Water Quality and Pollution

Water Quality and Pollution

Edited by Sheryl McMillan

SYRAWOOD
PUBLISHING HOUSE

New York

Published by Syrawood Publishing House,
750 Third Avenue, 9th Floor,
New York, NY 10017, USA
www.syrawoodpublishinghouse.com

Water Quality and Pollution
Edited by Sheryl McMillan

International Standard Book Number: 978-1-68286-432-6 (Hardback)

Cataloging-in-publication Data

Water quality and pollution / edited by Sheryl McMillan.
    p. cm.
Includes bibliographical references and index.
ISBN 978-1-68286-432-6
1. Water quality. 2. Water--Pollution. 3. Water--Pollution--Toxicology. 4. Water--Pollution--Physiological effect.
5. Water quality management. I. McMillan, Sheryl.
TD370 .W38 2017
628.161--dc23

Printed in the United States of America.

TABLE OF CONTENTS

Permissions

List of Contributors

Index

PREFACE

Water is a basic requirement for sustaining life on earth. It is required for different important processes like drinking, cooking, cleaning, washing, etc. Therefore, the depleting condition of water is a major concern in present times. It is extremely important to conserve the quality and quantity of water bodies on this planet. This book will give detailed information on the various causes responsible for water pollution and will provide different solutions to conserve water. The various studies that are constantly contributing towards advancing technologies and evolution of this field are examined in detail in the text. Those in search of information to further their knowledge will be greatly assisted by it. The text will prove to be immensely beneficial to students and researchers in this area.

This book is a result of research of several months to collate the most relevant data in the field.

When I was approached with the idea of this book and the proposal to edit it, I was overwhelmed. It gave me an opportunity to reach out to all those who share a common interest with me in this field. I had 3 main parameters for editing this text:

1. Accuracy – The data and information provided in this book should be up-to-date and valuable to the readers.

2. Structure – The data must be presented in a structured format for easy understanding and better grasping of the readers.

3. Universal Approach – This book not only targets students but also experts and innovators in the field, thus my aim was to present topics which are of use to all.

Thus, it took me a couple of months to finish the editing of this book.

I would like to make a special mention of my publisher who considered me worthy of this opportunity and also supported me throughout the editing process. I would also like to thank the editing team at the back-end who extended their help whenever required.

Editor

Highly Sensitive, Highly Specific Whole-Cell Bioreporters for the Detection of Chromate in Environmental Samples

Rita Branco[1,2], Armando Cristóvão[3,4], Paula V. Morais[1,4]*

1 IMAR, 3004-517 Coimbra, Portugal, 2 Escola Universitária Vasco da Gama, Mosteiro de S. Jorge de Milréu, Estrada da Conraria, Castelo Viegas – Coimbra, Portugal, 3 Center for Neuroscience and Cell Biology, University of Coimbra, Coimbra, Portugal, 4 Department of Life Sciences, FCTUC, University of Coimbra, Coimbra, Portugal

Abstract

Microbial bioreporters offer excellent potentialities for the detection of the bioavailable portion of pollutants in contaminated environments, which currently cannot be easily measured. This paper describes the construction and evaluation of two microbial bioreporters designed to detect the bioavailable chromate in contaminated water samples. The developed bioreporters are based on the expression of *gfp* under the control of the *chr* promoter and the *chrB* regulator gene of Tn*OtChr* determinant from *Ochrobactrum tritici* 5bvl1. pCHRGFP1 *Escherichia coli* reporter proved to be specific and sensitive, with minimum detectable concentration of 100 nM chromate and did not react with other heavy metals or chemical compounds analysed. In order to have a bioreporter able to be used under different environmental toxics, *O. tritici* type strain was also engineered to fluoresce in the presence of micromolar levels of chromate and showed to be as specific as the first reporter. Their applicability on environmental samples (spiked Portuguese river water) was also demonstrated using either freshly grown or cryo-preserved cells, a treatment which constitutes an operational advantage. These reporter strains can provide on-demand usability in the field and in a near future may become a powerful tool in identification of chromate-contaminated sites.

Editor: Melanie R. Mormile, Missouri University of Science and Technology, United States of America

Funding: This work was funded by Fundação para a Ciência e Tecnologia (FCT) under the PTDC/MAR/109057/2008 project. Rita Branco was supported by a grant, SFRH/BPD/48330/2008, from FCT. The funders had no role in study design, data collection and analysis, decision to publish, or preparation of the manuscript.

Competing Interests: The authors have declared that no competing interests exist.

* E-mail: pvmorais@ci.uc.pt

Introduction

Contamination of soils and water supplies with chromium compounds is considered a serious environmental issue. The extensive chromate utilization in industries has resulted in a large number of contaminated sites worldwide [1]. Nowadays, water-soluble heavy metal components, including chromate compounds, are of particular concern for drinking water and agricultural water quality [2]. Although many of these chromate compounds consist of complexes of low bioavailability and toxicity, they often persist in the environment and may be converted to more dangerous and bioavailable forms: the Cr(VI) free oxyanions.

Detection of the presence of low concentrations of chromate is indispensable to identify contaminated locations and to control the progress of remediation efforts. Analytic methods such as inductively coupled plasma atomic electron spectrometry (ICP/AES) or mass spectrometry (ICP/MS), flow injection atomic absorption (FIAAS) or electrochemical methods are the most used methods for quantification of total metals. These methodologies require the sample digestion with strong acids, which are pollutant, and are not able to distinguish between the bio-available and non-bio-available fractions of the metals. Therefore, the environmental risk for each metal is difficult to predict using these methodologies. [3,4,5]. On the other hand, the use of bacterial biosensors/bioreporters allows sensitive determinations of the bioavailable contaminant, without expensive equipment or specialized training. It provides a simple way of determining contaminants and some stress-induced biosensors report the mutagenic effects of samples with great sensitivity [6–8].

Microbial biosensors are increasingly being used as specific and sensitive sensing devices for measuring biologically relevant concentrations of pollutants [6,7,9]. Due to their low cost, lifespan, and range of fitting growth conditions (e.g. pH and temperature), microorganisms have been widely employed in the construction of biosensors.

In biosensor technology, in order to detect bioavailable environmental pollutants, the devices are constructed by fusing a pollutant-responsive promoter to a reporter gene. The reporter gene expression may be quantified and is a measure of the availability of a specific pollutant in the sample. Several field applications for these biosensors have been previously reported, such as quantification of polychlorinated biphenyls, alkanes, aromatic compounds, including polycyclic aromatic hydrocarbons (PAHs), biocides, antibiotics and heavy metals (reviewed in [10–12]).

In the case of heavy-metal detection, several bioreporters have been constructed to be used in general toxicity testing or in detection of specific metals, e.g. mercury, copper, lead, cadmium, nickel, zinc and arsenic [4,13–17]. Despite the different metal-specific bacterial reporters available, only a few have been drawn for the analysis of environmental samples and even fewer used as chromium (Cr(VI) and Cr(III))-specific bacterial sensors in direct testing of soils [5]. A recombinant luminescent bacterial sensor for chromate was constructed based

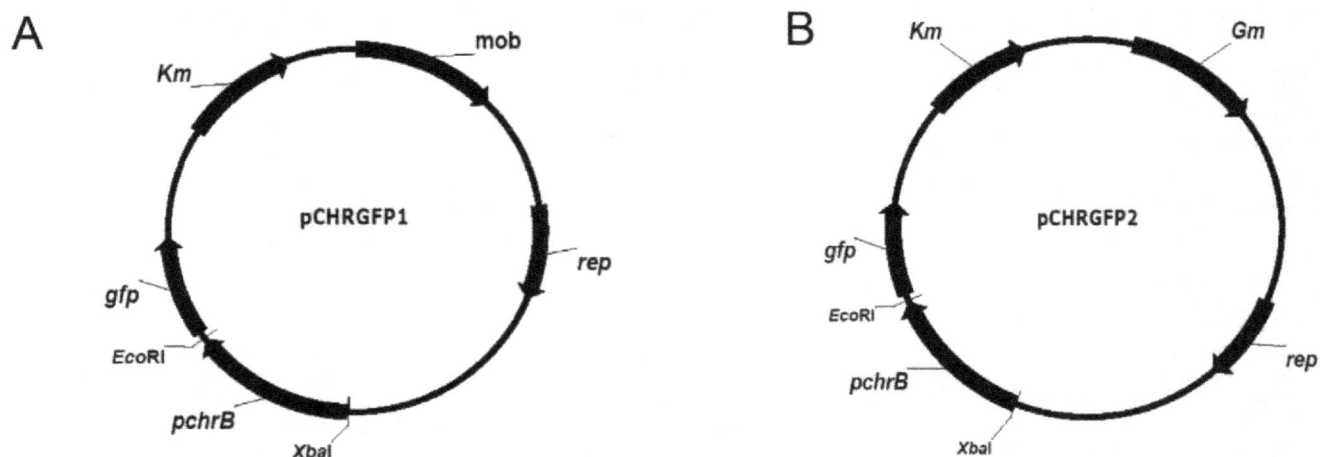

Figure 1. Schematic diagram of the plasmids engineered for this work. (A) Plasmid pCHRGFP1: pPROBE-NT containing the *chr* promoter and *chrB* gene upstream of *gfp*. (B) Plasmid pCHRGFP2: pCHRGFP1 with an additional gentamicin resistance gene cloned into *mob* gene. Arrows indicate the direction of transcription. The restriction enzymes indicated were those used in this work.

on a chromate resistance determinant located in megaplasmid pMOL28 of multiresistant *Ralstonia metallidurans* CH34 [18]. However, this bacterial reporter revealed to have its response affected by Cr(III) ions, and therefore, to be a system not completely selective to chromate.

Commonly used reporter systems include *lacZ* for β-galactosidase activity, *luxAB* and *lucFF* for bioluminescence measurements [19,20]. The gene for green fluorescent protein (GFP) is increasingly used to construct biosensors, because it allows for in situ assessment of elemental bioavailability and can report relevant concentrations of pollutants [21–23].

In the present study we describe the construction and characterization of a chromate-GFP bioreporter *E. coli*. This bioreporter was constructed by creating a transcriptional fusion between *gfp* gene and the genetic unit composed by the promoter and the chromate regulatory gene (*chrB*) from Tn*OtChr* of *Ochrobactrum tritici* 5bvl1. The characterized Tn*OtChr* element, carrying *chr* operon of *O. tritici* 5bvl1, contains a regulatory gene (*chrB*) and three additional chromate resistance genes (*chrA, chrC* and *chrF*) [24,25]. In order to have a bioreporter resistant to other toxics and able to detect chromate in multi-contaminated environmental samples, we also engineered the type strain of *O. tritici* to become fluorescent in the presence of toxic levels of

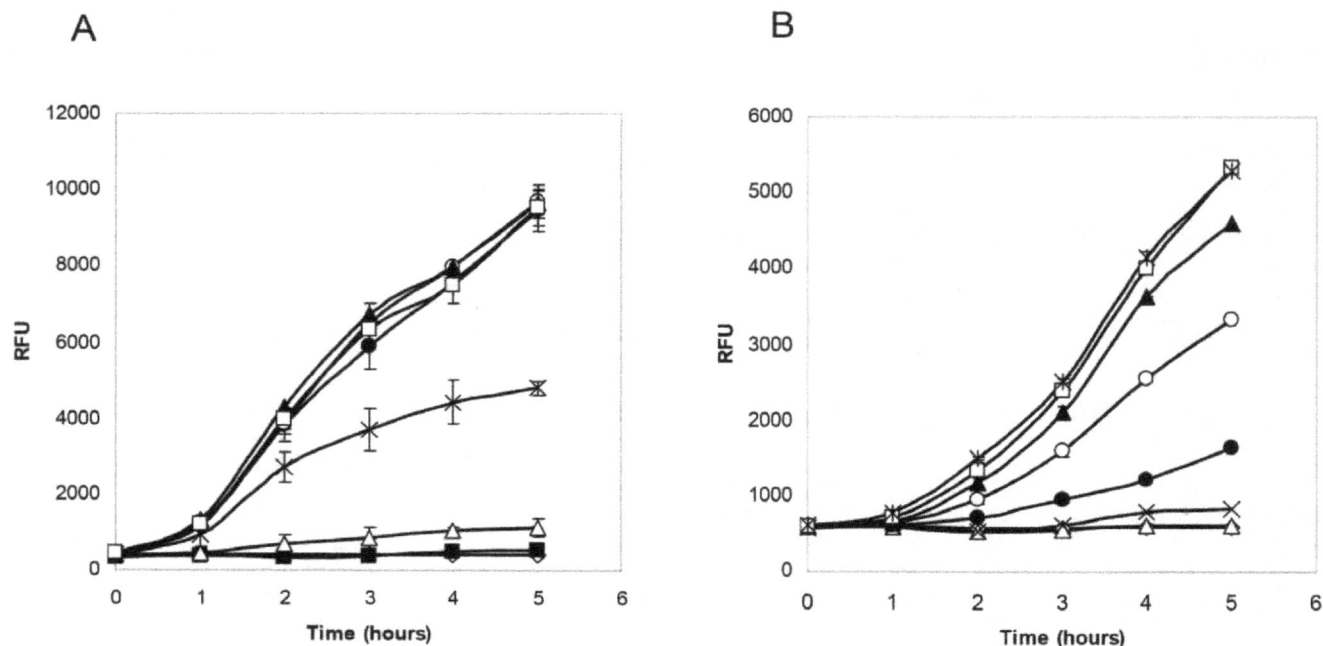

Figure 2. Time-dependent induction of pCHRGFP1 *E. coli* (A) and pCHRGFP2 *O. tritici* (B) reporters with chromate. Cells were exposed to increasing concentrations of chromate, 0 nM (◇), 10 nM (■), 100 nM (△), 1 μM (x), 5 μM (●) 10 μM (○), 25 μM (▲), 50 μM (□) e 100 μM (✳). The data are the mean values of three experiments with the standard deviations.

A

B

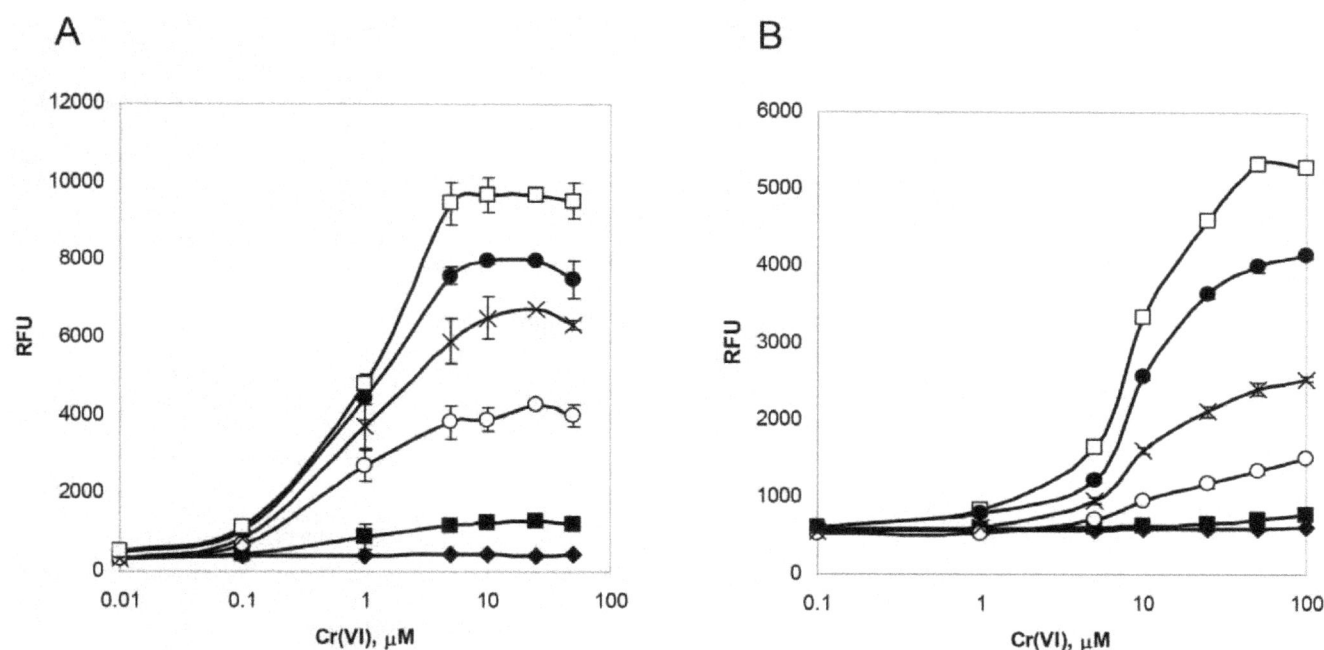

Figure 3. Chromate-fluorescence response of pCHRGFP1 *E. coli* (A) and pCHRGFP2 *O. tritici* (B) reporters measured after different exposure periods. 0 h chromate incubation (◆), 1 h chromate incubation (■), 2 h chromate incubation (○), 3 h chromate incubation (x), 4 h chromate incubation (●), 5 h chromate incubation (□). The data are the mean values of three experiments with the standard deviations.

chromate. This strain was chosen since it is an environmental strain [26] and its bacterial genus, *Ochrobactrum*, is widely distributed, being able to survive in different environments, including habitats where contamination with several compounds is present [27–29]. This great diversity, including habitats contaminated with chromate, suggests that environmental whole-cell bioreporter could be a great solution for chromate detection across a wide range of conditions.

Materials and Methods

Ethics Statement

No specific permits were required for the described field studies. The water samples were collected in a public river beach open to the public (Mondego) and in 3 other public stream waters (Malhou, Braças and Carritos). Only waters were used for analysis in the present study.

pCHRGFP1 *E. coli* Reporter Construction

The reporter plasmid was constructed by inserting a heavy metal responsive element from a natural resistance mechanism into a plasmid-containing promoterless green fluorescence protein (*gfp*) gene. Standard recombinant techniques were used for the construction of this chromate sensor plasmid [30]. The genomic region containing the *chr* promoter and *chrB* gene was amplified from strain *O. tritici* 5bvl1 DNA with primers Pchr1 (5′-CG***TCTAGA***GATTGCTTATTCCTATTGCCA-3′) and Pchr2 (5′-CG***GAATTC***TCATACGCTGAGGGTCCCTTT-3′), containing engineered *Xba*I and *Eco*RI recognition sites (restriction sites are italicized and in bold), following the standard PCR conditions. The resulting PCR product (1.11 Kbp) was gel purified prior to being cloned into pGEM-T plasmid (Promega). DNA sequencing was performed to verify that the correct amplification product was obtained prior to digestion by *Xba*I and *Eco*RI and

transference of the restriction digestion product into the multiple cloning site of pPROBE-NT [31]. This construct was then transformed into *Escherichia coli* DH5α resulting in a pCHRGFP1 *E. coli* chromate reporter (Figure 1A).

pCHRGFP2 *O. tritici* Reporter Construction

The previous reporter plasmid was engineered by introducing an additional antibiotic resistance cassette. After gentamicin resistance gene amplification from pBBR1MCS-5 plasmid [32], the PCR fragment was cloned downstream into *Not*I site of previous construct (Figure 1B). This plasmid was then inserted into *O. tritici* type strain by electroporation. Selection of plasmid-expressing clones was performed on LB plates with gentamicin (15 µg/ml). One clone was considered a pCHRGFP2 *O. tritici* reporter and was subjected to further analyses.

Reporter Activity Assays

The reporter strains pCHRGFP1 *E. coli* and pCHRGFP2 *O. tritici* were grown overnight, in a shaker (200 rpm) at 30°C in Luria–Bertani (LB) medium and used as inoculums. Cultures of the bioreporters used in the experiments (except when otherwise stated) were performed in TMM medium (Tris-buffered mineral salts medium [33] supplemented with glucose 0.5% and vitamins [24]), incubated at 37°C, and grown to an OD600nm of about 0.2–0.3, corresponding to cells on exponential phase of growth. Frozen cells at −80°C were prepared by resuspending exponential growing cells in TMM medium with 15% glycerol. To study the specificity and sensitivity of the bioreporters, these exponential cells were stressed by addition of the indicated concentration of chromate (as K_2CrO_4), chromium (III) chloride, arsenate, phosphate, sulphate, molybdate, tungstate, nickel, cobalt, cadmium, copper, and zinc. The stock metal solutions (1 M) were prepared by dissolving analytical grade (>99.0%) metal salts in ultra pure water sterilized by filtration. The stressed cultures were

Table 1. Induction of fluorescence in both bioreporters by several compounds.

Compound	Bioreporter	Concentration (µM)	Indution ratio
Ni²⁺	pCHRGFP1 E. coli	10	0.9±0.03
		100	1.2±0.12
	pCHRGFP2 O. tritici	10	1.2±0.08
		100	1.1±0.02
Co²⁺	pCHRGFP1 E. coli	10	0.7±0.02
		100	0.7±0.1
	pCHRGFP2 O. tritici	10	0.8±0.02
		100	0.7±0.04
Cd²⁺	pCHRGFP1 E. coli	10	0.8±0.01
		100	0.9±0.15
	pCHRGFP2 O. tritici	10	0.9±0.03
		100	0.8±0.07
Zn²⁺	pCHRGFP1 E. coli	10	0.7±0.03
		100	0.8±0.03
	pCHRGFP2 O. tritici	10	0.7±0.05
		100	0.8±0.03
Cu²⁺	pCHRGFP1 E. coli	10	0.9±0.03
		100	1.1±0.02
	pCHRGFP2 O. tritici	10	0.9±0.08
		100	1.0±0.02
MoO₄²	pCHRGFP1 E. coli	100	0.9±0.02
		1000	0.9±0.04
	pCHRGFP2 O. tritici	100	1.0±0.03
		1000	0.9±0.05
WO₄²⁻	pCHRGFP1 E. coli	100	1.0±0.01
		1000	1.0±0.04
	pCHRGFP2 O. tritici	100	1.1±0.01
		1000	0.9±0.08
Cr (III)	pCHRGFP1 E. coli	10	1.1±0.04
		100	1.4±0.01
		1000	1.3±0.01
	pCHRGFP2 O. tritici	10	1.0±0.01
		100	1.1±0.02
		1000	1.0±0.02

then incubated at 37°C, for the period of 5 hours. During this period, reaction volumes of 200 µl were transferred onto white 96-well plates and GFP fluorescence intensity was measured hourly and in triplicate.

The effect of the growth media composition in the ability of the pCHRGFP1 E. coli reporter to measure chromate was also tested using two different media. Exponential cells grown in the chemically defined medium TMM, and in the rich medium LB, were subjected to increasing chromate concentrations. To assess the effect of culture growth phase on bioreporter response, cells grown overnight, corresponding to cells at stationary phase, in LB or in TMM, were exposed to different concentrations of chromate and the results compared to the ones obtained by using cells in exponential growth phase.

Environmental analyses with Mondego river water were performed by mixing (1:1) exponential growing or frozen cells of pCHRGFP1 E. coli and pCHRGFP2 O. tritici reporters with natural river water (uncontaminated sample) or river water spiked with 1 µM and 10 µM of chromate (contaminated sample). Assays with environmental waters from Malhou, Braças and Carritos were performed as described above using only exponential growing cells.

Optical density and fluorescence were measured, in replicate experiments, during the incubation periods.

Spectrofluorometry

Fluorescence intensity was measured with a Gemini EM Fluorescence Microplate Reader (Molecular Devices), with emission, excitation and cutoff wavelengths at 510, 480 and 595 nm, respectively. Relative fluorescence unit (RFU) is defined as the culture fluorescence relative to culture biomass at OD600nm SpectraMax Plus384 Absorbance Microplate Reader (Molecular Devices). Induction ratio is defined as the RFU of an effector-exposed sample divided by the RFU of a no-effector control (0 µM).

Epifluorescent Microscopy

Bioreporter cells were visualized under a Leitz Laborlux K epifluorescent microscope (Leica) equipped with a 50 W mercury lamp, a BP 450–490 excitation filter and a LP 515 nm emission filter. Overnight chromate stressed cells were placed onto a microscope slide and examined under the microscope for GFP fluorescence expression.

Results

Genetic Description of the Bacterial Bioreporters

The chr promoter and the chrB gene of the chr resistance determinant of strain O. tritici 5bvl1 (TnOtChr) was cloned into the broad-host-range vector pPROBE-NT [31], upstream from the gfp gene, creating a pchrB_gfp transcriptional fusion. One bioreporter was created when E. coli DH5α cells were transformed with this construct, which resulted in the bioreporter designated by pCHRGFP1 E. coli (Fig. 1A). An additional gentamicin gene was inserted into mob gene of pCHRGFP1 resulting in pCHRGFP2 (Fig. 1B), which was then introduced into O. tritici type strain. This second bioreporter was named by pCHRGFP2 O. tritici.

Detection of Chromate by Bioreporter Strains

Cultures of pCHRGFP1 E. coli and pCHRGFP2 O. tritici were grown under a range of chromate concentrations (0.01 to 100 µM Cr(VI)), to assess whether functional GFP was produced and to measure the bioreporter sensitivity. The fluorescence data recorded from exponential growth cells of both strains in TMM, with and without chromate, under aerobic conditions are shown in Fig. 2 and Fig. 3. The results indicated that cells were able to produce functional GFP when subjected to chromate.

Time-dependent induction of the two bacterial reporters in response to chromate was determined by incubating periodically the cells with the metal ion. The induction of green fluorescence production by both bioreporters when exposed to chromate showed a time-dependence (Fig. 2). As shown in this figure, as the time of induction increased, there was an increase in the fluorescence emitted by bacterial strains. The profile of both bioreporters response showed that during the period of cell incubation, the specific fluorescence intensity continuously increased. Although the maximum signal for these reporters was reached after 5 h of exposure to chromate, the incubation

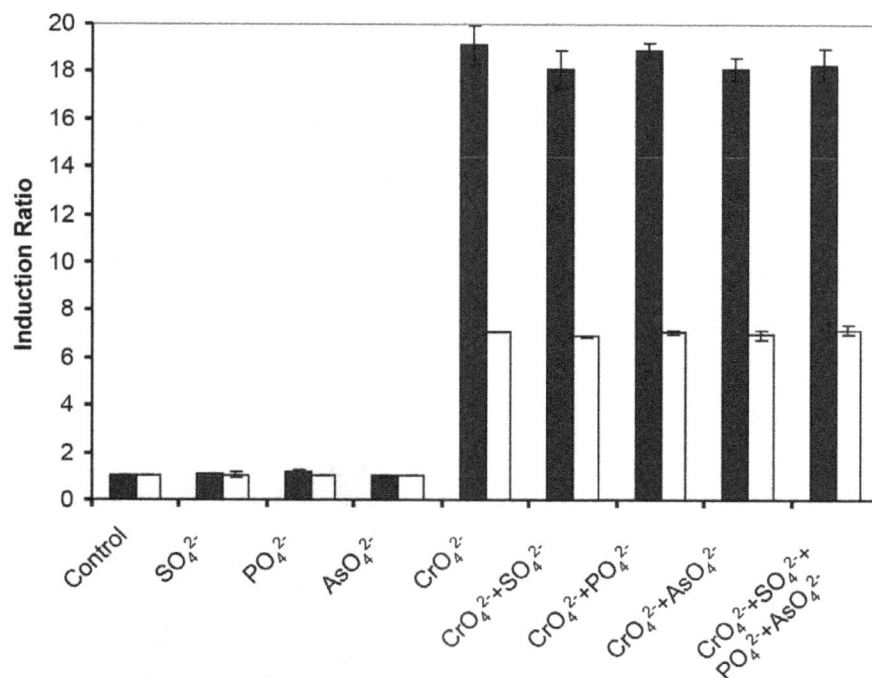

Figure 4. Selectivity of the pCHRGFP1 *E. coli* **and pCHRGFP2** *O. tritici* **reporters to different compounds.** Cells were treated individually with sulphate and phosphate in concentration of 1 mM and with arsenate and chromate in concentration of 0.1 mM or mixtures of the several compounds for 5 hours. Control refers to no-metal treatment reporter bacteria. The data are the mean values of three experiments with the standard deviations.

time could be certainly reduced to 3 h since the fluorescence signal obtained during this period of time was high enough to reach suitable conclusions, as it is possible to observe in Fig. 3.

The intensity of the GFP fluorescence emission of pCHRGFP1 *E. coli* and pCHRGFP2 *O. tritici* was found to be dependent on the concentration of chromate ions in the growth medium, given that the increase of chromate concentrations under the same incubation conditions resulted in increasing

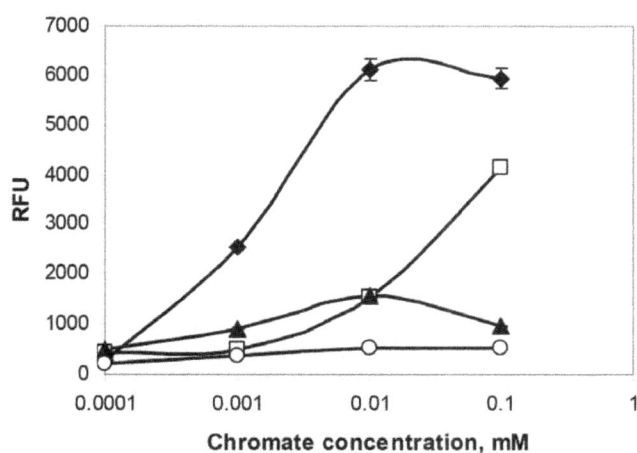

Figure 5. Effect of growth phase and culture medium on efficiency of pCHRGFP1 *E. coli* **reporter.** Exponential cells grown in TMM medium (◆) or LB medium (□) and stationary cells grown in TMM medium (▲) or LB medium (○) were incubated with the indicated chromate concentrations for 5 hours. The data are the mean values of three experiments with the standard deviations.

fluorescence (Fig. 2 and Fig. 3). The lowest concentration of chromate that caused a noticeable response by pCHRGFP1 reporter was 0.1 μM Cr(VI), with a fluorescence intensity four times higher than the control experiment after 5 h of incubation (Fig. 2A). The intensity of signal fluorescence increased along with the amount of chromate concentration, up to 10 μM (Fig. 3A), after which the fluorescence reached a plateau. When chromate concentration exceeded 50 μM, green fluorescence started to decrease slightly. This was possibly due to the toxic effect of chromium ions on the bacterial cells. The second bioreporter, pCHRGFP2 *O. tritici*, was not as sensitive to chromate as pCHRGFP1. Higher chromate concentrations (at least 1 μM) were needed to induce a clear green fluorescence production by pCHRGFP2 (Fig. 2B and Fig. 3B). These reporter assays also showed maximum fluorescence intensity values lower than those achieved by using the pCHRGFP1 *E. coli* bioreporter. Moreover, the saturation of fluorescence signal was achieved for higher chromate concentrations (>50 μM).

No GFP fluorescence emission could be detected in the parent strains *E. coli* DH5α and *O. tritici* carrying the empty pPROBE-NT plasmid (data not shown). In addition, Fig. 2 and Fig. 3 show that the background fluorescence exhibited by the untreated bioreporters (cells without chromate) did not change or oscillate during the incubation period.

Selectivity of the Chromate Bioreporters

The selectivity of the pCHRGFP1 *E. coli* and pCHRGFP2 *O. tritici* bioreporters to metal ions was also evaluated. The reporter cells were treated with increasing concentrations of the different divalent metal ions. The levels of fluorescence measured during incubation (5 h) are shown in Table 1. The metals Cd(II), Zn(II),

A

B

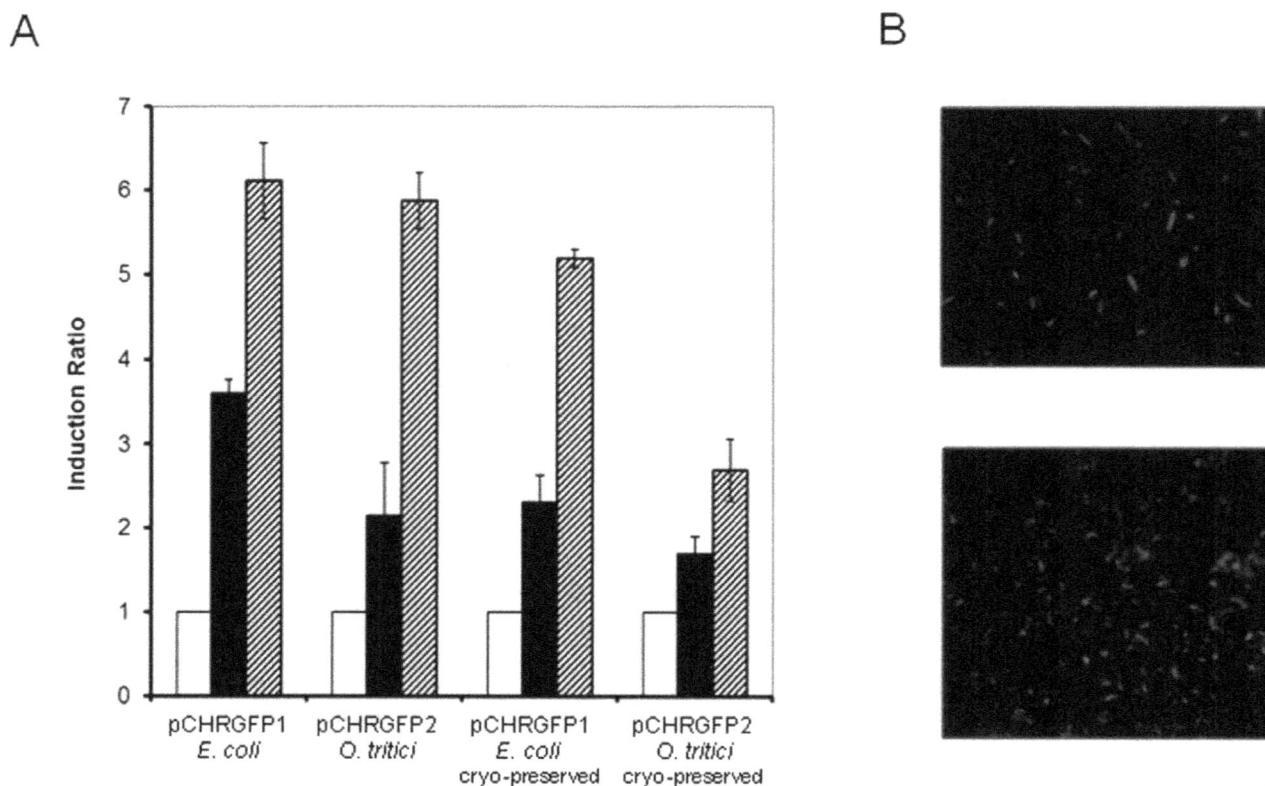

Figure 6. GFP reporter detection of chromate-contaminated environmental water samples. A) Cultures of fresh exponential-growing cells or cryo-preserved exponential cells of pCHRGFP1 *E. coli* and pCHRGFP2 *O. tritici* reporters were exposed to Mondego river water (white bars), Mondego river water spiked with 1 μM chromate (black bars) and 10 μM chromate (dashed bars). The fluorescence was assayed after 5 hours of incubation. The data are the mean values of three experiments with the standard deviations. **B)** Exponential grown cells of pCHRGFP1 *E. coli* (upper panel) and of pCHRGFP2 *O. tritici* (lower panel) were incubated overnight to Mondego river water spiked with 10 μM of chromate, after which green fluorescent protein expression in bacterial cells was visualized by using epifluorescence microscope. Magnification, ×1000.

Co(II), Ni(II) and Cu(II) did not induce significant changes (less than 1%) in green fluorescence compared to control.

Since structural configuration of chromate is comparable to arsenate, phosphate sulphate, molybdate and tungstate, both bioreporter cells were also treated with these compounds in order to evaluate their specificity to chromate. The concentrations of the compounds tested ranged from 10 μM to 10 mM, in order to ensure that, if the compound was an inducer, an induction concentration was tested. In the case of arsenate, lower concentrations (1 to 100 μM) were used because higher concentrations have demonstrated to be significantly inhibitory to cell growth. As it was observed in the tests using metals different from chromate, pCHRGFP1 *E. coli* or pCHRGFP2 *O. tritici* did not respond to any of the effectors, at any concentration tested (Table 1 and Fig. 4).

Although pCHRGFP1 *E. coli* and pCHRGFP2 *O. tritici* bioreporters did not respond to the several effectors analysed, the contaminated sites often contain mixtures of metal contaminants and other compounds. Thus, the effect of the presence of multiple effectors on reporters response was also assessed. The combinations of sulphate, phosphate or arsenate with chromate did not change the induction ratio observed in the individual test (with chromate only) (Fig. 4). Based on these induction results and despite of the lower chromate sensitivity of *O. tritici* reporter when compared with *E. coli* reporter, both bioreporter cells behaved as efficient chromate selective reporter systems.

Growth Phase and Medium Dependence on Bioreporter Cells

It is known that growth-phase and growth-medium can affect a reporter gene expression in a whole-cell biosensor. To assess this problem, the expression of *gfp* by pCHRGFP1 *E. coli* was studied under different bacterial growth conditions. In the experiments with exponential growth chromate-treated cells, the kinetics of GFP formation essentially matched those of biomass formation, with the net fluorescence of the culture increasing during the exponential phase of growth. Chromate induced *gfp* expression in cells grown in both media examined. However, the level of fluorescence was different between both growth media used: although the bioreporter did respond to chromate when cultivated in the rich medium, the induction ratio was only 25% of that obtained in TMM medium (Fig. 5). Cells obtained from overnight LB or TMM growth cultures (stationary phase) exhibited a fluorescence level much lower than the fluorescence signal obtained with exponential growth cells. These results confirmed that growth phase affects induction of the chromate bioreporter.

Chromate Bioreporters in Environmental Samples

Both bioreporter strains, pCHRGFP1 *E. coli* and pCHRGFP2 *O. tritici*, were used to determine chromate concentration in environmental water samples with differentchemical composition (Table 2). Waters differed mainly in their content of iron, chlorides, magnesium, selenium and barium. Portuguese river

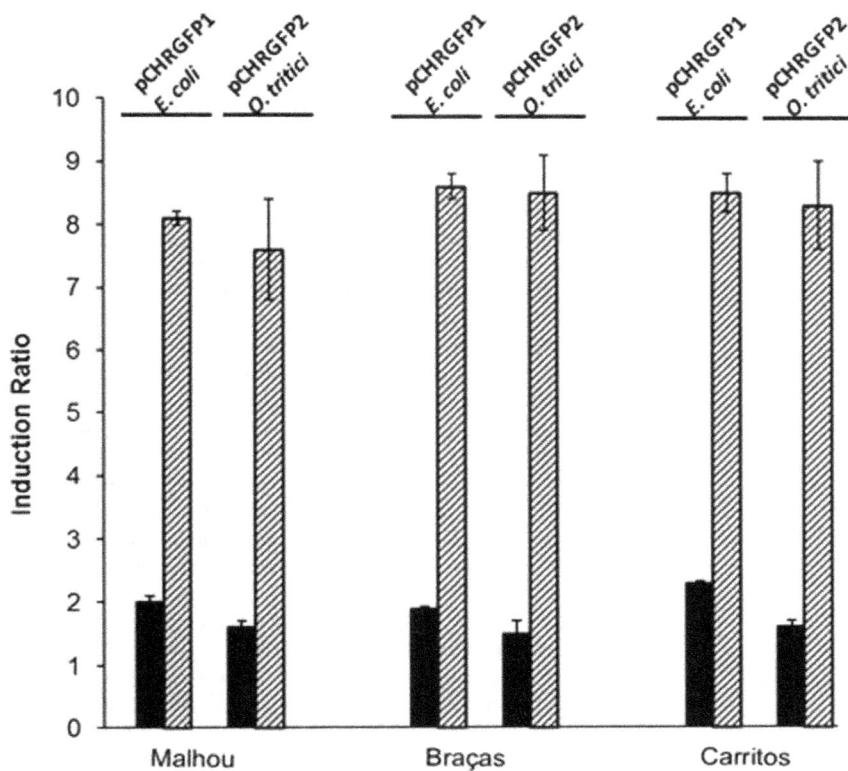

Figure 7. Fluorescence of bioreporter strains exposed to different chromate-contaminated environmental waters. Cultures of exponential-growing cells of pCHRGFP1 *E. coli* and pCHRGFP2 *O. tritici* were exposed to chromate spiked stream waters of Malhou, Braças and Carritos. These samples were contaminated with 1 µM chromate (black bars) and 10µM chromate (dashed bars). The fluorescence was assayed after 5 hours of incubation. The data are the mean values of three experiments with the standard deviations.

water samples, contaminated with 1 µM and 10 µM of chromate, induced GFP expression in exposed fresh cells and the green fluorescence intensity was time and dose dependent (Fig. 6 and Fig. 7). Additionally, cryo-preserved reporter cells were tested since they might be considered a useful for operational reasons, allowing their storage and use over time. In order to assess their activity, both bioreporter frozen cells were exposed to chromate contaminated environmental Mondego river water samples and showed to be able to produce GFP, when submitted to contaminated waters. Figure 6A shows that fresh exponential growing cells were more efficient than the frozen cells since the fluorescence levels were higher than those obtained by cryo-preserved cells. This data clearly demonstrates that strains distinguished chromate contaminated water of uncontaminated water even when used after freezing.

The different composition of the environmental waters tested did not affect the fluorescence signals (Fig. 7). Additionally, the only stream water, Malhou, which naturally contained 0.3 µM of total chromium, showed an increase of 1.5 fold comparatively to the other waters for the non-contaminated assays.

When examined by fluorescence microscopy, the image of the bacterial cultures of pCHRGFP1 *E. coli* and pCHRGFP2 *O. tritici* incubated overnight with chromate contaminated waters (10 µM chromate), clearly shows induction of fluorescence, revealed by the presence of green cells (Fig. 6B). Incubation of both reporter cells with non-contaminated environmental water resulted in cells without fluorescence and therefore no green color could be detected on images of epi-fluorescence (data not shown).

Discussion

In the present work, two whole-cell chromate reporters that can detect the presence of nanomolar amounts of chromate in laboratorial medium assays or micromolar concentrations of chromate in environmental samples were constructed. After the successful experiments with the pCHR*lacZ* fusion reporter [24], we concluded that ChrB could be used as a chromate-sensing system. To take advantage of this ability, a reporter construct was performed by placing a promoterless *gfp* gene under the control of a genetic fragment composed by a *chr* promoter and *chrB* gene of *O. tritici* 5bvl1.

Several bacterial bioreporters for different environmental pollutants have been developed using various reporter genes, such as *lacZ*, *luxAB*, and *lucFF* (reviewed in 13, 34). Although the colorimetric enzyme assay and bioluminescence have been very successful as reporters for detection of several compounds, such as metals, metalloids or aromatic compounds, these methods require the addition of exogenous substrates or cofactors. Moreover, these assays require additional experimental steps, such as centrifugation, cell lysis, pH adjustment, before enzyme activity measurements. Regarding biological chromate sensing, few efforts have been carried out to achieve an efficient chromate biosensor. The few cases reported are based on fusion of chromate resistance genes with *lux* reporter gene [5,18]. However, this system showed interactions among induction of the *chr* resistance determinant, chromate reduction, chromate accumulation, and sulfate concentration of the growth medium.

Table 2. Chemical analyses of environmental water samples.

Compound (mg/l)	Water Samples			
	Mondego	**Malhou**	**Braças**	**Carritos**
Nitrate	*<11	<11	<11	<11
Chloride	10	47	36.5	70.4
Phosphate	<0.14	<0.14	<0.14	<0.14
Sulfate	<15	<15	<15	<15
Phenol	<0.001	n.d.	<0.001	<0.001
Hydrocarbon	<0.01	<0.01	<0.01	<0.01
Fluoride	<0.3	0.73	<0.3	<0.3
Cyanide	<0.01	n.d.	<0.01	<0.01
Iron	0.09	<0.04	4.0	0.1
Magnesium	0.02	4.9	0.06	0.01
Copper	<0.01	n.d.	<0.01	<0.01
Zinc	<0.1	<0.1	<0.1	<0.1
Arsenic	<0.003	<0.003	<0.003	<0.003
Boron	<0.2	0.08	<0.2	<0.2
Cadmium	<0.0015	<0.0015	<0.0015	<0.0015
Chromium	<0.006	0.016	<0.006	<0.006
Lead	<0.006	<0.006	<0.006	<0.006
Selenium	0.001	0.0023	0.0026	0.001
Barium	0.02	n.d.	0.03	0.06
Mercury	<0.0004	<0.0004	<0.0004	<0.0004
pH	7.9	7.9	6.2	6.4
Dissolved oxygen (%)	97	80	41	84

*limit of detection.
n.d.: not determined.

The use of *gfp* as a reporter gene gives these sort of biosystems some advantages such as the ability to use a simple and clean method without the need for exogenous enzyme substrates or other chemical compounds. Additionally, the ability to use fluorometry and fluorescence microscopy to examine GFP expression allows assured results for assessing the effectors bioavailability [22,35].

Under uninduced conditions, fluorescence from pCHRGFP1 *E. coli* or pCHRGFP2 *O. tritici* was almost null, most probably because of the transcriptional shielding of the *pchrB-gfp* fusion. The ability to detect induction after moderately short incubation periods (3 hours) and at low inducer concentrations, nanomolar scale for pCHRGFP1 *E. coli* reporter and 1 µM for pCHRGFP2 *O. tritici* reporter, may reflect the fact that the *pchrB-gfp* fusion is a stable and well-performed construct, carried on an efficient plasmid that allows a high amplification of the signal, and also reflects the sensitivity of fluorescence detection instrumentation.

Both whole-cell reporters demonstrated ability to discriminate environmental water samples contaminated with chromium from uncontaminated samples. The notable ability to detect chromate levels lower than US or EU limit value for chromate concentrations in drinking water, 0.1 mg/l [36] or 0.05 mg/l [37] respectively make this system a hopeful successful tool to monitor the levels of chromate in environmental samples. Currently, many bacterial reporters or sensors have been reported and are

considered promising applications in the fields of biotechnology and environmental sciences [7]. The development of portable biosensors able to be applied on-site to monitor environmental pollution is an urgent demand. For example, arsenic contamination of groundwater is a serious problem in several countries and sensitive biosensors were already tested with arsenite contaminated waters from these countries [38,39]. A major characteristic of the whole-cell chromate biosensing systems developed in this study are their insensitivity to micromolar or millimolar levels of phosphate, sulphate, arsenate, molybdate, tungstate and diverse divalent metals. This is a noteworthy point, since environmental samples are composed of mixtures of compounds. In contaminated sites, the presence of many different chemicals other than the inducer compounds or environmental parameters, such as extreme salinity and pH values, could be toxic to or interfere with whole-cells, thereby causing inhibitory effects. Although it is impossible to determine the effect of all hypothetical components that could be present in an environmental sample, the bacterial reporters based on pCHRGFP represent a rapid, easy to perform, and inexpensive alternative to the conventional chemical method for chromate detection and measurement. It should also be noted that the chromate environmental bioassays required only small volumes of bacteria and the use of frozen cells did not impair its effectiveness. In the future, the optimization of this system, which can include the reconstitution of reporter strains from lyophilized powder, could greatly improve its applicability in the field as it has already been referred for other bioreporters [21,40]. These bacterial bioreporters or their improved versions could complement or even replace traditional analytical methods, providing with confidence data that could be useful in risk assessment and evaluation of the remediation needs of chromate contaminated sites.

Conclusions

The most innovative aspect of this paper is the demonstration of the high specificity and high sensitivity of two new chromate bioreporters constructed by our team. This paper includes the construction, laboratory characterization, and environmental sample testing of pCHRGFP1 *E. coli* and pCHRGFP2 *O. tritici* reporters for the detection of chromate. In fact, the pCHRGFP1 *E. coli* reporter has a very low detection limit enabling it to be used to evaluate chromate in potable waters. The most important advantage of these bioreporters is the fact it is a sensing system with a highly specific chromate selectivity, thus making possible to measure chromate among several compounds. In consequence, since field samples often contain unknown components, these bioreporters could be useful in the screening of chromate among unidentified contaminants in the environment. Moreover, chromium can be in the environment in different oxidation states which are not all bioavailable. The GFP-based bacterial reporter systems can provide information about the bioavailability of this metal, which is the most relevant information in environmental risk assessment and the potential biological impact of a contaminant [34].

Acknowledgments

We thank Dr. Steven Lindow for providing the pPROBE-NT plasmid.

Author Contributions

Conceived and designed the experiments: RB PVM. Performed the experiments: RB. Analyzed the data: RB. Contributed reagents/materials/analysis tools: AC PVM. Wrote the paper: RB PVM.

References

1. Cervantes C, Campos-Garcia J, Devars S, Gutierrez-Corona F, Loza-Tavera H, et al. (2001) Interaction of chromium with microorganisms and plants. FEMS Microbiol Rev 25: 335–347.
2. World Health Association (1993) Guidelines for drinking-water quality – chromium In: World Heath Association (Ed.). Guidelines for Drinking-Water Quality (vol.1, Recommendations). World Heath Association. Geneva. Switzerland. 45–46.
3. Tauriainen S, Virta M, Karp M (2000) Detecting bioavailable toxic metals and metalloids from natural water samples using luminescent sensor bacteria. Water Res 34: 2661–2666.
4. Corbisier P, van der Lelie D, Borremans B, Provoost A, de Lorenzo V, et al. (1999) Whole cell- and protein-based biosensors for the detection of bioavailable heavy metals in environmental samples. Anal Chim Acta 387: 235–244.
5. Ivask A, Virta M, Kahru A (2002) Construction and use of specific luminescent recombinant bacterial sensors for the assessment of bioavailabe fraction of cadmium, zinc, mercury and chromium in the soil. Soil Biol Biochem 34: 1439–1447.
6. Verma N, Singh M (2005) Biosensors for heavy metals. Biometals 18: 121–129.
7. Yagi K (2007) Applications of whole-cell bacterial sensors in biotechnology and environmental science. App Microbiol Biotechnol 73: 1251–1258.
8. Hansen LH, Sorensen SJ (2001) The use of whole-cell biosensors to detect and quantify compounds or conditions affecting biological systems. Microb Ecol 42: 483–494.
9. Ron E (2007) Biosensening environmental pollution. Curr Opin Biotechnol 18: 252–256.
10. Rodriguez-Mozaz S, de Alda MJL, Barceló D (2006) Biosensors as useful tools for environmental analysis and monitoring. Anal Bioanal Chem 386: 1025–1041.
11. Tecon R, van der Meer JR (2008) Bacterial biosensors for measuring availability of environmental pollutants. Sensors 8: 4062–4080.
12. Su L, Jia W, Hou C, Lei Y (2011) Microbial biosensors: A review. Biosens Bioelectron 26: 1788–1799.
13. Selifonova O, Burlage R, Barkay T (1993) Bioluminescent sensors for detection of bioavailable Hg(II) in the environment. Appl Environ Microbiol 59: 3083–3090.
14. Tauriainen S, Karp M, Chang W, Virta M (1998) Luminiscent bacterial sensor for cadmium and lead. Biosens Bioelectron 13: 931–938.
15. Tibazarwa C, Corbisier P, Mench M, Bossus A, Solda P, et al. (2001) A microbial biosensor to predict bioavailable nickel and its transfer to plants. Environ Pollut 113: 19–26.
16. Tom-Petersen A, Hosbond C, Nybroe O (2001) Identification of copper-induced genes in Pseudomonas fluorescens and use of a reporter strain to monitor bioavailable copper in soil. FEMS Microbiol Ecol 38: 59–67.
17. Riether K, Dollard M-A, Billard P (2001) lux and copAp::lux-based biosensors Appl Microbiol Biotechnol 57: 712–6.
18. Peitzsch N, Eberz G, Nies DH (1998) Alcaligenes eutrophus as a bacterial chromate sensor. App Environ Microbiol 64: 453–458.
19. Liu X, Germaine K, Ryan D, Dowling DN (2010) Whole-cell fluorescent biosensors for bioavailability and biodegradation of polychlorinated biphenyls. Sensors 10: 1377–1398.
20. Hakkila K, Maksimow M, Karp M, Virta M (2002) Reporter genes lucFF, luxCDABE, gfp, and dsred have different characteristics in whole-cell bacterial sensors. Anal Biochem 301: 235–242.
21. Hillson NJ, Hu P, Andersen GL, Shapiro L (2007) Caulobacter crescentus as a whole-cell uranium biosensor. App Environ Microbiol 73: 7615–7621.
22. Liao VH-C, Chien M-T, Tseng Y-Y, Ou K-L (2006) Assessment of heavy metal bioavailability in contaminated sediments and soils using green fluorescent protein-based bacterial biosensors. Environ Pollut 142: 17–23.
23. Roberto FF, Barnes JM, Bruhn DF (2002) Evaluation of a GFP reporter gene construct for environmental arsenic detection. Talanta 58: 181–188.
24. Branco R, Chung AP, Johnston T, Gurel V, Morais P. et al. (2008) The chromate-inducible chrBACF operon from the transposable element TnOtChr confers resistance to chromium(VI) and superoxide. J Bacteriol 190: 6996–7003.
25. Morais PV, Branco R, Francisco R (2011) Chromium resistance strategies and toxicity: what makes Ochrobactrum tritici 5bvl1 a strain highly resistant. Biometals 24: 401–410.
26. Lebuhn M, Achouak W, Schloter M, Berge O, Meier H, et al. (2000) Taxonomic characterization of Ochrobactrum sp. isolates from soil samples and wheat roots, and description of Ochrobactrum tritici sp. nov. and Ochrobactrum grignonense sp. nov. Int J Syst Evol Microbiol 50: 2207–2223.
27. Branco R, Alpoim MC, Morais PV (2004) Ochrobactrum tritici strain 5bvl1 - characterization of a Cr(VI)-resistant and Cr(VI)-reducing strain. Can J Microbiol 50: 697–703.
28. He Z, Gao F, Sha T, Hu Y, He C (2009) Isolation and characterization of a Cr(VI)-reduction Ochrobactrum sp. strain CSCr-3 from chromium landfill. J Hazard Mater 163: 869–873.
29. Liang B, Li R, Jiang D, Sun J, Qiu J, et al. (2010) Hydrolytic dechlorination of chlorothalonil by Ochrobactrum sp. CTN-11 isolated from a chlorothalonil-contaminated soil. Curr Microbiol 61: 226–233.
30. Sambrook J, Russell DW (2001) Molecular Cloning: A Laboratory Manual. 3rd ed. Ed. Cold Spring Harbor New York: Cold Spring Harbor Laboratory Press.
31. Miller WG, Leveau javascript:popRef('a1')JHJjavascript:popRef('a1'), Lindow SE (2000) Improved gfp and inaZ broad-host-range promoter-probe. Mol Plant Microbe Interact 13: 1243–1250.
32. Kovach ME, Elzer PH, Hill DS, Robertson GT, Farris MA, et al. (1995) Four new derivatives of the broad-host-range cloning vector pBBR1MCS carrying different antibiotic-resistance cassettes. Gene 166: 175–176.
33. Mergeay M, Nies D, Schlrgrl HG, Gerits J, Charles P, et al. (1985) Alcaligenes eutrophus CH34 is a facultative chemolithotroph with plasmid-bound resistance to heavy metals. J Bacteriol 162: 328–334.
34. Magrisso S, Erel Y, Belkin S (2008) Microbial reporters of metal bioavailability. Microb Biotechnol 1: 320–330.
35. Stiner L, Halverson LJ (2002) Development and characterization of a green fluorescent protein-based bacterial biosensor for bioavailable toluene and related compounds. App Environ Microbiol 68: 1962–1971.
36. U.S. Environmental Protection Agency (2002) National Primary Drinking Water Regulations – Consumer Fact Sheet on Chromium. Available: http://www.epa.gov/safewater/dwh/cioc/chromium.html.
37. E.C. (1998) Council Directive (98/83/EC) of 3 November 1998 on the quality of water intended for human consumption. Offic J Eur Commun L330.
38. Stocker J, Balluch D, Gsell M, Harms H, Feliciano J, et al. (2003) Development of a set of simple bacterial biosensors for quantitative and rapid measurements of arsenite and arsenate in potable water. Environ Sci Technol 37: 4743–4750.
39. Trang PTK, Berg M, Viet PH, Mui N, van der Meer JR (2005) Bacterial bioassay for rapid and accurate analysis of arsenic in highly variable groundwater samples. Environ Sci Technol 39: 7625–7630.
40. Philp J, French C, Wiles S, Bell J, Whiteley A, et al. (2004) Wastewater toxicity assessment by whole cell biosensor. In: The handbook of environmental chemistry. Vol. 5. Part I. Springer-Verlag Berlin Heidelberg. 165–225.

2

Sunscreen Products as Emerging Pollutants to Coastal Waters

Antonio Tovar-Sánchez[1]*, David Sánchez-Quiles[1], Gotzon Basterretxea[2], Juan L. Benedé[3], Alberto Chisvert[3], Amparo Salvador[3], Ignacio Moreno-Garrido[4], Julián Blasco[4]

1 Department of Global Change Research, Mediterranean Institute for Advanced Studies (UIB-CSIC), Esporles, Balearic Island, Spain, **2** Department of Ecology and Marine Resources, Mediterranean Institute for Advanced Studies (UIB-CSIC), Esporles, Balearic Island, Spain, **3** Department of Analytical Chemistry, Facultad de Química, Universitat de València, Burjassot, Valencia, Spain, **4** ICMAN-Instituto de Ciencias Marinas de Andalucía (CSIC), Puerto Real, Cádiz, Spain

Abstract

A growing awareness of the risks associated with skin exposure to ultraviolet (UV) radiation over the past decades has led to increased use of sunscreen cosmetic products leading the introduction of new chemical compounds in the marine environment. Although coastal tourism and recreation are the largest and most rapidly growing activities in the world, the evaluation of sunscreen as source of chemicals to the coastal marine system has not been addressed. Concentrations of chemical UV filters included in the formulation of sunscreens, such as benzophehone 3 (BZ-3), 4-methylbenzylidene camphor (4-MBC), TiO_2 and ZnO, are detected in nearshore waters with variable concentrations along the day and mainly concentrated in the surface microlayer (i.e. 53.6–577.5 ng L^{-1} BZ-3; 51.4–113.4 ng L^{-1} 4-MBC; 6.9–37.6 µg L^{-1} Ti; 1.0–3.3 µg L^{-1} Zn). The presence of these compounds in seawater suggests relevant effects on phytoplankton. Indeed, we provide evidences of the negative effect of sunblocks on the growth of the commonly found marine diatom *Chaetoceros gracilis* (mean $EC_{50} = 125 \pm 71$ mg L^{-1}). Dissolution of sunscreens in seawater also releases inorganic nutrients (N, P and Si forms) that can fuel algal growth. In particular, PO_4^{3-} is released by these products in notable amounts (up to 17 µmol PO_4^{3-} g^{-1}). We conservatively estimate an increase of up to 100% background PO_4^{3-} concentrations (0.12 µmol L^{-1} over a background level of 0.06 µmol L^{-1}) in nearshore waters during low water renewal conditions in a populated beach in Majorca island. Our results show that sunscreen products are a significant source of organic and inorganic chemicals that reach the sea with potential ecological consequences on the coastal marine ecosystem.

Editor: Wei-Chun Chin, University of California, Merced, United States of America

Funding: This work was funded by the ISUMAR project (CTM2011-22645) and CONCORDA project (384/2011) of the Spanish Ministries of Economy and Competitiveness and of Agriculture, Food and Environment. The funders had no role in study design, data collection and analysis, decision to publish, or preparation of the manuscript.

Competing Interests: The authors have declared that no competing interests exist.

* E-mail: atovar@imedea.uib-csic.es

Introduction

In spite of the fact that coastal tourism and recreation are becoming the largest and most rapidly growing activities in the world [1] and that sunscreen products have been used for nearly 80 years, the effect of sunscreens, as a source of introduced chemicals to the coastal marine system, has not yet been addressed. Sun protection cosmetics are composed of organic (para-aminobenzoates, cinnamates, benzophenones, dibenzoylmethanes, camphor derivatives and benzimidazoles, which absorb the UV radiations), and/or inorganic UV chemical filters (i.e. TiO_2 and ZnO) that reflect and scatter the UV radiation protecting human skin from direct radiation of sunlight [2,3]. There are around 45 UV chemical filters subjected to regulation in different countries [3,4]. In addition to these UV filters, sunscreen products contain other ingredients such as preservatives (e.g. parabens derivates) [5], coloring agents (e.g. ammonium sulphate, copper powder, ferric ammonium ferrocyanide, iron and zinc oxides, etc.) [6], film forming agents (e.g. acrylates and acrylamides) [7], surfactants, chelators, viscosity controllers (e.g. potassium cetyl phosphate, pentasodium ethylenediamine tetramethylene phosphonate among others) [8] and fragrances, etc.

Formulation and concentration of cosmetic ingredients in commercial sunscreens are varied, and legislated by local or international agencies (e.g. European Union Cosmetics Directive [9] or United States Food and Drug Administration [10]) to reach a compromise between adequate UV protection and minimal side effects for humans [2]. Studies conducted in lakes (i.e. Zurich and Hüttnersee Lakes, Swiss) suggest that UV filter removal processes from the water column are important, and can be mediated by biodegradation processes and/or absorption sedimentation [11]. Because of their lipophilicity, persistence and stability against biodegradation they have been shown to accumulate in the food chain [4,12].

Coastal tourism is considered one of the fastest growing forms of tourism in recent decades [1] being the Mediterranean one of the most important tourism regions in the world [13]. For decades, the Balearic Islands (Western Mediterranean Sea) have provided the traditional sun, sand and sea product. Tourism is the first economic activity in the Islands. The islands comprise a total surface area of 5040 km^2, 1428 km of coastline and have usually been considered in the literature as a typical example of a second-generation european mass tourist resort [14]. Majorca island (the largest of the Balearic archipelago), with about one million inhabitants, received 9.8 million international arrivals in 2010 [15], presenting one of the highest tourist rates per capita in the world [16].

In this study we estimate the potential effect of commercial sunscreen released in nearshore waters by beachgoers. We conduct field and laboratories studies to evaluate the presence of chemicals products released from sunscreens in coastal seawater and its effect on the marine phytoplankton. Particularly, (1) we present the results for UV chemical filters levels in different fractions of surface marine waters of three Majorca areas; (2) we evaluate the contribution of sunscreen products to the total dissolved P in nearshore waters of a populated beach in Majorca island; and (3) we test the effect of sunscreens on the growth rate of a marine diatom (i.e. *Chaetoceros gracilis*).

Materials and Methods

Ethics Statement

A permit for sampling in the bathing areas of Palmira and Santa Ponça beaches were obtained from the city hall of Calviá (Majorca Island). No specific permit was required for sampling in the Ses Salines Cape. The maritime area of Ses Salines is not private and protected for sampling. The study did not involve endangered or protected species.

Field Sampling

Surface nearshore waters of three beaches around Majorca Island were sampled in August-September 2011. Two areas corresponded to semi-enclosed and densely populated beaches in resort areas (maximum daily density of 3.5–4.5 users m^{-1} shoreline), and the third one, considered a control, was an open and scarcely used beach located in a pristine area (Figure 1).

Seawater Collection

Surface waters (microlayer and subsurface) were collected from a zodiac during August 20th, 21st in the swimming area of Palmira and Santa Ponça beaches, and in September 1st at Ses Salines Cape, respectively. Surface seawater (1 m depth) was collected using a peristaltic pump and pumped through acid-cleaned Teflon tubing coupled to a C-flex tubing (for the Cole-Parmer peristaltic pump head), filtered through an acid-cleaned polypropylene cartridge filter (0.22 μm; MSI, Calyx®) for the dissolved fraction, and collected in a 0.5 L low-density polyethylene plastic bottle [17]. Surface microlayer (SML) samples were collected using a glass plate sampler [18].

Chemical Analysis

Organic UV filters (i.e. BZ-3 and 4-MBC) in seawater were preconcentrated by dispersive liquid-liquid microextraction (DLLME). Stock solutions (500 ng mL^{-1}) of BZ-3 and MBC in ethanol were prepared. Then, a multicomponent aqueous stock solution (25 ng mL^{-1}) of BZ-3 and MBC was prepared from these, and it was used to prepare multicomponent aqueous working solutions (50–250 ng L^{-1} or 200–1000 ng L^{-1}). On the other hand, a stock solution (20 μg mL^{-1}) of deuterated benzophenone (benzophenone-d10 (BZ-d10)), which was used as surrogate, was prepared. Then, an aqueous working solution (65 ng mL^{-1}) was prepared from this.

Ten mL of each of the standard working solutions were pH adjusted with glacial acetic acid to 2–4, then 1–1.5 g of sodium chloride were added to reach a final content within 10–15% (m/v), and finally 25 μL of the working surrogate solution were added. Then, they were subjected to DLLME in 15-mL polyethylene centrifuge tubes, by rapidly injecting pre-mixed 940 μL of acetone with 60 μL of chloroform solutions. Once the cloudy solutions

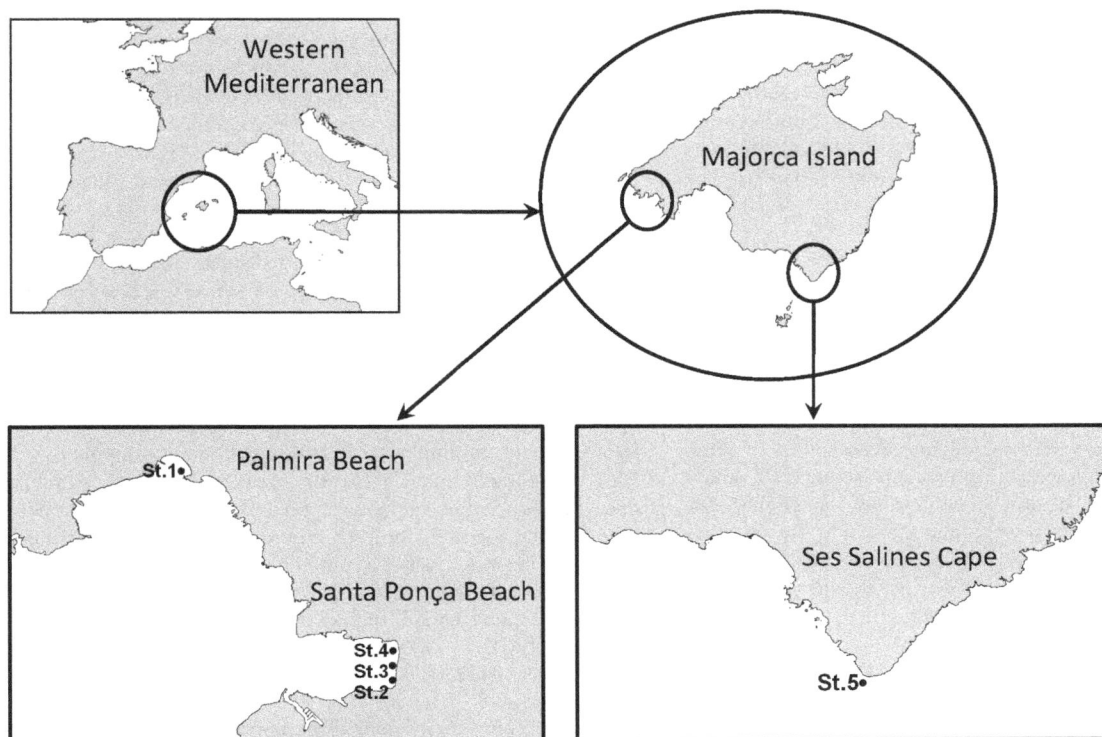

Figure 1. Sampling locations: St.1 (Palmira Beach); St.2, St.3, St.4 (Santa Ponça beach); St.5 (Ses Salines Cape).

were formed, they were centrifuged at 6000 rpm for 5 min. After centrifugation, approximately 25 µL of each of the organic sedimented phases were collected with the aid of a 50 µL syringe and transferred into 100 µL inserts placed inside 1.5 mL injection vials. Then, 2 µL were injected into the GC system coupled to a mass spectrometry (MS) detector operated in positive electron ionisation mode at ionisation energy of 70 eV and with a multiplier voltage set at 1400 V. The inlet temperature was 280 °C and the injection was accomplished in splitless mode (splitless time: 1 min). The separation was run at a 1 mL min^{-1} helium constant flow rate. The oven temperature program was: from 70 °C (1 min) to 170 °C at 10 °C min^{-1}, then to 200 °C at 2 °C min^{-1} and finally to 280 °C (6 min) at 10 °C min^{-1}. The transfer line and ion source temperatures were set at 280 and 250 °C, respectively. The chromatograms were recorded in selected ion monitoring (SIM) mode at the following mass/charge (m/z) ratios: 151, 227 (quantifier) and 228 for BZ3; 128, 211 and 254 (quantifier) for 4-MBC; and 82, 110 (quantifier) and 192 for BZ-d10.

Calibration was performed by plotting A_i/A_{sur} (where A_i is the peak area of the target analyte and A_{sur} that of the surrogate (i.e., BZ-d10), each one obtained by using its quantifier ion) versus target analyte concentration.

For the determination of the soluble fraction of organic UV filters, 10 mL of filtered water samples were pH adjusted, and sodium chloride and surrogate solution added as previously described. For the determination of the total content (i.e., soluble plus particulate fraction), 10 mL of unfiltered water samples were sonicated for 10 min, then they were filtered and treated as described before for the determination of the soluble fraction. Then, they were subjected to DLLME and injected into GC-MS system as previously described for standards. For each target analyte, A_i/A_{sur} was obtained and interpolated in the corresponding calibration line, and the concentration was finally obtained.

Titanium in seawater samples was analyzed with MSFIA-LWCC [19], after prior digestion with potassium peroxodisulfate. Recovery of spikes of Ti in seawater was 101.5±8.8%. Inorganic nutrients in seawater (i.e. PO_4^{3-}, NO_3^-, NO_2^-, SiO_2 and NH_4^+) were determined with an autoanalyzer (Alliance Futura) using colorimetric techniques [20]. The accuracy of the analysis was established using Coastal Seawater Reference Material for Nutrients (MOOS-1, NRC-CNRC), with recoveries of 100.7%, 100.5%, 97.4% and 86.8% for PO_4^{3-}, NO_3^-, NO_2^- and SiO_2, respectively. Zinc in seawater was preconcentrated by the APDC/DDDC organic extraction method [17] and analyzed by ICP-MS (PerkinElmer ELAN DRC-e). The accuracy of the analysis in seawater was established using Coastal water Reference Material for Trace Metals (CASS-4, NRC-CNRC) with recoveries of 95.6%. Metals in sunscreen were analyzed by ICP-AES (Perkin Elmer ICP-AES Optima 5300 DV) after previous chemical digestion [19]. Reference materials of sunscreen consist in three sunscreens with three different sun protection factors (SPF) made in our laboratory, with known concentrations for Ti and Zn. Recoveries were 108±6% for Ti and 101±1% for Zn. All sampling and analytical operations were performed following trace-metal clean techniques. All chemical analysis in samples and stocks solutions were analyzed by triplicate.

Nutrient Release Experiment

The experiment on the release of nutrients from sunscreen was carried out by dissolving 15 g of the sunscreen number 11 (Table S1) into 500 mL of artificial seawater (37 g of NaCl per L), and stirred up to five days in an orbital shaker (300 rpm) inside a culture chamber at 25°C. Dissolved phosphate estimations in

nearshore waters at Palmira beach (Figure S1) are based on the release kinetics experiment. The number of beachgoers was simulated from maximum midday direct counts and allowing for a sinusoidal variation during day length (13 h; Figure S2). Each hour, 25% of the beachgoers were assumed to swim, and only 10% of the sunscreen was dissolved in seawater. A median content of 0.08 µmolP g^{-1} sunscreen was used. Water renewal at the beach was calculated from a series of 6 hours averages of historical current measurements in the sampling site during the same season, obtained using moored Nortek ADCPs [21].

Microalgal Toxicity Biossays

Standard microalgal toxicity bioassays were carried out on ASTM [22] Substitute Ocean Water enriched with f/2 medium on 10^4 initial cellular density populations of Chaetoceros gracilis, obtained from the ICMAN marine culture collection, included in the BIOCISE index. Exposition was carried out at 20±1°C under continuous white light (35.2±1.1 µmol(quantum) m^{-2} s^{-1}) in a controlled culture chamber (Ibercex) in 50 mL of exposition media disposed in 125 mL borosilicate conical flasks topped with synthetic cotton -Perlon-. After a previous wide-range concentration experiment developed in order to get a narrower concentration interval [23], five concentrations plus a control were disposed by triplicate for each compose tested. After 72 hours exposition, cellular counts were performed under light microscopy on Neubauer counting chambers. Considering controls as 100% growth, percentage of growth inhibition was calculated for each cream concentration. Adjusting values of growth inhibition following Hampel et al., 2001 [24], EC50% 72h, it means, effective concentration of pollutant which inhibits microalgal population growth (biomass) at 72 hours [25] was calculated for each cream on this microalgal species.

Results and Discussion

Chemical analysis of the surface nearshore waters of three areas around Majorca Island showed that four of the main chemicals used in commercial sunscreens were detected in the surface waters, with the highest concentrations measured in the unfiltered fraction of the surface microlayer (SML) (i.e. BZ-3:580±50 ng L^{-1}; 4-MBC: 113±7 ng L^{-1}; and Ti: 38±7 µg L^{-1}; Zn: 10.8 µg L^{-1}; Table 1). Because of the lipophilic characteristic of these cosmetics [4], and the insolubility of many of their chemicals, sunscreen products tend to be more concentrated in the surface microlayer (SML) and to accumulate in soils and particles [26]. Levels of these chemicals co-varied throughout the day reaching the highest concentrations between 14:00 and 18:00 h (Figure 2A–D), a few hours after the beachgoers maximum numbers occurring around noon, and when sunlight radiation is maximum and sunscreen application is expected to be at the highest level of application [4]. In the case of the unfiltered fraction of the four compounds (i.e. BZ3, MBC, Zn and Ti) in the surface microlayer, midday concentrations exceeded between 60 and 90% background values (observed during night or early morning), suggesting a common source for these products. Even in the pristine beach of Ses Salines Cape (Figure 1) detectable concentrations of BZ-3 and 4-MBC (16±3 and 26±1 ng L^{-1}, respectively) were measured in the total fraction of SML, and in subsurface water (36±2 and 27±2 ng L^{-1}, respectively), whilst Ti was only detected in the total water fraction of the SML (23.7±1.7 ng L^{-1}). This suggests a high degree of alongshore connectivity due to the persistence of these products that results in their generalized impact around the island.

Our study also shows the release of some inorganic nutrients (i.e. PO_4^{3-}, NO_3^-, and NH_4^+) which may affect algal growth. A

Table 1. Midday concentration ± standard deviation (n = 3) of BZ-3 (ng L^{-1}), 4-MBC (ng L^{-1}), Ti (µg L^{-1}), Zn (µg L^{-1}) and nutrients (nmol.L^{-1}) in the unfiltered (Total) and filtered (<0.22 µm; Diss) fraction of the Surface microlayer (SML) and Subsurface (1 cm) seawater (SW) samples.

Beach	Sampling stations	Total SML				Diss SML								
		BZ-3	4-MBC	Ti	Zn	BZ-3	4-MBC	Ti	Zn	PO$_4^{3-}$	NO$_3^-$	NO$_2^-$	SiO$_2$	NH$_4^+$
Palmira	St.1	245.6±12.0	109.6±10.0	37.6±7.3	3.3	123±8.4	46.7±6.7	nd	2.0	70.5	718.8	37.8	5091.9	123.4
Santa Ponça	St.2	-	-	-	-	-	-	-	-	-	-	-	-	-
Santa Ponça	St.3	174.8±10.5	59.8±3.9	12.1±1.2	10.8	156.1±6.0	55.4±1.2	nd	7.7	8.5	452.5	3.9	5032.8	55.3
Santa Ponça	St.4	-	-	-	-	-	-	-	-	-	-	-	-	-
Ses Salines (control)	St.5	15.8±3.0	25.7±1.2	23.7±1.7	0.8	nd	nd	nd	0.5	153.5	933.0	38.5	969.6	66.5

Beach	Sampling stations	Total SW				Diss SW								
		BZ-3	4-MBC	Ti	Zn	BZ-3	4-MBC	Ti	Zn	PO$_4^{3-}$	NO$_3^-$	NO$_2^-$	SiO$_2$	NH$_4^+$
Palmira	St.1	143.6±7.4	62.5±4.5	nd	0.6	95.8±2.9	30.5±1.6	nd	0.4	14.2	161.2	nd	6653.5	129.5
Santa Ponça	St.2	76.2±4.6	65.0±5.0	nd	1.0	40.4±2.6	37.4±3.1	nd	0.9	nd	1218.1	nd	6209.9	81.9
Santa Ponça	St.3	155.4±5.6	51.4±2.5	nd	0.7	119.6±1.7	29.3±5.0	nd	0.7	nd	15.6	nd	5414.3	84.4
Santa Ponça	St.4	314.8±2.9	47.5±3.9	nd	0.7	241.9±5.2	14.7±4.0	nd	0.9	nd	299.0	100.3	6541.2	81.5
Ses Salines (control)	St.5	36.3±2.2	26.6 ± 2.0	nd	0.1	nd	nd	nd	0.1	95.1	2892.3	7.0	702.1	66.2

(-) not collected; nd: not detected.

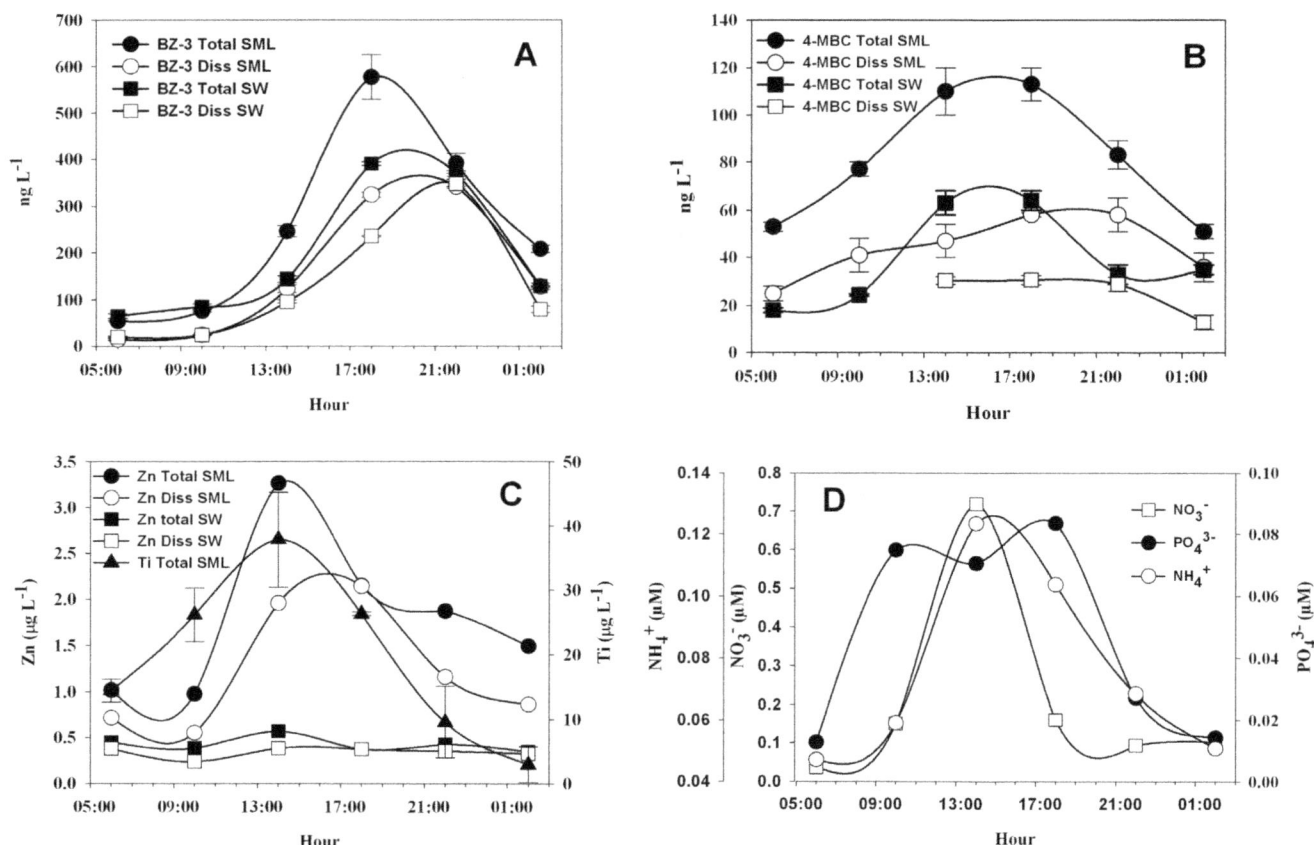

Figure 2. Concentration of BZ-3 (A), 4-MBC (B), Zn (C), Ti (C) and nutrients (D) in the unfiltered (Total) and filtered (<0.22 μm; Diss) fraction of the Surface microlayer (SML) and Subsurface (1 cm) seawater (SW) samples from Palmira Beach. Nutrients concentrations are only plotted in the dissolved fraction of the SML (D). Error bars represent the standard deviation (n = 3).

release kinetic experiment consisting of shaking a commercial sunscreen (num. 11 in Table S1) in artificial seawater (37g L^{-1} NaCl) showed rapid dissolution of silicate and nitrogen compounds (in the first 16 hours), and lower rates (if any) thereafter

(Figure 3). Conversely, the release dynamics of PO_4^{3-} was fairly more progressive and linear (0.01 μmol g^{-1} h^{-1}). Nutrient release kinetics varies depending on sunscreen composition, but since a majority of the tested products stabilized their release after 72 h

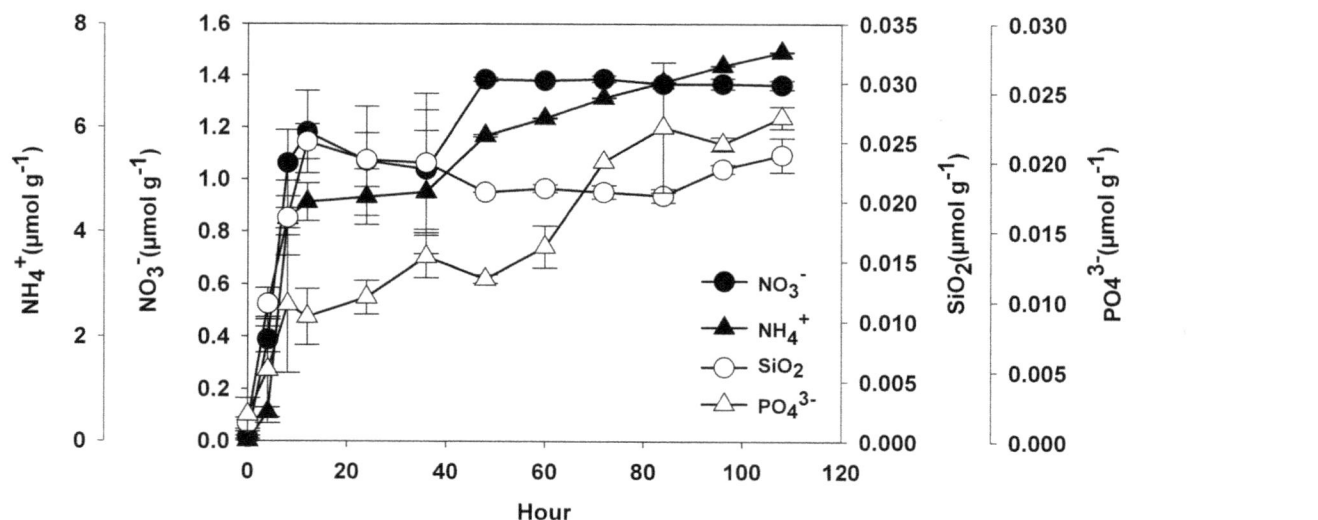

Figure 3. Kinetics of nutrient release from a commercial sunscreen in seawater (n = 3). NO_2^- was not detected.

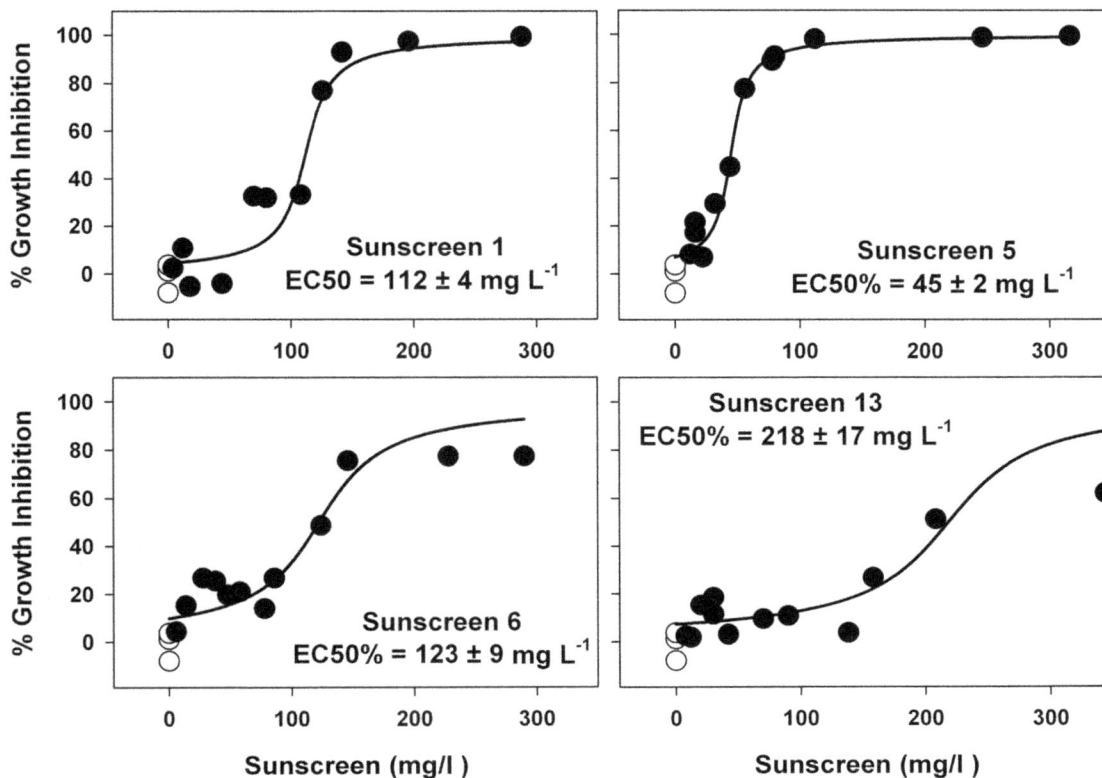

Figure 4. Growth inhibition rate for *Chaetoceros gracilis* exposed to different concentrations of commercial sunscreens after 72 hours culture. White circles represent controls.

we assumed that concentrations at this time to be indicative values for total content. The 13 commercial sunscreens tested provided final concentrations in water of 2 ± 5 µmol g^{-1} of PO_4^{3-}, 0.2 ± 0.4 µmol g^{-1} of NO_3^-, 0.001 ± 0.002 µmol g^{-1} of NO_2^-, 2 ± 2 µmol g^{-1} of SiO_2 and 0.02 ± 0.01 µmol g^{-1} of NH_4^+ (Table S1). It is particularly notable that on the average the release of PO_4^{3-} occurs in relatively high molar ratios compared to nitrogen forms. This high PO_4^{3-} mean concentration is nevertheless based on only a few products containing high phosphate concentrations (up to 17 µmol g^{-1}; Table S1), but its importance should not be ignored because they are widely consumed.

Inorganic nutrient inputs stimulate primary production in oligotrophic waters, as in the western Mediterranean Sea [27], so that recreational activities at some sites may represent a significant and previously overlooked nutrient source for the nearshore environment. Based on the estimated sunscreen dose (half of the recommended 2 mg cm^{-2} and ~36 g/adult person) [28] direct counts on the number of beachgoers, and water renewal estimations depicted from wind conditions, we conservatively estimated that on a calm day (low water renewal) sunscreens may increase by an average of 55% above the otherwise low PO_4^{3-} concentrations in Palmira Beach (Figure S1). Under low water renewal conditions, up to 0.12 µmol L^{-1}, this represents an approximate doubling of mean offshore concentrations. While other land sources are considered quantitatively more important, this modest, but significant contribution in a P-limited area could play an important role in the dynamics of nearshore phytoplankton. It has been demonstrated that concentrations of 20 nmol L^{-1} of P induce significant phytoplankton response in the coastal Mediterranean sea during summer [27]. Furthermore, in addition

to inorganic nutrients and chemical UV filters, sunscreens contain other constituents such as Al and Fe. These metals (together with Ti and Zn) were detected in at least one of the 13 commercial sunscreens analyzed (five of which are among the ten bestsellers in Spanish pharmacies in 2011, according to Sell Out database from IMS Health® [29]) (Table S2). Iron, together with P, is an essential micronutrient for phytoplankton growth and it is also suggested to limit primary production in the western Mediterranean [30].

We tested the effect of sunscreens on the growth rate of the marine phytoplankton *Chaetoceros gracilis*, which is a widespread species in the western Mediterranean [31]. The acute toxicity was measured by calculating half maximal effective concentration (EC_{50}) after 72h incubation, resulting in an average of 125 ± 71 mg L^{-1}. The sunscreen num. 5 (a solar spray; Table S1) induced the highest level of toxicity with a EC_{50} of 45 ± 2 mg L^{-1}, while sunscreen num. 13 (a solar milk) presented the lowest effects ($EC_{50}=218\pm17$ mg L^{-1}) (Figure 4). The concentrations of sunscreens at which EC50 occurs ($45–218$ mg L^{-1}) are higher than environmental concentrations measured in our studied areas. These amounts of sunscreen (at which EC50 occurs) are reflecting the threshold for acute toxicity. However, even at very low sunscreen concentrations certain inhibitory effect is observed (Figure 4). The higher toxicity of the spray versus cream or milk formats could be due to its higher content of hydrosoluble compounds, making them more bio-available to phytoplankton. Our results demonstrate the toxicity of the commercial sunscreen for marine phytoplankton, and confirm previous studies of toxicity carried out with individual organic and inorganic UV filters on marine organisms (including green algae, crustacean, phytoplankton and fishes) [32–38].

Conclusions

More than half of today's world population live in coastal areas, and estimates for the future suggest that in three decades from now nearly 75 percent of the world's population will live along coasts [39]. This fact, combined with data showing that sun protection products are one of the fastest growing products globally [40], points to sunscreens as a potential pollutant with implications for the coastal marine ecosystem. The results presented here suggest that sunscreens in coastal waters may produce deleterious effects in the coastal ecosystem, either, by inhibiting growth of some marine phytoplankton species or by adding essential micronutrients which may stimulate the growth of others.

Supporting Information

Figure S1   Modeled variations of PO_4^{3-} (µM) at Palmira Beach (Majorca) estimated from sinusoidal variations of beachgoers and 6 hours averaged current velocities. Dashed line indicates shelf water background concentration.

Figure S2   Diel variation of beach users at Palmira during a labor day (bars) and sinusoidal adjustment used in model simulations (line).

References

1. UNEP (2009) Sustainable Coastal Tourism - An Integrated Planning and Management Approach. ISBN: 978-92-807-2966-5. 133 p.
2. Giokas DL, Salvador A, Chisvert A (2007) UV filters: From sunscreens to human body and the environment. TrAC Trends in Analytical Chemistry 26: 360-374.
3. Salvador A, Chisvert A (2007) Analysis of cosmetic products. Amsterdam; London: Elsevier Science. 83 p.
4. Santos AJM, Miranda MS, Esteves da Silva JCG (2012) The degradation products of UV filters in aqueous and chlorinated aqueous solutions. Water Research 46: 3167-3176. doi:10.1016/j.watres.2012.03.057.
5. Peck AM (2006) Analytical methods for the determination of persistent ingredients of personal care products in environmental matrices. Anal Bioanal Chem 386: 907-939. doi:10.1007/s00216-006-0728-3.
6. Weisz A, Milstein SR, Scher AL (2007) 4.2. Colouring Agents in Cosmetic Products (Excluding Hair Dyes): Regulatory Aspects and Analytical Methods. Analysis of cosmetic products: 153.
7. Quartier S, Garmyn M, Becart S, Goossens A (2006) Allergic contact dermatitis to copolymers in cosmetics-case report and review of the literature. Contact dermatitis 55: 257-267.
8. Tønning K, Jacobsen E, Pedersen E (2009) Survey and Health Assessment of the exposure of 2 year-olds to chemical substances in Consumer Products. Danish Misnistry of the Environment. 327 p.
9. Council Directive 76/768/EEC of 27 July 1976 on the approximation of the laws of the Member States relating to cosmetic products (1976) and its successive amendments and adaptations. Official Journal L 262, 27/09/1976 P 0169-0200. Available: http://eur-lex.europa.eu/LexUriServ/LexUriServ.do?uri=CELEX:31976L0768:EN:HTML. Accessed 11 September 2012.
10. ACNielsen Global Services (2007) What's Hot around the Globe: Insights on Personal Care Products. Available: http://www.accessdata.fda.gov/scripts/cdrh/cfdocs/cfcfr/CFRSearch.cfm?CFRPart=70-82. Accessed 11 September 2012.
11. Poiger T, Buser HR, Balmer ME, Bergqvist PA, Müller MD (2004) Occurrence of UV filter compounds from sunscreens in surface waters: regional mass balance in two Swiss lakes. Chemosphere 55: 951-963.
12. Díaz-Cruz MS, Barceló D (2009) Chemical analysis and ecotoxicological effects of organic UV-absorbing compounds in aquatic ecosystems. TrAC Trends in Analytical Chemistry 28: 708-717. doi:10.1016/j.trac.2009.03.010.
13. Sundseth K (2009) Natura 2000 in the Mediterranean Region. Luxembourg: European Commission Environment Directorate General. ISBN 978-92-79-11587-5.10 p.
14. Knowles T, Curtis S (1999) AID-JTR135>3.0.CO;2-6.
15. Ministerio de Industria, Energía y Turismo (2012) Instituto de Estudios Turísticos. Available: http://www.iet.tourspain.es/es-es/estadisticas/frontur/paginas/default.aspx. Accessed 11 September 2012.
16. Garín-Muñoz T, Montero-Martín LF (2007) Tourism in the Balearic Islands: A dynamic model for international demand using panel data. Tourism Management 28: 1224-1235. doi:10.1016/j.tourman.2006.09.024.
17. Tovar-Sánchez A (2012) 1.17 - Sampling Approaches for Trace Element Determination in Seawater. In: Editor-in-Chief: Janusz Pawliszyn, editor.

Table S1   Concentration (average ± SDV) of nutrients in nmol g⁻¹ released from commercial sunscreens in seawater after 72 h shaking. Samples were analyzed by triplicate except for nutrients from sunscreen 4. SPF (Sun Protection Factor).

Table S2   Concentration of metals in µg g⁻¹ (average ± SDV, n = 3) in commercial sunscreens. SPF (Sun Protection Factor). Other elements such as Ag, Cd, Co, Cr, Cu, Mn, Mo, Ni, Si, Sr, Pb, Tl, V, and Zr were not detected.

Acknowledgments

We thank A. Massanet, P. Vidal, J. González, M.J. Alonso and J. Marcos for technical assistance. We thank Ajuntament de Calvià for the field support.

Author Contributions

Conceived and designed the experiments: AT-S DS-Q JLB AC AS IM JB. Performed the experiments: DS-Q JLB AC IM JB. Analyzed the data: AT-S DS-Q GB JLB AC AS IM JB. Contributed reagents/materials/analysis tools: AT-S DS-Q GB JLB AC AS. Wrote the paper: AT-S DS-Q GB AC AS.

Comprehensive Sampling and Sample Preparation. Oxford: Academic Press. 317-334. Available:http://www.sciencedirect.com/science/article/pii/B978012381373200017X. Accessed 1 October 2012.
18. Stortini AM, Cincinelli A, Degli Innocenti N, Tovar-Sánchez A, Knulst J (2012) 1.12 - Surface Microlayer. In: Editor-in-Chief: Janusz Pawliszyn, editor. Comprehensive Sampling and Sample Preparation. Oxford: Academic Press. 223-246. Available: http://www.sciencedirect.com/science/article/pii/B9780123813732000181. Accessed 1 October 2012.
19. Páscoa RNMJ, Tóth IV, Almeida AA, Rangel AOSS (2011) Spectrophotometric sensor system based on a liquid waveguide capillary cell for the determination of titanium: Application to natural waters, sunscreens and a lake sediment. Sensors and Actuators B: Chemical 157: 51-56. doi:10.1016/j.snb.2011.03.025.
20. Grasshoff K, Almgreen T (1976) Methods of seawater analysis. Verlag Chemie. 344 p.
21. Basterretxea G, Garcs E, Jordi A, Angls S, Mas M (2007) Modulation of nearshore harmful algal blooms by in situ growth rate and water renewal. Mar Ecol Prog Ser 352: 53-65. doi:10.3354/meps07168.
22. ASTM (2008) Practice for the Preparation of Substitute Ocean Water. ASTM International. p. Available: http://www.astm.org/Standards/D1141.htm. Accessed 12 September 2012.
23. Moreno-Garrido I, Lubián LM, Soares AMVM (2000) Influence of Cellular Density on Determination of EC50 in Microalgal Growth Inhibition Tests. Ecotoxicology and Environmental Safety 47: 112-116. doi:10.1006/eesa.2000.1953.
24. Hampel M, Moreno-Garrido I, Sobrino C, Lubián LM, Blasco J (2001) Acute Toxicity of LAS Homologues in Marine Microalgae: Esterase Activity and Inhibition Growth as Endpoints of Toxicity. Ecotoxicology and Environmental Safety 48: 287-292. doi:10.1006/eesa.2000.2028.
25. Nyholm N (1990) Expression of results from growth inhibition toxicity tests with algae. Arch Environ Contam Toxicol 19: 518-522. doi:10.1007/BF01059070.
26. Botta C, Labille J, Auffan M, Borschneck D, Miche H, et al. (2011) TiO2-based nanoparticles released in water from commercialized sunscreens in a life-cycle perspective: Structures and quantities. Environmental Pollution 159: 1543-1550. doi:16/j.envpol.2011.03.003.
27. Vaulot D, Lebot N, Marie D, Fukai E (1996) Effect of Phosphorus on the Synechococcus Cell Cycle in Surface Mediterranean Waters during Summer. Appl Environ Microbiol 62: 2527-2533.
28. Diffey B (2001) Sunscreen isn't enough. Journal of Photochemistry and Photobiology B: Biology 64: 105-108.
29. IMS (2012) IMS Health. Available: http://www.imshealth.com/portal/site/ims. Accessed 11 September 2012.
30. Bonnet S, Guieu C (2006) Atmospheric forcing on the annual iron cycle in the western Mediterranean Sea: A 1-year survey. J Geophys Res 111: C09010. doi:10.1029/2005JC003213.
31. Hernández-Almeida I, Bárcena MA, Flores JA, Sierro FJ, Sanchez-Vidal A, et al. (2011) Microplankton response to environmental conditions in the Alboran Sea (Western Mediterranean): One year sediment trap record. Marine Micropaleontology 78: 14-24. doi:10.1016/j.marmicro.2010.09.005.

32. Miller RJ, Lenihan HS, Muller EB, Tseng N, Hanna SK, et al. (2010) Impacts of Metal Oxide Nanoparticles on Marine Phytoplankton. Environmental Science & Technology 44: 7329–7334. doi:10.1021/es100247x.

33. Miller RJ, Bennett S, Keller AA, Pease S, Lenihan HS (2012) TiO2 Nanoparticles Are Phototoxic to Marine Phytoplankton. PLoS ONE 7: e30321. doi:10.1371/journal.pone.0030321.

34. Sieratowicz A, Kaiser D, Behr M, Oetken M, Oehlmann J (2011) Acute and chronic toxicity of four frequently used UV filter substances for Desmodesmus subspicatus and Daphnia magna. Journal of Environmental Science and Health, Part A 46: 1311–1319. doi:10.1080/10934529.2011.602936.

35. Kaiser D, Sieratowicz A, Zielke H, Oetken M, Hollert H, et al. (2012) Ecotoxicological effect characterisation of widely used organic UV filters. Environmental Pollution 163: 84–90. doi:10.1016/j.envpol.2011.12.014.

36. Chen J, Dong X, Xin Y, Zhao M (2011) Effects of titanium dioxide nanoparticles on growth and some histological parameters of zebrafish (Danio rerio) after a long-term exposure. Aquatic Toxicology 101: 493–499. doi:10.1016/j.aquatox.2010.12.004.

37. Coronado M, De Haro H, Deng X, Rempel MA, Lavado R, et al. (2008) Estrogenic activity and reproductive effects of the UV-filter oxybenzone (2-hydroxy-4-methoxyphenyl-methanone) in fish. Aquatic Toxicology 90: 182–187. doi:10.1016/j.aquatox.2008.08.018.

38. Kunz PY, Gries T, Fent K (2006) The ultraviolet filter 3-benzylidene camphor adversely affects reproduction in fathead minnow (Pimephales promelas). Toxicol Sci 93: 311–321. doi:10.1093/toxsci/kfl070.

39. Division on Earth & Life Studies (2012) Ocean - Reports: Academies' Findings - Division on Earth and Life Studies. Available: http://dels.nas.edu/Ocean/Coastal-Hazards/Reports-Academies-Findings. Accessed 12 September 2012.

40. Executive News Report from Nielsen Global Services (2007) What's Hot around the Globe: Insights on Personal Care Products. Available: http://kr.en.nielsen.com/reports/GlobalServiceStudies.shtml. Accessed 12 September 2012.

3

Application of Zero-Valent Iron Nanoparticles for the Removal of Aqueous Zinc Ions under Various Experimental Conditions

Wen Liang[1], Chaomeng Dai[1,2]*, Xuefei Zhou[1], Yalei Zhang[1]*

1 State Key Laboratory of Pollution Control and Resources Reuse, Tongji University, Shanghai, China, **2** College of Civil Engineering, Tongji University, Shanghai, China

Abstract

Application of zero-valent iron nanoparticles (nZVI) for Zn^{2+} removal and its mechanism were discussed. It demonstrated that the uptake of Zn^{2+} by nZVI was efficient. With the solids concentration of 1 g/L nZVI, more than 85% of Zn^{2+} could be removed within 2 h. The pH value and dissolved oxygen (DO) were the important factors of Zn^{2+} removal by nZVI. The DO enhanced the removal efficiency of Zn^{2+}. Under the oxygen-contained condition, oxygen corrosion gave the nZVI surface a shell of iron (oxy)hydroxide, which could show high adsorption affinity. The removal efficiency of Zn^{2+} increased with the increasing of the pH. Acidic condition reduced the removal efficiency of Zn^{2+} by nZVI because the existing H^+ inhibited the formation of iron (oxy)hydroxide. Adsorption and co-precipitation were the most likely mechanism of Zn^{2+} removal by nZVI. The FeOOH-shell could enhance the adsorption efficiency of nZVI. The removal efficiency and selectivity of nZVI particles for Zn^{2+} were higher than Cd^{2+}. Furthermore, a continuous flow reactor for engineering application of nZVI was designed and exhibited high removal efficiency for Zn^{2+}.

Editor: Vipul Bansal, RMIT University, Australia

Funding: This work was funded by the National Key Technologies R&D Program of China (No. 2012BAJ25B04), New Century Excellent Talents in University (NCET-11-0391), the Project of Shanghai Science and Technology Commission (No. 11QH1402600), and the National Natural Science Foundation of China (key program No. 21246001, 51138009, 41101480). The funders had no role in study design, data collection and analysis, decision to publish, or preparation of the manuscript.

Competing Interests: The authors have declared that no competing interests exist.

* E-mail: daichaomeng@tongji.edu.cn (CD); zhangyalei@tongji.edu.cn (YZ)

Introduction

Zinc is one of the trace elements closely related to human health. It is essential for living organisms [1]. But excessive amount of zinc in the environment is toxic to man, animals and plants. When the concentration of zinc increases above a limit, it may lead to acute gastroenteritis, peritonitis, growth retardation and even shock or death [2–4]. Zinc toxicity to aquatic organisms and ecosystems has been frequently reported [5,6]. Excessive zinc may lead to the death of fishes [7]. Irrigation water containing excessive zinc may cause poor crop growth and affect the health of the eaters [8,9]. The presence of zinc is mainly from industrial pollution, such as galvanizing plants, pigments, mine drainage, etc. Zinc is commonly detected in the aquatic environment with its widely use in industry [10]. Considering its toxicity and non-biodegradability, it is necessary to effectively remove zinc. Current main zinc removal techniques from aqueous solutions include physico-chemical precipitation, ion exchange, complexation, adsorption, electrodialysis, etc. [11–13].

Nanoscale zero-valent iron (nZVI) has been investigated as a new tool for the reduction of contaminated water and soil for more than 10 years, and the technology has been applied in many countries worldwide. The nZVI has been proven as a highly effective technology for the removal or degradation of various chemical pollutants, such as β-lactam and nitroimidazole based antibiotics [14], azo dyes [15], chlorinated solvents [16], chlorinated pesticides [17], organophosphates [18], nitroamines [19], nitroaromatics [16], alkaline earth metals [20], transition metals [21,22], post-transition metals [21,22], metalloids [21], actinides [21], etc. The successful application of nZVI in dissolved metals removal was explored and reported by many researchers [23].

The determined contaminant removal pathways of nZVI include adsorption, complexation, (co)precipitation and surface-mediated chemical reduction [24]. The removal mechanism by nZVI mainly involves adsorption/surface complexation for metal ions such as Zn^{2+} and Cd^{2+} which have the standard electrode potentials (E^0) for reduction to a metallic state that are very close to, or more negative than Fe^0 (-0.44 V). For metal ions such as Hg^{2+} and Cu^{2+} whose E^0 are much more positive than that of Fe^0, removal of metal ions is mainly realized via surface-mediated reductive precipitation in comparison. While metal cations are only slightly more electropositive than Fe^0, the removal is mainly realized via the adsorption with partial chemical reduction [22].

In this study, the removal mechanism of Zn^{2+} by nZVI was investigated based on the operation conditions, including nZVI solids loading, pH value and dissolved oxygen (DO). The X-ray Photoelectron Spectroscopy (XPS) of nZVI was performed to detect the valence of zinc and iron to determine whether chemical reaction happened. Furthermore, a continuous flow reactor was designed and applied to remove Zn^{2+} for evaluating the engineering application of nZVI.

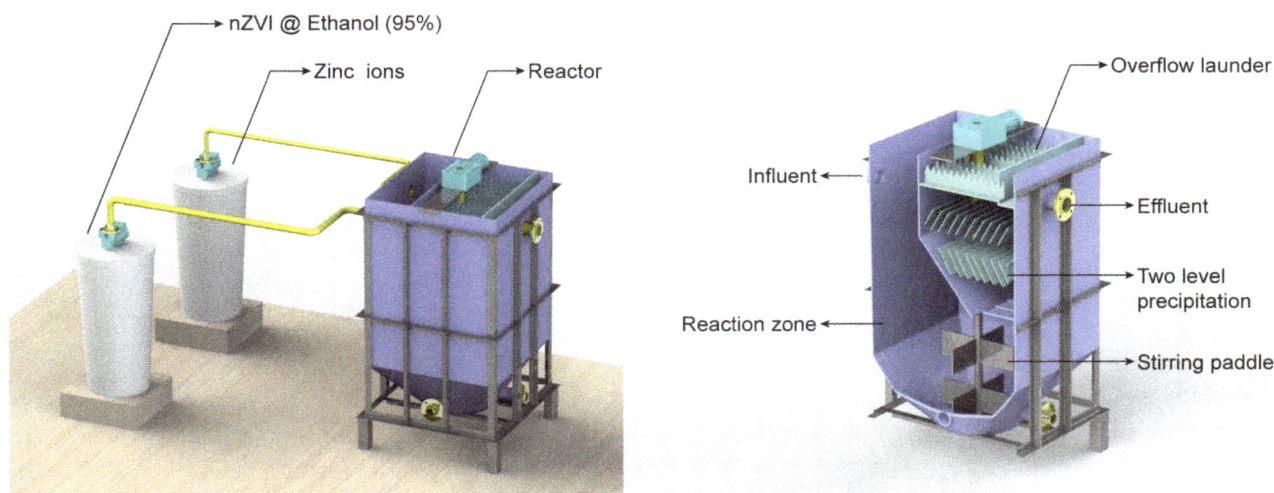

Figure 1. The continuous flow reactor. A continuous flow reactor was designed to realize the continuous removal of Zn^{2+} by nZVI.

Materials and Methods

Chemicals and Materials

Zinc chloride ($ZnCl_2$), analytic grade cadmium acetate ($Cd[CH_3COO]_2 \cdot 2H_2O$), sodium borohydride ($NaBH_4$, 98%) and ferric chloride anhydrous ($FeCl_3$) were purchased from Aladin (Shanghai, China). Hydrochloric acid (HCl), sodium hydroxide (NaOH), nitric acid (HNO_3) and anhydrous ethanol (C_2H_5OH) were obtained from Sinopharm Chemical Reagent Shanghai Co., Ltd. (Shanghai, China). All chemicals were used without further purification.

Deionized water was prepared with a Milli-Q water purification system (Millipore, Bedford, MA, USA). Microporous membranes (0.22 μm×50 mm) were obtained from CNW (Germany).

Synthesis of nZVI

The nZVI was synthesized according to the method of liquid-phase reduction of ferric trichloride by sodium borohydride [25]. The sodium borohydride ($NaBH_4$, 0.5 M) and ferric chloride anhydrous ($FeCl_3$, 0.1 M) with the volume ratio of 1:1 were vigorously reacted. Then the generated jet-black nZVI particles were collected through vacuum filtration and respectively washed with deionized water and anhydrous ethanol for three times. Finally, fresh nZVI particles were stored in anhydrous ethanol solution at 4°C in order to avoid oxidization prior to use.

Characterization of nZVI

Samples of nZVI were prepared by depositing a few droplets of ethanol-diluted nZVI solution onto a carbon-coated transmission electron microscopy (TEM) grid in an oxygen-limiting chamber. But the samples were exposed to air transitorily during transfer from the oxygen-limiting chamber to the microscope. The high-resolution TEM observation was performed using a JEOL JEM 2011 HR-TEM operated at 200 kV with an INCA EDS system.

The specific surface area of nZVI was measured by BET analysis.

The nZVI particles were dried in a refrigerated drying chamber and then kept under seal at 4°C for X-ray photoelectron spectroscopy (XPS) measurement and X-ray diffraction (XRD) measurement. The XPS spectra were obtained with a Perkin Elmer PHI 5000 ESCA System under Al Kα radiation at

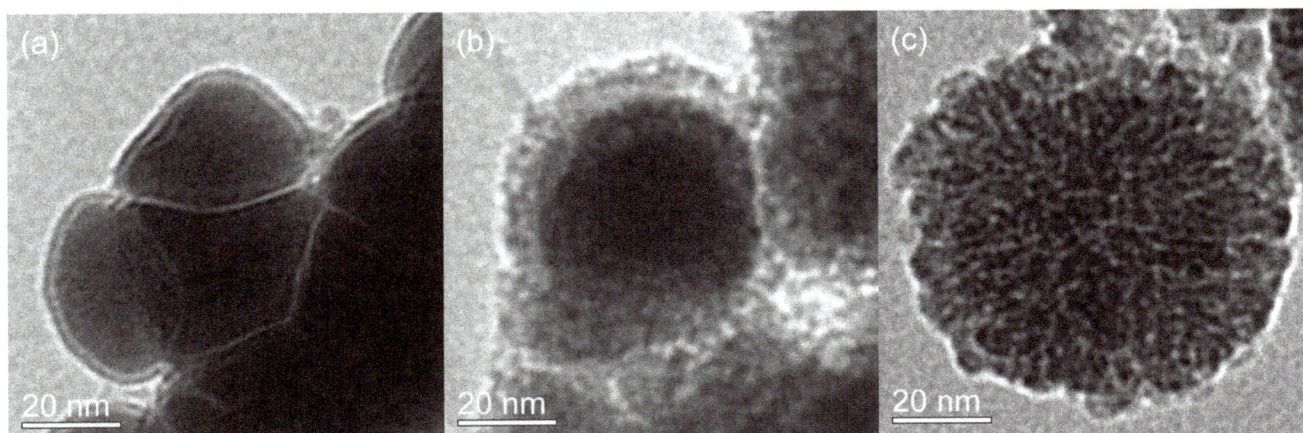

Figure 2. The TEM analysis of nZVI. Three kinds of nZVI particles were analyzed by TEM: (a) the fresh nZVI particles, (b) the stock nZVI after the reaction with Zn^{2+}, and (c) the stock nZVI as a blank control sample.

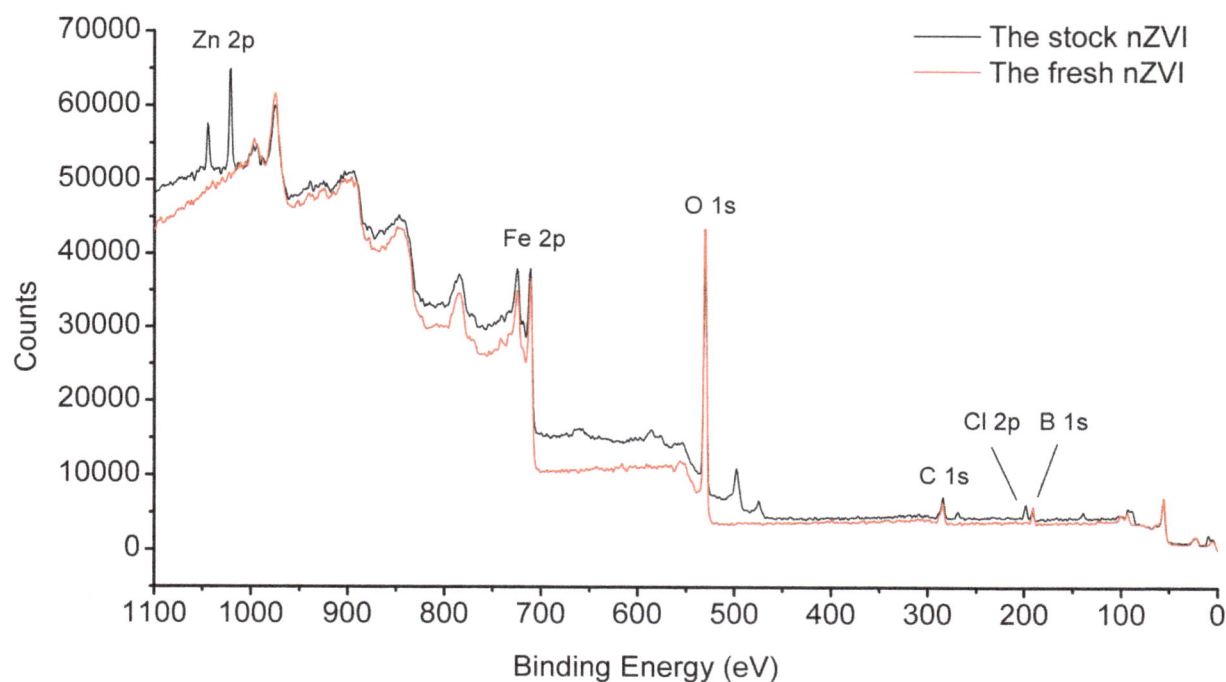

Figure 3. The XPS full scan analysis of nZVI.

1486.6 eV to study the conversion of the element contents and valence states on nZVI surface. The XRD was carried out on a Bruker X-ray D8 Advance diffraction instrument (Cu Kα) and the diffraction angle (2θ) from 10 to 90° was scanned.

Batch Experiments

A 100 mg/L stock solution of $ZnCl_2$ was prepared with deionized water. Uptake reactions were initiated by the addition of nZVI particles into 100 mL aliquots of Zn stock solution. The nZVI loading concentration in the solution was 0.1, 0.2, 0.3, 0.4, 0.6, 0.8, 1.0, 1.2, 1.4, 1.6, 1.8 and 2.0 g/L, respectively, at a zinc ion concentration of 100 mg/L. After mixing, the reactors were continuously shaken for 2 hours on an orbital shaker. The optimum loading of nZVI was obtained by comparing the results of the above experiments. All the experiments were performed in triplicate.

The effect of oxygen on Zn^{2+} removal by nZVI was investigated under the oxygen-limiting and oxygen-contained conditions with the optimum nZVI loading. The oxygen-limiting condition was established by flowing nitrogen over the solution. The initial solution pH value was controlled at 5. Reaction time was 5, 10, 15, 20, 25, 30, 40, 50, 60, 70, 80, 90 and 100 min, respectively. All the experiments were performed in triplicate.

To investigate the effect of solution pH on the Zn^{2+} removal by nZVI, the initial solution pH was adjusted from 3 to 5 with the initial Zn^{2+} concentration at 100 mg/L by small amounts of HCl or NaOH solution. Then water samples with different pH values were applied to 1 g/L nZVI. All the experiments were performed in triplicate.

To investigate the effect of cadmium on Zn^{2+} removal by nZVI, three different water samples were used. Sample 1 contained 100 mg/L Zn^{2+} solution. Sample 2 contained 100 mg/L mixture of Zn^{2+} Cd^{2+}. Sample 3 contained 100 mg/L Cd^{2+} solution. The uptake experiments were conducted with the optimum loading of nZVI for 2 h. All the experiments were performed in triplicate.

All solution samples were filtered with 0.22 μm membrane acidified with 4% ultrahigh purity HNO_3 before analysis. Zinc and iron in the sample were determined by inductively coupled plasma optical emission spectrometry (ICP-OES, PerkinElmer Optima 2100 DV, USA).

Experiment in Continuous Flow Reactor

A continuous flow reactor was designed to realize the continuous removal of Zn^{2+} by nZVI (Figure 1). The reactor

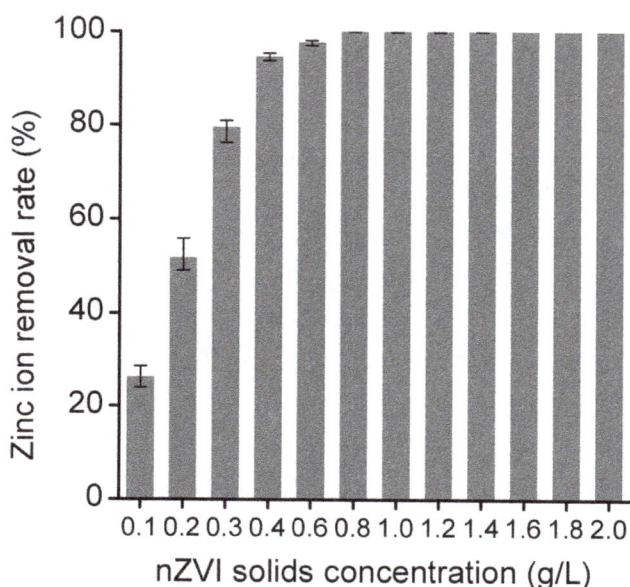

Figure 4. Effect of nZVI solids concentration on Zn^{2+} removal.

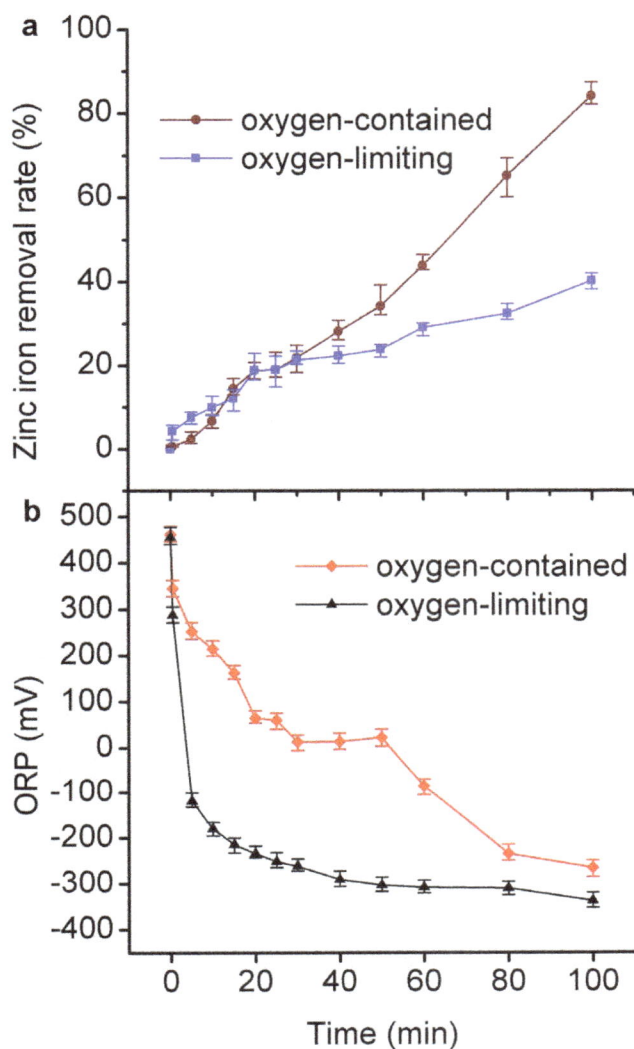

Figure 5. The Zn^{2+} removal and ORP in oxygen-contained and oxygen-limiting conditions.

Figure 6. Effect of pH value on Zn^{2+} removal.

was composed of reaction zone and precipitation zone, with the dimension of 0.2 m length, 0.2 m width, and 0.5 m height. The tank with inclined-plate could enhance the solid-liquid separation. The hydraulic retention time in the reaction zone was set to be 1 h and the nZVI solids concentration was set to be 1.0 g/L. The solution of Zn^{2+} with a concentration of 15 mg/L was stored in a reservoir tank and flowed to the reactor with a peristaltic pump at 120 mL/min. The design of two level precipitations made the nZVI particles settling to the reaction zone, so that the nZVI particles could be reused. The effluent was periodically sampled for analysis.

Statistical Analyses

One-way ANOVA was performed to assess the removal efficiency of Zn^{2+} by nZVI. Statistical significance was evaluated at $p<0.05$ level. All statistical analyses were performed with SPSS software (Ver 13.0; SPSS, Chicago, IL, USA). The experimental data were expressed as mean±standard deviation (SD).

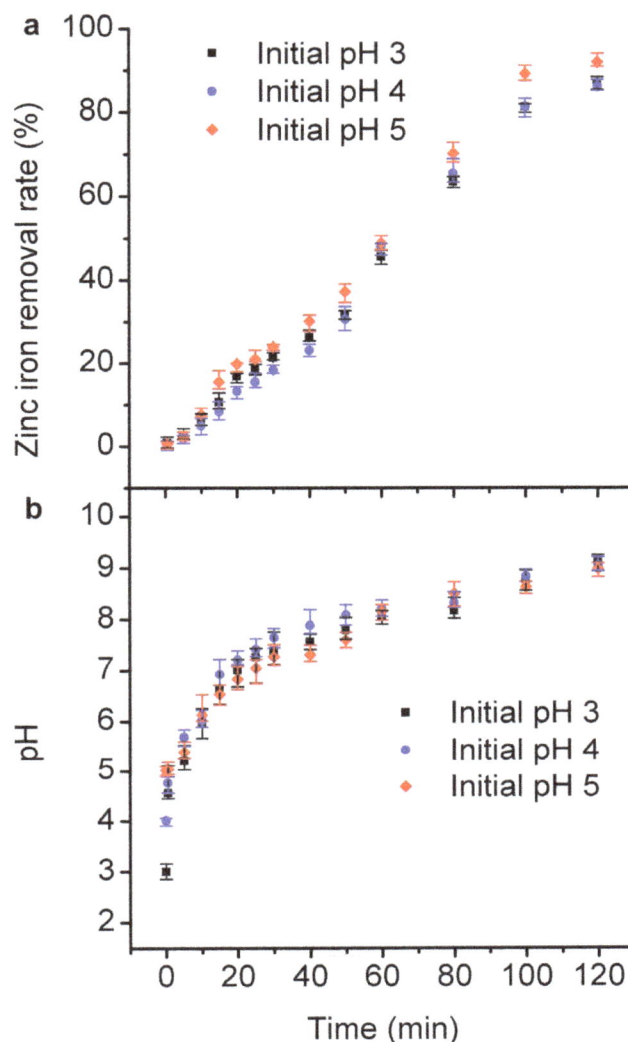

Results and Discussion

Characterization of nZVI

Three kinds of nZVI particles were analyzed by transmission electron microscopy (TEM). The fresh nZVI particles were shown in Figure 2(a). The smooth sphere indicated that little oxidation happened on the surface. The nZVI particles were typically less than 100 nm in diameter. As shown in Figure 2(b), the stock nZVI after the reaction with the concentration of 100 mg/L Zn^{2+} in 2 h, appeared a single particle composed of a dense core surrounded by a thin amorphous shell, which indicated that the reaction occurred on the surface of nZVI, and that a core-shell structure was formed during the reaction. As shown in Figure 2(c), the stock nZVI as the blank control showed a complete oxidation. The core structure disappeared due to corrosion, which indicated that the reaction on the surface of nZVI could protect Fe-core from further corrosion.

The specific surface area of the nZVI sample was measured by BET analysis. The analysis results indicated that the specific surface area of the nZVI sample was 18.9887 m^2/g, which was much higher than that of ZVI, 0.048 m^2/g [26]. The high specific surface area of nZVI demonstrated its high adsorption capacity.

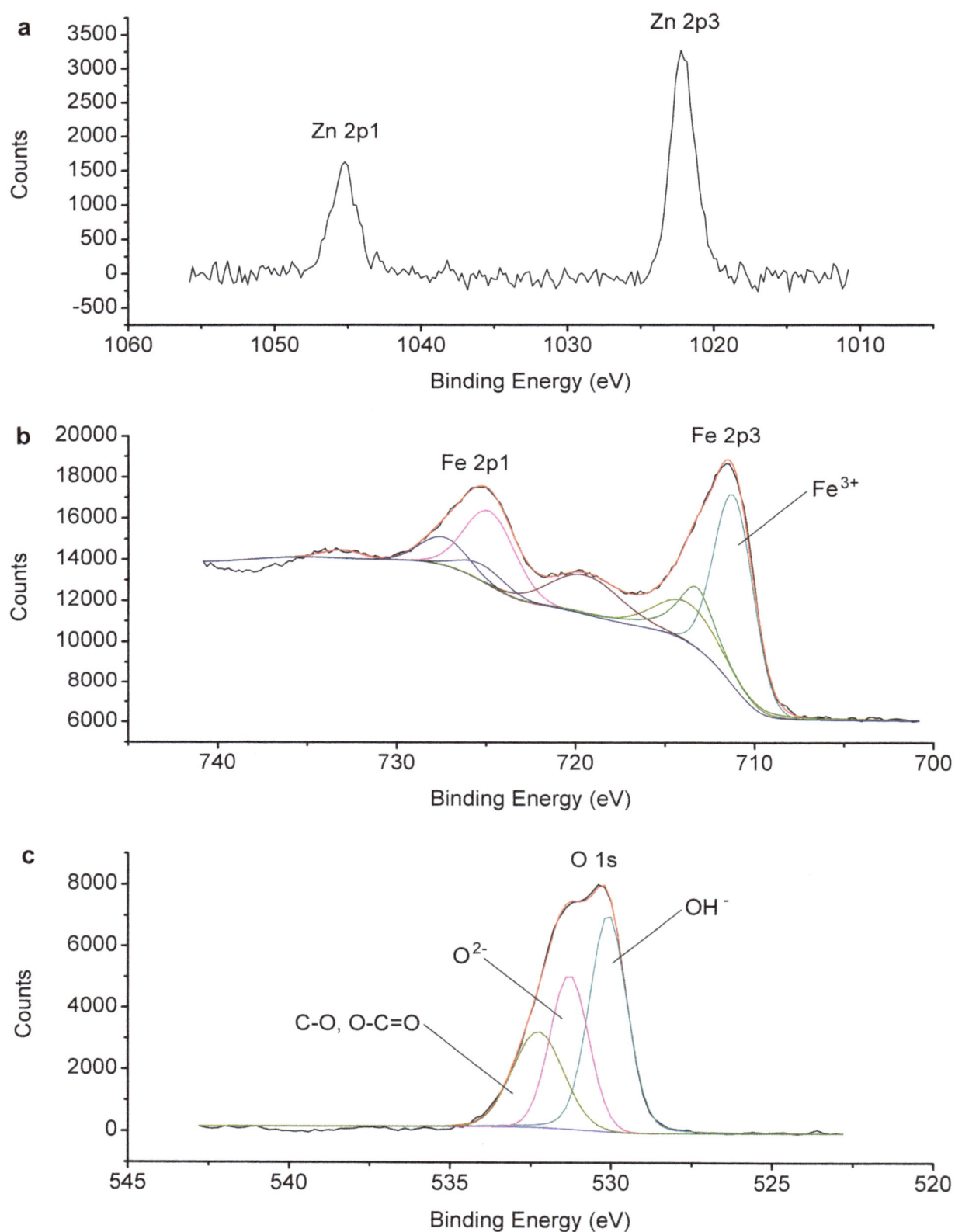

Figure 7. The XPS analysis of stock nZVI. The XPS narrow scan and curve fitting were analyzed: (a) the XPS narrow scan analysis of Zn 2p, (b) curve fitting analysis of Fe 2p, and (c) curve fitting analysis of O 1 s.

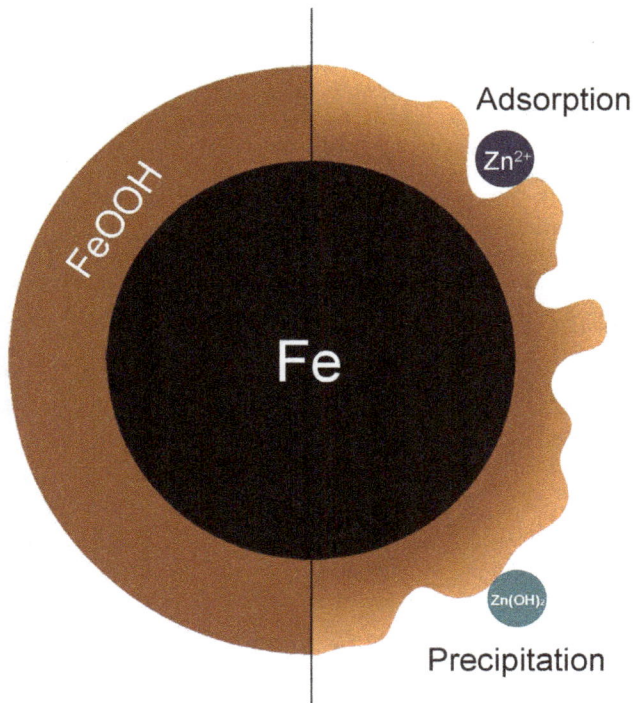

Figure 8. The structure of nZVI during reaction.

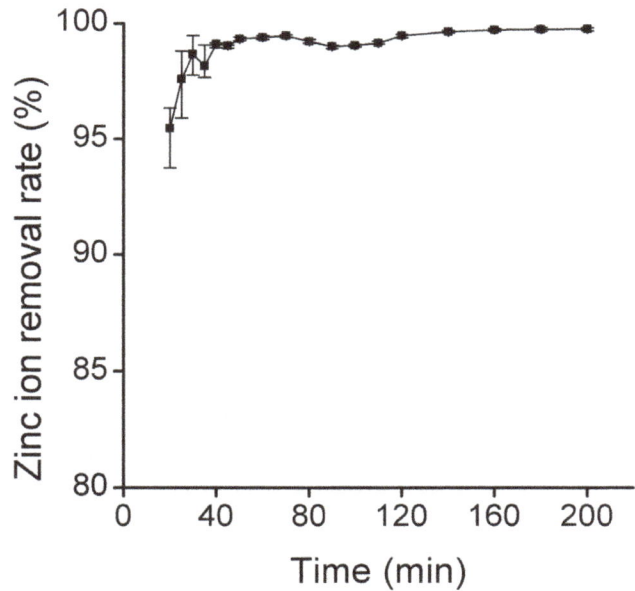

Figure 10. Simultaneous and individual removal of Zn²⁺ and Cd²⁺ by nZVI. Sample 1 contained 100 mg/L Zn²⁺ solution. Sample 2 contained 100 mg/L mixture of Zn²⁺ Cd²⁺. Sample 3 contained 100 mg/L Cd²⁺ solution.

The XPS spectra were also obtained to study the elements on the surface of nZVI. As shown in Figure 3, Zn and Cl were found on the surface of nZVI after the reaction.

Effect of nZVI Solids Concentration

The uptake experiments were conducted with 0.1 to 2 g/L nZVI at a zinc ion concentration of 100 mg/L, respectively for

2 h. As shown in Figure 4, the removal efficiency of Zn^{2+} was increased with the increase of nZVI loading. The added Zn^{2+} was completely removed under the nZVI loading of 0.8 g/L or higher. When nZVI loading was higher than 1.0 g/L, the removal efficiency of Zn^{2+} remained about 99%. Higher loading of nZVI could provide more surface area, which enhanced the Zn^{2+} removal efficiency by nZVI. Thus, we proposed that 1 g/L nZVI was the optimum solids concentration of nZVI required for complete removal of 100 mg/L Zn^{2+} under the examined

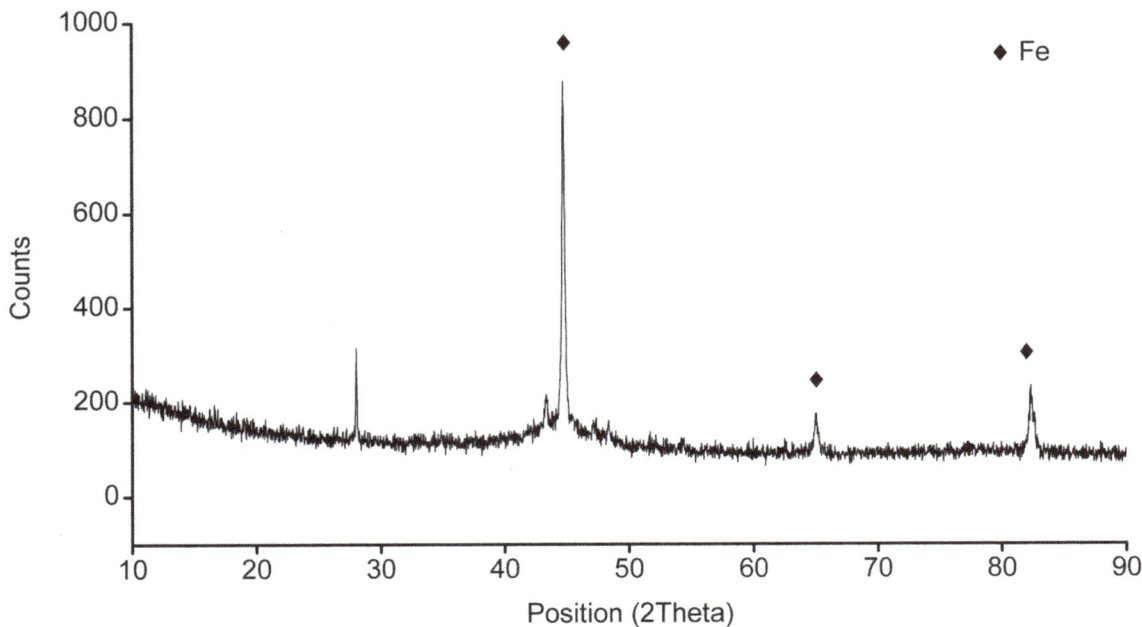

Figure 9. The XRD analysis of spent nZVI.

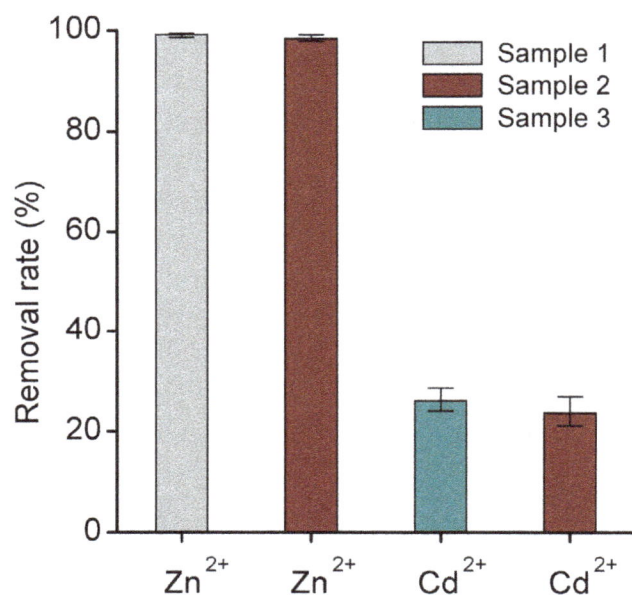

Figure 11. Removal of Zn^{2+} in continuous flow reactor.

experimental conditions. And the concentration ratio was adopted in the subsequent experiments.

The high removal efficiency of Zn^{2+} by nZVI had been proved by Weile Yan [27]. The determined removal pathways of the contaminant mainly included adsorption, complexation, (co) precipitation and surface-mediated chemical reduction [24]. The standard reduction potential of zinc E^0 (Zn^{2+}/Zn) is -0.76 v, while the standard reduction potential of iron E^0 (Fe^{2+}/Fe) is -0.44 v. The ionization tendency of zinc is higher than iron. So Zn^{2+} removal by nZVI is likely not caused by the surface-mediated chemical reduction.

Effect of Dissolved Oxygen

The effect of dissolved oxygen (DO) on Zn^{2+} removal was examined to further study the removal mechanism on the surface of nZVI particles. The freshly prepared 1.0 g/L iron particles were injected into the solution with a zinc ion concentration of 100 mg/L in two 3-neck flasks. One of the flasks had an oxygen-limiting condition with the DO concentration below 0.5 mg/L, but the other one had an oxygen-contained condition with the DO concentration above 5.0 mg/L. As shown in Figures 5(a) and 5(b), the removal extent of Zn^{2+} under oxygen-limiting condition was up to 40%, but a higher removal extent of 80% appeared under oxygen-contained condition. Under both conditions, the Zn^{2+} removal extent increased to 25% in the first 30 min. After that, the removal trend became different. Under oxygen-limiting condition, another 15% of Zn^{2+} removal extent was achieved. Under oxygen-contained condition, the removal extent reached 80%. Meanwhile, the range of oxidation-reduction potential (ORP) under the two conditions presented different processes. Under the oxygen-limiting condition, the value of ORP dropped rapidly, and then came to a gentle decline in the residual contact time. In contrast, under the oxygen-contained condition, the value of ORP decreased in the first 30 min, remained unchanged for the next 20 min, and then decreased in the residual contact time. The change processes of ORP indicated that complex redox reactions occurred. The initial drop of ORP was likely caused by the consumption of dissolved oxygen during the oxidation of the Fe

(0). The slower decline of ORP under oxygen-contained condition was caused by the supply of oxygen from atmosphere. The above results declared that DO was one of the important factors for the removal of Zn^{2+} by nZVI. As mentioned above, Zn^{2+} removal by nZVI is likely not caused by the surface-mediated chemical reduction. The result that nZVI particles were corroded and oxidized by DO might be interpreted as follows: the particles were covered by iron (oxy) hydroxide on the surface and the zinc ion was adsorbed on iron (oxy) hydroxide.

Under the oxygen-limiting condition, the predominant electron receptors should be water and the corrosion reaction could occur as follows [25,26]:

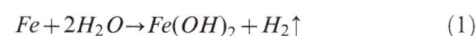

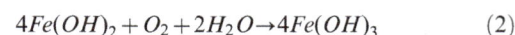

$$Fe + 2H_2O \rightarrow Fe(OH)_2 + H_2\uparrow \qquad (1)$$

$$4Fe(OH)_2 + O_2 + 2H_2O \rightarrow 4Fe(OH)_3 \qquad (2)$$

Under the oxygen-contained condition, the corrosion reaction could occur as follows [28]:

$$4Fe + 3O_2 + 2H_2O \rightarrow 4FeOOH \qquad (3)$$

Heavy metal could be adsorbed by $Fe(OH)_3$ and FeOOH. Nano/mico goethite has been proved to be efficient with variable capabilities in the removal of five metal ions including Zn^{2+} from aqueous solution [29]. Removal of arsenic from water with granular ferric hydroxide has been discussed [30,31]. The reason for low Zn^{2+} removal extent under the oxygen-limiting condition could be found in Equation (2). Under oxygen-limiting condition, the formation of $Fe(OH)_3$ was restricted by the oxygen supply, thus leading to the low removal efficiency.

The result that nZVI particles was corroded and oxidized into ferric ion by DO might be interpreted as follows: the particles were covered by iron hydroxide precipitation on their surface and the zinc ion was adsorbed on and/or co-precipitation on iron hydroxide.

Effect of Solution pH

The pH is an important factor for Zn^{2+} removal by nZVI. The freshly prepared 1.0 g/L iron particles were injected into the solution with a zinc ion concentration of 100 mg/L within 2 h. Uptake results at various pH conditions were shown in Figure 6(a). Some floccose sediment generated when the pH value of the stock zinc ions solution was above 8. Thus, the pH values of 3, 4 and 5 were selected. The higher removal efficiency was observed in the experiment with an initial pH value of 5. The percentage of uptake of Zn^{2+} rose gradually with an increase in pH. The similar effect of pH value has been elucidated by Kishimoto [26]. The variation of pH under three initial values was shown in Figure 6(b). The pH value rose with the contact time. The final pH values were above 8.5 in spite of different initial values. In most part of the corrosion process, the pH value was above 7, while the cementation, which was caused by the glued sediment particles, process was highly effective under acidic conditions in the absence of DO [23]. Thus, the cementation role was not the main effect in Zn^{2+} removal.

With the change in pH value, iron was corroded by acid and oxygen. During the corrosion of iron by acid, Equations (4), (5) and (6) could happen [25,26]. The H^+ inhibited the formation of iron (oxy)hydroxide, resulting in the low removal extent of Zn^{2+} by nZVI.

$$Fe + 2H^+ \rightarrow Fe^{2+} + H_2 \uparrow \tag{4}$$

$$2Fe + O_2 + 4H^+ \rightarrow 2Fe^{2+} + 2H_2O \tag{5}$$

$$4Fe^{2+} + O_2 + 4H^+ \rightarrow 4Fe^{3+} + 2H_2O \tag{6}$$

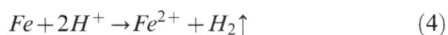

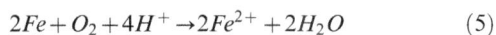

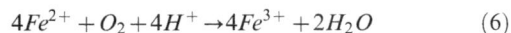

Equations (7), (8) and (9) could happen with the rise in pH value during oxygen corrosion [25,26,28]. Large iron (oxy)hydroxide could adsorb Zn^{2+} on the surface.

$$2Fe + O_2 + 2H_2O \rightarrow 2Fe(OH)_2 \tag{7}$$

$$4Fe(OH)_2 + O_2 + 2H_2O \rightarrow 4Fe(OH)_3 \tag{8}$$

$$4Fe + 3O_2 + 2H_2O \rightarrow 4FeOOH \tag{9}$$

Removal Mechanism

The heavy metal removal by nZVI generally involves redox, cementation, adsorption and precipitation. The standard reduction potential of zinc $E^0(Zn^{2+}/Zn)$ is -0.76 v, while the standard reduction potential of iron $E^0(Fe^{2+}/Fe)$ is -0.44 v. Therefore, it is not true that Zn^{2+} removal by nZVI is achieved due to the higher ionization tendency of zinc than that of iron. The cementation is usually effective under acidic pH without DO [23]. In this study, removal efficiency of Zn^{2+} was lower under acid condition. Accordingly, the high removal extent of Zn^{2+} by nZVI might be caused by adsorption and co-precipitation, which was proved by the effects of DO and pH as mentioned above. The formation of FeOOH on the surface of nZVI could be the main factor of Zn^{2+} removal because of its high adsorption affinity for aqueous solutes [32].

These phenomena could also be confirmed by XPS, as shown in Figure 7(a), (b) and (c). According to the spectra of Figure 7(a), the binding energy of Zn 2p3 was 1022.2 eV, the difference between Zn 2p3 and 2p1 was 23.2 eV, which declared the Zn chemical shift [33]. And it was one way to identify the change of valence. The binding energy of LMM transition, the sharpest auger peak of zinc, was 498.2 eV. Compared with the auger parameters and the strongest photoelectron peak of zinc in handbook of X-ray [33], it could be determined that the zinc on the surface of nZVI was divalent [34]. According to the curve fitting analysis of Fe 2p and O 1 s, the peaks of Fe^{3+}, OH^- and O^{2-} could be found [35], indicating the formation of FeOOH on nZVI surface. Accordingly, adsorption and co-precipitation are the most likely mechanism of Zn^{2+} removal by nZVI (Figure 8).

The nZVI is the core-shell structure: a single particle composed of a dense core surrounded by a thin amorphous shell exhibiting markedly less density than the interior core. As shown in Figure 2, the core-shell structure of nZVI could be found by TEM. The presence of Fe could be proved by XRD (Figure 9). The chemical composition of the passivated thin shell is believed to be a mixed Fe(II)/Fe(III) oxide phase [36,37]. When nanoscale iron particles are exposed to water media, they will obtain hydroxide groups and consequently an apparent surface stoichiometry in proximity to

FeOOH is formed [28]. The FeOOH-shell could enhance the adsorption. The H^+ inhibited the formation of iron (oxy)hydroxide, resulting in the low removal extent of Zn^{2+} by nZVI.

Effect of Cadmium

Simultaneous and individual removals of Zn^{2+} and Cd^{2+} by nZVI were shown in Figure 10. The removal extent of Zn^{2+} was much higher than that of Cd^{2+}. Simultaneous removals of Zn^{2+} and Cd^{2+} by nZVI were both lower than individual removal. The Zn^{2+} removal extent was 0.8% lower in sample 2 than sample 1. And the Cd^{2+} removal efficiency by nZVI was 2.4% lower in mixed contaminants of sample 2 than sample 3. Hardiljeet's research has proved that Cd^{2+} removal by nZVI was chemisorption [38]. So the selective behavior happened between Zn^{2+} and Cd^{2+} on nZVI. There could be the same adsorption sites for both Zn^{2+} and Cd^{2+}. The removal efficiency and selectivity of nZVI particles for Zn^{2+} were higher than Cd^{2+}.

Removal of Zn^{2+} in Continuous Flow Reactor

The high removal efficiency of Zn^{2+} by nZVI was demonstrated by jar-test. The removal extent of 100 mg/L Zn^{2+} could reach 85% by 1 g/L nZVI in 2 h, which should be adopted during reactor design and parameter control, such as influent flow, velocity, etc. The contact time of nZVI and Zn^{2+} should be 2 h or more and the nZVI solids concentration should be no less than 1 g/L when the concentration of Zn^{2+} was 100 mg/L. The results of Zn^{2+} removal in continuous flow reactor were shown in Figure 11. The maximum removal efficiency was up to more than 95%, and the removal efficiency was steady after a rapid increase in the first 30 min. Furthermore, it should be found that this experiment may provide an applicable purification approach for water polluted by heavy metal for this technology allowed the enhanced reactivity and the favorable field deployment capabilities without secondary pollution of nZVI particles.

Conclusions

This study demonstrated that the uptake of Zn^{2+} by nZVI was efficient. With the solids concentration of 1 g/L nZVI, more than 85% of Zn^{2+} could be removed within 2 h. The pH value and DO were the important factors of Zn^{2+} removal by nZVI. The DO enhanced the removal efficiency of Zn^{2+}. Under the oxygen-contained condition, oxygen corrosion gave the nZVI surface a shell of iron (oxy)hydroxide, and the removal efficiency reached 80%, which could show high adsorption affinity. In contrast, the removal efficiency of Zn^{2+} was only 40% under oxygen-limiting condition. The removal efficiency of Zn^{2+} increased with the increasing of the pH. Acidic condition reduced the removal efficiency of Zn^{2+} by nZVI because the existing H^+ inhibited the formation of iron (oxy)hydroxide. The higher removal efficiency was observed in the experiment with an initial pH value of 5. Adsorption and co-precipitation were the most likely mechanism of Zn^{2+} removal by nZVI. The FeOOH-shell could enhance the adsorption efficiency of nZVI. The removal extent of Zn^{2+} was much higher than that of Cd^{2+}. The removal efficiency and selectivity of nZVI particles for Zn^{2+} were higher than Cd^{2+}. Furthermore, a continuous flow reactor for engineering application of nZVI was designed and exhibited high removal efficiency for Zn^{2+}. The maximum removal efficiency was up to more than 95%, and the removal efficiency was steady after a rapid increase in the first 30 min.

Author Contributions

Conceived and designed the experiments: WL CD XZ YZ. Performed the experiments: WL. Analyzed the data: WL CD XZ YZ. Contributed

reagents/materials/analysis tools: CD XZ YZ. Wrote the paper: WL CD XZ YZ.

References

1. Salim R, Al-Subu M, Abu-Shqair I, Braik H (2003) Removal of zinc from aqueous solutions by dry plant leaves. Process Saf Environ 81: 236–242.
2. Committee SDW (1977) Drinking water and health: National Academy Press.
3. Emsley J (1989) The elements. Oxford: Oxford University Press.
4. EPA U (2005) Toxicological review of zinc and compounds. Washington, DC: US Environmental Protection Agency Report: 6.
5. Iwasaki Y, Kagaya T, Miyamoto Ki, Matsuda H (2009) Effects of heavy metals on riverine benthic macroinvertebrate assemblages with reference to potential food availability for drift-feeding fishes. Environ Toxicol Chem 28: 354–363.
6. Wang H, Liang Y, Li S, Chang J (2013) Acute toxicity, respiratory reaction, and sensitivity of three cyprinid fish species caused by exposure to four heavy metals. PLoS One 8: e65282.
7. Wagemann R, Barica J (1979) Speciation and rate of loss of copper from lakewater with implications to toxicity. Water Res 13: 515–523.
8. Giller KE, Witter E, Mcgrath SP (1998) Toxicity of heavy metals to microorganisms and microbial processes in agricultural soils: a review. Soil Bilo Biochem 30: 1389–1414.
9. Sadowski Z (2001) Effect of biosorption of Pb (II), Cu (II) and Cd (II) on the zeta potential and flocculation of Nocardia sp. Miner Eng 14: 547–552.
10. Gordon RB, Bertram M, Graedel TE (2006) Metal stocks and sustainability. Proc Natl Acad Sci U S A 103: 1209–1214.
11. Kwon J-S, Yun S-T, Kim S-O, Mayer B, Hutcheon I (2005) Sorption of Zn (II) in aqueous solutions by scoria. Chemosphere 60: 1416–1426.
12. Lu S, Gibb SW, Cochrane E (2007) Effective removal of zinc ions from aqueous solutions using crab carapace biosorbent. J Hazard Mater 149: 208–217.
13. Katsou E, Malamis S, Haralambous KJ (2011) Industrial wastewater pre-treatment for heavy metal reduction by employing a sorbent-assisted ultrafiltration system. Chemosphere 82: 557–564.
14. Fang Z, Chen J, Qiu X, Qiu X, Cheng W, et al. (2011) Effective removal of antibiotic metronidazole from water by nanoscale zero-valent iron particles. Desalination 268: 60–67.
15. Fan J, Guo Y, Wang J, Fan M (2009) Rapid decolorization of azo dye methyl orange in aqueous solution by nanoscale zerovalent iron particles. J Hazard Mater 166: 904–910.
16. Choe S, Lee SH, Chang YY, Hwang KY, Khim J (2001) Rapid reductive destruction of hazardous organic compounds by nanoscale Fe0. Chemosphere 42: 367–372.
17. Elliott DW, Lien H-L, Zhang W-X (2009) Degradation of lindane by zero-valent iron nanoparticles. J Environ Eng 135: 317–324.
18. Ambashta RD, Repo E, Sillanpää M (2011) Degradation of Tributyl Phosphate Using Nanopowders of Iron and Iron–Nickel under the Influence of a Static Magnetic Field. Ind Eng Chem Res 50: 11771–11777.
19. Naja G, Halasz A, Thiboutot S, Ampleman G, Hawari J (2008) Degradation of hexahydro-1,3,5-trinitro-1,3,5-triazine (RDX) using zerovalent iron nanoparticles. Environ Sci Technol 42: 4364–4370.
20. Celebi O, Uzum C, Shahwan T, Erten HN (2007) A radiotracer study of the adsorption behavior of aqueous Ba(2+) ions on nanoparticles of zero-valent iron. J Hazard Mater 148: 761–767.
21. Klimkova S, Cernik M, Lacinova L, Filip J, Jancik D, et al. (2011) Zero-valent iron nanoparticles in treatment of acid mine water from in situ uranium leaching. Chemosphere 82: 1178–1184.
22. Li X-q, Zhang W-x (2007) Sequestration of metal cations with zerovalent iron nanoparticles a study with high resolution X-ray photoelectron spectroscopy (HR-XPS). J Phys Chem C 111: 6939–6946.
23. Rangsivek R, Jekel MR (2005) Removal of dissolved metals by zero-valent iron (ZVI): kinetics, equilibria, processes and implications for stormwater runoff treatment. Water Res 39: 4153–4163.
24. Miehr R, Tratnyek PG, Bandstra JZ, Scherer MM, Alowitz MJ, et al. (2004) Diversity of contaminant reduction reactions by zerovalent iron: role of the reductate. Environ Sci Technol 38: 139–147.
25. Sun YP, Li XQ, Cao J, Zhang WX, Wang HP (2006) Characterization of zero-valent iron nanoparticles. Adv Colloid Interface Sci 120: 47–56.
26. Kishimoto N, Iwano S, Narazaki Y (2011) Mechanistic consideration of zinc ion removal by zero-valent iron. Water Air Soil Pollut 221: 183–189.
27. Yan W, Herzing AA, Kiely CJ, Zhang W-x (2010) Nanoscale zero-valent iron (nZVI): Aspects of the core-shell structure and reactions with inorganic species in water. J Contam Hydrol 118: 96–104.
28. Hoerlé S, Mazaudier F, Dillmann P, Santarini G (2004) Advances in understanding atmospheric corrosion of iron. II. Mechanistic modelling of wet–dry cycles. Corros Sci 46: 1431–1465.
29. Hafez H (2012) A study on the use of nano/micro structured goethite and hematite as adsorbents for the removal of Cr (III), Co (II), Cu (II), Ni (II), and Zn (II) metal ions from aqueous solutions. Int J Eng Sci 4: 3018–3028.
30. Guan X-H, Wang J, Chusuei CC (2008) Removal of arsenic from water using granular ferric hydroxide: Macroscopic and microscopic studies. J Hazard Mater 156: 178–185.
31. Badruzzaman M, Westerhoff P, Knappe D (2004) Intraparticle diffusion and adsorption of arsenate onto granular ferric hydroxide (GFH). Water Res 38: 4002–4012.
32. Otte K, Schmahl WW, Pentcheva R (2012) Density functional theory study of water adsorption on FeOOH surfaces. Surf Sci 606: 1623–1632.
33. Chastain J, King Jr RC (1992) Handbook of X-ray photoelectron spectroscopy. Perkin-Elmer Corporation 40: 221.
34. Liu F, Zhao Z, Qiu L, Zhao L (2009) Tables of peak positions for XPS photoelectron and auger electron peaks. Analysis and Testing Technology and Instruments 15: 1–17.
35. Yamashita T, Hayes P (2008) Analysis of XPS spectra of Fe^{2+} and Fe^{3+} ions in oxide materials. Appl Surf Sci 254: 2441–2449.
36. Signorini L, Pasquini L, Savini L, Carboni R, Boscherini F, et al. (2003) Size-dependent oxidation in iron/iron oxide core-shell nanoparticles. Phys Rev B 68: 195423.
37. Wang C, Baer DR, Amonette JE, Engelhard MH, Antony J, et al. (2009) Morphology and electronic structure of the oxide shell on the surface of iron nanoparticles. J Am Chem Soc 131: 8824–8832.
38. Boparai HK, Joseph M, O'Carroll DM (2011) Kinetics and thermodynamics of cadmium ion removal by adsorption onto nano zerovalent iron particles. J Hazard Mater 186: 458–465.

4

Acute Toxicity, Respiratory Reaction, and Sensitivity of Three Cyprinid Fish Species Caused by Exposure to Four Heavy Metals

Hongjun Wang*, Youguang Liang, Sixin Li, Jianbo Chang

Key Laboratory of Ecological Impacts of Hydraulic-Projects and Restoration of Aquatic Ecosystem of Ministry of Water Resources, Institute of Hydroecology, Ministry of Water Resources and Chinese Academy of Sciences, Wuhan, P.R. China

Abstract

Using 3 cyprinid fish species zebra fish, rare minnow, and juvenile grass carp, we conducted assays of lethal reaction and ventilatory response to analyze sensitivity of the fish to 4 heavy metals. Our results showed that the 96 h LC_{50} of Hg^{2+} to zebra fish, juvenile grass carp, and rare minnow were 0.14 mg L^{-1}, 0.23 mg L^{-1}, and 0.10 mg L^{-1}, respectively; of Cu^{2+} 0.17 mg L^{-1}, 0.09 mg L^{-1}, and 0.12 mg L^{-1} respectively; of Cd^{2+} 6.5 mg L^{-1}, 18.47 mg L^{-1}, 5.36 mg L^{-1}, respectively; and of Zn^{2+} 44.48 mg L^{-1}, 31.37 mg L^{-1}, and 12.74 mg L^{-1}, respectively. Under a 1-h exposure, the ventilatory response to the different heavy metals varied. Ventilatory frequency (Vf) and amplitude (Va) increased in zebra fish, juvenile grass carp, and rare minnows exposed to Hg^{2+} and Cu^{2+} (P<0.05), and the Vf and Va of the 3 species rose initially and then declined when exposed to Cd^{2+}. Zn^{2+} had markedly different toxic effects than the other heavy metals, whose Vf and Va gradually decreased with increasing exposure concentration (P<0.05). The rare minnow was the most highly susceptible of the 3 fish species to the heavy metals, with threshold effect concentrations (TEC) of 0.019 mg L^{-1}, 0.046 mg L^{-1}, 2.142 mg L^{-1}, and 0.633 mg L^{-1} for Hg^{2+}, Cu^{2+}, Cd^{2+}, and Zn^{2+}, respectively. Therefore, it is feasible to use ventilatory parameters as a biomarker for evaluating the pollution toxicity of metals and to recognize early warning signs by using rare minnows as a sensor.

Editor: Z. Daniel Deng, Pacific Northwest National Laboratory, United States of America

Funding: This study was supported by the National Science and Technology Major Project of Water Pollution Control and Management of China (Grant No. 2009ZX07528-003), the transforming program of agricultural scientific and technological achievements of China (Grant NO.2011GB23320006) and National Natural Science Foundation of China (Grant 51209150). The funders had no role in study design, data collection and analysis, decision to publish, or preparation of the manuscript.

Competing Interests: The authors have declared that no competing interests exist.

* E-mail: whj103392@163.com

Introduction

With the development of industry and agriculture, numerous heavy metal pollutants have been released into water bodies by various means, resulting in serious water pollution. In polluted waters, exposure of fish to heavy metals leads to interactions between these chemicals and biological systems and causes biochemical disturbances[1–2]. Mercury (Hg), copper (Cu), cadmium (Cd), and zinc (Zn) are the 4 most common heavy metals. Hg bioaccumulates in organisms, especially in carnivorous fish at high trophic levels of the food chain, and its concentration can become tens of thousands-fold greater than that in water. In the early 1960s, Songhua River (China) was seriously polluted, and it became a water body that typified Hg pollution from discharge of industrial wastewater [3]. Cd is a nonessential heavy metal of high environmental concern due to its toxicity, common usage, industrial production, and emissions from fossil fuel combustion. Cd is on some governments'priority-substance lists (e.g., Canadian Environmental Protection Act, 1994), and in environments impacted by man, concentrations may reach values of micrograms per liter or higher [4]. Cu is an essential element, required by all living organisms for several physiological functions and biochem-

ical reactions [5]. Zn is also an essential trace element, but excess of both Cu and Zn is poisonous to organisms.

Using fish as biological indicators is advantageous because changes in their behaviors (e.g., avoidance responses, swimming patterns, and breathing) in response to environmental changes can be measured directly. Indeed, behavior has been used as an integral parameter of physiological activity and as a robust biological warning indicator of the quality of water supplies and effluents[6–7]. Fish ventilatory parameters that are known to be sensitive to toxins include ventilatory rate (opercular movement), depth or amplitude of ventilation, and coughing or gill purge rate [8].

Over the past 40 years, an early warning system was successfully developed by using fish as bioindicators and the ventilatory parameters of fish were used as biomarkers for online monitoring of water pollution[9–10]. Numerous studies found that this warning system was sensitive to the resource water and to emission of pollutants [11]. The system was used in Europe where it has been successfully applied in several rivers since 1990[12–13], and has been extensively applied in South Africa [14]. In response to the September 11, 2001 terrorist attacks in the United States (US) and a subsequent anthrax bacteria event, the US began to pay attention to the research and application of biological early

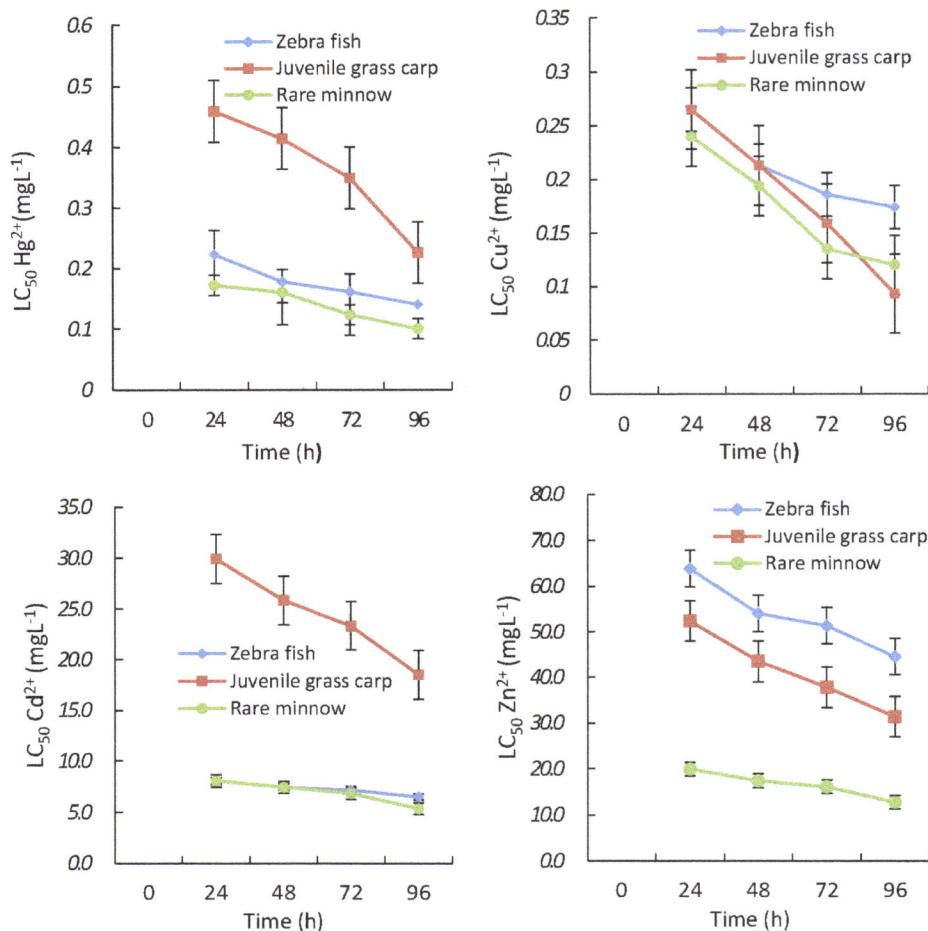

Figure 1. LC$_{50}$ values (mgL^{-1}) over time for zebrafish (*Brachydanio rerio*), rare Minnow (*Gobiocypris rarus*) and juvenile grass carp (*Ctenopharyngodon idellus*) exposed to 4 kinds of heavy metals.

warning systems [6]. However, such studies are still lacking in China.

Related research on fish behavior toxicology indicated that ventilatory behavior was sensitive to heavy metals. Many researchers primarily used bluegill sunfish[8–9] and rainbow trout [11] to conduct toxicological experiments. A review summarizing the methodology for measurement and interpretation of fish ventilatory patterns as early warning signals of water quality deterioration and incipient toxicity was available [15]. In the work presented here, we used standard experimental fish – zebra fish, a rare local minnow, and grass carp – for a ventilatory toxicology experiment examining 4 heavy metals. We analyzed the sensitivity of 4 heavy metals in experimental fish and considered how the respiratory parameters of local rare minnow could provide basic data for monitoring and giving early warning of heavy metal pollution in water bodies.

Materials and Methods

Experimental Animals

The 3 cyprinid fish species used in the present study included 2 native, local Chinese species: juvenile grass carp (Ctenopharyngodon idellus) and rare minnow (Gobiocypris rarus), and one introduced, standard international experimental fish: zebra fish (Brachydanio rerio). The mean lengths and weight (± standard deviation) of zebra fish, juvenile grass carp, and rare minnow were 2.3±0.3 cm and 0.22±0.05 g, 4.1±0.2 cm and 1.84±0.33 g, and 2.6±0.7 cm and 0.25±0.06 g, respectively. G. rarus and B. rerio were obtained from Institute of Hydrobiology, Chinese Academy of Sciences, China, and C. idellus was obtained from a fish hatchery in Hubei Province, China. The death rate of experimental fish, which were raised for 14 days and fasted for 24 hours before the experiment, was less than 2%. All procedures and animal handling were in accordance with the guideline approved by Chinese Association For Laboratory Animal Sciences. The study was approved by the animal ethics committee of the Institute of Hydroecology, Ministry of Water Resources and Chinese Academy of Sciences (protocol number: IHE20110525).

After the experiment, fish that had gone through the heavy metal treatment and were less active or stressed were euthanized using overdose of Benzocaine. The surviving fish that did not have symptoms of poison were put in clean fresh water and closely monitored. No fish died during the ventilatory monitoring experiments.

Dilution Water and Heavy Metals

The tap water used in the experiment was aerated, dechlorinated, and charged with oxygen for more than 48 h. Dissolved oxygen, pH, conductivity, total hardness, and water temperature

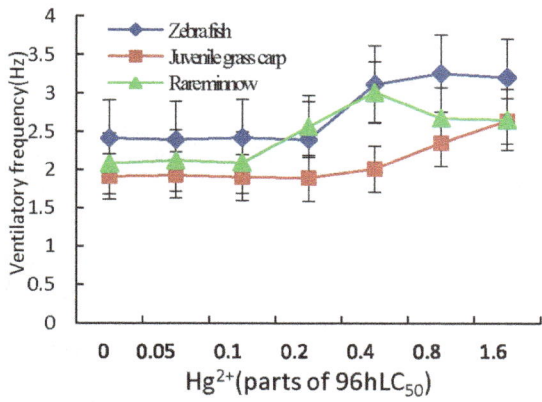
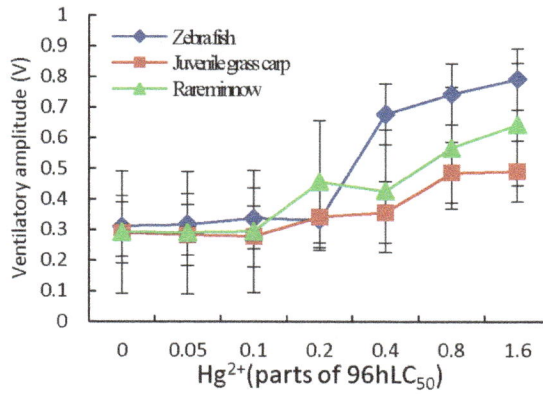
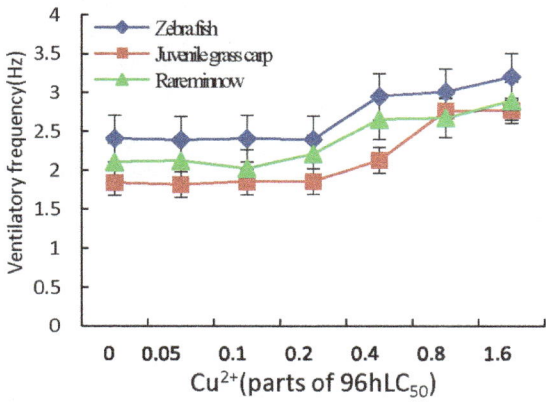
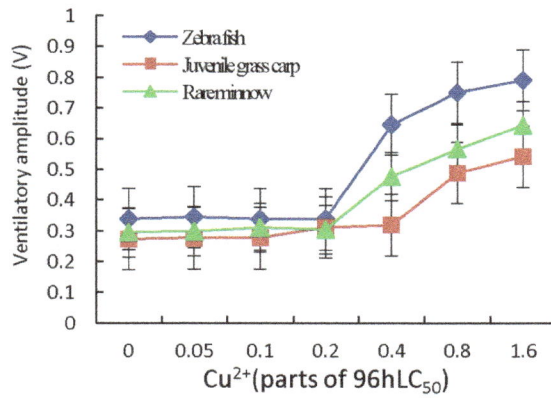
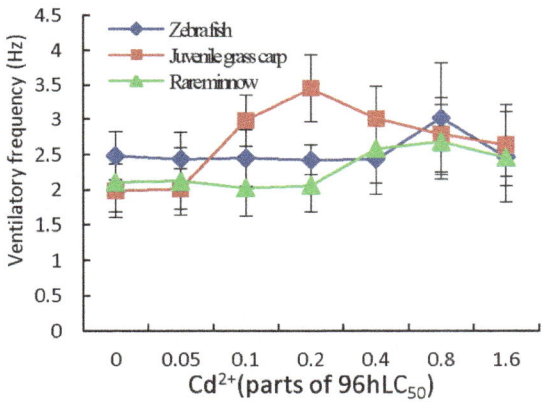
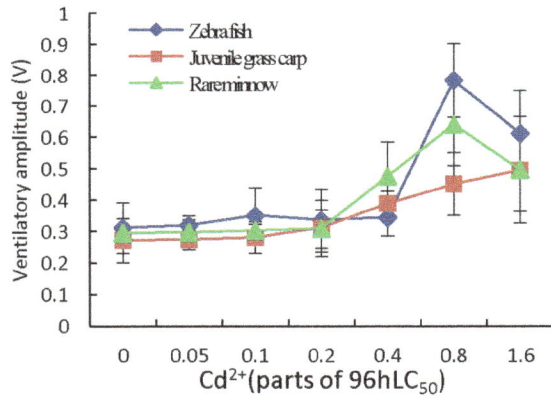
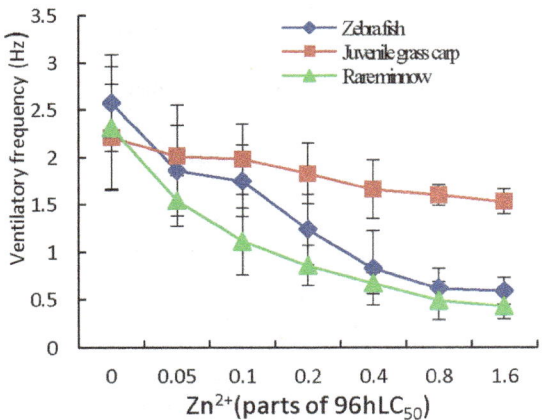
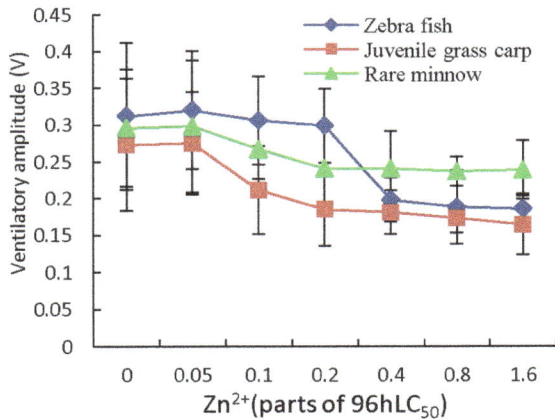

Figure 2. Ventilatory responses over time for zebrafish (*Brachydanio rerio*), **rare minnow** (*Gobiocypris rarus*) **and juvenile grass carp** (*Ctenopharyngodon idellus*) **exposed to 4 kinds of heavy metals.**

were 7.5–7.8 mg L^{-1}, 7.7–7.9, 130–290 μScm^{-1}, 120 mg L^{-1} (as $CaCO_3$), and 22–24°C, respectively. Concentrations of Hg, Cu, Cd, and Zn were not detected in the dilution water. All chemical reagents used in the experiment were of analytical grade, and the purity of $HgCl_2$, $CuSO_4 \cdot 5H_2O$, $Cd(NO_3)_2 \cdot 4H_2O$, and $ZnCl_2$ were 99.50%, 99.0%, 99.0%, and 98.0%, respectively.

Testing Equipment

A biological early warning system, manufactured by Biological Monitoring, Inc., USA (model: Bio-sensor 7008), was used. The main principles of its use are to use fish as indicator organisms, monitor ventilatory frequency and amplitude of fish by sensors, and to provide a warning when the aquatic environment changes according to changes in the ventilatory parameters. The early warning system consists of 5 parts: (1) ventilatory monitoring sensor (Bio-Sensor), (2) signal filter and amplifier (Bio-Amp), (3) computer data processing and display system, (4) YSI water quality analyzer, and (5) automatic alarm device and water sampler. Operculum respiration and other neuromuscular activities of fish generate several microvolt bioelectrical signals, the strongest of which is a respiratory signal. The signal is received by the sensor in the respiratory chamber and is then sent to signal filter and amplifier (Bio-Amp) before being transferred to a computer. The computer makes judgments as to outliers according to default statistical methods and sends out early warning signals, providing continuous, online monitoring of water pollution[16–17]. The system automatically collects water samples after an alarm and determines water quality by chemical analyses.

Testing Procedures

LC_{50} assessment. Before the official experiment, we conducted a preliminary experiment by using a hydrostatic method. To do this, we prepared a wide concentration series of experimental liquids, recorded the quantity of dead fish for each concentration every 6 h, and promptly removed dead fish. Each experiment lasted 96 h, and we selected the minimum lethal concentration for the experimental fish (24 h LC_{100}), and the maximum tolerated concentration (96 h LC_0). The series concentration of the official experiment ranged from 24 h LC_{100} to 96 h LC_0 with geometric series as interval, set up 5 groups of concentration gradients, and established 3 series and 1 blank

control in each group. The experimental liquid volume was 15 L, and there were 10 fish per tank. Fish were not fed during the experiment; poisoning symptoms, time of death time, and number of deaths were recorded. The fish were observed continuously for 96 additional hours, with dead fish and metabolites being promptly removed. The standard to determine death in experimental fish was a lack of reaction upon prodding the fish's tail with a glass bar [18].

Ventilatory responses. The experimental method was according to ASTM [15]. Experimental fish were put into 8 monitoring chambers after 14 days of accommodation, to conduct a respiratory adaptability experiment for 4 days in control water. A standard solution of heavy metals was configured for a series of concentration gradients. The ratio of heavy metal concentration to the corresponding 96 h LC_{50} was defined as U. U-values for each metal were 0, 0.05, 0.1, 0.2, 0.4, 0.8, and 1.6, respectively. The tested standard solution was introduced into the monitoring chambers at a constant flow (100 $mLmin^{-1}$) by using a metering pump, and respiration reaction experiments were conducted for every concentration for 1 h. The signal values of fish respiratory reactions (ventilatory frequency Vf and ventilatory amplitude Va) were recorded for the conditions of control water and heavy metal exposure.

Data Analysis

We calculated the median lethal concentration (LC_{50}) of the acute toxicity experiment and 95% confidence intervals with probability unit regression by using the PROBIT function of SPSS 16.0 software (SPSS Inc., USA). VF and VA data were analyzed with one-way analysis of variance (ANOVA) to determine whether significant differences existed among experimental groups (SPSS 16.0). If a difference was significant, Duncan's test of multiple comparison was applied. P values of <0.05 were considered significant. Early fish biological statistical algorithms using the moving average method were applied to set up an evaluation interval of 8 min with 6 statistical calculation samples [10]. The relative sensitivity of the behavioral responses was evaluated by comparing their threshold values. Threshold effect concentration (TEC) was estimated by defining the geometric mean between the lowest observed effect concentration (LOEC) and the no observed effect concentration (NOEC) [19].

Table 1. Correlation analysis among the toxicity of the 4 heavy metals to the 3 Cyprinid Fish species.

Species		Zebra fish	Juvenile grass carp	Rare minnow
Zebra fish	Pearson Correlation	1	0.977**	0.960**
	Sig. (2-tailed)		0.000	0.000
	N	20	20	20
Juvenile grass carp	Pearson Correlation	0.977**	1	0.985**
	Sig. (2-tailed)	0.000		0.000
	N	20	20	20
Rare minnow	Pearson Correlation	0.960**	0.985**	1
	Sig. (2-tailed)	0.000	0.000	
	N	20	20	20

**Correlation is significant at the 0.01 level (2-tailed).

Table 2. Analysis of the acute toxicity, respiratory reaction and sensitivity of different fish species.

	Species	NOEC	LOEC	TEC	96 hLC$_{50}$	Parts of 96 hLC$_{50}$	Hardness (mg/L CaCO$_3$)	pH	Reference
Hg^{2+}	zebra fish	0.05	0.056	0.053	0.14	0.38	120	7.8	This paper
	rare Minnow	0.018	0.02	0.019	0.1	0.19	120	7.8	This paper
	grass carp	0.041	0.046	0.043	0.23	0.19	120	7.8	This paper
	largemouth bass		0.01					7	[29]
	fathead minnow				0.168		45	7.4	[22]
Cu^{2+}	zebra fish	0.062	0.068	0.065	0.17	0.38	120	7.8	This paper
	rare Minnow	0.045	0.048	0.046	0.12	0.39	120	7.8	This paper
	grass carp	0.066	0.072	0.069	0.09	0.77	120	7.8	This paper
	largemouth bass		0.048					7	[29]
	Bluegill				1.1		45	7.5	[23]
	brook charr				0.1		45	7.5	[24]
Cd^{2+}	zebra fish	2.5	2.6	2.55	6.5	0.39	120	7.8	This paper
	rare Minnow	2.14	2.144	2.142	5.36	0.4	120	7.8	This paper
	grass carp	1.84	1.847	1.843	18.47	0.1	120	7.8	This paper
	rainbow trout		0.064				82	7.8	[32]
	largemouth bass		0.15					7	[29]
	Bluegill				20.4		200	7.7	[25]
	fathead minnow				7.2		200	7.7	[26]
	flag fish				2.5		44	7.5	[27]
Zn^{2+}	zebra fish	2.2	2.224	2.212	44.48	0.05	120	7.8	This paper
	rare Minnow	0.629	0.637	0.633	12.74	0.05	120	7.8	This paper
	grass carp	3.12	3.137	3.128	31.37	0.1	120	7.8	This paper
	fathead minnow				9.20		203	7.7	[28]
	rainbow trout		2.55				50	7.8	[30]
	Brook charr		1.39				45	7.5	[31]

Results

Effect of Acute Lethal Toxicity on 3 Cyprinid Fish Species Exposed to Heavy Metals

During the experiment, no fish in the control group died, and the acute lethal effect caused by the 4 heavy metals on the 3 species of fish are shown in Figure 1. The 96 h LC$_{50}$ (and 95% confidence intervals) values of zebra fish, grass carp, and juvenile rare minnow were 0.14 (0.068–0.268) mg L^{-1}, 0.23 (0.171–0.295) mg L^{-1}, and 0.10 (0.076–0.131) mg L^{-1}, respectively, when exposed to Hg^{2+}. The 96 h LC$_{50}$ values of zebra fish, grass carp, juvenile rare minnow were 0.17 (0.102–0.29) mg L^{-1}, 0.09 (0.065–0.124) mg L^{-1}, and 0.12 (0.075–0.175) mg L^{-1}, respectively, when exposed to Cu^{2+}. The 96 h LC$_{50}$ values caused by Cd^{2+} were 6.5 (6.15–6.83) mg L^{-1}, 18.47 (13.55–25.09) mg L^{-1}, and 5.36 (1.58–9.12) mg L^{-1}, respectively. The 96 h LC$_{50}$ values caused by Zn^{2+} were 44.48 (36.7–52.7) mg L^{-1}, 31.37 (25.74–38.27) mg L^{-1}, and 12.74 (4.16–23.9) mg L^{-1}, respectively. The fitting equation of acute toxicity showed a good linear relationship among the zebra fish, grass carp, and juvenile rare minnow for the 4 heavy metals, and the related coefficients were 0.90–0.99, 0.88–0.99, and 0.92–0.99, respectively, which showed that the acute lethal effect exhibited a significant dose-effect relationship. Figure 1 indicates that LC$_{50}$ decreased with increasing exposure time and that the toxicity of heavy metals to the fish also demonstrated a significant time-effect relationship.

The Effect of Heavy Metals on the Respiratory Behaviors of Fish

Compared to the control, the mean Vf and Va of the 3 species of cyprinid fish exposed to Hg^{2+} generally increased with increasing concentration (Fig. 2). It can be seen from Figure 2 that with an increase in exposure concentration, Hg^{2+} will stimulate the Vf and Va of each fish species. When the ratios of Hg^{2+} concentration to its 96 h LC$_{50}$ (U) were 0.05, 0.1, and 0.2, the Vf and Va of zebra fish tended to increase, although not significantly. When the U-value increased to 0.4, 0.8, and 1.6, respectively, Vf and Va of zebra fish increased significantly (P<0.05). When the U-value was 0.8 (i.e., the concentration of Hg^{2+} was 0.112 mg L^{-1}), Vf reached its highest level, increasing from 2.41 Hz in the control to 3.26 Hz, with 35.2% increase. When the U-value was 1.6 (i.e., the concentration of Hg^{2+} was 0.224 mg/L), Va reached its highest level, increasing from 0.31 V in the control to 0.79 V, with 155.6% increase.

When exposed to Hg^{2+} with an U-value of 0.8, the Vf of juvenile grass carp increased significantly (P<0.01); when the U-value was 0.2, its Va increased significantly (P<0.05). When the U-value exceeded 0.1, Vf and Va of the rare minnow increased significantly (P<0.01) compared to the control group.

When exposed to Cu^{2+}, zebra fish, juvenile grass carp, and rare minnow had similar respiratory responses as when exposed to Hg^{2+}. When U-values were less than 0.2, the values of Vf and Va showed no significant difference; when the U-value was greater than 0.4, the values of Vf and Va increased significantly (P<0.01).

With the rise of concentration in Cd^{2+} exposure, the values of Vf and Va for each fish species increased initially, followed by a decrease. When the U-value of juvenile grass carp exceeded 0.05, the Vf value increased prominently (P<0.01), reaching a maximum at U = 0.2 and subsequently decreasing. Nevertheless, Vf was still significantly higher than that of the control group (P<0.05). Compared to Vf, Va had a lagged effect; zebra fish and rare minnows showed a similar response, but with different exposure response concentrations.

Compared to the other heavy metals, Zn^{2+} had different toxic effects. The Vf of all fish species decreased with the increase in exposure concentration (P<0.01); the Vf value of rare minnows had an obvious dose-effect relationship with exposure concentration ($Y = 2.758\,e^{-0.28\,x}$, $R^2 = 0.98$). With the increase in Zn^{2+} exposure concentration, the Va values of all fish were significantly reduced (P<0.05); the Va value of rare minnows declined most dramatically (P<0.05) with U-values greater than 0.05, while it tended to be stable with a U-value of 0.2.

Discussion

In this study, we analyzed the sensitivity of 4 heavy metals in experimental fish and have shown that the respiratory parameters of rare minnow can provide basic data for monitoring heavy metal pollution in water bodies. Different species of fish have been used to monitor the pollution of water bodies [6,8,11] However, These fish species reported in the literature are hard to find or hard to raise in China. In this study, we used two widespread Chinese native species and one international standard species to monitor the water body pollution. We have shown that these species are sensitive to the 4 heavy metals, which are the major pollutants.

Evidence has shown that the toxicity of pollutants is correlated among different aquatic organisms. Firth [20] used different kinds of biology (Rainbow-Trout and Ceriodaphnia) to determine the toxicity of wastewater discharged by a paper mill and found that the toxicity of different kinds of biology had good linear relativity. On the basis of experimental studies, Jiang Min et al. [21] discovered that the influence of heterocyclic nitrogen on the toxicity of luminous bacteria fit that of its effect on zebra fish and daphnia magna, with a correlation coefficient of greater than 0.99. In this study, we selected zebra fish, rare minnow, and grass carp as the standard fish for the experiment. Grass carp is one of 4 Chinese carp species. The rare minnow is a rare cyprinid fish in China that represent small- and medium-sized fish, respectively, from different aquatic habitats. As shown in Table 1, the 4 heavy metals yielded similar acute toxicity values among the 3 fish species (P<0.05), with R values of 0.977, 0.960, and 0.985, respectively. This correlation analysis indicated that the toxic effect of each heavy metal on a given cyprinid fish species was similar and that there is little effect of habitat and size at maturity on heavy metal toxicity. As a result, reliable toxicity data can be obtained on the basis of acute toxicity in early life stages.

The lethal effect of Hg^{2+}, which was highest on zebra fish, was higher than that reported for fathead minnow [22]. As for Cu^{2+}, its toxicity was highest on juvenile grass carp, and its 96 h LC_{50} was 0.09 mg L^{-1}, far lower than 1.1 mg L^{-1} reported for bluegill [23] and also lower than 0.1 mg L^{-1} reported for brook charr [24]. Among the 3 cyprinid fishes, both Cd^{2+} and Zn^{2+} produced the highest mortality rate in the rare minnow. The 96 h LC_{50} of Cd^{2+} for rare minnow was lower than that for bluegill [25] and fathead minnow [26] but higher than that for flag fish [27]. The 96 h LC_{50} of Zn^{2+} for rare minnow was 12.74 mg L^{-1}, which was higher than 9.20 mg L^{-1} for fathead minnow [28].

Our study has found that, compared with the acute lethal effect, the heavy metal ion exhibited a greater effect on fish breathing. Gills are the major target organ for water-borne pollutants, and they are the site for metal uptake [5]. As shown in Table 2, of the 3 cyprinid species studied, rare minnow was the most sensitive to Hg^{2+}, Cu^{2+}, and Zn^{2+}, while the TEC of Hg^{2+} on rare minnow was 0.019 mg L^{-1} (i.e., 19% of its 96 h LC_{50}), which was close to the sensitivity of the breathing effect on largemouth bass [29]. The response time of the 3 cyprinid fish was between 32 and 56 min. The Cu^{2+} LOEC was the same as that for largemouth bass, but the sensitivity to Zn^{2+} was higher than that of both rainbow trout [30] and brook charr [31]. Therefore, it was very appropriate to select the rare minnow as China's native experimental fish for the study on respiratory toxicology. Meanwhile, we observed that the ventilation frequency of rare minnow was noticeably related to the dosage of Zn^{2+}, and the relationship between dosage and effect was expressed as $Y = 2.7588e^{-0.28x}$, $R^2 = 0.986$. Gintaras Svecevicius [17] exposed rainbow trout to hexavalent chromium and found a noticeable relationship between dosage and effect, i.e., $Y = 98.388/(1+0.152e^{6.9844x})$, $R^2 = 0.98$, P<0.05. In addition, the TEC was 0.2 mg CrL^{-1}, only 7% of the 96 h LC_{50} coefficient, leading Svecevicius to believe that ventilation frequency could be used as a biomarker in the standard test of the toxicity of water.

Moreover, the 4 heavy metal ions stimulated different types of respiratory reactions in the experimental fish. Hg^{2+} and Cu^{2+} were both noticeably stimulative on Vf and Va of each of the fish species; when fish were exposed to an increased concentration of ions. Both Vf and Va of Cd^{2+} initially rose and then gradually dropped. Zn^{2+} had an obviously different toxicity effect from that of the other heavy metal ions, as both its Vf and Va dropped gradually, which is similar to the conclusion drawn by Diamond et al. for bluegill [32]. Different types of heavy metal ions have different effects on fish respiration. The mechanism by which these toxic substances exert their influence on fish respiration is still unclear and needs to be explored.

Conclusions

The acute toxicities of the 4 heavy metal ions on zebra fish, juvenile grass carp, and rare minnow were linearly correlated, indicating clear relationships between dosage and effect as well as between time and effect. The respiratory activity of the fish was highly sensitive to heavy metal pollution, and this sensitivity was much greater than the lethal reaction. Among the 3 cyprinid species studied, rare minnow was most sensitive to heavy metals. It is feasible to use the respiratory parameter of the rare minnow as a biomarker for evaluating the toxicity of heavy metals and to utilize this species as a sensor to monitor and predict heavy metal pollution.

Author Contributions

Conceived and designed the experiments: HJW YGL JBC. Performed the experiments: HJW SXL. Analyzed the data: HJW SXL JBC. Contributed reagents/materials/analysis tools: HJW SXL. Wrote the paper: HJW SXL.

References

1. Gail M, Dethloff, Daniel Schlenk (1999) Alteration in physiological Parameters of Rainbow Trout (*Oncorhynchus mykiss*) with Exposuer to Copper and Copper/Zinc Mixtures. Ecotoxicology and Environmental Safety 42: 253–264.
2. Ibrahim Orun, Zeliha Selamoglu Talas (2008) Antioxidative role of Sodium selenite against the toxic effect of heavy metals(Cd^{2+},Cr^{3+}) on some Biochemical and Hematological Paramenters in the Blood of Rainbow trout (Oncorhynchus mykiss Walbaum,1792). Fresenius Environmental Bulletin 17: 1242–1246.
3. Clarkson TW, Magos L, Myers GJ (2003) The toxicology of mercury-current exposures and clinical manifestations. The New England Journal of Medicine 349: 1731–1737.
4. Felten V, Charmantier G, Mons R, Geffard A, Rousselle P, et al. (2008) Physiological and behavioural responses of Gammarus pulex(*Crustacea:Amphipoda*) exposed to cadmium. Aquatic Toxicology 86: 413–425.
5. Rafael Mendonca Duarte, Ana Vristina Leite Menezes, Leonardo da Silveira Rodrigues, Vera Maria Fonseca de Almeida, Adalberto Luis Val (2009) Copper sensitivity of wild ornamental fish of the Amazon. Ecotoxicology and Environmental Safety 72: 693–698.
6. Van der Schalie WH, Sheed TR, Knechtgnes PL, Widder MW (2001) Using higher organisms in biological early warning systems for real-time toxicity detection. Biosens Bioelectron 16: 457–465.
7. Sérgio Reis Cunha, Renata Gonçalves, Sérgio Rui Silva, Ana Dulce Correia (2008) An automated marine biomonitoring system for assessing water quality in real-time. Ecotoxicology 17: 558–564.
8. Shedd TR, van der Schalie WH, Widder MW, Burton DT, Burrows EP (2001) Long-term Operation of an Automated Fish Biomonitoring system for Continuous Effluent Acute Toxicity Surveillance. Bull.Environ.Contam.Toxicol 66: 392–399.
9. Sloof W (1979) Dection Limits of a Biological Monitoring System Based on Fish Respiration. Bull.Environ.Contam.Toxicol 23: 517–523.
10. David Gruber, John Cairns Jr (1981) Data acquisition evaluation in biological monitoring systems. Hydrobiologia 83: 387–393.
11. Kramer KJM, Botterweg J (1991) Aquatic biological early warning systems: an overview. In: Jeffrey DJ, Madden B, editors. Bioindicators and Environmental Management. London: Academic Press. 95–126.
12. Gerhardt A (1999) Recent trends in online biomonitoring for water quality control. In: Gerhardt A, editors. Biomonitoring of Pulluted Water. Utikon-Zurich: Trans Tech Publications Ltd. 95–118.
13. Peter Diehl, Thomas Gerke, Jeuken AD, Jaqueline Lowis, Ruud steen, et al. (2006) Early Warning Strategies and Practices Along the River Rhine. Hdb Env Chem 5: 99–124.
14. Morgan WSG, Kuhn PC, Allais B, Wallis G (1982) An appraisal of the performance of a continuous automatic fish biomonitoring system at an industrial site. Water Sci Technol 14: 151–161.
15. ASTM (1995(Reapproved 2008)) Standard guide for ventilatory behavioral toxicology testing of freshwater fish. E 1768–95. West Conshohocken: ASTM International. 1–10.
16. Gruber D (2007) Automated Biological Monitors: Has Their Time Arrived? IN: AWWA, editors. Proceedings Good Laboratory Practices Conference. Charlottesville, VA. 1–35.
17. States S, Stoner M, Westbrook C (2007) Development of a Contamination Warning System for the Pittsburgh Water and Sewer Authority. IN: AWWA, editors. Proceedings of the Water Technology Conference. Alexandria, VA. 16.
18. USEPA (2002) Short-time methods for estimating the chronic toxicity of effluent and receiving water to freshwater organisms. NW Washington DC: Environmental Protection Agency Office of Water. 135 p.
19. Gintaras Svecevicius (2009) Use of Behavioral Responses of Rainbow Trout Oncorhynchus mykiss in Identifying Sublethal Exposure to Hexavalent Chromium. Bull.Environ.Contam.Toxicol 82: 564–568.
20. Firth BK, Backman CJ (1990) Comparison of Microtox Testing with Rainbow-Trout (Acute) and Ceriodaphnia (Chronic) Bioassays in Mill Wastewaters. Tappi Journal 73: 169–174.
21. Jiang M, Li YM, Gu GW (2005) Study on toxicity of nitrogenous heterocyclic compounds to aquatic organisms. Acta Scientiae Circumstantiae 25: 1253–1258.
22. Snarski VM, Olson GF (1982) Chronic toxicity and bioaccumulation of mercuric chloride in the fathead minnow, Pimephales promelas. Aquat. Toxicol 2: 143–156.
23. Benoit DA (1975) Chronic effects of copper on survival, growth, and reproduction of the bluegill (*Lepomis macrochirus*). Trans. Amer. Fish. Soc 104: 353–358.
24. McKim JM, Benoit DA (1971) Effects of long-term exposures to copper on survival, growth, and reproduction of brook trout (*Salvelinus fontinalis*). J. Fish. Res. Board Can 28: 655–662.
25. Eaton JG (1974) Chronic cadmium toxicity to the bluegill (*Lepomis macrochirus*). Trans. Amer. Fish. Soc 103: 729–735.
26. Pickering QH, Gast MH (1972) Acute and chronic toxicity of cadmium to the fathead minnow (*Pimephales promelas*). J. Fish. Res. Board Can 29: 1099–1106.
27. Spehar RL (1976) Cadmium and zinc toxicity to flagfish, Jordanella floridae. J. Fish. Res. Board Can 33: 1939–1945.
28. Brungs WA (1969) Chronic toxicity of zinc to the fathead minnow, Pimephales promelas Rafinesque. Trans. Amer. Fish. Soc 98: 272–279.
29. Morgan WSG (1979) Fish locomotor behavior patterns as a monitoring tool. J. Water Pollut. Control Fed 51: 580–589.
30. Cairns J Jr, Sparks RE (1971) The use of bluegill breathing to detect zinc. In: US EPA, editors. Water Pollut. Control Res. Ser. 18050 EDQ 12/71. Washington, D.C: Publications Branch(Water), Research Information Division, R&M, Environmental Protection Agency.
31. Drummond RA, Carlson GW (1977) Procedures for measuring cough (gill purge) rates of fish. In: US EPA, editors. Ecological Research Series. 600/3–77–133. Duluth, Minnesota: Environmental Research Laboratory-Duluth. 1–47.
32. Diamond JM, Parson MJ, Gruber D (1990) Rapid detection of sublethal toxicity using fish ventilatory behavior. Environ. Toxicol. Chem 9: 3–11.

5

Are *In Vitro* Methods for the Detection of Endocrine Potentials in the Aquatic Environment Predictive for *In Vivo* Effects? Outcomes of the Projects SchussenAktiv and SchussenAktiv*plus* in the Lake Constance Area, Germany

Anja Henneberg[1]*, Katrin Bender[3], Ludek Blaha[2], Sabrina Giebner[3], Bertram Kuch[4], Heinz-R. Köhler[1], Diana Maier[1], Jörg Oehlmann[3], Doreen Richter[5], Marco Scheurer[5], Ulrike Schulte-Oehlmann[3], Agnes Sieratowicz[3], Simone Ziebart[3], Rita Triebskorn[1]

1 Animal Physiological Ecology, University of Tübingen, Tübingen, Germany, **2** Faculty of Science, RECETOX, Masaryk University, Brno, Czech Republic, **3** Department Aquatic Ecotoxicology, University of Frankfurt am Main, Frankfurt am Main, Germany, **4** Institute for Sanitary Engineering, Water Quality and Solid Waste Management, University of Stuttgart, Stuttgart, Germany, **5** Water Technology Center Karlsruhe, Karlsruhe, Germany

Abstract

Many studies about endocrine pollution in the aquatic environment reveal changes in the reproduction system of biota. We analysed endocrine activities in two rivers in Southern Germany using three approaches: (1) chemical analyses, (2) *in vitro* bioassays, and (3) *in vivo* investigations in fish and snails. Chemical analyses were based on gas chromatography coupled with mass spectrometry. For *in vitro* analyses of endocrine potentials in water, sediment, and waste water samples, we used the E-screen assay (human breast cancer cells MCF-7) and reporter gene assays (human cell line HeLa-9903 and MDA-kb2). In addition, we performed reproduction tests with the freshwater mudsnail *Potamopyrgus antipodarum* to analyse water and sediment samples. We exposed juvenile brown trout (*Salmo trutta* f. *fario*) to water downstream of a wastewater outfall (Schussen River) or to water from a reference site (Argen River) to investigate the vitellogenin production. Furthermore, two feral fish species, chub (*Leuciscus cephalus*) and spirlin (*Alburnoides bipunctatus*), were caught in both rivers to determine their gonadal maturity and the gonadosomatic index. Chemical analyses provided only little information about endocrine active substances, whereas the *in vitro* assays revealed endocrine potentials in most of the samples. In addition to endocrine potentials, we also observed toxic potentials (E-screen/reproduction test) in waste water samples, which could interfere with and camouflage endocrine effects. The results of our *in vivo* tests were mostly in line with the results of the *in vitro* assays and revealed a consistent reproduction-disrupting (reproduction tests) and an occasional endocrine action (vitellogenin levels) in both investigated rivers, with more pronounced effects for the Schussen river (e.g. a lower gonadosomatic index). We were able to show that biological *in vitro* assays for endocrine potentials in natural stream water reasonably reflect reproduction and endocrine disruption observed in snails and field-exposed fish, respectively.

Editor: John A. Craft, Glasgow Caledonian University, United Kingdom

Funding: The project SchussenAktiv was funded by the Ministerium für Umwelt, Naturschutz und Verkehr Baden-Württemberg (UVM), and the foundation "Natur und Umwelt" of the Landesbank Baden-Württemberg (LBBW). We acknowledge support by Deutsche Forschungsgemeinschaft and Open Access Publishing Fund of Tubingen University. The project SchussenAktivplus is funded by the Federal Ministry for Education and Research (BMBF) and cofounded by the Ministry of Environment Baden-Wü rttemberg. In addition, the city of Ravensburg, the AZV Mariatal and the AV Unteres Schussental financially contribute to the project. The commercial companies (Jedele & Partner GmbH and Ö konsult GbR) provided funding to this study. SchussenAktivplus is connected to the BMBF action plan "Sustainable water management (NaWaM)" and is integrated in the BMBF frame programme "Research for sustainable development FONA". Contract period: 1/2012 to 12/2014, Funding number: 02WRS1281A. All funders had no role in study design, data collection and analysis, decision to publish, or preparation of the manuscript.

* E-mail: anja.henneberg@uni-tuebingen.de

Introduction

Endocrine disruptors (EDs) are substances which can affect the endocrine system by imitating or repressing body's own hormones. Chemicals with endocrine potentials form a very diverse group and the number of chemicals known to cause endocrine effects in organisms is constantly increasing. This group includes for example synthetic estrogens, bioflavonoids, organochlorine pesticides, dioxins, furans, phenols, alkylphenols, polychlorinated biphenyls, phthalates, and brominated flame retardants. Also, naturally produced steroid hormones like 17β-estradiol (E2), estrone (E1), or testosterone, as well as phytohormones have the potential to affect endocrine systems in other organisms. However, natural endocrine-active chemicals are often less persistent than synthetic EDs [1].

Recently, a growing number of scientists, in particular toxicologists and ecologists, have pointed out the hazardous effects that different endocrine-active chemicals may have on the environment and animal and human health [2]. For example, many EDs are suspected to contribute to the development of breast cancer in women and prostate and testicular cancers in men, to reduce male fertility and to interact with the immune system [3,4]. Disruptions of endocrine functions also occur in wildlife. Reduced fertility, abnormal development of embryos, feminization, and demasculinization are reported for birds, reptiles, mammals, and fish, while defeminization and masculinization are reported for gastropods (summarized in [5]). A number of distinct characteristics make EDs especially problematic. First, the wide range of effects caused by EDs makes it difficult to identify all hazardous effects. Second, low exposure levels are sufficient to cause serious consequences. For example, 17α-ethinylestradiol (EE2) is considered to be a very potent estrogen for fish; its lowest observed effect concentration for vitellogenesis in rainbow trout is 0.1 ng/L [6]. Therefore, already concentrations of estrogens and their mimics that are currently observed in freshwaters may impact the sustainability of wild fish populations [5,7], even though direct evidence to relate endocrine disruption to wildlife population decline is rare [8,9]. Third, many EDs are highly persistent, which often leads to long-term exposure. Once released into the environment, EDs may affect biota over many years, and it is difficult to assess these long-term effects with regards to the whole ecological community. Fourth, mixtures of EDs can interact, and thus either enhance or counteract the action of single substances. Studies on mixture toxicity offer increasing evidence that joint effects can occur when all mixture components are below levels at which individual chemicals cause observable effects [10,11].

A main source for ED chemicals is the discharge of waste water treatment plants (WWTPs) into recipient waters. River pollution through waste water is especially relevant in areas with industry, high human population density, and/or intensive agriculture. Today, most waste water is treated in developed countries, but often endocrine disrupting chemicals cannot be completely removed by routine waste water treatment, and additional techniques to improve waste water purification are necessary [12]. Even in highly developed countries untreated waste water may be dumped into rivers when the capacity of WWTPs and stormwater overflow basins is exceeded during heavy rain events [13].

Given the evident relevance of EDs and the importance of WWTPs for their discharge into the environment, the present study assesses the effects of WWTPs on the water quality of two tributaries of Lake Constance, the Schussen and Argen rivers, as part of the "SchussenAktiv" and "SchussenAktiv*plus*" projects. As a first step, these projects examine the current ecological state in Schussen and Argen rivers. After different types and sizes of WWTPs at the Schussen are technically improved, these projects will then evaluate the effects of improved waste water treatment [14]. The present study reports the results on the water quality before the technical improvement of the examined WWTPs and consists of three main parts: chemical analyses of endocrine-active substances, a set of *in vitro* bioassays, and *in vivo* tests. These tests are employed to investigate estrogenic, anti-estrogenic, and anti-androgenic potentials and effects (and their temporal variability and trends) in the Schussen and Argen rivers and were jointly applied in view to elucidate the predictive value of chemical

analyses or biological *in vitro* assays for organism-level endocrine effects in field-exposed biota.

Using chemical analyses, we focused on the identification of endocrine-active substances in surface waters and sediments. Previous chemical analyses detected up to 82 micropollutants, including EDs, in tributaries of Lake Constance. Thirty-five of these substances were found at ecotoxicologically relevant concentrations, for which effects on mortality, development, health, and reproduction of aquatic organisms cannot be excluded [15]. During the whole project we will analyse more than 150 micropollutants in waste water, surface water, sediments, and tissue samples [14].

Importantly, chemical analyses alone often provide very little information on the biological effects and do not take into account interactions among individual chemicals in mixtures. Therefore, we applied various bioassays to provide complementary information on biological potencies. Specifically, we use *in vitro* reporter gene assays detecting estrogen receptor (ER) or androgen receptor (AR) activation, and cell proliferation assays like the E-screen. These assays seem to be promising with respect to their mechanistic nature, relative simplicity, and potential high throughput [16–18]. Several field studies have demonstrated the diagnostic potential of bioassays, including studies with contaminated water and sediment samples [19–25].

However, sometimes results from *in vitro* assays are imprecise estimates for effects observed *in vivo* (see, e.g. [26]). For example, in a study on zebrafish [7], the relative estrogenic potency of EE2 that was observed was about 25 times more potent in *in vivo* than could be expected based on the *in vitro* results. Therefore, we complement our *in vitro* assays by using *in vivo* tests with mudsnails and fish. For investigations of native water and sediment samples in the laboratory assessing reproduction disrupting potentials, we used the freshwater mudsnail *Potamopyrgus antipodarum*, which has been shown to be a sensitive test organism responding to reproduction disrupting chemicals, including estrogens and their mimics. Such effects can be assessed by quantifying embryo numbers in the brood pouch [27]. As a second *in vivo* test for assessing endocrine effects, we evaluated expression of the egg yolk precursor protein vitellogenin (vtg) in juvenile brown trout. Normally, only female fish produce vitellogenin, which is estrogen-dependent. However, estrogenic xenobiotics can also act on the hepatic receptors to induce synthesis of vitellogenin in males and juveniles [28]. Therefore, vitellogenin levels in male and juvenile trout can be used as a biomarker of exposure to estrogen active substances in the environment [6,28–32].

In addition, we examined feral fish (chub and spirlin) to determine their gonadal development and to assess if there are indications for endocrine disorders in the feral fish population.

In contrast to large parts of extant literature, in this study we combined chemical analyses with *in vitro* assays and *in vivo* tests (Fig. 1). Thus, it was our aim to obtain a more precise and complete evaluation of endocrine activities at the Schussen and Argen rivers; in particular to investigate whether symptoms of endocrine disruption in field-living individuals are reflected by signals from *in vitro* laboratory assays or by the results derived from a detailed chemical monitoring programme.

Materials and Methods

1 Study Sites, Bypass Systems and Exposure Experiments

As a model region for a densely populated area, we investigated the Schussen river, a major tributary of Lake Constance. A total of 20 WWTPs and more than 100 stormwater overflow basins are connected to the Schussen [14]. Sampling site S 0 was upstream

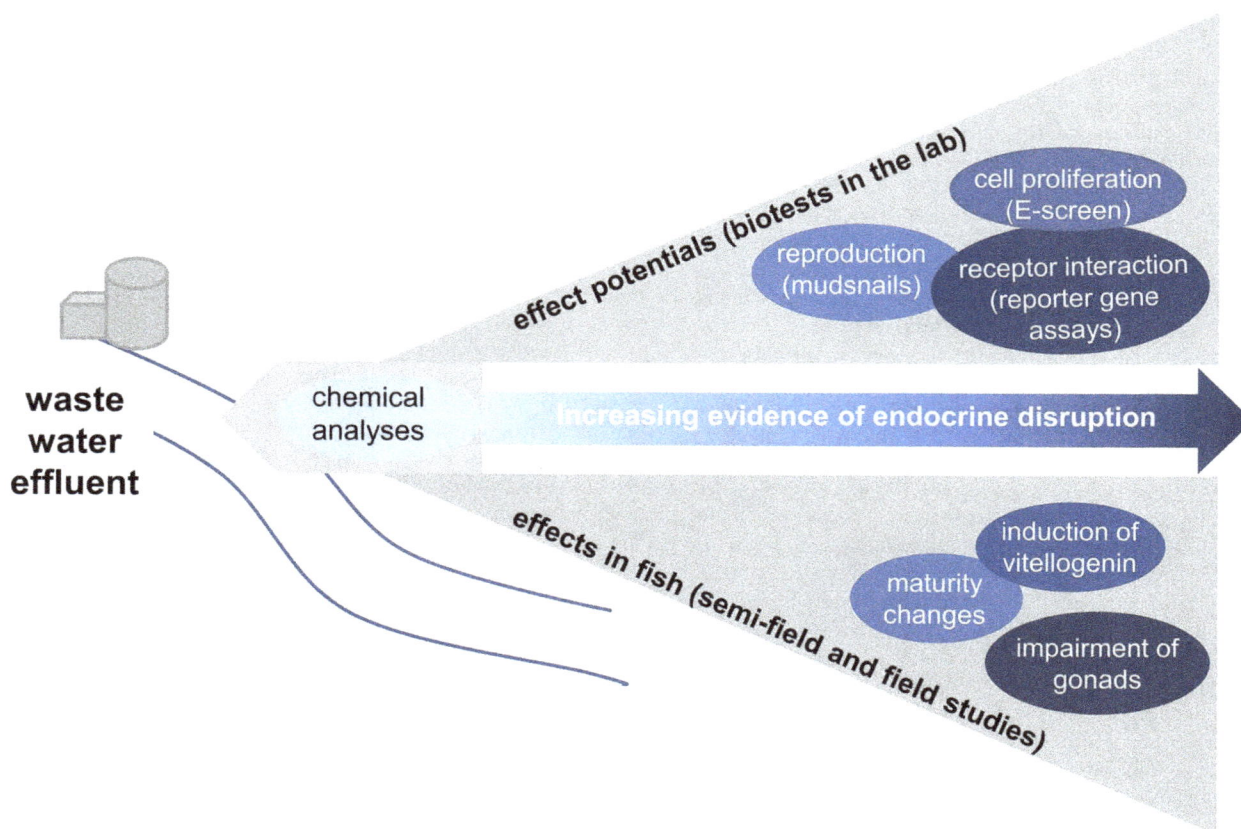

Figure 1. Model of the study design. This figure gives an overview of the study design and all performed analyses. Based on their results, we arranged the tests according to their evidence for endocrine disruption.

from one of the major waste water treatment plants (WWTP Langwiese) and a stormwater overflow basin, and site S 1 was located downstream from the stormwater overflow basin, but upstream from the WWTP Langwiese. Site S 3 was several kilometres downstream from the WWTP Langwiese, and S 6 was situated nearby the river mouth area at Lake Constance. Since a literature review by Triebskorn and Hetzenauer [15] showed less pollution at the Argen river, a reference sampling site, called S 4, was examined there. The location and sampling sites are shown in figure 2.

We collected water and sediment samples from all sampling sites. In addition, we analysed waste water (WW) from the WWTP Langwiese, which is one of the largest WWTP in the catchment area of the Schussen river (170,000 population equivalents). This WWTP has been upgraded with an active charcoal filter in autumn 2013. Table 1 shows all the sampling campaigns that we conducted from 2009 to 2013 (named from A to N).

Two feral fish species, chub (*Leuciscus cephalus*) and spirlin (*Alburnoides bipunctatus*), were caught at sampling sites S 3 (Schussen) and S 4 (Argen) using electrofishing. In addition, we built bypass systems at both rivers, one downstream WWTP Langwiese at the Schussen and one at the Argen to simulate semi-field conditions (see figure 2 for the locations). These flow-through-systems were situated near the rivers, and river water was continuously passed through 250 L aquaria by a pump. At both bypass systems, we installed a sediment trap to guarantee similar concentrations of suspended particles. Technical supervision of water temperature, oxygen content, conductivity, and flow-through volume was

carried out every 10 minutes, and failures were immediately reported by a short message. In these semi-field test systems, we performed exposure experiments with brown trout (*Salmo trutta* f. *fario*). The bypass systems allowed us to keep fish under controlled conditions that were close to their natural conditions (for a detailed description of the bypass systems, see [14]). As a negative control, we kept fish in 250 L aquaria under laboratory conditions in climate chambers at the University of Tübingen. Details for the exposure conditions of fish and catching procedure are described in 4.1. and 4.2.

Ethic statement. This study was carried out in strict accordance with German legislation (animal experiment permit nos. ZO 1/09 and ZP 1/12, field sampling permit AZ 35/9185.82–2, District Magistracy of the State of Baden-Württemberg).

2 Chemical Analysis of Endocrine-active Compounds

We analysed effluent samples from the WWTP Langwiese, surface water, and sediment samples from all sampling sites at different times (see Table 2). Immediately after extracting, 1 L of surface water sample and 0.2 L of WWTP effluent were preconcentrated by solid phase extraction (SPE) with a polymeric sorbent (Strata X, Phenomenex, Aschaffenburg, Germany) using an automated enrichment system (Autotrace, ThermoScientific). 4-n-nonylphenol and 17-α-methyltestosterone were added as surrogate standards prior the extraction process. We used 4-n-nonylphenol as a standard because literature did not describe its occurrence in aqueous environmental samples. The eluted samples

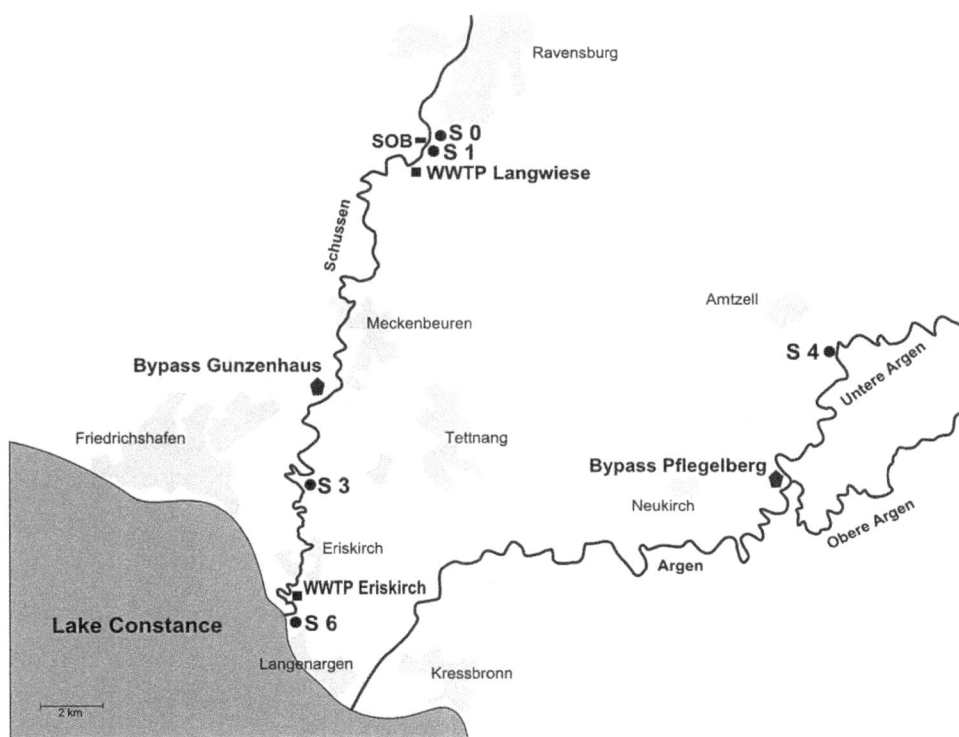

Figure 2. Location of the sampling sites and bypass systems at the Schussen and Argen rivers in Southwest Germany. Waste water treatment plant (WWTP) Langwiese and Eriskirch, as well as the storm water over-flow basin (SOB) at the Schussen. Geographic coordinates: S 0 = N47° 45′ 29.40″, E9° 35′ 21.78″, S 1 = N47° 45′ 19.22″, E9° 35′ 25.35″, S 3 = N47° 39′ 16.09″, E9° 31′ 53.35″, S 6 = N47° 37′ 4.73″, E9° 31′ 50.33″S 4 = N47° 44′ 20.46″, E9° 53′ 42.78″, bypass Gunzenhaus = N47° 40′ 44.00″, E9° 32′ 24.77″, and bypass Pflegelberg = N47° 39′ 11.21″, E9° 44′ 30.80″.

were completely dried and derivatised by adding n-methyl-n-tri-methylsilyltrifluoracetamid (MSTFA) + trimethyliodosilane (TMJS) reagent. The analytical method is based on gas chromatography separation coupled to mass spectrometry detection (GC – MS, Agilent). Measurements were carried out in the laboratories of the Water Technology Center Karlsruhe (TZW, Karlsruhe, Germany). The procedures for sample preparation and analysis are based on DIN EN ISO 18857–1 (February 2007).

Sediment samples were also analysed by GC/MS. The sediment samples (1 g) were fortified with surrogate standards and extracted twice with 10 ml of acetone/cyclohexane (1:10) in an ultrasonic bath for 15 minutes. Subsequently, the samples were centrifuged and the extracts were combined. The extracts were blown down to dryness and derivatised by adding MSTFA + TMJS reagent. Separation of the analytes was achieved by a Rxi - 5 Sil MS column (30 m x 0.25 mm, 0.25 μm) purchased from Restek (Fuldabrück, Germany). Transfer line temperature was 290°C. Temperature programme started with 120°C with holding time of 1 min was then ramped to 180°C with 15°C/min with no hold and then further ramped to 290°C with 5°C/min and 10 min hold. For the analysis a gas chromatograph 6890 coupled to a

mass spectrometer 5973 (both Agilent Technologies, Waldbronn, Germany) were used.

3 Detection of Endocrine Potentials – *In vitro* and *In vivo*

3.1 *In vitro* - E-screen assay. With the E-screen assay, we analysed effluent samples from the WWTP (Langwiese) and surface water samples from all sampling sites. The assay is based on the enhanced proliferation of human breast cancer cells (MCF-7) in the presence of estrogen active substances in the samples. The cell proliferation assay was developed by Soto et al. [17], optimized by Körner et al. [18,33], and modified by Schultis (2005, unpublished data). To determine the estrogenic activity, the acidified (pH 2.5 – 3) water samples (1 L) were solid phase extracted (C18-cartridges, Varian Mega Bond Elut, 1 g). After drying the cartridges overnight by lyophilization and elution with methanol (2 x 5 mL), dimethylsulfoxid (DMSO, 50 μL) was added as a keeper to prevent loss of volatile substances. The MCF-7 cells were stored humidified (37°C, 5% CO2) in Dulbecco's modified Eagle's medium (DMEM) with fetal bovine serum and phenol red as buffer tracer (culture medium) and passed weekly. To accomplish the E-screen assay the cells were trypsinized and the

Table 1. Dates of the sampling campaigns.

Code	A	B	C	D	E	F	G	H	J	K	L	M	N
Month	July	Oct.	June	Aug.	Oct.	May	July	Sept.	Oct.	May	July	Sept.	May
Year	2009		2010			2011				2012			2013

Table 2. Chemical analysis of water and sediment samples.

WWTP (Langwiese)	2010	2011	2012
	C, D, E	F, G, H, J	K, L
Site S 0	C		K, L, M
Site S 1	C		K, L, M
Site S 3	C	F	K, L, M
Site S 4	C	F	K, L, M

culture medium was replaced by phenol red free DMEM with charcoal dextran treated fetal bovine serum (experimental medium). The cell suspension (75 µL, approx. 2300 cells/well) was plated into 96-well plates (Sarstedt, Newton, USA) and stored in the incubator for 24 h. For assaying the samples, dilution series were prepared (9 concentrations per sample) and added to the cells (8 wells per concentration). For providing a positive control (standard dose-response curve) the cells were exposed to a dilution series of 17β-estradiol (2.5·10–14 mol/L–2.5·10–10 mol/L). Neat experimental medium served as negative control (8 wells per plate). The E-screen assay was terminated after a five-day incubation time by removing the medium, washing the cells with phosphate buffered saline buffer and fixing them with trichloroacetic acid. After incubation (30 min; 4°C) the trichloroacetic acid was removed by washing the plates under a gentle stream of cold water. After drying the plates at 40°C the cell protein was stained with sulforhodamin B. After incubation (10 min) the dye was washed off with aqueous acetic acid (1%) and the plates were dried again at 40°C. The cell attaching dye was resuspended with tris-buffer and incubated (20 min; 4°C). The extinction was measured at 550 nm using a microtiter plate reader (MRX, Dynatech laboratories, Virginia, USA). Analysis of the dose-response curve was performed using the software Table Curve 2D (Jandel, San Rafael, CA).

The resulting estrogenic activity reflects a sum parameter over all estrogen active substances present in the samples and is expressed in concentration units of the reference substance E2 (17β-estradiol equivalent concentration, EEQ). The assessment of cytotoxicity in cells exposed to the investigated samples is important, because a high toxicity can overlay the estrogenic response. For example, if a water sample is both highly cytotoxic and estrogenic, the exposed cells should be triggered to proliferate but will not be able to do so because the cytotoxicity represses the cell proliferation. As a result, one will get an undersized "estrogenic response" from the test. Cytotoxicity was indirectly detected using different dilutions of the concentrated samples. The EC50 TOX value is the concentration of the examined sample in which 50% of the cells are able to grow. For illustration, we calculated the reciprocal values of the EC50 TOX values; high 1/EC50 TOX values represent a high cytotoxicity in the sample.

3.2 In vitro - Cellular reporter gene assays for estrogens and androgens. With the reporter gene assays, we analysed effluent samples from the WWTP Langwiese and sediment samples from the sampling sites S 3 (Schussen) and S 4 (Argen). For effluents, one litre of each sample was filtered through a glass fiber filter using vacuum and extracted by SPE with SDB Waters Oasis (500 mg; columns were activated by 6 ml of methanol and equilibrated by 8 mL of distilled water, maximum backpressure was −30 kPa, and the flow rate did not exceed 10 mL/min). After SPE, the columns were dried, eluted with 6 mL methanol (no backpressure used), and concentrated by a nitrogen stream to final

volumes which corresponded to 1200-times concentrated effluents. Sediment samples from the Schussen (S 3) and the Argen (S 4) were dried by freeze-drying (Christ lyophilization instrument), sieved through a 2 mm sieve, and 10 g were extracted for 1 h in 150 mL dichloromethane (automatic extractor Büchi System B-811). Extracts were concentrated by a nitrogen stream to the last drop and then dissolved in methanol. All extracts were stored at −80°C until testing.

To determine estrogenicity and antiestrogenicity, the human cell line HeLa-9903 was used according to the slightly modified protocol of US EPA [34]. Cells were grown in DMEM-F12 without phenol red (Sigma Aldrich, USA), containing 10% fetal calf serum, at 5% CO_2 and 37°C. Once the cells reached about 80% confluence, they were trypsinized and seeded into a sterile 96-well plate at density 20 000 cells/well. For experiments, cells were grown in medium containing fetal calf serum treated with dextran-coated charcoal (which strongly reduces concentrations of natural steroids in the serum). After 24 h, the cells were exposed to the dilution series of the tested samples (6 different concentrations of each sample were tested), to the reference estrogen E2 (dilution series 1–500 pM E2) for the calibration, and to the blank and solvent controls (0.5% v/v methanol). To test for antiestrogenicity, the samples were co-exposed simultaneously with 33 pM E2, and the inhibitions of E2-induced responses were recorded. We used ICI 182,780 (7α,17β-[9-[(4,4,5,5,5-Pentafluoropentyl)sulfinyl]nonyl]estra-1,3,5(10)-triene-3,17-diol) as positive control. After the exposure, intensity of the luminescence was measured using Promega Steady Glo Kit (Promega, Mannheim, Germany). Effects on androgen receptor (AR) were evaluated with MDA-kb2 human breast cancer cell line [35]. Exposures were conducted in Leibowitz L-15 medium supplemented with 5% (v/v) stripped FCS at 37°C without added CO_2. For testing antiandrogenicity, cells were seeded into 96-well plates (15,000 cells/well) in medium supplemented with 1 nM dehydrotestosterone (DHT) and exposed to a dilution series of extracts. After 24 h exposure, lysis buffer was added and luminescence measured after 30 min using 100 µL of substrate for luciferase according to Wilson et al. [35]. In all experiments, the solvent (methanol or DMSO) concentration did not exceed 0.5% v/v. Exposures were conducted for 24 h at 37°C.

3.3 In vivo - Reproduction in potamopyrgus antipodarum. Potamopyrgus antipodarum (GRAY 1843), the mudsnail, originates from New Zealand. It can be found on soft sediments of standing or slowly flowing water bodies as well as in estuarine areas on the coasts at salinities up to 15‰ [36]. European populations consist almost entirely of female snails reproducing parthenogenetically. In Europe, male snails are found only very rarely [37,38] and were never observed in our own laboratory culture. Although reproduction occurs throughout the year, the maximum offspring production occurs in spring and early summer, while the minimum is from autumn to early winter [39]. P. antipodarum performs a very distinct kind of brood care, termed ovovivipary [40]. The eggs develop in the anterior part of the oviduct, which is transformed into a brood pouch. After removing the shell of the snail, embryos can be accurately seen through the epithelia. By opening the brood pouch and subsequently removing the embryos and counting them, the reproduction success of each female is easy to determine.

Mudsnails for the testing of Schussen and Argen samples were taken from the laboratory culture of the Department Aquatic Ecotoxicology at Goethe University Frankfurt am Main, Germany. Tests were conducted according to the Standard Operating Procedure (SOP Part III: Reproduction test using sediment exposure) [41] and an OECD guideline proposal [42]. We measured mortality and the number of embryos in the brood pouch after 28 days of exposure.

Sediments from the two field sites S 3 and S 4, and from the effluent of WWTP Langwiese were analysed. Samples from the field sites, stored frozen (-23°C) until the start of testing, were obtained in seven independent sampling campaigns (C, D and E 2010, F, G, H and J 2011).

Samples were thawed at room temperature before testing and individual sediments were mixed with a stainless steel spatula. An aliquot of 100 g sediment (wet weight) was transferred into the test vessels (1 L screw-cap borosilicate glass). WW samples were thawed and 800 mL transferred into 1 L screw-cap borosilicate glass vessels. For the negative control (C) and the positive control (PC) an artificial sediment consisting of 95% quartz sand (grain size 50–200 μm) and 5% dried and fine-grounded beech leaves (*Fagus sylvatica*) was used per replicate. For the PC, the artificial sediment was spiked with a nominal concentration of 30 μg/kg of 17α-ethinylestradiol (EE2) in order to verify the estrogen-sensitivity of the test organisms. All sediment and WW samples were tested with two replicates, while four replicates were used for control groups (C and PC). All sediment samples, including C and PC, were covered with 800 mL of fully reconstituted water according to OECD [42]. Test vessels were aerated via a Pasteur pipette. Twenty adult snails with a shell height of 3.5 to 4.3 mm were used for each replicate vessel (static system, light-dark rhythm of 16:8 h, 16±1°C, pH 8.0±0.5, oxygen content >8 mg/L, oxygen saturation >80% and conductivity 770±100 μS/cm). Only the WW samples were characterized by a slightly higher conductivity (797–1166 μS/cm). Water parameters were checked for each replicate at the beginning and end of the experiment and once a week during the experiment. Animals were fed three times a week with fine-grounded TetraPhyll® (0.2 mg dry weight per snail). After 28 days, all surviving snails were removed from the sediment and narcotized (2.5% magnesium chloride hexahydrate). The shell and aperture height were measured. The embryos were then removed from the pouch and counted, whereby shelled and unshelled embryos were distinguished.

4 Detection of Endocrine Effects – *In vivo*

4.1 Vitellogenin detection in brown trout. Juvenile brown trout (*Salmo trutta f. fario*) were used as test animals for the active exposure experiments in 2011 and 2012. Freshly fertilized brown trout eggs were bought from a hatchery (2011: Störk, Bad Saulgau, Germany and 2012: Schindler, Alpirsbach, Germany) and exposure started 4 hours after fertilization in three different treatments (laboratory, bypass station at the Schussen and at the Argen). In each bypass station, 300 eggs were exposed in an aquarium with a constant flow-through rate of 12 l/min of water from the streams. As laboratory control, 300 eggs were held in an aquarium at 8°C in filtered tap water with a filter (Co.: JBL 1500e). A third of the water volume was exchanged once per week and, after the eying of the embryos, the light/dark photoperiod

simulated field conditions. After hatching juvenile trout were fed by food for fry (Co.: BioMar, Biomar Inicio plus) and exposure continued till sampling (2011/12 exposure time: 99 days post fertilisation; 2012/13 exposure time: 111 days and 124 days post fertilisation). For vitellogenin analyses, larvae from each treatment were killed with an overdose MS-222 (tricaine mesylate, Sigma-Aldrich, St. Louis, USA), and the region between head and pectoral fin from each individual was placed in Eppendorf tubes, snap-frozen, and stored at −80°C.

All the following steps were undertaken on ice. Homogenates of juvenile trout were prepared by adding homogenization buffer (4-times the sample weight; PBS+2 TIU Aprotinin, C. Roth, Germany), mixing with a plastic pestle, centrifuging (10 min, 4°C, 20000×g (Eppendorf 5810R)) [31] and storing the supernatants at −80°C. As recommended by the provider of the test kit, a minimum of 1:20 dilution was used. Each sample was tested in duplicate. In 2012/2013, the semi-quantitative ELISA test kit, which is recommended for vitellogenin analyses of salmonides, was used (Biosense Laboratories AS, Bergen, Norway; V01002402: Semi-quantitative vitellogenin Salmonid (Salmoniformes) bio-marker ELISA kit). The enzyme activity (absorbance) which is measured in the assay is proportional to the concentration of vitellogenin in the sample (Automated Microplate Reader Elx 8006, Bio-Tek Instruments, INC., Winooski, Vermont, USA). Purified vitellogenin from Atlantic salmon (*Salmo salar*) was used as a positive control within every assay run as recommended by Biosense.

In 2011/12, we used a quantitative kit with a rainbow trout-specific antibody against vitellogenin (Biosense Laboratories AS, Bergen, Norway; V01004402: rainbow trout (Oncorhynchus mykiss) vitellogenin ELISA kit). As a pre-test to check the cross-reaction between rainbow trout antibody and brown trout vitellogenin, we analysed juvenile brown trout which we exposed for 16 days either to 40 ng/L EE2 or to clean water. Results of control fish showed 0 ng/L vitellogenin and EE2 exposed brown trout showed 2377±285 ng/L vitellogenin (each treatment: n = 6). This test showed that we are able to detect brown trout vitellogenin by using the rainbow trout specific antibody (rainbow trout kit).

4.2 Maturity stage and gonadosomatic index (GSI) of feral fish. In the field, at sites S 3 (downstream from WWTP Langwiese, Schussen) and S 4 (Argen) two feral fish species, chub (*Leuciscus cephalus*) and spirlin (*Alburnoides bipunctatus*), were caught by electrofishing (for caught fish numbers see in the result section). Fish were killed with an overdose of MS-222 (tricaine mesylate, Sigma-Aldrich, St. Louis, USA), weighed, and measured length-wise. The gonads were removed, weighed, and a small part of the middle part of the gonad was fixed in 2% glutaraldehyde in 0.1 M cacodylic acid for histological analyses. After embedding the fixed parts of the gonads in paraffin and cutting them in 3 μm slices, the slices were stained using two different methods (hematoxylin-eosin staining and alcianblue-PAS staining). Per fish 6 slices in three cell layers were evaluated by light microscopy and classified in 3 maturity stages according to Nagel et al. [43].

Female gonads:

Stage 1: Only oogonia or 90 to 100% previtellogenic or early perinucleolar oocytes present, <10% vitellogenic oocytes or yolk vesicle stadia

Stage 2: >10% vitellogenic oocytes or yolk vesicle stadia present, <50% mature oocytes with yolk and/or lipid

Stage 3: >50% mature oocytes with yolk and/or lipid present

Male gonads:

Stage 1: >80% spermatogonia, no spermatozoa present

Stage 2: <30% spermatozoa, residual spermatogonia, spermatocytes, and spermatids present.

Stage 3: >30% spermatozoa, residual spermatocytes, and spermatids present.

All statements refer to percentages of areas in the histological sections. The gonadosomatic index (GSI) was calculated according to Kang et al. [44]:

$$GSI = (weight\ of\ gonads * 100)/total\ weight$$

5 Statistical Analyses

5.1 In vitro tests. The samples applied to the E-screen assay were quantified via the dose-response curve of the reference substance 17β-Estradiol (E2) and the curve of a dilution series of a sample extract. The estrogenic activity of the sample was calculated as the ratio of the EC50-values of 17β-estradiol (E2; positive control) and the dilution curve:

$$17\beta - estradiol\ equivalent\ concentration(EEQ) =$$

$$EC_{50(E2,ng/L)}/EC_{50(sample)}$$

The limit of detection (LOQ) was defined as EC_{10} of the sample extract curve in comparison to the standard curve of E2. The LOQs depended on the individual concentration factor being used for the samples and were in the range of 0.01 ng/L–0.1 ng/L.

All samples analysed in the cellular reporter gene assays were tested in at least five different concentrations against each endpoint. Each treatment was performed in three replicates. The luminescence values measured in the estrogenicity and androgenicity assays were expressed as percentages of the maximum effect by subtracting the solvent control response and relating the values to the maximal response of standard ligand (E2$_{max}$ for estrogenicity or DHT (dehydrotestosterone) $_{max}$ for androgencity). Maximum induction values as well as the shape of the curve differed among samples, thus equal efficacy or parallelism of the dose–response curves could not be assumed [45]. Final EEQ values (17-beta-estradiol equivalents) or DHT-equivalents were based on relating the amount of model ligand (E2 or DHT) causing 25% of the E2$_{max}$ response (EC$_{25}$) to the amount of sample causing the same response (determined from regression analysis). The EC values were calculated by nonlinear logarithmic regression of dose–response curve of calibration standard and samples in Graph Pad Prism (GraphPad Software, San Diego, USA). Assays enabled detecting estrogenic activity higher than 0.5 ng EEQ/L of effluent or 6 ng EEQ/kg of sediment. Antiestrogenicity and antiandrogenicity were expressed as the sample concentration that caused 25% inhibition of luminescence (IC$_{25}$, g/ml) in the presence of competing ligand E2 (for antiestrogenicity) or DHT (antiandrogenicity). The IC values were determined on the basis of the linear regression models. The reciprocal value of IC$_{25}$ is presented as 1/EC$_{25}$ of the studied sample.

5.2 In Vivo Tests. The statistical analysis of data of the reproduction test with *P. antipodarum* was performed using Prism®, version 4.03 software (GraphPad Software, San Diego, CA, USA). Normally distributed data (D'Agostino-Pearson test) with equal variances (Bartlett test) were tested with a one-way ANOVA with Dunnett's post test for significant differences to the negative control (K). In all other cases, the nonparametric Kruskal-Wallis with Dunn's post test was used. Mortalities, expressed as quantal data, were analysed using Fisher's exact test.

Statistical analyses, which addressed the results of *in vivo* tests with fish, were performed with JMP 10.0 (SAS Systems, USA). Data were tested for normality using the Shapiro-Wilk W-test. If data were normally distributed the t-test was conducted, otherwise the Wilcoxon test or Steel-Dwass-test was used.

Results and Discussion

1 Chemical Analysis

A total of more than 150 micropollutants, including endocrine-active chemicals, were analysed in more than 75 water and sediment samples. The following substances were always below their detection limits: 4-iso-nonylphenol, iso-nonylphenoldiethoxylat (detection limits: 25 ng/L) and all analysed polybrominated diphenyl ethers (BDE-100, −138, −153, −154, −183, −209, −28, −47, −66, −85, and −99; detection limits: 10 ng/L). Highly potent steroid hormones like 17α-ethinylestradiol and 17β-estradiol were not detected (detection limits: 1 ng/L). Our detection limits are high, and due to the fact that EE2 is biologically active in concentrations of 1 ng/L [46], biological effects of EE2 could be present although EE2 was not detected by our chemical analyses. In few samples, estrone was detectable but only in low concentrations up to 0.8 ng/L at S 3.

The phytohormone β-sitosterol was detectable in 5 out of 7 WW samples (max. 990 ng/L), in 1 out of 2 water samples of S 3 (360 ng/L) and in 2 out of 2 water samples of S 4 (max. 1.2 µg/L). 4-tert.-Octylphenol (in 3 out of 7) and bisphenol A (in 4 out of 7) were measurable in low concentrations in WW samples (detection limit: 5 ng/L). In the past, octylphenol occurred in surface water of the Schussen in concentrations up to 0,098 µg/L [15], which were close to the suggested target value of 0,1 µg/L for endocrine disrupting chemicals [47].

Sediment samples were analysed from campaigns C and F, and only low concentrations of β-sitosterol were found at all examined sampling sites. o,p-DDT, p,p-DDD, p,p-DDE and p,p-DDT were not detectable in any sediment samples (detection limit of 2 µg/kg dry weight). Analysed sediment samples of campaigns K, L and M showed a temporary occurrence of BDE-209 (max. 0.2 µg/kg) and di(n-butyl) phthalate (DBP) (max. 66 µg/kg) at sampling sites at the Schussen. Concentrations of perfluorooctanesulfonate (PFOS) and perfluorobutanoate (PFBA) were detectable only in few samples with concentrations up to 3.26 µg/kg.

In summary, the chemical analyses showed only few endocrine active substances in all investigated compartments. The phytohormone β-sitosterol was found in µg/L concentrations, but compared with synthetic or natural hormones, it is considered to be less potent by a factor 10^4 [48]. This indicates that the risk of causing endocrine effects in animals living in the Schussen and Argen seems to be low. The fact that only few highly potent endocrine disrupting chemicals were found was unexpected (especially for waste water samples), because other studies (summarized in [15]) showed that there are detectable endocrine active substances, especially in the Schussen river.

2 Endocrine Effect Potentials

2.1 E-screen assay. Figures 3 and 4 show means of EEQ and toxicity from all samples of the campaigns in 2010 (sampling C, D, E), 2011 (F, G, H, J), 2012 (K, L, M), and 2013 (N). The highest estrogenic activity was measured in the WW samples with a mean of 3.1 ng/L EEQ. At the sampling sites downstream from the WWTP (S 3 and S 6), EEQs of about 0.8 ng/L were detected.

The lowest estrogenic activity was measured at the Argen (S 4) with 0.04 ng/L EEQ. Variability of the estrogenicity caused by seasonal or event-triggered effects assume to the average EEQs. Despite of these variations the results clearly showed a higher pollution of the river Schussen. The results of the cytotoxicity tests correlated with the results of the E-screen assay. Highest toxicities were observed in the WW samples and we had to exclude 5 of 9 samples in the E-screen because the high cytotoxic activity compromised the sensitivity of the E-screen assay. Similarly, samples of S 3 (5 out of 11) and S 6 (6 out of 11) showed high cytotoxicity and were also excluded. In contrast, samples of S 0, S 1, and S 4 had no evidence of cytotoxicity. Therefore, the estrogenic activity at Argen (S 4) and at two sampling sites at the Schussen (S 0 and S 1) could be assessed as low, whereas the WW clearly showed the highest observed estrogenic effects. The sampling sites downstream from the WWTP (S 3 and S 6) were charged less with estrogenic compounds compared to the WW. Due to an overlay of hormone action by cytotoxic effects, it is likely that the estrogenic potential in our samples from WW, S3 and S6 was actually higher than what our results suggest. Previous studies have found estrogenic activities in upper ranges as the one we measured with the E-screen assay: for WW samples (6–11 ng/L EEQ in [23,49]) and for rivers (4 ng/L EEQ in [49]). The EEQ values determined by E-screen in Schussen samples are clearly indicative of expected significant field effects as it was recently proposed [50]. The mean value of 3.1 ng EEQ/L is above the E-screen-specific Estrogenic Limits (ELs) suggested (higher than 2 ng EEQ/L [50]).

2.2 Reporter gene assays. Estrogenicity: In the effluent samples studied, no or only low estrogenicity was detected (one sample in campaign D with 0.88 ng/L of E2 equivalents, see Table 3). Nevertheless, the value determined with this reporter gene assay may indicate effects in vivo as it is within the range (or above) the Estrogenic Limits recently suggested. A number of research studies provide information on the estrogenicity of contaminated effluents and waters. These include a recent EU-wide study of 75 WWTP effluents [51], which has demonstrated that 27 of the analysed WW samples show estrogenic activity above the detection limit of 0.5 ng/L EEQ and that, in positive samples, estrogenicity varies from 0.53 to 17.9 ng/L EEQ.

For sediment samples, the HeLa bioassay shows a low estrogenic potential, referring to absolute values. However, the trend between localities is clear - much weaker effects were

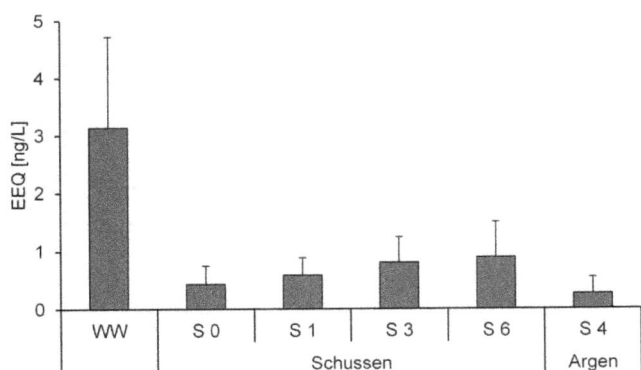

Figure 4. E-screen assay (cytotoxicity). Results of the E-screen assay regarding the cytotoxicity of the analysed samples. Expressed in $1/EC_{50}$ Tox (concentration in which 50% of the cells are able to grow) units; means and standard deviation. WW (Waste water of WWTP Langwiese) n = 9, S 0 n = 5, S 1 n = 4, S 3 n = 11, S 6 n = 11 and S 4 n = 11.

apparent at S 4 (Argen; only 2 positive samples, maximum 14 pg/g EEQ) in comparison to S 3 (Schussen; maximum up to 55 pg/g EEQ), compare Table 3. Comparable estimates for sediment samples for other studies are relatively rare. For Czech sediments, median values measured using MVLN cells were around 100 pg/g EEQ (with maxima around 500 pg/g) [20] and 4.7–22 pg/g [52]. In various European sediments (ESP, DE, CZ) values about 75–669 pg/g EEQ [53], in rivers in France up to 200–6430 pg/g EEQ [54] and in four Italian rivers (7 sites) values between 15.600±7.300 pg/g EEQ [55] were reported. In comparison with the absolute values of these studies, our data are within the range or lower.

Antiestrogenicity: In effluent samples - similar to estrogenicity – we recorded weak antiestrogenic effects: only a single sample shows a measurable effect (campaign J - antiestrogenic index 0.4 $[g/ml]^{-1}$). With respect to sediments, antiestrogenic effects were observed in several samples. Similar to estrogenicity, more pronounced effects were detected in the Schussen river (S 3; maxima up to 840 of the antiestrogenicity index $[g/ml]^{-1}$) in comparison to the Argen river (S 4; maxima up to 485 $[g/ml]^{-1}$). Antiestrogenicity showed seasonal dynamics with lower levels in spring and higher ones in autumn (Table 3). Previously, seasonal dynamics were reported in antiestrogenicity as well, with values in sediments ranging from 35–153 $[g/ml]^{-1}$ during spring to 250–1000 $[g/ml]^{-1}$ during autumn [20]. There are only few studies assessing antiestrogenicity in sediments: in Italian and Tunisian sediments no antiestrogenic effects were found, whereas in 3 rivers from an agricultural area in Nebraska (USA) a strong inhibition of E2-induced effects was reported [54,56].

Antiandrogenicity: For **effluents**, none of the samples showed antiandrogenicity up to the highest equivalent concentration that was tested (i.e. 12-times concentrated). To our knowledge, only few studies investigated antiandrogenicity of surface waters or effluents, and the values reported previously were highly variable. Previous works reported 438 µg/L of antiandrogen flutamide equivalents (FluEq) for a river in Italy [57] and in Chinese surface water antiandrogenicity ranged from 20 to 935 µg/L FluEq [58]. Statistical modelling of the 30 WWTPs from UK waters predicted antiestrogenicity in FluEq values ranging 0–100 µg/L (with median and average of 10 and 20 µg/L, respectively) indicating that chemical cocktails of both estrogens and antiandrogens may contribute to the wild fish feminization [59].

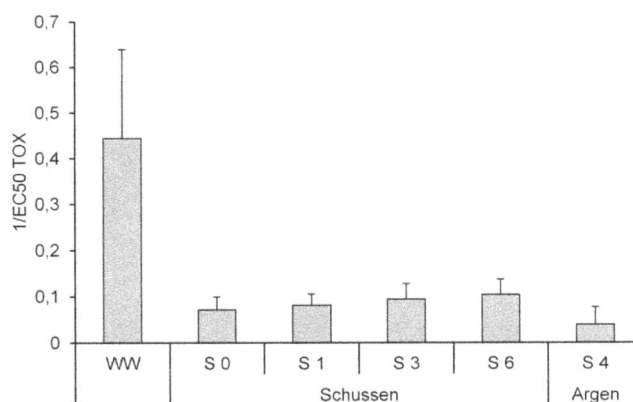

Figure 3. E-screen assay (estrogenic activity). Results of the E-screen assay expressed in 17β-estradiol equivalents (EEQ) in ng/L; means and standard deviation. Only data of samples which showed a low cytotoxicity (see figure 4) were used. WW (Waste water of WWTP Langwiese) n = 4, S 0 n = 5, S 1 n = 4, S 3 n = 7, S 6 n = 6 and S 4 n = 11).

Table 3. Summary results of mammalian cell reporter gene assays.

SEDIMENT SAMPLES

	2010		2011				
	C	D	E	F	G	H	J
Estrogenicity - [EEQ - pg E2 equivalent/g dw]							
Site S 3	18,0	40,8	54,5	n.e.	n.e.	n.e.	49,7
Site S 4	14,08	6,13	n.e.	n.e.	n.e.	n.e.	n.e.
Antiestrogenicity index [g/ml]$^{-1}$							
Site S 3	511	-	645	602	437	840	719
Site S 4	408	-	n.e.	412	210	485	198
Antiandrogenicity index [g/ml]$^{-1}$							
Site S 3	19,9	n.e.	n.e.	8,2	6,6	25,1	51,0
Site S 4	4,4	n.e.	n.e.	8,5	8,4	19,5	13,3

EFFLUENTS (WWTP, Langwiese)

	2010		2011				
	C	D	E	F	G	H	J
Estrogenicity [EEQ - ng/L]	n.e.	0,878	n.e.	n.e.	n.e.	n.e.	n.e.
Antiestrogenicity index [1/IC25]	n.e.	n.e.	n.e.	n.e.	n.e.	n.e.	0,4
Antiandrogenicity [1/IC25]	n.e.	n.e.	n.e.	n.e.	n.e.	n.e.	n.e.

n.e. = no effect up to the highest tested concentration, i.e. 0.5 g sediment dw/ml or equivalent of 12x concentrated water.
- = no samples analysed.

In sediments (see Table 3), several samples always showed stronger anti-androgenic effects at S 3 at the Schussen compared to S 4 at the Argen. No anti-androgenic effects were observed during two campaigns (D and E). In general, higher effects were observed at S 3. Nevertheless, all values were lower in comparison to contaminated river sediments studied before [20]. Because the LOEC for fish is 63–651 µg/L FluEq as summarized by Runnalls et al. [60], we rarely expect antiandrogenic effects of the tested water in fish. Antiandrogenicity of sediment samples was also determined in previous studies, but the reported effects cannot be directly compared due to the use of different expressions/units: in sediments from the Czech Republic, antiandrogenicity was observed but not quantified [61,62]; in Italian sediments a maximum inhibition of - 20% of dehydrotestosterone was reported [63], and in French sediments 1.1–32.5 µg/g flutamide equivalents were measured [54].

2.3 Comparison of *in vitro* assays.

Effluents of the WWTP Langwiese showed a higher estrogenic activity in the E-screen (four samples with mean 3.1 ng/L EEQ; Fig. 3) than in the reporter gene assay (estrogenicity detected only in one sample: 0.88 ng/L of EEQ; table 3). Therefore, the five day proliferation E-screen test seems to be more sensitive for the estrogenic assessment in comparison with the 24-h gene activation assays. Due to the high cytotoxicity observed in effluents, at S 3, and S 6 in the E-screen, we contend that the real estrogenic pollution is higher than 3.1 ng/L for effluents of the WWTP Langwiese (similarly for sampling sites S 3 and S 6). We used the reporter gene assay to analyse sediment samples, but not for surface water. Similar to the water sample results (measured with the E-screen), sediments from the Schussen (S 3; maximum 55 pg/g EEQ) showed higher estrogenic activities than those from the Argen (S 4; maximum 14 pg/g EEQ).

When comparing our results for sediment and water samples, it was obvious that the sediment samples showed a higher estrogenic activity than the water samples. Note that measurements of surface water (by E-screen) and sediment samples (by reporter gene assay) are not directly comparable due to different endpoints (growth vs gene transactivation) as well as origin of the cell lines used (MCF-7 vs HeLa-9903 [17,34]). Previous work showed that the reporter gene assay with HGELN cells (which are derived from the HeLa cells used in the present study) may be less sensitive than the E-screen (with MCF-7 cells) when indicidual compounds are considered [64]. However, interpretation of tests with complex mixture samples (as performed in the present study with effluents, waters and sediments) may be more complicated depending on the actual composition of the studied samples. For example, simultaneous presence of both estrogens and antiestrogens may induce different responses (both estrogenic and antiestrogenic, depending on the concentration ranges and ratios). In the present study, high antiestrogenicity was detected in studied sediments being systematically higher at the S3 site in Schussen river. These results suggest that estrogenicity could be underestimated, and might be even higher than measured by the reporter gene assay. This is in line with results of Peck et al. [65], who have suggested that riverine sediments are a major sink and a potential source of persistent estrogenic contaminants. A study at the Upper Danube River in Southern Germany with *in vitro* assays also showed that endocrine disrupting potentials were elevated in selected sediments and confirmed an accumulation of endocrine active substances in sediments [66].

To summarize, our *in vitro* assays showed apparent endocrine disruptive potentials at the Schussen and Argen. These potentials varied over time, and were more pronounced at the Schussen. The presence of cytotoxic and antiestrogenic potentials implies that direct estrogenic potentials at the Schussen might be underestimated.

2.4 Reproduction in *potamopyrgus antipodarum*.

In order to assess the relevance of in vitro bioassays for the in vivo situation, we investigated reproduction in the mudsnail Potamopyrgus antipodarum. The overall mortality during the tests was quite low with a mean value of 5.8% and 9.5% for the negative and positive control, respectively. Although the mortality was nominally higher in the WWTP effluent samples (mean: 22.4%) and in sediments from the two field sites, S 3 and S 4 (15.2% and 13.7%, respectively), this increase was neither statistically significant when merging the values from all sampling campaigns nor for the single sampling campaigns (Fisher's exact test, p>0.05). As the number of embryos in the brood pouch of *P. antipodarum* is positively correlated with shell height, all test animals were taken from a defined size class (3.5 to 4.3 mm shell height) at the start of the experiment. At the end of the experiment, differences in shell height between the treatment groups were very low (maximum difference of mean shell height: 4.01% between negative control and sediment from S 3 in August 2010) and not statistically significant (ANOVA, p>0.05). The average number of embryos in the brood pouch of females in the negative control group was 8.92, while females in the positive control group had a mean of 14.4 embryos in the brood pouch. This represents a highly significant increase of 74.5% (p<0.01, Fig. 5).

The mean embryo number of 8.67 in mudsnails that were exposed for four weeks to the WWTP effluent was not statistically significant different from the negative control. In contrast, the total number of embryos in female snails which have been exposed to the two field sediments from S 3 and S 4 was significantly higher than in the negative control with mean values of 16.9 and 17.0, respectively (Fig. 5). This increase by 104%–105% was even well above the level of the positive control (ANOVA with Dunnett's post test, p<0.01). There was no significant difference in embryo numbers between females from the two field sediments.

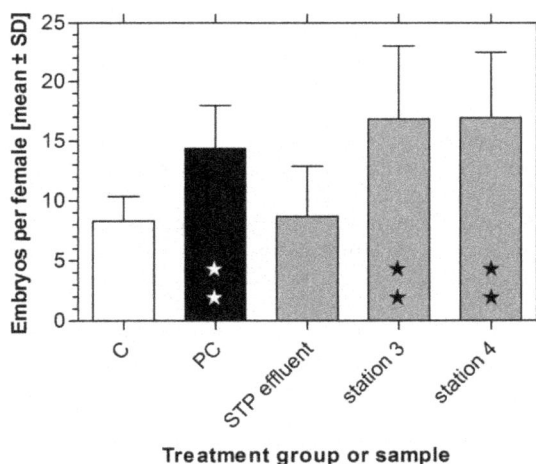

Figure 5. Reproduction test with the mudsnail. Means and standard deviation of the reproduction test with *Potamopyrgus antipodarum*. Total embryo number per female in negative (C) and positive controls (PC), in effluent water from the waste water treatment plant Langwiese (STP effluent) and in the two field sediments from sampling sites S 3 at the Schussen river and S 4 at the Argen river (station 3 and station 4) over the seven sampling campaigns. Asterisks indicate significant differences vs. C (one-way ANOVA with Dunnetts multiple comparison test; p<0.01).

It remains controversial as to whether reproduction in snails is regulated by an estrogen signalling pathway, homologous to vertebrates. Although there is broad empirical evidence that an exposure of caenogastropods and bivalves to estrogens and their mimics alters sexual differentiation and reproductive parameters, in some cases even at environmentally-relevant concentrations [27,42], the observed effects on embryo numbers in *P. antipodarum* cannot univocally be attributed to estrogen signalling. This is because the endocrine systems of molluscs are insufficiently characterised and the precise mode(s) of action of endocrine active chemicals, including estrogens and their mimics are not fully understood. However, the significant increase of embryo production observed in the field sediments S 3 and S 4 is a clear indication for reproductive disruption with obvious potential for population level consequences [27,67,68].

The apical effects of an exposure to endocrine active chemicals in *P. antipodarum* have been reviewed by Duft et al. [27]. Exposure to various xeno-estrogens (BPA, octylphenol, nonylphenol, EE2) resulted in increased embryo numbers in the brood pouch of mudsnails. In the case of BPA, a stimulation of the reproductive output was noted in a sediment test with an EC_{50} of 5.67 µg/kg and an EC_{10} of 0.19 µg/kg after four weeks [69]. Exposure to BPA and EE2 via water was investigated by Jobling et al. [70], again resulting in a stimulated embryo production, with significant effects at a concentration of 5 µg BPA/L (NOEC 1 µg BPA/L) and 25 ng EE2/L (NOEC 5 ng EE2/L), respectively. A reproduction-disrupting effect of EE2 in *P. antipodarum* was confirmed by Sieratowicz et al. with a LOEC of 50 ng/L and a NOEC of 25 ng/L [39]. Most of the observed concentration-response relationships for both compounds, however, were biphasic, with an inverted U-shaped curve [39,70]. This is important for the interpretation of results from tests with reproduction disrupting chemicals or environmental samples with *P. antipodarum* because at very high concentrations, the stimulation of reproductive performance declines, and may even fall back to the level of the negative control. Corresponding observations have been made in several other studies with snails [67,69,71–73]. They can be explained by a dominant stimulating effect of these reproductive disrupting test compounds at low concentrations and a decrease in embryo production due to their general toxicity at higher concentrations.

Therefore, the significantly enhanced embryo numbers in mudsnails exposed to the field sediments from S 3 and S 4 indicate the presence of reproductive disrupting compounds. The effects at both rivers are higher than the effects in the positive control with a concentration of 30 µg EE2/kg, which indicates severe pollution by reproductive-disrupting compounds in the sediments of both rivers. In contrast, the lack of significant differences in embryo numbers between the WWTP effluent and the negative control is not necessarily evidence for a lack of such compounds in the waste water. In complex environmental samples, the presence of reproduction-toxic substances may compensate for the effects of estrogens and other disruptive compounds on embryo production in a way that stimulating effects can be completely masked. It is also possible that, at high concentrations of reproductive-disrupting compounds in waste water, the number of embryos is again reduced to the negative control level due to the already discussed biphasic curve of the concentration-effect relationship.

Galluba & Oehlmann [24] applied the *in vivo* reproduction test with *P. antipodarum* and the yeast estrogen screen (YES) as an *in vitro* assay in parallel for 50 sediments from smaller rivers and creeks. It was shown that 54% of the sediments exhibited a promoting effect on snail reproduction and also showed an estrogenic activity in the YES while 82% of the samples which were active in the YES

caused an increased snail reproduction. Despite this coincidence, the Spearman correlation between EEQs and embryo number in the snails was not significant because sediments with the highest EEQs in the YES caused no or little increase of embryo numbers. The lack of a significant correlation between the two systems may reflect the difference by which estrogens are acting in the yeast cells compared to how they are acting in the snail. Alternatively, it may be an indication that embryo numbers had returned to control levels at very high exposure to reproductive-disrupting compounds, reflecting the biphasic concentration response of the snails.

Galluba & Oehlmann [24] also discussed the possibility that lower embryo numbers in the artificial control sediment may reflect sub-optimal conditions for the development and reproduction of the snails. However, if embryo numbers in the tested field sediments are not compared to the artificial control sediment but to a natural reference sediment with no measurable estrogenic activity in the YES, an identical number of sediments turned out to exhibit significantly more embryos. This shows that reproduction in *P. antipodarum* is almost identical in natural sediments without estrogenic activity and in artificial sediments so that alternative explanations for enhanced embryo numbers such as the supply of more or better suited food can be ruled out.

Previous studies have pointed out that an increase in reproductive output in snails can have an adverse effect on the population [67,69,72]. A stimulation of reproductive output outside the main reproduction period may result in oviduct malformations as shown by Oehlmann et al. for *Marisa cornuarietis* [73]. Furthermore, the stimulation of reproduction outside of the breeding season is a waste of an organism's energy reserves because offspring face less favourable environmental conditions for survival and growth during these periods [68]. Further possible consequences are a reduced somatic growth of adults and a decreased reproductive performance during the actual breeding season [71].

3 Endocrine Effects in Fish

3.1 Vitellogenin detection in brown trout. In 2011/2012, juvenile brown trout, which were exposed at the bypass stations for 99 days after fertilization, showed higher average vitellogenin levels at the Schussen bypass compared to the Argen bypass and the negative control (Fig. 6). However, the differences were not significant. We analysed the samples with a kit that is specific for rainbow trout. Auxiliary tests indicate that the antibody cross-reacts more weakly with brown trout vitellogenin. Therefore we exposed juvenile rainbow and brown trout for 16 days to 40 ng EE2/L. After the exposure, we measured an average vitellogenin level of 2377 ng/L in the brown trout but found a higher average vitellogenin level of 279988 ng/L in the rainbow trout (while we analysed six brown trout samples, we were only able to analyse two rainbow trout samples because the others showed a strong reaction that exceeded the allowed extinction level of the assay). Given the difference in the ways the antibody binds with vitellogenin in brown and rainbow trout, we conjecture that the actual vitellogenin levels in juvenile brown trout were higher than shown in figure 6. Estrogen active compounds in the Schussen are likely causes for the increased vitellogenin levels. Vitellogenin levels in trout exposed at the Argen were lower compared to those from the Schussen, but not significantly so (p = 0,4030). This might result from the lower anthropogenic pollution of the Argen river [15]. Trout exposed at the Argen showed vitellogenin levels comparable to those of the negative control (p = 1,00).

In 2012/2013, vitellogenin analyses in 111 day-old juvenile brown trout showed no significant differences between trout

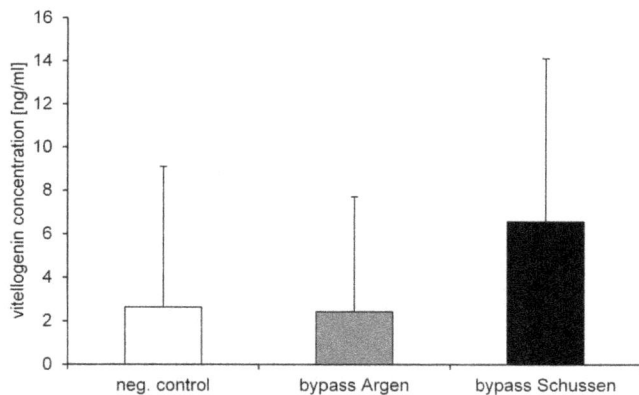

Figure 6. Vitellogenin in juvenile brown trout. Vitellogenin levels in homogenates of juvenile brown trout 99 days post fertilization in 2011/2012, means and standard deviation. Analysed by Biosense rainbow trout vitellogenin ELISA kit. Samples: Neg. control n = 6 (1 out of 6 pos. result, bypass Argen n = 10 (2 out of 10 showed a pos. result), bypass Schussen n = 10 (5 out of 10 showed a pos. result). No significant differences (Steel-Dwass-test: neg. control- bypass Argen p = 1,00, neg. control- bypass Schussen p = 0,5787 and, bypass Schussen- bypass Argen p = 0,4030).

exposed at the Schussen bypass and at the Argen bypass (Fig. 7, sampling March 2013). However, the values recorded for the negative control were significantly lower than those of trout exposed at the bypass stations. For the analyses, we used the semi-quantitative ELISA optimized for salmonids. The cross-reaction of the monoclonal antibody, BN-5, with brown trout vitellogenin is strong and recommended for vitellogenin analyses with brown trout [74]. Given that the negative control showed significantly lower levels (Steel-Dwass-test: neg. control- bypass Schussen p = 0,0159 and neg. control- bypass Argen p = 0,0221), the vitellogenin production in our juvenile brown trout is likely caused

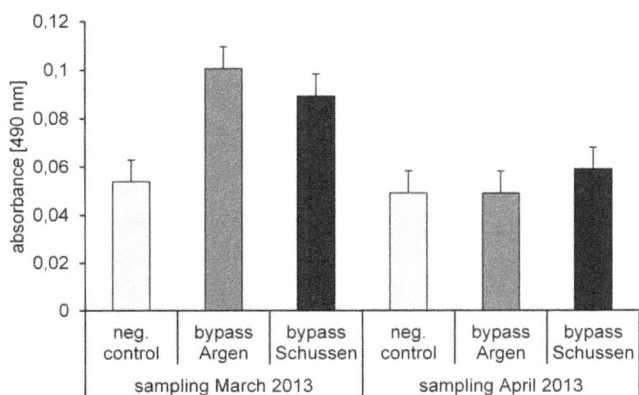

Figure 7. Semi-quantitative vitellogenin detection in juvenile brown trout. Absorbance measured in homogenates of juvenile brown trout 111 days post fertilization and 124 days after fertilization exposed in 2012/2013; means and SD. Each sampling analysed with one semi-quantiative vitellogenin salmonid (Salmoniformes) biomarker ELISA kit (enzyme activity = colour intensity is proportional to the concentration of vitellogenin in the sample). Samples March 2013: Neg. control n = 5, bypass Schussen n = 7, bypass Argen n = 6. Significant differences with Steel-Dwass-test: neg. control- bypass Schussen p = 0,0159 and neg. control- bypass Argen p = 0,0221; * = p<0.05. Samples April 2013: Neg. control n = 12, bypass Schussen n = 12, bypass Argen n = 12. No significant differences with Steel-Dwass-test.

by estrogen-like substances occurring in the Schussen and Argen. However, analyses of vitellogenin in juvenile brown trout from a second sampling (124 days of exposure; see figure 7, sampling April 2013) did not show any significant differences between all three treatments, and the vitellogenin levels were all in the range of the negative control.

A previous study conducted by Stalter et al. [31], showed a significant increase in the vitellogenin concentration (nearly 70 ng/mL compared to less than 10 ng/mL in the control) in yolk-sac rainbow trout which were directly exposed to WWTP effluents for 60 days. Other studies that examined WWTP effluents using sexually immature or male trout also showed a correlation between vitellogenin levels and WWTP effluents [6,75,76]. Another reason for the increased vitellogenin levels could be an immune response caused by pathogens occurring in the river water [77]. However, Zhang et. al [77] argued that juvenile fish are probably not able to produce vitellogenin as an immune response. Hence, we conjecture that mainly estrogens are responsible for the increased vitellogenin levels.

Overall, the vitellogenin levels we have detected were rather low compared to previous studies. However, these studies either exposed trout directly to WWTP effluents [6,32,78] or examined older feral trout [76,79,80]. We interpret our results as showing that an estrogenic pollution might be present in both rivers, but that concentrations apparently have varied and were able to induce vitellogenin production only in some cases.

3.2 Gonadal maturity and gonadosomatic index of feral fish. Generally, the gonadal maturity levels (Fig. 8) we observed in chub were higher in summer than in autumn, which is due to the spawning season (April to June). After the spawning season, the gonadal maturity normally decreases until females generate new eggs and males build new spermatozoa. Female chub caught at the Argen showed an increased gonadal maturity compared to chub from the Schussen (Fig. 8), potentially reflecting a higher estrogenicity in the Schussen river or anti-estrogenic effects at the Argen river. We did not observe any differences in the gonads between male chub caught at the Schussen and Argen.

In female spirlin from the Schussen and Argen rivers, differences in the maturity of gonads were low in summer (Fig. 9). In autumn, female spirlin caught at the Schussen showed a higher gonadal maturity than those from the Argen. Similar to the results obtained for male chub, we did not observe any differences in the maturity of male gonads between Schussen and Argen spirlin (Fig. 9). Because the spawning season for spirlin and chub is from April to July, it was expected that in autumn no spermatozoa would be detectable in the gonads of males and the maturity would be lower [81–83]. We did not find evidence for endocrine effects on male maturity in both rivers. Contrary to our results, a study on wild roach living in rivers receiving high amounts of effluents showed a progression of spermatogenesis mainly in males, whereas the females appeared to be less affected [84].

Female chub and spirlin reacted contrary to one another at the Schussen, whereas no difference between the two species could be observed at the Argen. At the Schussen, female chub (Fig. 8) showed a lower gonadal maturity but female spirlin (Fig. 9) a higher gonadal maturity compared to their respective conspecifics from the Argen. One possible reason for the observed differences is that the two species react differently to substances occurring in the Schussen. Although the water temperature at the Schussen is slightly higher than at the Argen in general, this is not a likely explanation for the observed differences. Higher temperatures could lead to faster gonadal growth and higher gonadal maturity [85,86], and hence, cause a higher gonadal maturity of fish at the Schussen. However, as a higher maturity was only observed for

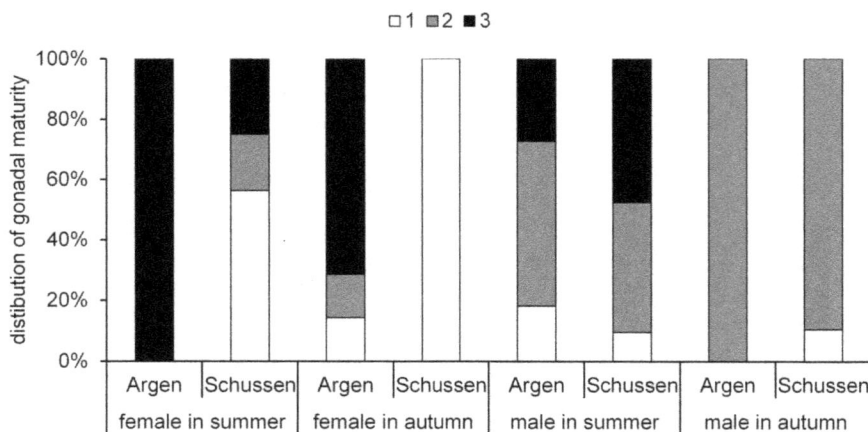

Figure 8. Maturity of chub. Distribution of gonadal maturity (stage 1 = immature; stage 2 = intermediate and, stage 3 = mature) of feral chub. 2009–2011. Females: summer Argen n = 2, summer Schussen n = 16, autumn Argen n = 7, autumn Schussen n = 12. Males: summer Argen n = 11, summer Schussen n = 21, autumn Argen n = 10, autumn Schussen n = 19.

female spirlin, the temperature is less likely to be the main cause for the observed effect.

In spirlin we only determined the gonadal maturity because in the field it was technically not possible to weight small gonads exactly. In summer, we did not observe any differences in the GSI values for chub between the Argen and Schussen (results not shown). In autumn, female and male chub caught at the Argen showed a significantly higher GSI than chub from the Schussen (Fig. 10). Also, female chub from the Schussen showed a distinctly lower GSI than the lowest value reported for chub by Mert et al. [87]. This could be the result of substances and stress factors in the Schussen which hinder the development of the gonads and cause a delayed maturity. The fact that both sexes show a reduced GSI could be explained either by the simultaneous presence of anti-estrogenic, androgenic, and estrogenic substances or by a general worse health status of fish at the Schussen compared to fish at the Argen.

This is in line with several studies, which showed a reduced gonad growth in fish caught at polluted areas [88–90]. Investiga-

tions in brown trout also showed lower GSI values and vitellogenin production for trout caught downstream of WWTPs compared to trout caught upstream of WWTPs [91,92]. A study about the interaction between 17β-trenbolone (TB) and EE2 in relevant environmental concentrations observed a decrease of the GSI of male eelpout after 21 days of exposure to EE2 alone or in combination with TB compared to controls [93].

4 Comparisons

Our *in vivo* tests revealed endocrine potentials/effects at the Schussen as well as at the Argen. The reproduction tests with *P. antipodarum* showed an equal increase in the number of embryos at both rivers, which were even higher than in the positive control (with a concentration of 30 μg EE2/kg). The vitellogenin levels we observed in juvenile brown trout also were increased at both rivers. Data of Jobling et al. [70] indicate that both, the nature of the response and the relative sensitivities to environmental estrogens, are comparable for *P. antipodarum* and rainbow trout. In concordance with this observation, our results for mudsnails were

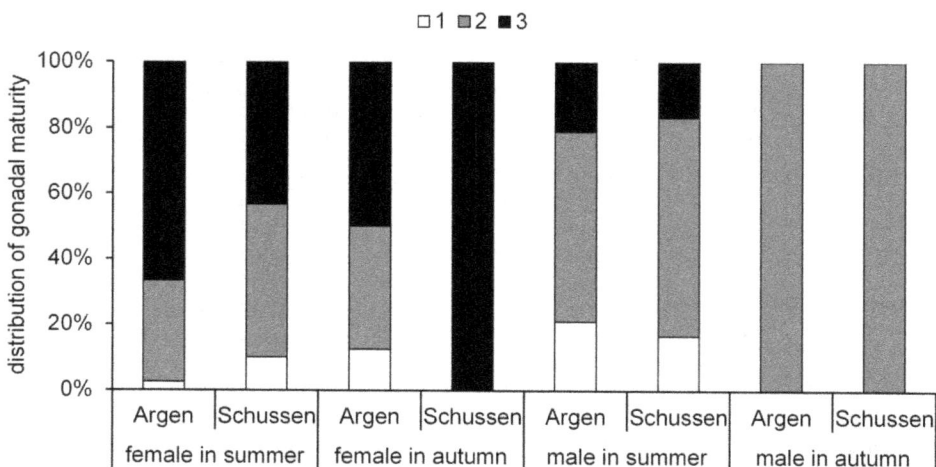

Figure 9. Maturity of spirlin. Distribution of gonadal maturity (stage 1 = immature; stage 2 = intermediate and, stage 3 = mature) of feral spirlin. 2009–2011. Females: summer Argen n = 35, summer Schussen n = 30, autumn Argen n = 16, autumn Schussen n = 7. Males: summer Argen n = 19, summer Schussen n = 3, autumn Argen n = 19, autumn Schussen n = 8.

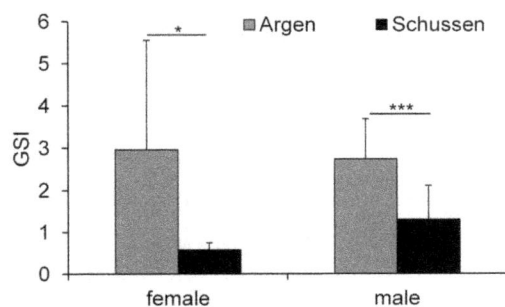

Figure 10. Gonadosomatic Index (GSI). Gonadosomatic Index of female and male chub caught in autumn 2010–2012 (sampling campaign E, J, and M); means and SD. Females: Argen n = 5 and Schussen n = 10. Males: Argen n = 12 and Schussen n = 16. Asterisks indicate significant differences between Schussen and Argen (* = p< 0.05 and *** = p<0.001).

qualitatively in line with those for brown trout. We performed the tests with *P. antipodarum* with sediments only for 4 weeks, whereas the trout were exposed directly after their fertilization to the river water for several months. The results were stronger for mudsnails, despite the fact that exposure time were much longer for trout. A potential explanation for this is that sediments (used for mudsnails) showed high estrogenic and antiandrogenic activities (as indicated by the reporter gen assay), whereas in the surface water, which we used for the trout tests, only low estrogenic activities were detected (as revealed in the E-screen). While *in vitro* and *in vivo* (mudsnails and vitellogenin production) tests provided qualitatively comparable perceptions of the endocrine-disruptive activity, the results of the chemical analyses did not reveal the presence of endocrine substances at effect concentrations, probably because not even the broad range of substances analysed in this study could represent the plethora of potentially endocrine-active compounds which are supposedly present in the environment. Moreover, mixture effects might be important: even if individual compounds were not detected, a combination of substances at lower-than-detectable levels could cause an effect. The gonadal maturity examinations in feral chub and spirlin did not provide clear indications for the presence of endocrine active substances. Nonetheless, chub of both sexes caught at the Schussen showed reduced GSI values compared to those caught at the Argen. A mechanistic interaction of endocrine-active (androgenic and/or estrogenic) and toxic compounds, as indicated by the *in vitro* assays, could explain the reduced GSI values at the Schussen river.

When analysing effluents of the WWTP Langwiese, all our tests revealed temporary endocrine activities. However, chemical analyses revealed only low concentrations of chemicals like estrone, β-sitosterol, octylphenol, and bisphenol A, which fluctuated over time. We conclude that constant presence, but concentrations below the limit of detection, possibly, a variety of compounds were the reason why our chemical analyses did not succeed in detecting high numbers of potent endocrine disrupting substances. In addition, chemical analyses only reflect snap-shots of pollution (single sample from the field or 24 h sample of the WWTP effluent) whereas fish were exposed for several weeks (trout) or for their lives (chub, spirlin). Our *in vitro* assays indicated that the aggregate estrogenic potential was relatively low (0.9 to 3 ng/L EEQ), but high cytotoxicity (as indicated by the E-screen) and the existence of antiestrogenic potentials (as indicated by reporter gene assays) could probably lead to an underestimation of estrogenic potentials. Notably, mudsnails exposed to effluents

showed no increase in the number of embryos compared to the negative control, but it is likely that estrogenic activities were masked by toxic substances, as indicated by increased mortality rates of mudsnails exposed to waste water, however, they were not significant higher. Our results suggest that the waste water has both estrogenic and toxic potentials.

Conclusion

Using a biological and chemical monitoring programme at two German rivers, we investigated whether symptoms of endocrine disruption in feral animals are reflected by results obtained in biological *in vitro* assays and by chemical analyses. In our case, chemical analyses provided only little information about the occurrence of endocrine active substances. In contrast, the results of our *in vitro* assays showed endocrine-disruptive activities for most of the analysed samples, indicating that the discharge of treated waste water results in elevated endocrine-disruptive potentials. Similar results were obtained *in vivo* using mudsnail reproduction tests and measuring GSI values of feral fish. In contrast, vitellogenin levels of trout and the maturity of feral fish showed only a slight indication of estrogenic activities.

Our multiple testing approach revealed that the E-screen assay reports higher estrogenic activities compared to the reporter gene assay (for waste water samples), which suggests that the E-screen assay was more sensitive in our analyses. Furthermore, it showed that *in vivo* tests with mudsnails alone would have led to an underestimation of the estrogenic activity of the waste water samples.

Our results imply that an interpretation of individual test results can be questionable, because different conclusions could be drawn from the results (e.g., as toxic effects might overlay endocrine effects), and an over- or underestimation of the endocrine pollution might result. We therefore propose a combination of *in vitro* and *in vivo* tests supported by advanced targeted instrumental analyses to assess endocrine pollution in rivers. The individual test results of the present study provide varying degrees of evidence for endocrine-mediated effects in fish that were due to possible interactions of toxic and endocrine impacts (Fig. 1). Nonetheless, the proposed combination of in vitro and in vivo tests overall strongly supports the plausibility of endocrine disruption in the test river, which results from chemicals that were not detected or detected only in low concentrations by our chemical analyses.

Acknowledgments

The technical help of Martin Steček and Martin Benišek (RECETOX) with laboratory analyses using mammalian cell reporter gene assays is acknowledged. We thank the fishers, B. Engesser and colleagues, for excellent work. Many thanks are due to M. Weyhmüller for the maintenance of the bypass systems as well as to the staff from the department of Animal Physiological Ecology (A. Dietrich, M. Di Lellis, K. Peschke, A. and V. Scheil, P. Thellmann, K. Vincze, and especially S. Krais) for help with the field work and sampling. Many thanks go also to J. R. Kielhofer, University of Arizona for valuable comments on the language of the manuscript.

Author Contributions

Conceived and designed the experiments: AH KB LB SG BK HRK DM JO DR MS USO AS RT SZ. Performed the experiments: KB AH BK DM DR MS AS SZ. Analyzed the data: KB LB SG AH BK DM JO DR MS USO AS SZ. Contributed reagents/materials/analysis tools: KB LB SG AH BK DM JO DR MS USO AS SZ. Wrote the paper: LB AH BK HRK DM JO DR MS RT.

References

1. Dickson RB, Eisenfeld AJ (1981) 17 Alpha-ethinyl estradiol is more potent than estradiol in receptor interactions with isolated hepatic parenchymal cells. Endocrinology 108: 1511–1518.

2. Jobling S, Tyler CR (2006) Introduction: the ecological relevance of chemically induced endocrine disruption in wildlife. Environmental health perspectives 114: 7–8.

3. Kavlock RJ, Ankley GT (1996) A Perspective on the Risk Assessment Process for Endocrine-Disruptive Effects on Wildlife and Human Health. Risk analysis 16: 731–739.

4. Sharpe RM, Irvine DS (2004) How strong is the evidence of a link between environmental chemicals and adverse effects on human reproductive health? BMJ 328: 447–451.

5. Colborn T, vom Saal FS, Soto AM (1993) Developmental effects of endocrine-disrupting chemicals in wildlife and humans. Environmental Health Perspectives 101: 378–384.

6. Purdom CE, Hardiman PA, Bye VVJ, Eno NC, Tyler CR, et al. (1994) Estrogenic Effects of Effluents from Sewage Treatment Works. Chemistry and Ecology 8: 275–285.

7. Van den Belt K, Berckmans P, Vangenechten C, Verheyen R, Witters H (2004) Comparative study on the in vitro/in vivo estrogenic potencies of 17β-estradiol, estrone, 17α-ethynylestradiol and nonylphenol. Aquatic Toxicology 66: 183–195.

8. Köhler H-R, Triebskorn R (2013) Wildlife ecotoxicology of pesticides: can we track effects to the population level and beyond? Science 341: 759–765.

9. Kidd KA, Blanchfield PJ, Mills KH, Palace VP, Evans RE, et al. (2007) Collapse of a fish population after exposure to a synthetic estrogen. Proceedings of the National Academy of Sciences 104: 8897–8901.

10. Kortenkamp A (2007) Ten years of mixing cocktails: a review of combination effects of endocrine-disrupting chemicals. Environmental Health Perspectives 115: 98.

11. Schwarzenbach RP, Escher BI, Fenner K, Hofstetter TB, Johnson CA, et al. (2006) The challenge of micropollutants in aquatic systems. Science 313: 1072–1077.

12. Bolong N, Ismail A, Salim MR, Matsuura T (2009) A review of the effects of emerging contaminants in wastewater and options for their removal. Desalination 239: 229–246.

13. Heinz B, Birk S, Liedl R, Geyer T, Straub KL, et al. (2009) Water quality deterioration at a karst spring (Gallusquelle, Germany) due to combined sewer overflow: evidence of bacterial and micro-pollutant contamination. Environmental Geology 57: 797–808.

14. Triebskorn R, Amler K, Blaha L, Gallert C, Giebner S, et al. (2013) SchussenAktivplus: reduction of micropollutants and of potentially pathogenic bacteria for further water quality improvement of the river Schussen, a tributary of Lake Constance, Germany. Environmental Sciences Europe 25: 1–9.

15. Triebskorn R, Hetzenauer H (2012) Micropollutants in three tributaries of Lake Constance, Argen, Schussen and Seefelder Aach: a literature review. Environmental Sciences Europe 24: 1–24.

16. Janosek J, Hilscherova K, Blaha L, Holoubek I (2006) Environmental xenobiotics and nuclear receptors-interactions, effects and in vitro assessment. Toxicology In Vitro 20: 18–37.

17. Soto AM, Sonnenschein C, Chung KL, Fernandez MF, Olea N, et al. (1995) The E-SCREEN assay as a tool to identify estrogens: an update on estrogenic environmental pollutants. Environmental Health Perspectives 103: 113–122.

18. Körner W, Hanf V, Schuller W, Kempter C, Metzger J, et al. (1999) Development of a sensitive E-screen assay for quantitative analysis of estrogenic activity in municipal sewage plant effluents. The Science of the total environment 225: 33–48.

19. Hilscherova K, Kannan K, Holoubek I, Giesy JP (2002) Characterization of estrogenic activity of riverine sediments from the Czech Republic. Archives of Environmental Contamination and Toxicology 43: 175–185.

20. Hilscherova K, Dusek L, Sidlova T, Jalova V, Cupr P, et al. (2010) Seasonally and regionally determined indication potential of bioassays in contaminated river sediments. Environmental Toxicology and Chemistry 29: 522–534.

21. Jarosova B, Blaha L, Vrana B, Randak T, Grabic R, et al. (2012) Changes in concentrations of hydrophilic organic contaminants and of endocrine-disrupting potential downstream of small communities located adjacent to headwaters. Environment International 45: 22–31.

22. Houtman CJ, Booij P, Jover E, Pascual del Rio D, Swart K, et al. (2006) Estrogenic and dioxin-like compounds in sediment from Zierikzee harbour identified with CALUX assay-directed fractionation combined with one and two dimensional gas chromatography analyses. Chemosphere 65: 2244–2252.

23. Körner W, Bolz U, Süßmuth W, Hiller G, Schuller W, et al. (2000) Input/output balance of estrogenic active compounds in a major municipal sewage plant in Germany. Chemosphere 40: 1131–1142.

24. Galluba S, Oehlmann J (2012) Widespread endocrine activity in river sediments in Hesse, Germany, assessed by a combination of in vitro and in vivo bioassays. Journal of Soils and Sediments 12: 252–264.

25. Brander SM, Connon RE, He G, Hobbs JA, Smalling KL, et al. (2013) From 'Omics to Otoliths: Responses of an Estuarine Fish to Endocrine Disrupting Compounds across Biological Scales. PLoS ONE 8: e74251.

26. Folmar LC, Hemmer M, Denslow ND, Kroll K, Chen J, et al. (2002) A comparison of the estrogenic potencies of estradiol, ethynylestradiol, diethylstilbestrol, nonylphenol and methoxychlor in vivo and in vitro. Aquatic Toxicology 60: 101–110.

27. Duft M, Schmitt C, Bachmann J, Brandelik C, Schulte-Oehlmann U, et al. (2007) Prosobranch snails as test organisms for the assessment of endocrine active chemicals—an overview and a guideline proposal for a reproduction test with the freshwater mudsnail Potamopyrgus antipodarum. Ecotoxicology 16: 169–182.

28. Kime DE, Nash JP, Scott AP (1999) Vitellogenesis as a biomarker of reproductive disruption by xenobiotics. Aquaculture 177: 345–352.

29. Ackermann GE, Schwaiger J, Negele RD, Fent K (2002) Effects of long-term nonylphenol exposure on gonadal development and biomarkers of estrogenicity in juvenile rainbow trout (Oncorhynchus mykiss). Aquatic Toxicology 60: 203–221.

30. Tyler CR, van Aerle R, Hutchinson TH, Maddix S, Trip H (1999) An in vivo testing system for endocrine disruptors in fish early life stages using induction of vitellogenin. Environmental Toxicology and Chemistry 18: 337–347.

31. Stalter D, Magdeburg A, Weil M, Knacker T, Oehlmann J (2010) Toxication or detoxication? In vivo toxicity assessment of ozonation as advanced wastewater treatment with the rainbow trout. Water Research 44: 439–448.

32. Sumpter JP, Jobling S (1995) Vitellogenesis as a biomarker for estrogenic contamination of the aquatic environment. Environmental Health Perspectives 103: 173.

33. Körner W, Hanf V, Schuller W, Kempter C (1999) Development and testing of a simple screening system for estrogen-like acting environmental chemicals.: University of Tübingen and University Women's Hospital Ulm. PUGU 95 004.

34. Environmental Protection Agency US (2011) ESTROGEN RECEPTOR TRANSCRIPTIONAL ACTIVATION (Human CELL LINE – HeLa - 9903). Standard Evaluation Procedure (SEP). ENDOCRINE DISRUPTOR SCREENING PROGRAM U.S. Environmental Protection Agency Washington, DC 20460, September 2011.

35. Wilson VS, Bobseine K, Lambright CR, Gray LE, Jr. (2002) A novel cell line, MDA-kb2, that stably expresses an androgen- and glucocorticoid-responsive reporter for the detection of hormone receptor agonists and antagonists. Toxicological Sciences 66: 69–81.

36. Jacobsen R, Forbes VE (1997) Clonal variation in life-history traits and feeding rates in the gastropod, Potamopyrgus antipodarum: performance across a salinity gradient. Functional Ecology 11: 260–267.

37. Wallace C (1979) Notes on the occurrence of males in populations of Potamopyrgus jenkinsi. Journal of Molluscan Studies 45: 383–392.

38. Ponder WF (1988) Potamopyrgus antipodarum - a molluscan coloniser of Europe and Australia. Journal of Molluscan Studies 54: 271–285.

39. Sieratowicz A, Stange D, Schulte-Oehlmann U, Oehlmann J (2011) Reproductive toxicity of bisphenol A and cadmium in Potamopyrgus antipodarum and modulation of bisphenol A effects by different test temperature. Environmental Pollution 159: 2766–2774.

40. Fretter V, Graham A (1994) British prosobranch molluscs. Their functional anatomy and ecology. Journal of the Marine Biological Association of the United Kingdom 74: 985–985.

41. Schmitt C, Duft M, Brandelik C, Schulte-Oehlmann U, Oehlmann J (2008) SOP for testing of chemicals: Reproduction test with the prosobranch snail Potamopyrgus antipodarum for testing endocrine active chemicals. Part III: Reproduction test using sediment exposure. Goethe University Frankfurt am Main. Department Aquatic Ecotoxicology.

42. OECD (2010) Detailed Review Paper on Mollusc Life-Cycle. Organisation for Economic Co-operation and Development, Paris. (= OECD Series on Testing and Assessment No. 121). 182.

43. Nagel R, Ludwichowski K-U, Oetken M, Schmidt J, Jackson P, et al. (2004) Ringtest zur Validierung der Prüfrichtlinie Fish Life-Cycle Test mit dem Zebrabärbling (Danio rerio): Forschungs und Entwicklungsvorhaben des Umweltbundesamtes; F+E- Vorhaben 200 67 411.

44. Kang IJ, Yokota H, Oshima Y, Tsuruda Y, Shimasaki Y, et al. (2008) The effects of methyltestosterone on the sexual development and reproduction of adult medaka (Oryzias latipes). Aquatic Toxicology 87: 37–46.

45. Villeneuve DL, Blankenship AL, Giesy JP (2000) Derivation and application of relative potency estimates based on in vitro bioassay results. Environmental toxicology and chemistry 19: 2835–2843.

46. Thorpe KL, Cummings RI, Hutchinson TH, Scholze M, Brighty G, et al. (2003) Relative Potencies and Combination Effects of Steroidal Estrogens in Fish. Environmental Science & Technology 37: 1142–1149.

47. Brauch H-J (2011) Organische Spurenstoffe in Gewässern. Vorkommen und Bewertung. Gwf-Wasser/Abwasser 12: 1206–1211.

48. Körner W, Bolz U, Triebskorn R, Schwaiger J, Negele R-D, et al. (2001) Steroid analysis and xenosteroid potentials in two small streams in southwest Germany. Journal of Aquatic Ecosystem Stress and Recovery 8: 215–229.

49. Bicchi C, Schiliro T, Pignata C, Fea E, Cordero C, et al. (2009) Analysis of environmental endocrine disrupting chemicals using the E-screen method and stir bar sorptive extraction in wastewater treatment plant effluents. The Science of the total environment 407: 1842–1851.

50. Jarošová B, Bláha L, Giesy JP, Hilscherová K (2014) What level of estrogenic activity determined by in vitro assays in municipal waste waters can be considered as safe? Environment International 64: 98–109.

51. Loos R, Carvalho R, António DC, Comero S, Locoro G, et al. (2013) EU-wide monitoring survey on emerging polar organic contaminants in wastewater treatment plant effluents. Water Research 47: 6475–6487.

52. Vondráček J, Machala M, Minksová K, Bláha L, Murk AJ, et al. (2001) Monitoring river sediments contaminated predominantly with polyaromatic hydrocarbons by chemical and in vitro bioassay techniques. Environmental Toxicology and Chemistry 20: 1499–1506.

53. Schmitt C, Balaam J, Leonards P, Brix R, Streck G, et al. (2010) Characterizing field sediments from three European river basins with special emphasis on endocrine effects – A recommendation for Potamopyrgus antipodarum as test organism. Chemosphere 80: 13–19.

54. Kinani S, Bouchonnet S, Creusot N, Bourcier S, Balaguer P, et al. (2010) Bioanalytical characterisation of multiple endocrine- and dioxin-like activities in sediments from reference and impacted small rivers. Environmental Pollution 158: 74–83.

55. Viganò L, Benfenati E, Cauwenberge Av, Eidem JK, Erratico C, et al. (2008) Estrogenicity profile and estrogenic compounds determined in river sediments by chemical analysis, ELISA and yeast assays. Chemosphere 73: 1078–1089.

56. Sellin Jeffries MK, Conoan NH, Cox MB, Sangster JL, Balsiger HA, et al. (2011) The anti-estrogenic activity of sediments from agriculturally intense watersheds: Assessment using in vivo and in vitro assays. Aquatic Toxicology 105: 189–198.

57. Urbatzka R, van Cauwenberge A, Maggioni S, Vigano L, Mandich A, et al. (2007) Androgenic and antiandrogenic activities in water and sediment samples from the river Lambro, Italy, detected by yeast androgen screen and chemical analyses. Chemosphere 67: 1080–1087.

58. Zhao JL, Ying GG, Yang B, Liu S, Zhou LJ, et al. (2011) Screening of multiple hormonal activities in surface water and sediment from the Pearl River system, South China, using effect-directed in vitro bioassays. Environmental Toxicology and Chemistry 30: 2208–2215.

59. Jobling S, Burn RW, Thorpe K, Williams R, Tyler C (2009) Statistical modeling suggests that antiandrogens in effluents from wastewater treatment works contribute to widespread sexual disruption in fish living in English rivers. Environ Health Perspect 117: 797–802.

60. Runnalls TJ, Margiotta-Casaluci L, Kugathas S, Sumpter JP (2010) Pharmaceuticals in the Aquatic Environment: Steroids and Anti-Steroids as High Priorities for Research. Human and Ecological Risk Assessment: An International Journal 16: 1318–1338.

61. Mazurová E, Hilscherová K, Šídlová-Štěpánková T, Köhler H-R, Triebskorn R, et al. (2010) Chronic toxicity of contaminated sediments on reproduction and histopathology of the crustacean Gammarus fossarum and relationship with the chemical contamination and in vitro effects. Journal of Soils and Sediments 10: 423–433.

62. Mazurová E, Hilscherová K, Triebskorn R, Köhler H-R, Maršálek B, et al. (2008) Endocrine regulation of the reproduction in crustaceans: Identification of potential targets for toxicants and environmental contaminants. Biologia 63: 139–150.

63. Louiz I, Kinani S, Gouze ME, Ben-Attia M, Menif D, et al. (2008) Monitoring of dioxin-like, estrogenic and anti-estrogenic activities in sediments of the Bizerta lagoon (Tunisia) by means of in vitro cell-based bioassays: Contribution of low concentrations of polynuclear aromatic hydrocarbons (PAHs). The Science of the total environment 402: 318–329.

64. Gutendorf B, Westendorf J (2001) Comparison of an array of in vitro assays for the assessment of the estrogenic potential of natural and synthetic estrogens, phytoestrogens and xenoestrogens. Toxicology 166: 79–89.

65. Peck M, Gibson RW, Kortenkamp A, Hill EM (2004) Sediments are major sinks of steroidal estrogens in two United Kingdom rivers. Environmental Toxicology and Chemistry 23: 945–952.

66. Grund S, Higley E, Schönenberger R, Suter M-F, Giesy J, et al. (2011) The endocrine disrupting potential of sediments from the Upper Danube River (Germany) as revealed by in vitro bioassays and chemical analysis. Environmental Science and Pollution Research 18: 446–460.

67. Schmitt C, Oetken M, Dittberner O, Wagner M, Oehlmann J (2008) Endocrine modulation and toxic effects of two commonly used UV screens on the aquatic invertebrates Potamopyrgus antipodarum and Lumbriculus variegatus. Environmental Pollution 152: 322–329.

68. Giesy JP, Pierens SL, Snyder EM, Miles-Richardson S, Kramer VJ, et al. (2000) Effects of 4-nonylphenol on fecundity and biomarkers of estrogenicity in fathead minnows (Pimephales promelas). Environmental Toxicology and Chemistry 19: 1368–1377.

69. Duft M, Schulte-Oehlmann U, Weltje L, Tillmann M, Oehlmann J (2003) Stimulated embryo production as a parameter of estrogenic exposure via sediments in the freshwater mudsnail Potamopyrgus antipodarum. Aquatic Toxicology 64: 437–449.

70. Jobling S, Casey D, Rodgers-Gray T, Oehlmann J, Schulte-Oehlmann U, et al. (2004) Comparative responses of molluscs and fish to environmental estrogens and an estrogenic effluent. Aquatic Toxicology 66: 207–222.

71. Oehlmann J, Schulte-Oehlmann U, Bachmann J, Oetken M, Lutz I, et al. (2006) Bisphenol A induces superfeminization in the ramshorn snail Marisa cornuarietis (Gastropoda: Prosobranchia) at environmentally relevant concentrations. Environ Health Perspect 114 Suppl 1: 127–133.

72. Weltje L, vom Saal FS, Oehlmann J (2005) Reproductive stimulation by low doses of xenoestrogens contrasts with the view of hormesis as an adaptive response. Human & Experimental Toxicology 24: 431–437.

73. Oehlmann J, Schulte-Oehlmann U, Tillmann M, Markert B (2000) Effects of endocrine disruptors on prosobranch snails (Mollusca: Gastropoda) in the laboratory. Part I: Bisphenol A and octylphenol as xeno-estrogens. Ecotoxicology 9: 383–397.

74. Nilsen BM, Berg K, Arukwe A, Goksøyr A (1998) Monoclonal and polyclonal antibodies against fish vitellogenin for use in pollution monitoring. Marine Environmental Research 46: 153–157.

75. Vajda AM, Barber LB, Gray JL, Lopez EM, Woodling JD, et al. (2008) Reproductive Disruption in Fish Downstream from an Estrogenic Wastewater Effluent. Environmental Science & Technology 42: 3407–3414.

76. Bjerregaard P, Hansen PR, Larsen KJ, Erratico C, Korsgaard B, et al. (2008) Vitellogenin as a biomarker for estrogenic effects in brown trout, Salmo trutta: laboratory and field investigations. Environmental Toxicology and Chemistry 27: 2387–2396.

77. Zhang S, Wang S, Li H, Li L (2011) Vitellogenin, a multivalent sensor and an antimicrobial effector. The International Journal of Biochemistry & Cell Biology 43: 303–305.

78. Harries JE, Sheahan DA, Jobling S, Matthiessen P, Neall P, et al. (1997) Estrogenic activity in five United Kingdom rivers detected by measurement of vitellogenesis in caged male trout. Environmental Toxicology and Chemistry 16: 534–542.

79. Bjerregaard LB, Madsen AH, Korsgaard B, Bjerregaard P (2006) Gonad histology and vitellogenin concentrations in brown trout (Salmo trutta) from Danish streams impacted by sewage effluent. Ecotoxicology 15: 315–327.

80. Burki R, Vermeirssen EL, Körner O, Joris C, Burkhardt-Holm P, et al. (2006) Assessment of estrogenic exposure in brown trout (Salmo trutta) in a Swiss midland river: integrated analysis of passive samplers, wild and caged fish, and vitellogenin mRNA and protein. Environmental toxicology and chemistry 25: 2077–2086.

81. Bless R (1996) Reproduction and habitat preference of the threatened spirlin (Alburnoides bipunctatus Bloch) and soufie (Leuciscus souffia Risso) under laboratory conditions (Teleostei: Cyprinidae). In: Kirchhofer A, Hefti D, editors. Conservation of Endangered Freshwater Fish in Europe: Birkhäuser Basel. 249–258.

82. Türkmen M, Haliloglu H, Erdogan O, Yıldırım A (1999) The growth and reproduction characteristics of chub Leuciscus cephalus orientalis (Nordmann, 1840) living in the River Aras. Turkish Journal of Zoology 23: 355–364.

83. Koç HT, Erdoğan Z, Tinkci M, Treer T (2007) Age, growth and reproductive characteristics of chub, Leuciscus cephalus (L., 1758) in the İkizcetepeler dam lake (Balıkesir), Turkey. Journal of Applied Ichthyology 23: 19–24.

84. Jobling S, Beresford N, Nolan M, Rodgers-Gray T, Brighty GC, et al. (2002) Altered sexual maturation and gamete production in wild roach (Rutilus rutilus) living in rivers that receive treated sewage effluents. Biology of Reproduction 66: 272–281.

85. Economou AN, Daoulas C, Psarras T (1991) Growth and morphological development of chub, Leuciscus cephalus (L.), during the first year of life. Journal of Fish Biology 39: 393–408.

86. Stenseth NC (2004) Marine ecosystems and climate variation: the North Atlantic; a comparative perspective. Oxford [u.a.]: Univ. Press.

87. Mert R, Bulut S, Solak K (2011) Some biological properties of the Squalius cephalus (L.1758) population inhabiting Apa Dam Lake in Konya (Turkey). Afyon Kocatepe University Journal of Science 6 (2): 1–12.

88. Andersson T, Förlin L, Härdig J, Larsson Å (1988) Physiological Disturbances in Fish Living in Coastal Water Polluted with Bleached Kraft Pulp Mill Effluents. Canadian Journal of Fisheries and Aquatic Sciences 45: 1525–1536.

89. Adams SM, Bevelhimer MS, Greeley MS, Levine DA, Teh SJ (1999) Ecological risk assessment in a large river-reservoir: 6. Bioindicators of fish population health. Environmental Toxicology and Chemistry 18: 628–640.

90. Munkittrick KR, McMaster ME, Portt CB, Kraak GJVD, Smith IR, et al. (1992) Changes in Maturity, Plasma Sex Steroid Levels, Hepatic Mixed-Function Oxygenase Activity, and the Presence of External Lesions in Lake Whitefish (Coregonus clupeaformis) Exposed to Bleached Kraft Mill Effluent. Canadian Journal of Fisheries and Aquatic Sciences 49: 1560–1569.

91. Kobler B, Lovas R, Stadelmann P (2004) Ökologische und fischbiologische Untersuchungen der Ron oberhalb und unterhalb der Kläranlagen Rain und Hochdorf (Kanton Luzern): Schlussbericht 1999–2002.

92. Bernet D (2003) Biomonitoring in Fliessgewässern des Kantons Bern: Synthesebericht. Teilprojekt Fischnetz Nr. 99/16. 1–57.

93. Velasco-Santamaría YM, Bjerregaard P, Korsgaard B (2010) Gonadal alterations in male eelpout (Zoarces viviparus) exposed to ethinylestradiol and trenbolone separately or in combination. Marine Environmental Research 69, Supplement 1: S67–S69.

6

Heavy Metal Accumulation by Periphyton Is Related to Eutrophication in the Hai River Basin, Northern China

Wenzhong Tang[1], Jingguo Cui[2], Baoqing Shan[1]*, Chao Wang[1], Wenqiang Zhang[1]

1 State Key Laboratory on Environmental Aquatic Chemistry, Research Center for Eco-Environmental Sciences, Chinese Academy of Sciences, Beijing, China, **2** Beijing Sound Environmental Engineering Co., Ltd., Beijing, China

Abstract

The Hai River Basin (HRB) is one of the most polluted river basins in China. The basin suffers from various types of pollutants including heavy metals and nutrients due to a high population density and rapid economic development in this area. We assessed the relationship between heavy metal accumulation by periphyton playing an important role in fluvial food webs and eutrophication in the HRB. The concentrations of the unicellular diatoms (type A), filamentous algae with diatoms (type B), and filamentous algae (type C) varied along the river, with type A dominating upstream, and types B then C increasing in concentration further downstream, and this was consistent with changes in the trophic status of the river. The mean heavy metal concentrations in the type A, B and C organisms were Cr: 18, 18 and 24 mg/kg, respectively, Ni: 9.2, 10 and 12 mg/kg, respectively, Cu: 8.4, 19 and 29 mg/kg, respectively, and Pb: 11, 9.8 and 7.1 mg/kg respectively. The bioconcentration factors showed that the abilities of the organisms to accumulate Cr, Ni and Pb decreased in the order type A, type B, then type C, but their abilities to accumulate Cu increased in that order. The Ni concentration was a good predictor of Cr, Cu and Pb accumulation by all three periphyton types. Our study shows that heavy metal accumulation by periphyton is associated with eutrophication in the rivers in the HRB.

Editor: Fanis Missirlis, CINVESTAV-IPN, Mexico

Funding: This work was supported by the National Natural Science Foundation of China (No. 21107126), the One-Three-Fiver Program of Research Center for Eco-Environmental Sciences (No. YSW2013B02), and the National Water Pollution Control Program (No. 2012ZX07203-002). The funders had no role in study design, data collection and analysis, decision to publish, or preparation of the manuscript.

Competing Interests: JGC is employed by Beijing Sound Environmental Engineering Co., Ltd. There are no patents, products in development or marketed products to declare. All other authors have declared that no competing interests exist.

* E-mail: bqshan@rcees.ac.cn

Introduction

Heavy metal pollution in aquatic ecosystems has been a serious global environmental problem for a long time [1,2,3]. Heavy metals are persistent in aquatic environments because of their resistance to decomposition under natural conditions [4,5]. One of the greatest problems associated with the persistence of heavy metals is the potential for them to bioaccumulate and biomagnify, potentially resulting in long-term implications for human and aquatic ecosystem health [6,7,8,9]. Periphyton is an important aquatic resource and is a significant component in river ecosystems [10,11], it plays an important role in fluvial food webs [12,13]. Periphyton is highly sensitive to environmental stressors [14,15], and is often used as a pollution indicator to assess water quality [13,16,17]. Heavy metal accumulation by periphyton has a strong effect on river ecosystems because of these factors.

The Hai River Basin (HRB) is one of the most polluted river basins in China. It is in an area that has a high population density and that is undergoing rapid economic development. The HRB suffers from various types of pollution [18]. Heavy industrial development and rapid urbanization have caused significant pollution of the rivers in the HRB, and the main pollutants include nitrogen (N), phosphorus (P) and heavy metals. Many rivers in this area have also degenerated into shallow streams because of excessive water extraction, and this has caused periphyton (mainly benthic algae) to become the primary producer. The ecological degradation of a river can cause enormous changes in the periphyton community, and the periphyton species composition is routinely used as an indicator of heavy metal pollution in a river [19], because periphyton can accumulate metals from the ambient water and from sediments [20]. Therefore, it is important to understand the relationship between heavy metal accumulation by periphyton and various other pollutants (including N and P) in the HRB, in order to provide a reference point for the future control and management of heavy metals in the river ecosystem.

Periphyton is an important food source for invertebrates and some fish, and can be an important accumulator of heavy metals [21]. These accumulated metals may be transferred from periphyton to the consuming organisms [22]. We studied heavy metal accumulation by periphyton in the HRB rivers, and our specific objectives were to: 1) assess the effects of eutrophication (i.e., increased N and P concentrations) on periphyton community succession; 2) determine the degree of heavy metal accumulation in different periphyton community types; 3) assess the effects of eutrophication on heavy metal accumulation by periphyton.

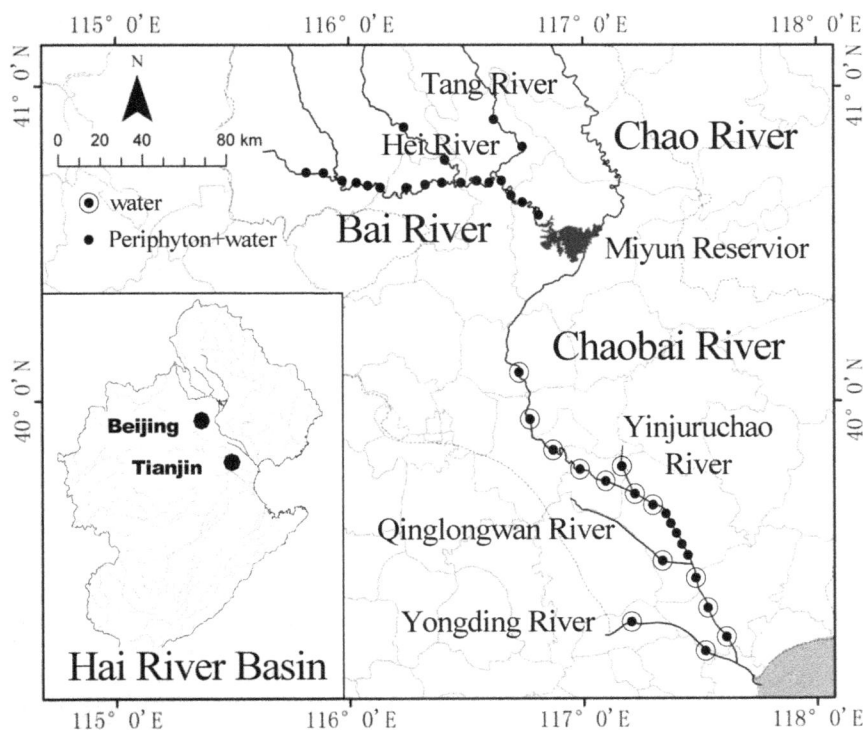

Figure 1. Location of the Chaobai River Watershed in the Hai River Basin and the distribution of the sampling sites.

Materials and Methods

Ethics Statement

No specific permits were required for the field studies described here. The study area is not privately-owned or protected in any way, and the field studies did not involve endangered or protected species.

Study Area

The HRB is mainly within Hebei Province, and includes Beijing, Tianjin and parts of Inner Mongolia, Shanxi, Henan and Shandong provinces. The HRB has an area of 318,000 km^2 and a temperate continental monsoon climate. The mean annual precipitation is 527 mm. The HRB is one of several major river basins managed by the Chinese Ministry of Water Resources. Heavy industrial development and rapid urbanization have caused significant pollution in the aquatic environments in this region. Water resources are in high demand, and deteriorating water quality has exacerbated the shortage of water resources. The HRB has, therefore, attracted much attention from the Chinese government and it has become one of the most important basins

in the national 11th and 12th five-year plans for water pollution control.

Chaobai River Watershed (CRW), which is one of the most important watersheds in the HRB, was selected as the study area (Fig. 1). This watershed is a significant source of drinking water for Beijing. Miyun reservoir, which is 100 km northeast of Beijing, feeds the Chao and Bai rivers. Drought and water overuse in recent years have caused some downstream parts of the Chao River to have a decreased flow, and the Bai River has turned into a subalpine shallow stream with a gravel and cobble riverbed. Arable fields lie on either side of the Bai River, and the runoff from these fields produces agricultural non-point-source pollution. The Bai and Chaobai rivers and their tributaries (the Hei, Tang, Yinjuruchao, Qinglongwan and Yongding rivers) in the CRW were selected as the main study sites because the water quality in these systems covers all of the trophic classes (oligotrophic, mesotrophic, eutrophic and hypereutrophic), allowing us to study periphyton responses. Water and periphyton samples were collected from 38 study sites in May and June.

Table 1. TN, TP and heavy metal concentrations in the water from the studied rivers in the Chaobai River Watershed (each concentration is the mean ± the standard deviation).

Site	TN	TP	Cr	Ni	Cu	Pb
	mg/L	μg/L	μg/L	μg/L	μg/L	μg/L
Bai River and its tributaries	2.4±1.4	17±11	1.2±1.1	1.2±1.1	3.2±4.1	1.0±1.0
Chaobai River and its tributaries	6.5±4.3	608±952	6.8±7.9	6.3±4.2	21±63	1.3±0.8

Table 2. Main species found in the three periphyton types found in the rivers in the Chaobai River Watershed and the eutrophication conditions (TN and TP concentrations) in the ambient water.

	Type A	Type B	Type C
	Fragilaria	*Cladophora+Tabellaria*	*Cladophora*
	Cymbella	*Cladophora+Cocconeis*	*Oscillatoria*
	Navicula	*Cladophora+Cocconeis+Tabellaria*	*Spirogyra*
Main species	*Gomphonema*	*Cladophora+Tabellaria+Fragilaria*	*Ulothrix*
	Cocconeis	*Cladophora+Tabellaria+Gomphonema*	*Oedogonium*
		Ulothrix+Cocconeis+Fragilaria	
		Oedogonium+Cocconeis+Fragilaria	
		Spirogyra+Navicula +Fragilaria+Cymbella	
TN concentration (mg/L)	1.1±0.36	1.8±1.1	4.9±1.8
TP concentration (µg/L)	12±4.1	16±15	292±440

Chemical Analysis of Water Samples

Water samples (three replicates) were acidified to below pH 2 and analyzed in the laboratory within 72 h of collection. A potassium persulfate digestion was used to prepare the samples before the total N (TN) and total P (TP) concentrations were determined [23]. The TN concentrations were determined using an automated chemical analysis instrument (Smart Chem 200, Westco, Italy), and the detection limit was 0.001 mg/L. Samples for heavy metal analysis were collected in clean acid-washed glass bottles, acidified with concentrated HNO_3, stored at 4°C, and analyzed within 72 h of collection. TP and heavy metals were all analyzed by inductively coupled plasma-mass spectrometry (ICP-MS) (7500a, Agilent, USA), which had a detection limit 0.001 µg/L for each analyte.

Periphyton Sampling and Elemental Analysis

Our field investigations revealed great differences in the periphyton community structures present in the HRB, from small epilithic/epiphytic diatoms to large filamentous algae. Sampling sites with similar conditions were selected to avoid physical and hydrological parameters (such as light, shade, substrata, water depth and flow velocity) affecting the results. Periphyton was collected from rock surfaces at each site using a nylon-bristled brush. If there was enough periphyton available, at least three samples were collected at each site each month. Each sample was divided into three aliquots for analysis. The first aliquot was

preserved in 3–5% glutaraldehyde solution in the field [24], and the community structure was identified using a microscope (Olympus BX51, Japan) in the laboratory. The second aliquot was frozen in the field, stored at −16°C in a cooler (Mobicool, BC55 DC, USA), freeze dried in the laboratory, then ground into a powder and passed through a 100 mesh sieve. The sample powder was then digested using a microwave digestion system (CEM, Matthews, NC, USA) and analyzed for heavy metals (Cr, Cu, Ni and Pb) by ICP-MS (7500a, Agilent, USA). The recoveries varied but were all within the range 90–95%, and the precision was good, with a relative standard deviation (RSD) of less than 3%. The last aliquot was stored for use as a backup in case the analysis of either of the other aliquots failed.

Bioconcentration Factor

The bioconcentration factor (BCF) was used to assess the accumulation efficiency of the heavy metals, by comparing the concentration in the biota with the concentration in the external medium [25]. We calculated the BCF using Eq. 1

$$BCF = \frac{C_P (mg/kg)}{C_W (\mu g/L)} \qquad (1)$$

where C_P and C_W are the heavy metal concentrations in the biota (mg/kg) and the water (µg/L), respectively. A BCF greater than 1

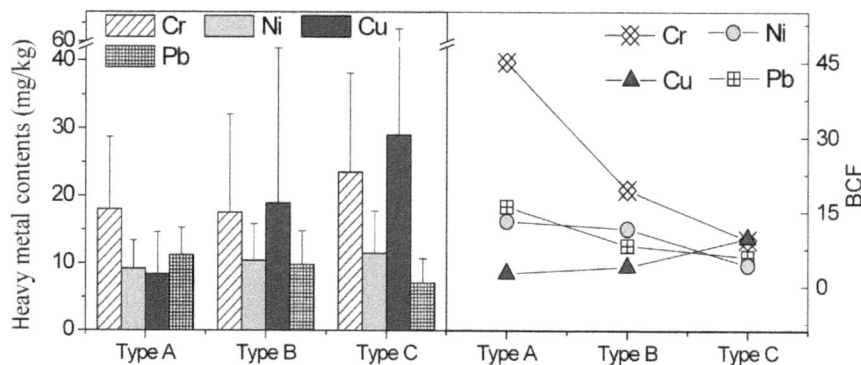

Figure 2. Heavy metal concentrations in the three types of periphyton and their bioconcentration factors (BCFs).

indicates that the periphyton enriched heavy metals from the water.

Statistical Analysis

The experimental data were analyzed using SPSS 17.0 for Windows. Spearman correlation was used to assess the relationships between the heavy metal concentrations in the three periphyton types. A one-sample t-test ($p \leq 0.05$) was used to analyze the variance. Origin Pro 8.0 was used to plot the experimental data.

Results and Discussion

Chemical Characteristics of the River Water

Table 1 shows the TN, TP and heavy metal concentrations in the water from the rivers in the CRW that were studied. Overall, the TN and TP concentrations clearly increased moving from the upstream part of the study area (the Bai River and its tributaries) to the downstream part (the Chaobai River and its tributaries). The Cr, Ni and Cu concentrations were about six times higher in

the upstream than in the downstream river water, but the Pb concentrations were almost the same. Defining the trophic state is more difficult in rivers than in lakes [26], but the US Environmental Protection Agency has suggested boundaries for the trophic classification of rivers [27], and these are that the oligotrophic–mesotrophic boundary is at $TN = 0.70$ mg/L and $TP = 25$ µg/L and that the mesotrophic–eutrophic boundary is at $TN = 1.5$ mg/L and $TP = 75$ µg/L. Using these criteria we classified the trophic status of the rivers studied, from upstream to downstream in the CRW, as ranging from oligotrophic, through mesotrophic to eutrophic, and some downstream parts were even classified as hypereutrophic. These results show that there are strong anthropogenic interferences (industrial and agricultural activities, urbanization) in the rivers in the CRW, and these interferences have caused the river water quality to deteriorate severely.

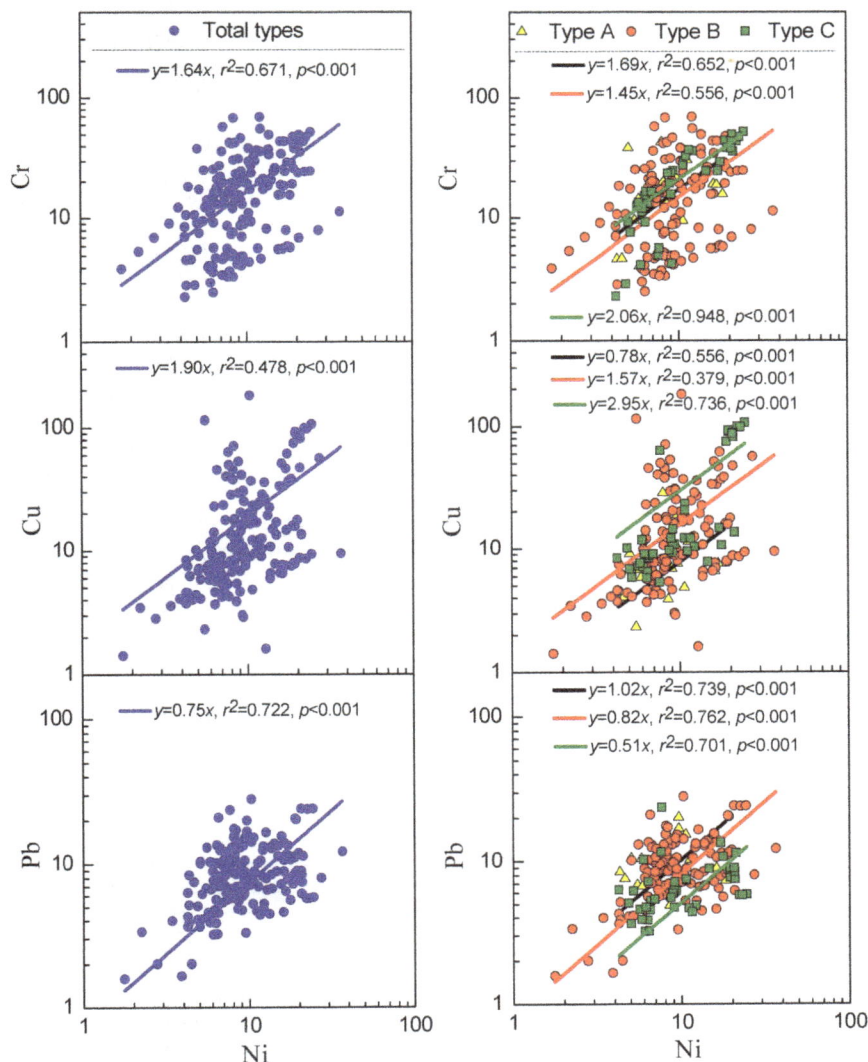

Figure 3. Linear regressions between the Ni concentration and the Cr, Cu and Pb concentrations (mg/kg) in the three periphyton types (as a total on the left and separately on the right). The concentrations were log-transformed.

Table 3. Spearman correlation coefficients for the heavy metal concentrations in the periphyton.

	Type	Cr	Ni	Cu	Pb
Cr	A	1	0.278	0.863[a]	0.595[a]
	B	1	0.341[a]	−0.184[b]	0.365[a]
	C	1	0.904[a]	0.711[a]	0.188
Ni	A		1	0.331	0.251
	B		1	0.453[a]	0.371[a]
	C		1	0.799[a]	0.466[a]
Cu	A			1	0.598[a]
	B			1	0.194[b]
	C			1	0.476[a]
Pb	A				1
	B				1
	C				1

[a]Correlation is significant at the 0.01 level (2-tailed).
[b]Correlation is significant at the 0.05 level (2-tailed).

Effect of Eutrophication Conditions (N and P Concentrations) on the Periphyton Community

We found great morphological differences in the periphyton in the HRB, indicating that there were wide variations in the periphyton communities in different areas. Algae are the main contributors to the periphyton, and their conformations differ greatly, from unicellular diatoms to filamentous algae. Many methods for classifying periphyton types have been published previously. For example, Maltalis and Vincent divided periphyton into black crust, brown film, green crust, and green filaments by their colors and other characteristics [28]. The method used in a particular study is generally selected based on the specific requirements of the research. In our study, classifying the periphyton by morphology will allow us to identify eutrophication conditions (high N and P concentrations), and help to assess the water quality.

Periphyton transitions from unicellular diatoms to filamentous algae were observed along the river flow direction in the CRW, and these transitions were consistent with the observed changes in trophic status. The periphyton was divided into three different types, based on their conformations (Table 2). Type A was primarily epilithic unicellular diatoms, such as *Fragilaria* and *Cymbella*. Type B was filamentous algae with epiphytic diatoms (a hybrid type). Typical type B periphyton was *Cladophora* with *Tabellaria* attached. Type C was filamentous algae, such as *Cladophora* and *Oscillatoria*, with few (a biomass <5% of the total) or no unicellular diatoms.

Many factors affect the periphyton community, including light, temperature and availability of nutrients in the water [24]. We selected sampling sites with similar physical and hydrological conditions so that the effects of eutrophication (high N and P concentrations) could be identified. Areas that had type B periphyton communities had slightly higher TN and TP concentrations in the river water than the areas with type A periphyton communities (Table 2), but areas with type C periphyton communities had much higher TN and TP concentrations in the water. We concluded that increasing nutrient concentrations travelling downstream were the main factors causing the changes in the periphyton communities, and this agrees with the results of a

study conducted by Kelly [29]. As has been shown previously, seasonal variations in the concentrations of some metals in river periphyton can be related to variations in the periphytic algae and cyanobacteria species present [30]. Therefore, we believe that the changes in the periphytic assemblages we found may have been connected with heavy metal accumulation by the periphyton.

Heavy Metal Accumulation by Different Periphyton Types

There were great variations in the amounts of the heavy metals accumulated by the different periphyton types (Fig. 2). The Cr concentrations were 18 ± 11 and 18 ± 15 mg/kg in periphyton types A and B, respectively, but 24 ± 15 mg/kg in type C. Cr was enriched in the periphyton in the decreasing order type A (BCF = 45)>type B (BCF = 19)>type C (BCF = 9.4). The Ni concentrations in the three periphyton types were not significantly different (type A 9.2 ± 4.2 mg/kg, type B 10 ± 5.4 mg/kg, type C 12 ± 6.2 mg/kg), even though the Ni concentration in the ambient water increased moving downstream (Table 1). The Ni BCF values, therefore, decreased in the order type A (13)>type B (12)>type C (4.3). The excretion of extracellular organic matter by diatoms enables them to scavenge metals, and may provide the diatoms with more resistance to Ni toxicity [31]. The high Ni BCF for type A (unicellular diatoms) suggested that they have a strong ability to accumulate Ni, which is consistent with previous findings. The Pb BCF values (type A 16, type B 8.2, type C 5.9) explained the decrease in Pb concentration moving from type A (11 ± 4.1 mg/kg), through type B (9.8 ± 5.0 mg/kg), to type C (7.1 ± 3.6 mg/kg).

Unlike the Cr, Ni and Pb concentrations, the Cu concentrations increased clearly in the order type A (8.4 ± 6.2 mg/kg)<type B (19 ± 23 mg/kg)<type C (29 ± 35 mg/kg). The Cu BCF values also increased in the order type A (2.6)<type B (3.9)<type C (9.7). These results indicate that the filamentous algae (type C) sampled could accumulate more Cu than the unicellular diatoms (type A) could. The presence of Cu could influence the rate of photosynthesis in the diatoms more than in the filamentous algae [32]. In our study, the filamentous algae gradually became the dominant species moving from type A to type C, so type C would be the most resistant to the effects of Cu, and this may have caused the stronger ability of type C to enrich Cu.

Overall, the abilities of the periphyton to accumulate Cr, Ni and Pb decreased but the ability to accumulate Cu increased moving from type A to type C in the CRW rivers. Previous studies have shown that filamentous algae, such as *Cladophora*, are generally the best bioindicators for heavy metals in aquatic ecosystems [25]. In our study, the changes in periphyton types found (from type A to type C) caused a significant increase in the ecological risk associated with Cu pollution, but a decrease in the ecological risks associated with Cr, Ni and Pb pollution.

Correlation Analysis of the Heavy Metal Concentrations in Periphyton

The relationships between the heavy metal concentrations in the three periphyton types were analyzed using Spearman's correlations (Table 3). Strong positive correlations ($p \leq 0.01$) were found between the Cr and Cu concentrations, the Cr and Pb concentrations, and the Cu and Pb concentrations in type A periphyton. Except for between the Cr and Cu concentrations in type B and the Cr and Pb concentrations in type C, all four heavy metals positively correlated with each other in periphyton types B and C. There was a considerable increase in the significance of the correlations between Ni and the other three metals (Cr, Cu and Pb) moving from periphyton types A through to C. Therefore, we produced scatter plots of the Ni concentration and the concentrations of the other three metals, for each of the periphyton types, and fitted a line to each (Fig. 3). There tend to be correlations between the accumulations of different heavy metals in periphyton [33,34]. In our study, the Cr, Cu and Pb concentrations all increased with the Ni concentration in all three periphyton types. The Cr:Ni ratios decreased in the order type C>type A ≈ type B and the Cu:Ni ratios decreased in the order type C>type B>type A. However, the Pb:Ni ratios gradually decreased moving from type A through to type C. The Ni concentration was a good predictor of the Cr, Cu and Pb accumulation in periphyton types A, B and C, and the differences in Ni (and Cr, Cu and Pb) accumulation between the filamentous algae and unicellular diatoms may be caused by different co-accumulation effects. Periphyton type A had a different ability to accumulate heavy metals from type C (Figs. 2 and 3). The increasing TN and TP concentrations resulted in the changes seen in the periphyton community types in the studied rivers. Therefore, we concluded that eutrophication (increased N and P concentrations) were important factors affecting the accumulation of heavy metals by periphyton in the rivers in the CRW.

Conclusions

Changes in the periphyton assemblages, from unicellular diatoms to filamentous algae, were observed moving along the flow direction in the rivers studied, and these changes were consistent with changes in the trophic status. The periphyton were classified into type A (unicellular diatoms), type B (filamentous algae with diatoms), and type C (filamentous algae), based on their morphologies. The changes in periphyton types caused a significant increase in the ecological risk associated with Cu pollution but decreased risks associated with Cr, Ni and Pb pollution. The Ni concentration was a good indicator of the Cr, Cu and Pb accumulation in the periphyton. These results provide information that will be useful for designing further research and for managing the rivers in the study area, and will help enable heavy metal pollution controls to be established in the HRB.

Author Contributions

Conceived and designed the experiments: WZT JGC BQS CW WQZ. Performed the experiments: WZT JGC CW WQZ. Analyzed the data: WZT JGC. Contributed reagents/materials/analysis tools: WZT JGC. Wrote the paper: WZT JGC.

References

1. Jiang M, Zeng GM, Zhang C, Ma XY, Chen M, et al. (2013) Assessment of heavy metal contamination in the surrounding soils and surface sediments in Xiawangang River, Qingshuitang District. PLoS One 8: e71176.
2. Gao XL, Chen CTA (2012) Heavy metal pollution status in surface sediments of the coastal Bohai Bay. Water Res 46: 1901–1911.
3. Tang WZ, Shan BQ, Zhang H, Mao ZP (2010) Heavy metal sources and associated risk in response to agricultural intensification in the estuarine sediments of Chaohu Lake Valley, East China. J Hazard Mater 176: 945–951.
4. Suresh G, Sutharsan P, Ramasamy V, Venkatachalapathy R (2012) Assessment of spatial distribution and potential ecological risk of the heavy metals in relation to granulometric contents of Veeranam lake sediments, India. Ecotox Environ Safe 84: 117–124.
5. Huang HJ, Yuan XZ, Zeng GM, Zhu HN, Li H, et al. (2011) Quantitative evaluation of heavy metals' pollution hazards in liquefaction residues of sewage sludge. Bioresource Technol 102: 10346–10351.
6. Qiu YW, Lin DA, Liu JQ, Zeng EY (2011) Bioaccumulation of trace metals in farmed fish from South China and potential risk assessment. Ecotox Environ Safe 74: 284–293.
7. Rainbow PS (2007) Trace metal bioaccumulation: Models, metabolic availability and toxicity. Environ Int 33: 576–582.
8. Hu B, Li G, Li J, Bi J, Zhao J, et al. (2013) Spatial distribution and ecotoxicological risk assessment of heavy metals in surface sediments of the southern Bohai Bay, China. Environ Sci Pollut R 20: 4099–4110.
9. Chakravarty R, Banerjee PC (2012) Mechanism of cadmium binding on the cell wall of an acidophilic bacterium. Bioresource Technol 108: 176–183.
10. Sterner RW (1993) Daphnia growth on varying quality of Scenedesmus: mineral limitation of zooplankton. Ecology 74: 2351–2360.
11. Bowes MJ, Ings NL, McCall SJ, Warwick A, Barrett C, et al. (2012) Nutrient and light limitation of periphyton in the River Thames: Implications for catchment management. Sci Total Environ 434: 201–212.
12. Barbour MT, Gerritsen J, Snyder B, Stribling J (1999) Rapid bioassessment protocols for use in streams and wadeable rivers. Washington, DC: USEPA.
13. Hauer FR, Lamberti GA (2011) Methods in stream ecology. San Diego: Academic Press.
14. Cairns JJ, Pratt JR, Niederlehner B, McCormick P (1986) A simple, cost-effective multispecies toxicity test using organisms with a cosmopolitan distribution. Environ Monit Assess 6: 207–220.
15. Kelly M, Penny C, Whitton B (1995) Comparative performance of benthic diatom indices used to assess river water quality. Hydrobiologia 302: 179–188.
16. Moreira-Santos M, Soares AM, Ribeiro R (2004) A phytoplankton growth assay for routine in situ environmental assessments. Environ Toxicol Chem 23: 1549–1560.
17. Vis C, Hudon C, Cattaneo A, Pinel-Alloul B (1998) Periphyton as an indicator of water quality in the St Lawrence River (Quebec, Canada). Environ Pollut 101: 13–24.
18. Aji K, Tang C, Song X, Kondoh A, Sakura Y, et al. (2008) Characteristics of chemistry and stable isotopes in groundwater of Chaobai and Yongding River basin, North China Plain. Hydrol Process 22: 63–72.
19. Gold C, Feurtet-Mazel A, Coste M, Boudou A (2002) Field transfer of periphytic diatom communities to assess short-term structural effects of metals (Cd, Zn) in rivers. Water Res 36: 3654–3664.
20. Morin S, Duong T, Dabrin A, Coynel A, Herlory O, et al. (2008) Long-term survey of heavy-metal pollution, biofilm contamination and diatom community structure in the Riou Mort watershed, South-West France. Environ Pollut 151: 532–542.
21. Newman MC, McIntosh AW (1989) Appropriateness of aufwuchs as a monitor of bioaccumulation. Environ Pollut 60: 83–100.
22. Hill WR, Stewart AJ, Napolitano GE (1996) Mercury speciation and bioaccumulation in lotic primary producers and primary consumers. Can J Fish Aquat Sci 53: 812–819.
23. China EPA (2002) Monitoring and analysis methods for water and waste water (4th). Beijing: China Environmental Science Press.
24. Biggs B, Kilroy C, editors (2000) Stream periphyton monitoring manual. Christchurch: NIWA.
25. Doshi H, Seth C, Ray A, Kothari I (2008) Bioaccumulation of heavy metals by green algae. Curr Microbiol 56: 246–255.
26. Dodds WK (2006) Eutrophication and trophic state in rivers and streams. Limnol Oceanogr 51: 671–680.
27. Buck S, Denton G, Dodds W, Fisher J, Flemer D, editors (2000) Nutrient criteria technical guidance manual: rivers and streams. Washington, DC: USEPA.
28. Maltais M-J, Vincent WF (1997) Periphyton community structure and dynamics in a subarctic lake. Can J Bot 75: 1556–1569.
29. Kelly MG (1998) Use of community-based indices to monitor eutrophication in European rivers. Environm Conserv 25: 22–29.
30. Anishchenko OV, Gladyshev MI, Kravchuk ES, Ivanova EA, Gribovskaya IV, et al. (2010) Seasonal variations of metal concentrations in periphyton and taxonomic composition of the algal community at a Yenisei River littoral site. Cent Eur J Biol 5: 125–134.
31. Sekar R, Nair K, Rao V, Venugopalan V (2002) Nutrient dynamics and successional changes in a lentic freshwater biofilm. Freshwater Biol 47: 1893–1907.
32. Nielsen ES, Wium-Andersen S (1971) The influence of Cu on photosynthesis and growth in diatoms. Physiol Plantarum 24: 480–484.
33. Cui JG, Shan BQ, Tang WZ (2012) Effect of periphyton community structure on heavy metal accumulation in mystery snail (Cipangopaludina chinensis): A case study of the Bai River, China. J Environ Sci 24: 1723–1730.
34. Wang HK, Wood JM (1984) Bioaccumulation of nickel by algae. Environ Sci Technol 18: 106–109.

7

Microbial Diversity of Emalahleni Mine Water in South Africa and Tolerance Ability of the Predominant Organism to Vanadium and Nickel

Ilunga Kamika, Maggie N. B. Momba*

Department of Environmental, Water and Earth Sciences, Faculty of Science, Tshwane University of Technology, Pretoria, Gauteng, South Africa

Abstract

The present study aims firstly at determining the microbial diversity of mine-water collected in Emalahleni, South Africa and secondly isolating and characterizing the most dominant bacterial species found in the mine water in terms of its resistance to both V^{5+} and Ni^{2+} in a modified wastewater liquid media. The results revealed a microbial diversity of 17 orders, 27 families and 33 genera were found in the mine-water samples with *Marinobacteria* (47.02%) and *Anabaena* (17.66%) being the most abundant genera. Considering their abundance in the mine-water samples, a species of the *Marinobacter* genera was isolated, identified, and characterised for metal tolerance and removal ability. The MWI-1 isolate (*Marinobacter* sp. MWI-1 [AB793286]) was found to be closely related to *Marinobacter goseongensis* at 97% of similarity. The isolate was exposed to various concentrations of Ni^{2+} and V^{5+} in wastewater liquid media and its tolerance to metals was also assessed. The MWI-1 isolate could tolerate V^{5+} and Ni^{2+} separately at concentrations (in terms of MIC) up to 13.41±0.56 mM and 5.39±0.5 mM at pH 7, whereas at pH 3, the tolerance limit decrease to 11.45±0.57 mM and 2.67±0.1 mM, respectively. The removal of V^{5+} and Ni^{2+} in liquid media was noted to gradually decrease with a gradual increase of the test metals. A significant difference (p<0.05) between V^{5+} and Ni^{2+} removal was noted. *Marinobacter* sp. MWI-1 achieved the maximum permissible limit of 0.1 mg-V^{5+}/L prescribed by UN-FAO at 100 mg/L, while at 200 mg/L only V^{5+} was removed at approximately 95% and Ni^{2+} at 47%. This study suggests that mine-water indigenous microorganisms are the best solution for the remediation of polluted mine water.

Editor: Vipul Bansal, RMIT University, Australia

Funding: The authors are grateful to the National Research Foundation (NRF) for the funding of this project (Grant number: M590). The funders had no role in study design, data collection and analysis, decision to publish, or preparation of the manuscript.

Competing Interests: The authors have declared that no competing interests exist. This submission has been approved by all the authors, the responsible authority where the work was carried out (Tshwane University of Technology) and also by the Sponsors.

* E-mail: mombamnb@tut.ac.za

Introduction

Mine water remains one of the major problems of concern, not only in South Africa, but also worldwide. This is due to its environmental, socio-economic and public health impacts [1]. It is mostly characterised by extreme pH (acidity or alkanity), high salinity levels, high concentrations of SO_4^{2-}, Al and several other toxic metals such as Fe, Cd, Co, Cu, Mo, Zn, Ni, V and sometimes even radionuclide [2]. In South Africa in particular, mining activities have a long history and have played a major role in both economic development and environmental pollution countrywide [3]. Although significant progress has recently been made in South Africa to address mine water management, environmental pollution due to the disposal of untreated mine water still remains [1]. Microorganisms, due to their ubiquitousness, have been viewed as one of the best ways to deal with this problem. Due to their ability to survive, grow and reproduce in such harsh environments, an interest in microorganisms was aroused among researchers worldwide [4]. Nevertheless, their presence in extreme environments such as mine water affects their species diversity [5]. Wang and co-workers [5] have pointed out that extreme conditions can be defined by levels of environmental factors, the effects of which pose difficulties for the survival of specific taxa or

all taxa. In addition, both Johnson and Hallberg [6] and Imarla et al. [7] have also reported that microbial community composition is largely bound to geochemical parameters such as pH and metal ion concentrations. As a result, microorganisms isolated from such environments are considered to be a valuable tool in the treatment of highly polluted mine water. A number of techniques such as culture-dependent and culture independent techniques have been developed and used to study microbial diversity of water, wastewater, soil and air [8,9]. While several microorganisms are not able to readily grow in pure culture, the culture-independent approach has seen its apogee for the simple reason that this method has the advantage of directly profiling microbial populations present in specific ecosystems straight from the environmental samples [10,11]. Although metal pollution is a major concern worldwide and also in South Africa, the microbial diversity of the mine water in the latter country has not been fully examined. This study is one of a few to describe the microbial diversity present in the water at the vanadium mine in Mpumalanga, South Africa. Moreover, the discovery of new microorganisms in harsh environments has provided some knowledge on the understanding of microbial biosynthetic processes which enhance the bioremediation of contaminated environments [12]. The present study aims at firstly assessing the bacterial diversity of mine water collected from a vanadium mine in South Africa and secondly isolating and characterizing the most dominant bacterial species found in the mine water in terms of its resistance to both V^{5+} and Ni^{2+} in a modified wastewater liquid media.

Materials and Methods

Study area and mine water sample collection and preparation

Mine water samples of 500 mL (a total of 48 samples) were collected in a sterile plastic sampling bottle (500 mL) on a weekly basis (with 4 samples per week) between September and November 2012 from the effluent of the vanadium mine in Emalahleni, Mpumalanga, South Africa (25°50''26.4' and 29°09'' 09.9'). No specific permit was needed for the collection of the wastewater samples in the described sample area and this study did not involve endangered or protected species. However, an official letter from the University was submitted to Mr. Ajith Ramnarain (Process Development Manager) to assist us with the collection of wastewater samples. Samples were kept in a cooler box (4°C) while being conveyed to the laboratory for microbial and physicochemical analysis. For microbial analysis, 10 mL of samples (a total of 12 samples) collected at the second week of each month were centrifuged at 10 000 ×g for 10 min and the pellets were thereafter kept under −20°C until analysis, for no more than 4 h. In preparation for the chemical analysis process, samples collected throughout the experiment were allowed to settle for 2 h prior to filtration and No.1 filter papers (Whatman) were used. The profile of the filtered samples was determined in terms of the chemical oxygen demand (COD), dissolved oxygen (DO), pH and chemical contents (metals, semi-metals and non-metals). The COD concentration was measured using closed reflux methods as described in the standard methods of APHA [13], while the pH and the DO were analysed using a pH probe (Model: PHC101, HACH) and DO probe (Model: LDO, HACH), respectively. Chemical elements were determined using the Inductively Couple Plasma Optical Emission Spectrometer (ICP-OES) (Spectro Arcos, Kleve Germany). The limits of detection (LOD) varied between 10–60 µg/L depending on the elements.

DNA extraction, PCR amplification of purified DNA and pyrosequencing

The cell pellets [10 mL] harvested by centrifuging the mine water were re-suspended in a 1× TE buffer (pH 8.0). An aliquot [10 µL] was examined for protozoan species under a light microscope (Axiovert S100, Carl Zeiss, Germany) at ×100 to ×400 magnification. Bacterial DNA was extracted with the ZR Fungal/Bacterial DNA Kit™ (ZYMO Research, Pretoria, South Africa) according to the procedures provided by the manufacturer. The PCR reaction was performed on the extracted DNA samples using universal primers 27F [5'AGRGTTTGATCMTGGCT-CAG3'] and 1492R [5'GGTTACCTTGTTACGACTT3'] [14]. A PCR reaction mixture of a total volume of 50 µL, containing 19 µL Nuclease-free water, 25 µL 2× Dream Taq™ PCR master mix (10×Dream Taq™ buffer, 2 µM dNTP mix and 1.25 U Dream Taq™ polymerase), 2 µL of each PCR primer [10 µM] (synthesised by Inqaba Biotechnologies Industry, Pretoria) and 2 µL of genomic DNA [50 ng/µL] was prepared in a 200 µL PCR tube. The amplification was carried out in a thermal cycler (MJ Mini™ Personal Thermal Cycler, Biorad, SA) and consisted of 30 cycles of 1 min each at 94°C of denaturation, 30 s at 58°C of annealing step, 1 min of extension step at 72°C followed by the final extension step of 72°C for 10 min and cooling to 4°C.

The PCR product [10 µL] was analysed using 1% (m/v) agarose gel (Merck, SA) stained with 5% of 10 mg/mL ethidium bromide (Merck, SA) and the correct band size (approximately 1500 bp) was excised. To amplify variable regions (V1-3) of the bacterial 16S rRNA gene, the DNA was recovered from the gel slices by using the GeneJET™ gel extraction kit (Fermentas); thereafter, it was re-amplified with primers A1.4 [5'**CGTA-TCGCCTCCCTCGCGCCATCA**tctctatgcgAGRGTTTGATC-MTGGCTCAG3'] and B1 [5'**CTATGCGCCTTGCCAGC-CCGCTCAGG**TATTACCGCGGCTGCTG3'*] [14]. These primers contained the appropriate adaptor and barcode sequences that were necessary for running the samples on the GS-FLX-Titanium (Roche). The PCR reaction was analysed as described previously, but with an annealing temperature of 50°C as reported by Tekere et al. [8]. The entire PCR product was loaded onto a 1% agarose gel and the correct band size (500 to 600 bp) was excised from the gel and subsequently purified as previously mentioned. The DNA concentrations were quantified by using a Nanodrop spectrophotometer (Nanodrop2000, Thermo Scientific, Japan). The samples were pooled at equal concentrations of the filtration and biofilm samples. The pooled samples were sequenced on the GS-FLX-Titanium series (Roche) at Inqaba Biotechnology Industries, South Africa. Sequences of not less than 200 pb were classified using the online Ribosomal Database Project (RDP) naive Bayesian Classifier, Version 2.4, December 2012, which is assigned to the taxonomical hierarchy: RDP training set 10, based on nomenclatural taxonomy and Bergey's Manual with a confidence threshold of 95%.

Tolerance limits of test organisms to vanadium and nickel

Microbial isolation. Due to the abundance of *Marinobacter* species in the collected mine water samples, the *Halomonas elongata* (HMC) medium was used [15]. The medium was prepared as follows: 7.5 g/L casamino acids, 5 g/L peptone, 1 g/L yeast extract, 50 g/L NaCl, 20 g/L MgSO₄. 7H₂O, 3 g/L sodium citrate, 0.5 g/L K₂HPO₄, 0.05 g/L FeSO₄(NH₄)₂SO₄ either with or without 2% agar and pH adjusted at 7.2±0.2. After the medium preparation, it was autoclaved and incubated overnight at 37°C to check for any contamination. Only media which showed no growth were used. The bacteria isolation was achieved by inoculating 1 mL of mine water into the HMC broth for 48 h and thereafter placed on HMC agar. A single colony was selected and serially streaked onto the HMC agar for purification. Pure isolates (MWI-1) were inoculated into the HMC broth to obtain the required bacterial concentration used during the experiment.

Modified wastewater liquid media preparation. Domestic wastewater samples were also collected from the effluent (before disinfection) of the Daspoort wastewater treatment plant in Pretoria. This was firstly screened and only samples with very low metal concentrations [<0.01 mg/L] were used for the preparation of the modified wastewater liquid media. D-glucose anhydrate [2.5 g/L], MgSO₄·7H₂O [0.5 g/L] and KNO₃ [0.18 g/L] were added to the filtrate to serve as a carbon source and nutrient supplement for the culture media [16]. The test metal used in the experimental study was of analytical grade, purchased from Sigma Aldrich (Cape Town, South Africa). Sodium meta-vanadate anhydrous ($NaVO_3$) and nickel nitrate [$Ni(NO_3)_2$] were used as a source of V^{5+} and Ni^{2+} ions, respectively. The stock solution of $V^{5+}$and Ni^{2+} at a concentration of 2 000 mg/L were prepared using deionised water. From this solution, aliquots of specific volume, corresponding to the final V^{5+} and Ni^{2+} concentrations ranging from 50 mg/L to 800 mg/L (increasing at a geometric scale of 50 mg/L), were added to a 250 mL flask containing the wastewater-mixed liquid medium so as to obtain a final volume of 150 mL. As the optimal growth of Marinobacter goseongensis has been reported occurring at pH 7.5 [17], the pH of the mixed liquid media was adjusted at 7.3 ± 0.3 using 1.0 M HCl and 1.0 M NaOH (Merck, SA). Additionally, a parallel experiment using V^{5+} and Ni^{2+} (1:1, 2:1, 1:2, v:v) concomitantly, was performed. ICP-OES was used to confirm the metal concentrations in the wastewater-liquid media. The culture medium was autoclaved at 121°C for 15 min and cooled down to room temperature prior to use. To check the sterility of this medium, 1 mL aliquot was plated onto the sterile bacteriological agar and incubated at 37°C for 24 h; media indicating any microbial growth were not included in the experimental study. Only flasks containing the sterile media were inoculated with a known population of the respective test organism isolates. In order to mimic the environmental condition, a parallel experiment was carried out at a pH value of 3.

Temperature and metal-tolerance characterisation. Prior to assessing the metal tolerance ability of the isolates, the optimum growth temperature of MWI-1 [AB793286] was determined by incubating the isolates at various temperatures (25°C, 30°C and 35°C) in the HMC broth. Afterwards, a series of experiments were conducted in 250 ml Erlenmeyer flasks containing 150 mL of the modified wastewater-liquid media. These series of experiments had one positive control and one negative control. The positive control flask contained the free metals-liquid media medium, while the negative control contained the wastewater-liquid media with the highest concentration [1 000 mg/L] of either V^{5+} or Ni^{2+}. Sample flasks as well as those containing the positive controls were inoculated with the isolates (approximately 100 cfu/mL). All the inoculated flasks and the controls were initially incubated at 30°C±2°C and aliquots were taken every day for 4 days. The median lethal concentration (LC_{50}) of the test metal for each of the test microbial isolates was determined as described by previous investigators [18–21]. The minimum inhibitory concentration (MIC) of the test metal (referring to the smallest concentration of an antimicrobial agent necessary to inhibit growth of microorganisms) was determined according to Shirdam et al. [22]. MIC values were noted when the isolates failed to grow on the plates. After incubation, the microbial isolates were classified as sensitive or tolerant to Ni^{2+} according to the inhibition of growth cells.

Effect of pH on the tolerance limits of test organisms. To check the effect of pH on the tolerance limits of bacterial isolate to nickel or vanadium, the isolate was inoculated in mixed liquid media containing the smallest concentration of Ni^{2+} or V^{5+} necessary to inhibit their growth (MIC). The experimental study was conducted at a constant temperature of 30°C in a shaking incubator at a speed of 100 rpm. During each sampling regime, aliquot samples were taken every 24 h for 4 days for microbial estimation.

Molecular characterization of the isolate for metal-tolerance ability. In order to assess the ability of the bacterial isolate to tolerate Ni^{2+} and V^{5+} toxicity, its molecular character-isation on metal-tolerance ability was performed by the amplifi-cation of the cnrB2, van2, nccA and smtAB genes that encode for cobalt-nickel-cadium-vanadium resistance as well as, using specific primers (Table 1). The PCR amplification of target genes were done in a thermal cycler (MJ MiniTM Personal Thermal Cycler, Biorad SA) using 200-µL PCR tubes and a reaction mixture volume of 50 µL. The reaction mixture consisted with 25 µl 2× Dream Taq^{TM} PCR master mix (10× Dream Taq^{TM} buffer, 2 µM dNTP mix and 1.25 U Dream Taq^{TM} polymerase), 2 µl of each PCR primer [10 µM] (synthesised by Inqaba Biotechnologies Industry, Pretoria, South Africa) and 2 µl of genomic DNA [50 ng/µl] and was made up 50 µl with ultra-pure nuclease-free

water [19 µL]. The following amplification conditions were used: denaturation of template DNA at 94°C for 2 min, followed by 30 cycles of denaturation at 94°C for 1 min, annealing of template DNA for 30 s at specific temperature (Table 1) and an extension time of 1 min at 72°C for the primers. After the last cycle the samples were kept at 72°C for 10 min to complete the synthesis of all the strands and a cooling temperature of 4°C was applied. The PCR product [10 µL] was analysed using 1% (m/v) agarose gel (Merck, SA) stained with 5% of 10 mg/mL ethidium bromide (Merck, SA) and electrophoresed to determine the product size, which was visualised under UV light in an InGenius L Gel documentation system (Syngene).

Sequencing 16S rRNA. Prior to sequencing, the bacterial DNA was extracted and amplified using the universal primers as stated above. The amplified PCR products of approximately 1 500 bp were purified using a DNA clean and concentrator-25 kit (Zymo Research, SA). The concentrated DNA samples were then stored under −20°C and dispatched to Inqaba Biotechnology Industries (Pretoria, South Africa) for sequencing. The GS Junior-454 Sequencer (Roche) was used to sequence the DNA sample. To carry the phylogenetic analysis, a partial 16S rRNA gene sequence (16S rDNA) from the MWI-1 isolate was compared with other Alteromonadaceae that were available in the database. The 16S rDNA sequences were aligned using CLUSTAL2X and then the phylogenetic tree was constructed with the use of the MEGA 5 computed package. The confidence level of the phylogenetic tree topology was evaluated with a generation of 100 bootstrap sets.

Effect of metal ions on cell morphology. To determine the effect of V^{5+} and Ni^{2+} on the cell morphology, the MWI-1 incubated for 24 h in metal solution of V^{5+} or Ni^{2+} [50 mg/L] and the negative control (MWI-1 not exposed to metal ions) were centrifuge sample [10 mL] at 8, 000×g at 4°C for 5 minutes. Microbial cell was washed twice with 0.1 M phosphate buffered saline and fix overnight in 2% glutaraldehyde [prepared in 0.1 M PBS]. Pellets were dehydrated through an ethanol series from 10% to absolute, and for each series samples were held for 30 minutes. Samples were placed on a brass stub, sputter-coated with gold and examined by SEM.

For the Infrared spectra of the treated and untreated biomass were obtained using a Fourier transform infrared spectrometer (FTIR BOMEM MB 104). Prior to analysis, the biomass was centrifuged [10 mL] at 8, 000×g at 4°C for 5 minutes. The pellets were washed twice with 0.1 M phosphate buffered saline, dehydrated through an ethanol series starting from 50% to absolute and thereafter dried. After drying, 5 mg of bacterial cells were encapsulated in 150 mg of KBr. For analysis, all the infrared spectra were recorded over the range of 4000 to 500 cm^{-1}.

Statistical Analyses

The data were statistically analysed using the Stata computer software (version: STATA V12, STATA Corp. LP, 2012). Analysis of variance (ANOVA) was performed to compare the average percentage abundance between different bacteria. The Tukey HSD pairwise comparisons were also performed to see which bacteria where different in terms of the average percentage. The non-parametric Kruskal Wallis test was performed to compare the rank sum of the tolerance rate of the bacteria between V^{5+} and Ni^{2+}. The ordinary linear regression and hierarchical linear model were used to compare the average percentage of metal removed by the bacterium between the three treatments, which are mixed, V^{5+} and Ni^{2+}. Another statistical multivariate analysis was performed to check the difference between die-off rates at different pH-values when considering the type of test metals and incubation time. Every statistical

Table 1. Primers targeting some genes encoding metal-resistance in microbes.

Gene name	Sequence forward (5′–3′)	Sequence reverse (5′–3′)	Annealing temperature
nccA	ACGCCGGACATCACGAACAAG	CCAGCGCACCGAGACTCATCA	57°C
van2	CAAGTTCGTCGTCAACTT	CACTCGAGACAGGTATCA	30°C
smtAB	GAT CGA CGT TGC AGA GAC AG	GAT CGA GGG CGT TTT GAT AA	56°C
cnrB2	TACTGGCGATGTACTCGC	GAAGGTATTACGGGTGGC	55°C

analysis performed assumed that the observations were dependent between each other. For each parameter, data analysis was repeated assuming the observations are independent. The interpretation was performed at a two-sided 95% confidence limit. Each experimental study or analysis was performed in triplicate except for metal uptake and tolerance limits which were performed in quintuplicate.

Nucleotide sequence accession number

The 16S rRNA gene sequence of the isolate MWI-1 (*Marinobacter* sp. MWI-1) has been deposited in DNA Data Bank of Japan (DDBJ) Nucleotide sequence database and is available from the above database (http://www.ddbj.nig.ac.jp) as well as GenBank nucleotide database at the NCBI website (http://www.ncbi.nlm.nih.gov) under accession number AB793286.

Results

Mine water profile

The chemical profile of the mine water samples is presented in Table 2. Several chemical elements were found in the samples with Zn, Cu, Mn, U, Hg, nitrate and phosphorus not exceeding 10 mg/L. Besides sodium (3429.08–6426.48 mg/L), which showed the highest concentration, Ni, V, Fe, Ca, Mg and K were also found to be the most abundant chemical elements in the mine water samples and their concentrations ranged between 11.37 mg/L and 437.46 mg/L. The mine water samples indicated high acidity with pH ranging from 2.34 to 3.78, whereas the temperature, the conductivity and the COD had values ranging between 12.17°C and 13.80°C, 397 μ/Sm and 645 μ/Sm and 529.27 mg/L and 661.54 mg/L, respectively. Although chemical contents in the mine water throughout the sampling period appeared to be different in terms of concentrations, no significant difference ($p > 0.05$) was noted. In contrary, there was only significant difference in the mean concentration between Na and Ni ($p < 0.05$) when adjusting for the time, and the interaction of the time and the chemical. Specifically, Na was on average 5021.8 mg/L (4493.601 to 5549.972) more concentrated than Ni.

Microbial diversity in mine water

The microbial community structure of mine water samples was determined using the 16S rRNA gene amplicon pyrosequencing method which targeted one DNA region (V1–3); respective sequences are summarised in Table 3. A total of 2 047 sequences were identified in the mine water and only sequences with a similarity of 95% to 100% to the species in the database were used. To determine the abundance of each taxon, the number of sequences from a particular taxon was plotted against the total number of sequences used. The bacterial phylum and classes in the South Africa mine water appeared to be less diverse (Figure 1). A total of 6 phyla with 10 classes were identified in the mine water

samples. Among these, it is evident that the phyla Proteobacteria (58.33%) and Cyanobacteria (36.25%) were the most abundant, followed by the Bacteriodetes (3.33%), *Firmicutes* (0.83%), *Actinobacteria* (0.83%) and *Chloroflexi* (0.42%). For bacterial classes, *Gammaproteobacteria* (59.67%) was the most predominant in the mine water sample and this was followed by *Cyanobacteria* (17.28%), *Alphaproteobacteria* (12.76%), *Betaproteobacteria* (5.35%) and *Flavobacteria* (2.88%). The rest of the bacterial classes represented a percentage of abundance of less than 0.5%.

In the 6 phyla and 10 classes, a high diversity in terms of bacterial orders, families and genera were observed in the mine water samples (Figure 2), with a total of 17 orders, 27 families and 33 genera being found. Of the orders, *Alteromonadales* (59.18%) was the most abundant followed by *Nostocales* (14.29%) and the rest that revealed less than 10% each. In addition, *Alteromonadaceae* (58.82%) and *Nostocaceae* (17.66%) were the most abundant families present with *Marinobacteria* (47.02%) and *Anabaena* (17.66%) being the most abundant genera. Of the total genera found in the mine water samples, unclassified bacteria (15.67%) were also observed in great abundance when compared to other genera. This was confirmed by a statistical test performing analysis of variance to compare the average percentage abundance between different bacteria. The test is significantly different at a 0.05 significance level. Furthermore, the Tukey HSD pairwise comparisons were also performed to see which bacteria where different in terms of the average percentage abundance and revealed that *Marinobacteria* was on average 31% more abundant than *Anabaena* in the water samples. Similar observation was noted with other bacteria when compared to *Marinobacter* (average ranging from 38–46%).

Tolerance limits and metal-removal ability of bacterial species

Growth curve of the mine water isolates (MWI-1) at different temperatures in metal-free media. Due to their predominance, the genera *Marinobacter* were isolated from the mine water samples and assessed for their possible ability to resist V^{5+} and Ni^{2+}. Figure 3 illustrates the growth curves of the MWI-1 in a metal-free medium (HMC broth). The MWI-1 revealed a lag-phase between time 0 to 2 followed by an exponential phase from time 2 h to 8 h when inoculated at 25°C, 30°C and 35°C with bacterial counts of 7 logCFU/mL, 8 logCFU/mL and 7 logCFU/mL, respectively. At 30°C, MWI-1 indicated a second exponential growth [10 logCFU/mL] from time 16 h to the end of the experiment, whereas a death phase was observed at 25°C and 35°C, repectively.

As MWI-1 grew very well at 30°C, its tolerance to V^{5+} and Ni^{2+} in the modified liquid media was tested at the said temperature and at pH 7.2±0.2. Figure 4 illustrates the growth performance of this isolate in the modified liquid media containing either V^{5+} or Ni^{2+} or both at two different concentrations to highlight the difference of the toxic effect between the two metals. In general,

Table 2. Profile of mine water samples collected from the vanadium mine, South Africa (n = 3).

	Ni [mg/L]	V [mg/L]	Zn [mg/L]	Cu [mg/L]	Mn [mg/L]	U [mg/L]	Fe [mg/L]	Hg [mg/L]
September	31.99±4.25	434.45±67.30	0.24±0.04	5.53±0.03	2.19±0.12	1.58±0.56	13.21±2.38	3.95±0.27
October	22.07±1.68	437.46±53.28	4. 41±1.08	1.51±0.01	6.34±0.18	1.57±0.09	18.11±1.69	4.64±1.28
November	19.79±3.25	420.94±17.64	2.25±0.67	7.52±0.67	3.12±0.05	1.60±0.21	11.37±3.94	2.76±0.06
	Ca [mg/L]	Mg [mg/L]	Na [mg/L]	K [mg/L]	Cr [mg/L]	Co [mg/L]	As [mg/L]	Pb [mg/L]
September	518.37±37.29	217.65±1.25	5413.85±97.69	216.32±51.69	0	0	0	0
October	507.91±17.29	213.46±4.98	6426.48±43.69	240.93±1.39	0	0	0	0
November	496.28±29.64	206.80±57.37	3429.08±131.5	238.4±15.37	0	0	0	0
	DO [mg/L]	NO$_3^-$ [mg/L]	PO$_4^{3-}$ [mg/L]	pH [pH units]	Temperature [°C]	Conductivity [μ/Sm]	COD [mg/L]	
September	5.67±0.22	4. 14±1.98	2.17±0.05	2.3±0.02	13.40±0.07	533.00±2.59	661.54±67.11	
October	5. 73±0.11	2.08±0.15	5.12±0.01	3.8±0.25	12.17±0.05	397.00±1.15	583.61±13.51	
November	5.63±0.04	1.48±0.37	3.07±0.12	3.2±0.02	13.80±0.01	645.00±2.22	529.27±53.25	

the growth of the isolates decreased with the increases of the metal concentrations. The MWI-1 was able to significantly grow in the presence of V^{5+} at 100 mg/L [9 logCFU/mL] and 200 mg/L [8 logCFU/mL], while in the presence of Ni^{2+} MWI-1 could only grow at 100 mg/L. Concomitantly, Ni^{2+} toxicity was able to inhibit the growth in all the volume ratios. Statistical evidence revealed a significant difference ($p<0.05$) in terms of growth performance between the positive controls and those samples treated with Ni^{2+} while no significant difference ($p>0.05$) was indicated for MWI-1 treated with V^{5+}. Another statistical analysis shows a significant difference for positive controls when compared with those samples treated with both metals concomitantly.

A general observation indicated that the MWI-1 isolate was more tolerant to V^{5+} than to Ni^{2+} in the modified liquid media (Table 4). The MWI-1 isolate could resist V^{5+} (MIC) up to 13.41 mM [approximately 700 mg/L] and only reached 5.39 mM [approximately 250 mg/L] in the presence of Ni^{2+} at pH 7. When exposed to test metals (V^{5+} and Ni^{2+}) at acidic medium (pH 3), the results showed a decrease of tolerance ability of the isolate with 24 h-LC50 ranging from 5.89–6.87 mM for V^{5+} and 2.04–2.22 mM for Ni^{2+} and MIC ranging from 10.8–11.78 mM for V^{5+} and 2.56–2.73 mM for Ni^{2+}. The tolerance or sensitivity of the MWI-1 isolate was also revealed by its ability to remove V^{5+} and Ni^{2+} at pH 7 (Table 4). In the presence of V^{5+} or Ni^{2+} separately (pH 7), the MWI-1 was able to remove up to 99.95% of 100 mg-V^{5+}/L and 86.42% of 100 mg-Ni^{2+}/L. When pH was adjusted at 3, the removal of both metals decreased to a range of 76.35–59.69 and 63.53–38.27 for V^{5+} and Ni^{2+}, respectively. In consortium, Ni^{2+} toxicity appeared to disturb the removal of V^{5+}, none of the metals (V^{5+}/Ni^{2+}) was removed at a percentage of over 30%, with the exception of V^{5+} (30.15%) that

Table 3. Summary of pyrosequencing data from mine water samples.

Sequence	V 1–3
Number of sequence	2047
Total length of sequences [bp]	769 890
Average length of sequences [bp]	400

was removed at a ratio of 1:1 [100 mg/L/100 mg/L]. The non parametric Kruskal Wallis test was performed to compare the rank sum of the tolerance rate of the bacteria between V^{5+} and Ni^{2+}. The test was significant at a 0.05 significance level with a p-value of 0.0037. The rank sum of the tolerance level of the bacteria was higher in V^{5+} than Ni^{2+}. Another statistical analysis revealed a significant difference ($p<0.05$) in terms of tolerance ability to Ni^{2+} between the two experiments (Experiment at pH 7 and pH 3), whereas no significant difference ($p>0.05$) in terms of V^{5+}-tolerance was shown between the two experiments. This implies that at the pH (pH value 3) of the industrial wastewater (polluted environment), the isolate (MWI-1) ability to resist to Ni^{2+} decreases, while for V^{5+} there is no decrease in terms of tolerance ability when compared to the tolerance at pH 7.

During the course of the study, the ordinary linear regression and hierarchical linear model were used to compare the average percentage of metal removed by the bacteria between the three treatments, which are mixed, V^{5+} and Ni^{2+}. Both adjusted and unadjusted models were fitted, and the treatments were found to be significantly different for both adjusted and unadjusted model. The treatments were found to be significantly different ($p<0.05$) (Table 5). In particular, V^{5+} and Ni^{2+} treatments were significantly different from mixed ($p<0.05$). Precisely on average V^{5+} used separately was removed 80.5%(35.206 to 58.110) more than mixed, and nickel was removed 46.7% (35.206 to 58.110) more than mixed. Similar statistical observation was noted after adjusting for concentration and individual elements with on average V^{5+} removed 61.3%(52.230 to 70.375) more than mixed, and Ni^{2+} removed 27.5.6% (18.405 to 36.550) more than mixed. This statistical evidence revealed that V^{5+} is less toxic then Ni^{2+}, and there is interaction between the two metals which increase their toxicity and rending their mixture more toxic then individual elements.

Effect of pH on the tolerance limits of test organisms. To check the effect of pH on the tolerance limit of the isolate to V^{5+} and Ni^{2+}, the MWI-1 inoculated in wastewater liquid media containing separately the test metal at a concentration able to kill the 50% of the test isolate after 24 h (LC$_{50}$) and the experimental study was conducted at 30°C in a shaking incubator at a speed of 100 rpm. Table 6 shows an increase on percentage die-off rate over the decrease of pH throughout the study period. Furthermore, the percentage die-off rate of the isolate (MWI-1) also increased over time of exposure.

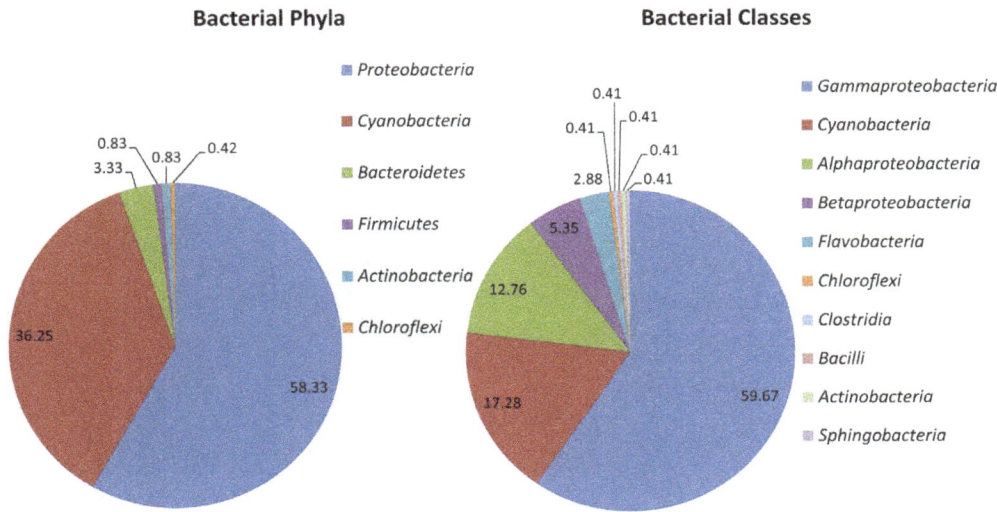

Figure 1. Relative abundance and diversity of bacterial phylum and classes in South African mine water.

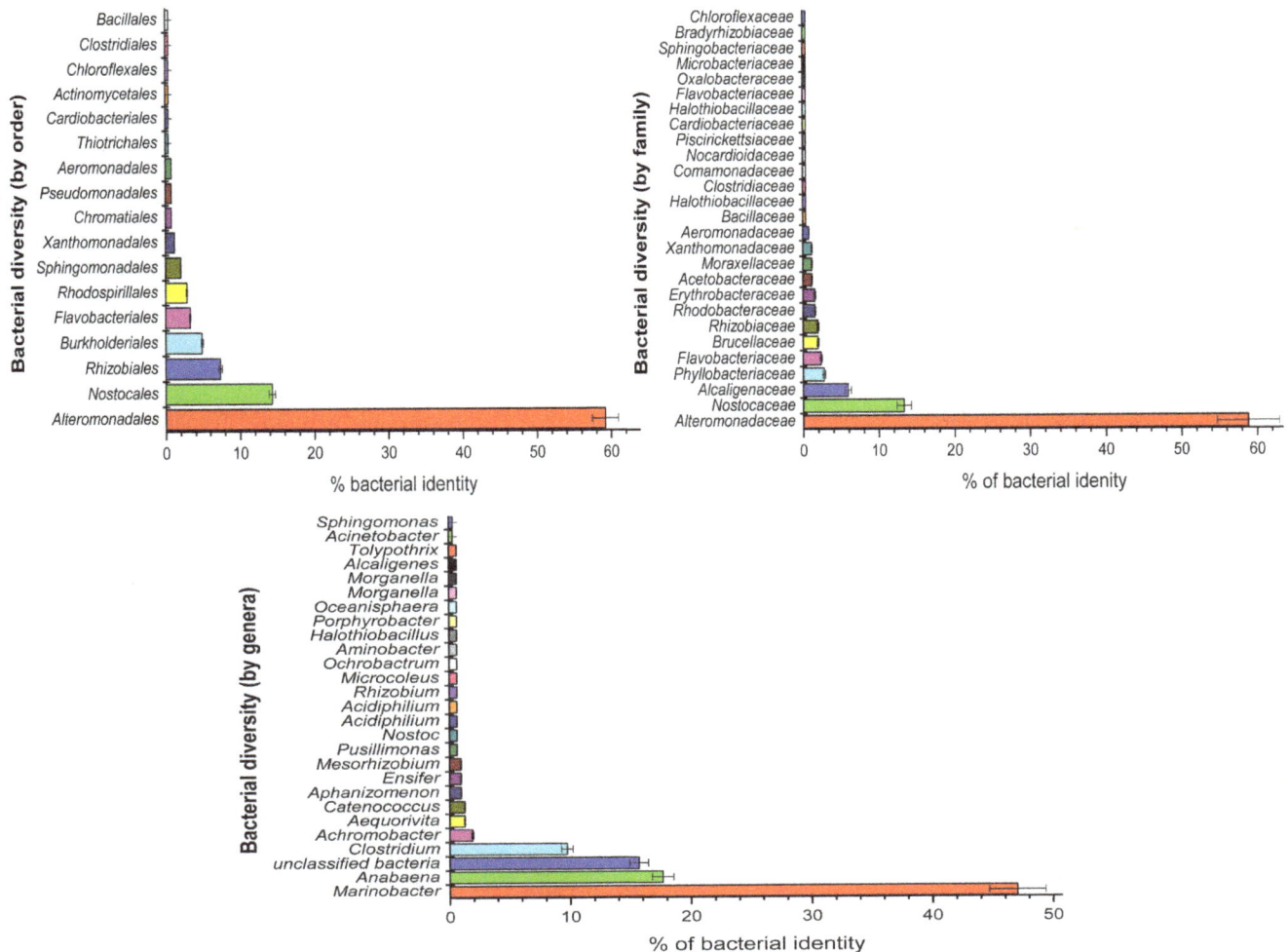

Figure 2. Composition of the bacterial orders, family and genera detected in the mine water with sequences of the variable region V1–3 of the 16S rRNA genes.

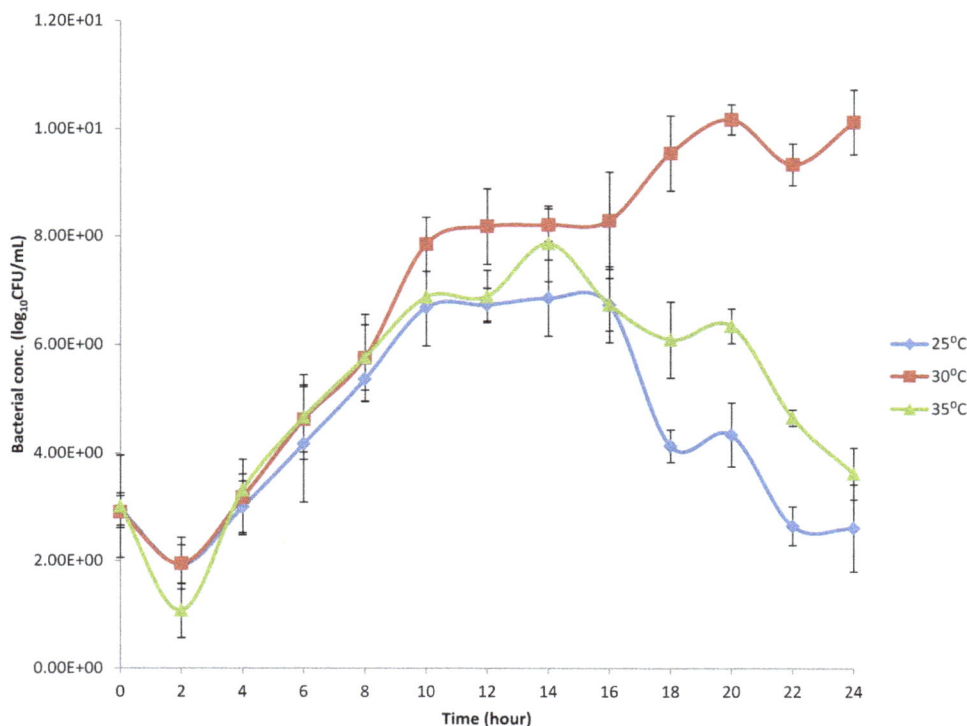

Figure 3. Growth curve of MWI-1 in a metal-free medium (HMC broth) inoculated at different temperatures (25°C, 30°C, 35°C) at pH 7.2±0.2 for 24 h.

When stressed with Ni^{2+}, the percentage die-off rate of the isolate ranged from 39.25% (pH 8) to 85.15% (pH 1) after 24 h of exposure, whereas in the presence of V^{5+}, the isolate die-off rate could not research 60% after 24 h of exposure.

Subsequently, a statistical multivariate analysis was performed to check the difference between die-off rates at different pH-values when considering the type of test metals. Table 7 indicates that the effect of pH level on the die-off rate was significant at a 0.05 significance level. Specifically, an increase of 0.5 in the pH level decreases the die-off rate by 2.5% (-3.009 to -2.083). After adjusting for the type of metals, adjusting for the incubation period, adjusting for the type of metal and incubation period, significant difference ($p<0.05$) was also found between die off rates at different pH level. Similarly to the above, an increase of 0.5 in the pH level decreases the die-off rate by 2.5% (-3.009 to -2.083). However, when adjusting for the type of metal, incubation period and the interaction between the type of metal and the pH level was significant at a 0.05 significance level with an increase of 0.5 in the pH level decreases the die-off rate by 3% (-3.338 to -2.598). This implies that changes on pH level affect the growth of the isolate when exposed to V^{5+} or Ni^{2+} and this is also function on the incubation period. In addition, the statistical test also revealed that the optimal pH of the MWI-1 isolate in the modified wastewater liquid media ranges between 7 to 8 pH values.

SEM, FTIR and Molecular analyses. To assess the effect of metal ions on cell morphology of MWI-1 in wastewater culture medium, a scanning electron microscopy (SEM) was used (Figure 5). The bacterial cells exposed to metal ions were significantly changed and showed major damage, characterized aglomeration and lyse of microbial cells when compared to isolate not exposed to metal ions. Microbial cells exposed to Ni^{2+}

(Figure 5b) revealed more damage than those exposed with V^{5+} (Figure 5c).

The Fourier transform infrared spectroscopy (FTIR) analysis was perfomed to investigate the role of extracellular polymer in V^{5+} and Ni^{2+} bioadsorption. The FTIR spectroscopy of untreated and treated MWI-1 with V^{5+} and Ni^{2+} separately is shown in Figure 6. The result revealed the effects of metal ions (V^{5+} and Ni^{2+}) on the functional group of membrane wall elucidating their biosorption onto the membrane. The bacterial biomass' spectrum (Blank) revealed different peaks representing amid group (C = 0, N-H, C-N), carboxylic group (C-0) and phosphate group (P = 0), reflecting the complex nature of the bacterial wall. A significant difference was found between untreated biomass and V^{5+}-treated biomass, whereas the Ni^{2+}-treated biomass did not show major difference when compared to the untreated biomass. The spectrum pattern of V^{5+}-treated biomass showed changes of certain bands in the region of 870–1626.91 cm^{-1} and 2921.8–3271.7 cm^{-1} compared to untreated biomass. The spectrum pattern of Ni^{2+}-treated biomass however showed changes in the region of 1044.7–1301.8 cm^{-1} compared to the untreated biomass' spectrum. This revealed that MWI-1 could adsorb more V^{5+} than Ni^{2+} in the liquid media.

The phylogenetic analysis using the neighbour-joining method with a bootstrap value of 100 replicates indicated that an MWI-1 isolate belonged to the genus *Marinobacter* and was most closely related to *Marinobacter goseongensis* strain En6 [EF660754.1 and NR044340.1] at a similarity of approximately 97% (Figure 7). Molecular study investigated the resistance ability of the microbial isolates in a gene level, to check whether the heavy-metal removal ability of test isolate is linked to specific genes such as *nccA* (Ni, Co, Cd-resistance), *cnrB2* (Ni and Co-resistance), *van2* (V-resistance) and *smtAB* (gene encoding synechococcal metallothioneins) using

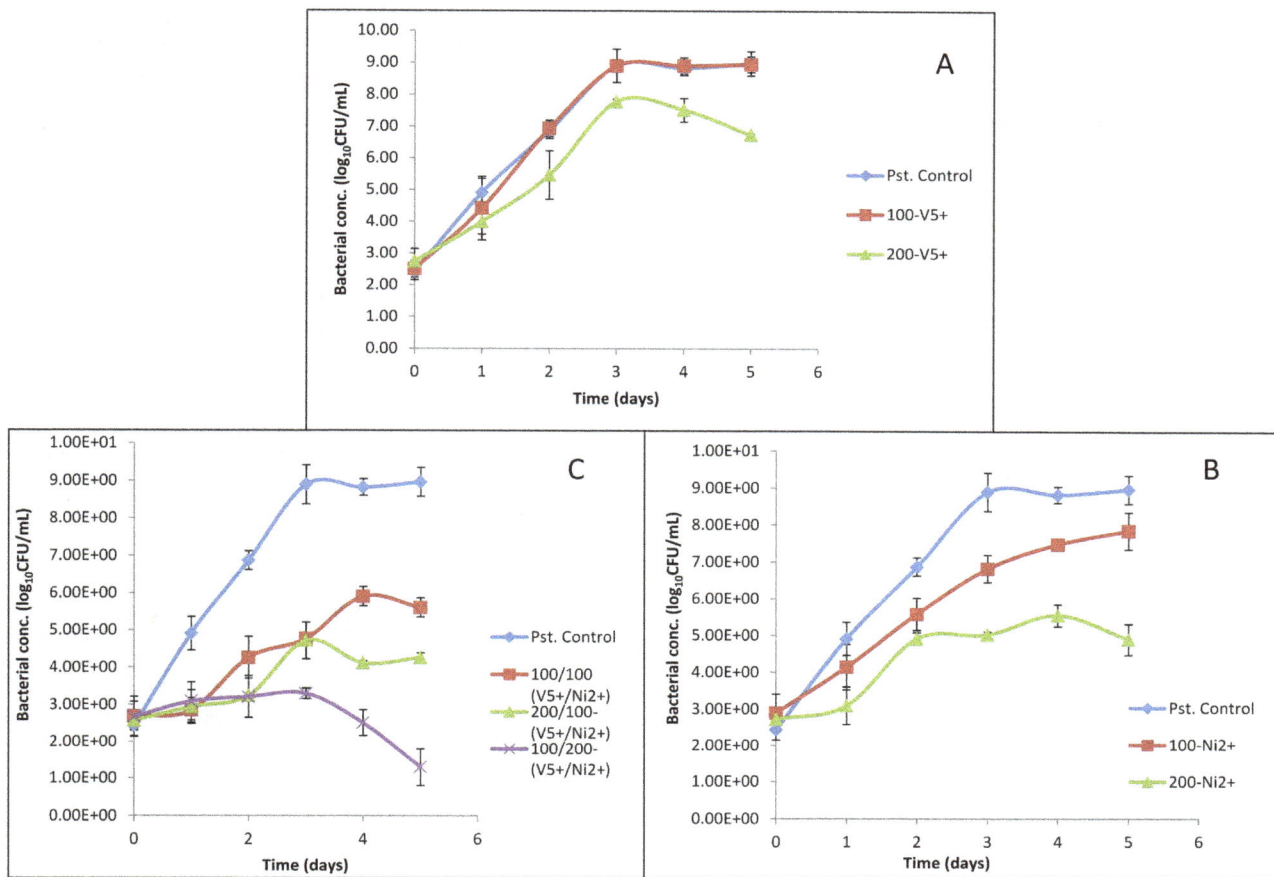

Figure 4. Growth performance of MWI-1 in a medium containing either V^{5+} (A) or Ni^{2+} (B) or both (B) at 100 mg/L and 200 mg/L, 30°C, pH 7.2±0.2.

the conventional PCR (Figure 8). The *smtAB* gene was targeted as it encodes the production of metallothionein involved in the resistance to several metal ions (such as Zn, Se, Cd, Hd, Ag, As, etc.) in several bacterial species. Of all the genes targeted in the gDNA of MWI-1 isolate, *nccA*, *cnrB2* and *smtAB* showed positive amplification. An amplified products of approximately 400 bp,

500 bp and 1141 bp revealing the presence of *cnrB2*, *smtAB* and *nccA* genes were reproductively detected, whereas, the vanadium-resistant gene *van2* was not found.

Table 4. MWI-1 isolate tolerance limits (MIC and 24 h LC_{50}) to V^{5+} and Ni^{2+}, and removal ability in the modified liquid media (n = 5).

	Percentage removal (Single metal), pH 7		Percentage removal (Single metal), pH 3	
	100mg/L	200mg/L	100mg/L	200mg/L
V^{5+} [%]	99.23±0.57	94.6±2.19	76.32±4.29	59.69±2.15
Ni^{2+} [%]	82.42±3.43	43.67±3.76	63.59±6.71	38.27±1.65
	Percentage removal (V5+/Ni2+, v/v) (mixed metals), pH 7			
	100/100 [mg/L]	200/100 [mg/L]	100/200 [mg/L] (v/v)	
V^{5+} [%]	30.15±7.1	17.17±0.87	9.51±1.97	
Ni^{2+} [%]	23.96±2.6	11.48±1.48	6.04±2.82	
	Tolerance limit at pH 7		Tolerance limit at pH 3	
	24h-LC50	MIC	24h-LC50	MIC
V^{5+} [mM]	10.15±0.59	13.41±0.56	6.54±0.56	11.45±0.57
Ni^{2+} [mM]	3.68±0.49	5.39±0.5	2.11±0.16	2.67±0.1

Table 5. Unadjusted and adjusted hierarchical regression model-Percentage removal.

Covariates	A	B	C
V^{5+}	80.48(69.54–91.42)	61.30(52.23–70.38)	69.61(65.03–74.19)
Ni^{2+}	46.66(35.72–57.60)	27.48(18.40–36.55)	19.17(14.59–23.75)
200/100 mg/L		−12.73(−20.35–−5.109)	−12.73(−16.28–−9.185)
100/200 mg/L		−19.28(−26.90–−11.66)	−19.28(−22.83–−15.73)
100 mg/L		22.14(14.52–29.76)	38.75(33.73–43.76)
Metal Code		5.117*(−1.107–11.34)	5.117(2.221–8.013)
V*100 mg/L			−33.22(−40.31–−26.12)
Constant	16.39(10.92–21.85)	24.50(18.27–30.72)	24.50(21.60–27.39)

Confidence Interval in parentheses; Note: p value for all the parameters was less than 0.01.
*$p < 0.05$, n = 30.
A: Main effect, **B**: Adjusting for concentration and individual elements, **C**: Adjusting for concentration, interaction between the treatment and the concentration, and individual elements.
Note: Each model includes the coefficient and their confidence interval in bracket.

Discussion

Mine water pollution negatively impacts the chemical and microbiological quality of both surface and groundwater [23]. Being a country with extensive industrialisation and major historical mine activities, water pollution by metal ions has emerged as one of the serious challenges currently faced by water service authorities in South Africa. As a result, the National Water Act of 1998 has been signed to make provision for the legal requirements, registration and authorisation to discharge wastewater into water sources [24]. Hence, this study firstly focused on the chemical characteristics of the vanadium mine located in Mpumalanga, South Africa. The results revealed that the mine water samples used in this study were highly acidic at a range below the permissible minimum limit of 5.5 pH unit set by the South African Water Act, No. 36 of 1998. If discharged into the receiving water bodies, this might be identified as a major factor in the speciation and diversity of microbial populations [25]. The electrical conductivities of the mine water varied significantly ($p < 0.05$) and were found to be at a range above the maximum permissible limit of 250 µ/Scm for effluent discharged into the receiving water bodies [26]. Several other pollutants found in the mine water samples had concentrations which far exceeded the maximum recommended limits of 0.2 mg-Ni/L, 0.1 mg-Mn/L, 0.1 mg-V/L, 0.01 mg-Cu/L, 0.1 mg-Zn/L, 0.005 mg-Hg/L, 0.03 mg-U/L, 0.3 mg-Fe/L, 15 mg-NO_3^-/L, 10 mg-PO_4^{3-}/L and 75 mg-COD/L, prescribed by the UN-FAO, EPA and South Africa Water Act, No. 36 of 1998 [24,27,28,29].

This study secondly focused on determining the bacterial diversity in the mine water samples and assessing their tolerance to high V^{5+} and Ni^{2+} concentrations in modified wastewater liquid

Table 6. Percentage die-off rate of *Marinobacter* sp. MWI-1 stressed with V^{5+} and Ni^{2+} over various pH in wastewater liquid media (n = 3).

pH	Nickel				Vanadium			
	Day 1	Day 2	Day 3	Day 4	Day 1	Day 2	Day 3	Day 4
1	85.15	99.5	99.5	99.5	52.8	89.2	99.5	99.5
1.5	77.11	99.6	99.6	99.6	52.4	85.3	99.6	99.6
2	75.31	99.59	99.59	99.59	47.7	72.5	85	97.4
2.5	65.12	96.51	99.61	99.61	52.5	72.6	86.5	96.9
3	44.92	95.34	99.58	99.58	47.8	76.7	84.7	95.2
3.5	58.49	91.32	99.62	99.62	36.2	75.3	88.5	93.2
4	59.73	83.22	93.96	99.66	55.1	60.7	81.6	91.8
4.5	49.77	69.41	94.98	99.54	51.1	55.6	80.1	88.7
5	48.82	77.56	94.09	99.61	47.4	53	77.1	86.9
5.5	41.75	73.3	93.69	99.51	50.2	52.9	80	84.7
6	48.98	74.49	94.39	98.98	49.5	49.5	71.8	75.5
6.5	51.02	79.59	95.92	97.14	52.4	55.2	75	78.2
7	45.96	68.98	79.8	87.37	28.6	47.3	70.1	74.1
7.5	43.98	58.59	60.65	88.43	55.1	55.1	71.9	76
8	39.25	57.94	60.62	83.64	35.1	44.3	67.9	70.2

Table 7. Unadjusted and adjusted hierarchical regression model-pH effect vs die-off rate, n = 30.

Covariates	pH_code	V	Day	Metal*pH level	Day*pH level	Constant
A	−2.546 (−3.009−−2.083)					81.64(69.28−94.00)
B	−2.546 (−3.009−−2.083)	−12.03(−16.03−−8.029)				87.66(82.99−92.32)
C	−2.546 (−2.814−−2.278)		22.18(19.86−24.49)			70.55(58.51−82.59)
D	−2.546(−2.814−−2.278)	−12.03(−14.35−−9.713)	22.18(19.86−24.49)			76.57(73.63−79.51)
E	−2.968(−3.338−−2.598)	−18.79(−23.54−−14.03)	22.18(19.92−24.44)	0.844(0.321−1.368)		79.95(76.40−83.49)
F	−1.769(−2.113−−1.425)	−12.03(−14.13−−9.930)	34.60(30.18−39.03)		−1.554(−2.040−−1.067)	70.35(67.06−73.65)
G	−2.191(−2.599−−1.783)	−18.79(−23.07−−14.50)	34.60(30.32−38.89)	0.844(0.373−1.316)	−1.554(−2.025−−1.082)	73.73(70.02−77.44)

Confidence Interval in parentheses; Note: p value for all the parameters was less than 0.01.
A: Model Main effect, **B**: Adjusting for metal, **C**: Adjusting for incubation period (Time), **D**: Adjusting for metal and time, **E**: Adjusting for metal, time and metal*pH interaction, **F**: Adjusting for metal, time and time* pH interaction, **G**: adjusting for metal, time, metal*pH interaction and and time* pH interaction.
Note: Each model includes the coefficient and their confidence interval in bracket.

media. With the use of a culture-independent method, a diverse bacterial population was found in the mine water samples as presented in Figure 1 and Table 3 with *Marinobacteria* (47.02%), indicating a predominance among other families. It is known that uncultured bacteria constitute the largest portion of the population in the environment [30]. In this study, uncultured bacteria were excluded and only approximately 15% of bacteria were unclassified. Although several studies have revealed evidence that free-living microorganisms exhibit non-random distribution patterns across diverse habitats at various spatial scales, the geochemical parameters also determine the type of microbial population and microbial diversity of a specific habitat [25,31]. Therefore, it proved to be difficult to compare the present result on microbial diversity to those previously published on other habitats. Nevertheless, it has been reported that *Proteobacteria* are present in extreme environments [32]. A study by Raji et al. [33] revealed the predominance of *Proteobacteria* (53%) and other uncultured bacteria (41%) in deep mines in South Africa. Other studies on microbial diversity in hot water springs in South Africa, reported a predominance of *Proteobacteria, Bacteriodetes* and *Planctomycetes* at a percentage of 54.43%, 10.74% and 13.03%, respectively, with an abundance of *Xanthomonadales* belonging to the *Proteobacteria*. Another study conducted in China revealed that 60.62% of the

microbial population in the mine water samples was affiliated to *Proteobacteria* phylum with a proportion of 3.10% being *Alphaproteobacteria*, 24.78% *Betaproteobacteria*, 31.41% *Gammaproteobacteria* and 1.33% *Deltaproteobacteria* [32]. Kuang et al. [25], when assessing the microbial diversity of acid mine drainage samples from southeast China, reported a predominance of *Proteobacteria*, with the *Betaproteobacteria*, which were linked to an acidic environment, at pH>2.4.

Due to their abundance, the *Marinobacter* genus of the *Proteobacteria* phylum was isolated in the present study and assessed for their tolerance to and removal of V^{5+} and Ni^{2+} in the modified liquid media. The isolate (*Marinobacter* sp. MWI-1) was firstly sequenced and found to be closely related to *Marinobacter goseongensis*. This *Marinobacter* sp. MWI-1 showed very good growth at 30°C in the HMC broth, while at 25°C and 35°C a prompt die-off was noted when inoculated in a metal-free media. Findings of this study corroborated to those of Roh et al. [17] who reported that the optimum temperature for *Marinobacter goseongensis* sp. nov. should be between 25°C and 30°C. When inoculated in modified wastewater liquid media, culture media containing V^{5+} and Ni^{2+}, separately or combined and incubated at 30°C, reveals a significant growth (p<0.05) of the *Marinobacter* sp. MWI-1 in the media with V^{5+} when compared to the media with Ni^{2+}.

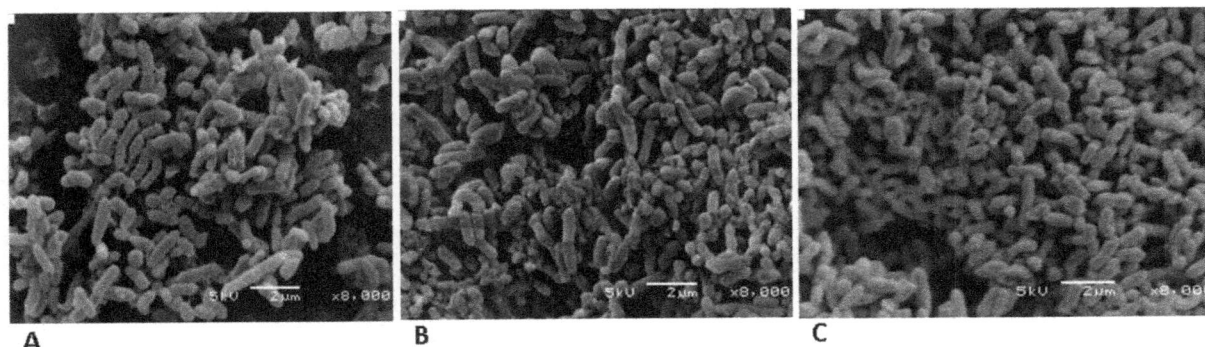

Figure 5. Scanning electron micrograph of strain MWI-1 grown on wastewater liquid media without test metal stressed (A), with vanadium stress (B) and with nickel stress (C).

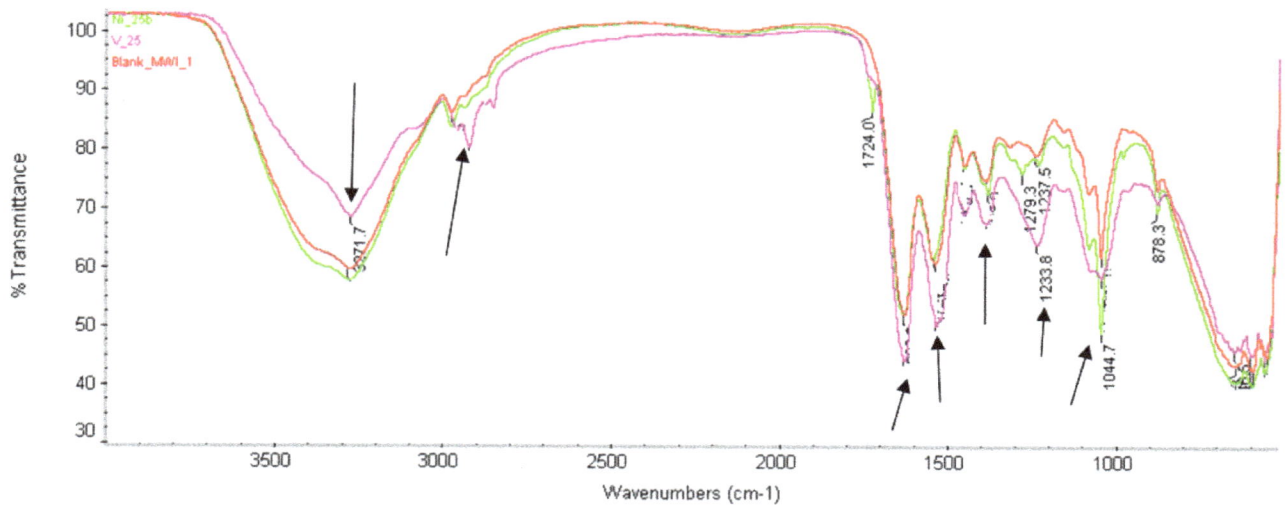

Figure 6. FTIR spectra of Marinobacter sp. MWI-1 before and after exposure to test metals.

Marinobacter sp. MWI-1 was more tolerant towards V^{5+} than towards Ni^{2+} (Table 4). The toxicity of the test metals to *Marinobacter* sp. MWI-1 appeared to have a relatively negative effect on the metal-removal ability of the test isolate in the modified wastewater liquid media with V^{5+} indicating the highest level of removal regardless of the pH value of the culture media (Table 4). It has been reported that bacterial strains can be characterised as being tolerant towards metal such as Ni^{2+}, if it is

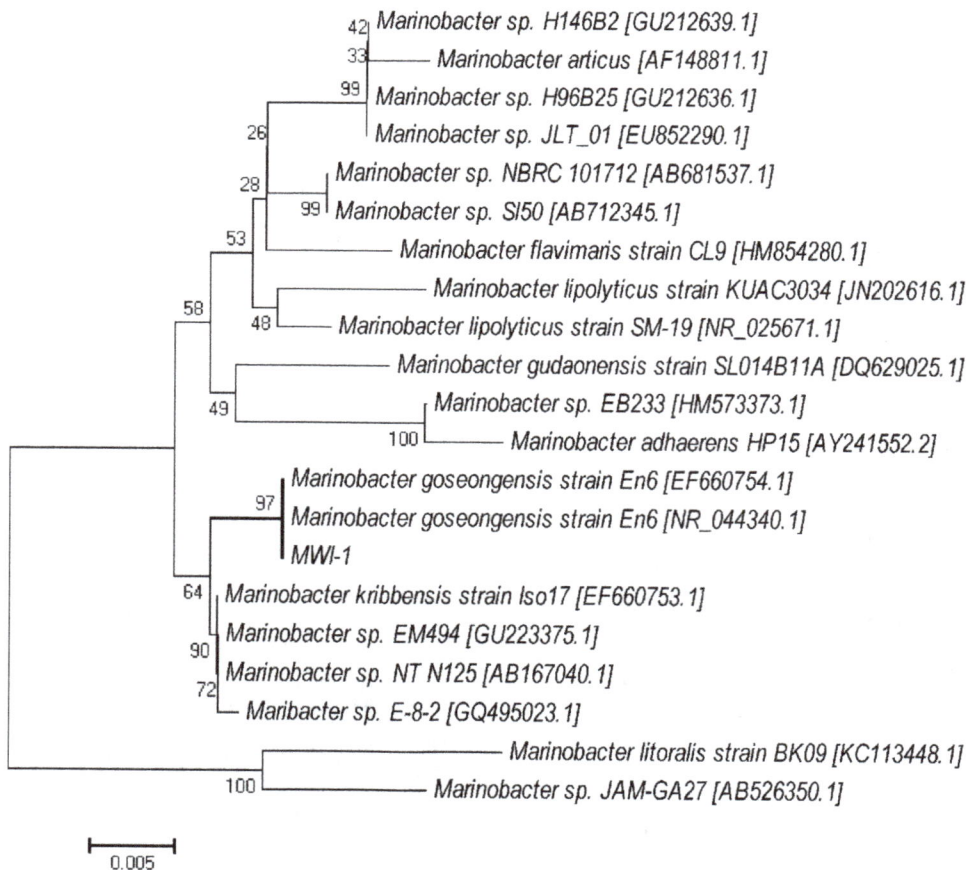

Figure 7. Phylogenetic tree using the neighbour-joining method, constructed and based on the bacterial 16S rRNA gene sequence detected in the present study along with similar sequences detected from the NCBI and RDP databases.

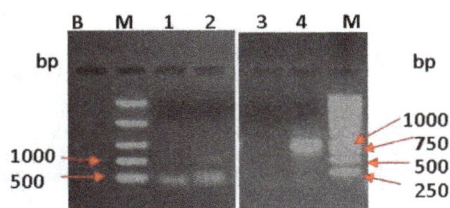

Figure 8. Agarose gel electrophoresis of PCR products of total genomic DNAs with primers targeting gene nccA (Lane: 4), van2 (lane: 3), smtAB (lane: 2) and cnrB2 (lane: 1). Lanes: M: DNA ladder (Marker) and B: Negative (No template DNA).

capable of expressing growth at concentrations higher than 100 mg/L of the metal [34]. This indicated that although the toxic effect of Ni^{2+} on the isolate when compared to the effect of V^{5+}, the isolate was still tolerant towards the metal (Ni^{2+}). However, the percentage removal decreased when both metals were concomitantly used revealing a synergistic action between the two metals. The tolerance ability of the test isolate towards Ni^{2+} was confirmed by a positive amplification of genes encoding the resistance of Ni^{2+}, whereas the gene *van2* encoding vanadium resistance was not found. The present study could not provide sufficient evidence of *Marinobacter* sp. MWI-1 V^{5+} resistance ability. The tolerance to V^{5+} by the *Marinobacter* sp. MWI-1 could be explained by the presence of a gene (*smtAB*) encoding the production of metalothionein which is a family of cysteine-rich proteins capable to bind metals and are suspected to be also involved in providing metal-resistant ability to bacteria [35].

The microbial species have been reported to be highly resistant to heavy metals and also having a very high ability to remove heavy metals. Zucconi et al. [36] stated that isolating microorganisms from extreme environments represent an appropriate practice to select metal-resistant strains that could be used for heavy metal removal and bioremediation purposes. Since most of the species of the genus *Marinobacter* have been reported to be halophilic, heterotrophic neutrophiles and living under extreme environmental conditions such as pH and high salinity, they have been isolated from several habitats such as seawater, petroleum refineries, oil-refineies, and so forth [37]. Due to their ability to grow under extreme habitats, strains found in *Marinobacter* could have the ability to remove metal in the environment. Researchers have reported that several strains of *Marinobacter* spp. such as *Marinobacter aquaeolei* have iron transport capabilities and are also capable of oxidising iron [38]. It has also been revealed that the abundance of genes involved in phosphonate metabolism in *Marinobacter* spp. could serve as binding agents for metals and furthermore be used as a heavy metal defence by these species [39]. According to Liao et al. [40], the *Marinobacter* species can be able to tolerate high concentrations of metal and metalloids. The authors revealed that *Marinobacter* sp. MnI7-9 isolated in a deep Indian sea could grow well at Mn concentration of 10 mM and also oxidize the metal at the same concentration.

Previous studies have well illustrated the importance of pH in microbial growth and activities which tend to be seen as one of the

major limiting parameter for the performance of wastewater treatment systems [18]. In addition, acid-tolerant microorganisms are viewed as being beneficial for the treatment of highly polluted wastewater from the mines or industry [40]. However, by investigating the effect of pH variations on bacterial activity in the wastewater liquid media treated with V^{5+} or Ni^{2+} (at concentrations reaching their specific 24 h-LC_{50}), the present study showed that the tolerance capability of *Marinobacter* sp. MWI-1 was significantly dependent on the pH and temperature of the wastewater liquid media (Table 4).

When assessing the biosorption of the test metal ions by biomass (Figure 7), it has been found that V^{5+} was more active on membrane wall implying high adsorption by *Marinobacter* sp. MWI-1 compared to Ni^{2+}. Shifts in bands and peaks have also been reported by Lameiras et al. [41] and Tunali et al. [42] and these was reported to be indicative of bondage of functional groups (e.g. hydroxyl group and –NH stretching peak), indicative of Stretching of functional groups (C-N and aromatic-CH) and elucidating metal biosorption.

During the course of the study, a multivariate statistical analysis was performed supposing that the observations were dependent between each other. For each parameter, data analysis was repeated assuming the observations are independent and the results were the same as for the dependent case, with differences only in the width of the confidence interval.

In conclusion, the MWI-1 isolate (*Marinobacter* sp. MWI-1 [AB793286]), closely related to *Marinobacter goseongensis*, demonstrates high tolerance to both V^{5+} and Ni^{2+}. This indicates that mine water is a reservoir of novel microbial species which can sufficiently be used for the removal of heavy metals in highly polluted effluents. The tolerance and removal ability of the isolate (*Marinobacter* sp. MWI-1) was found to be pH and temperature dependent with 7-8 pH values and 30°C as an optimum, respectively. Further studies on geochemistry and microbial diversity need to be conducted to unveil how the chemistry of the effluent from the vanadium mine in South Africa can affect the microbial diversity of the environment. Furthermore, studies carried out on microbial diversity from extreme environments such as mine water are needed in order to isolate novel hypertolerant microbial species for heavy metal removal. The results in the present study constitute one of the first regarding the bacterial diversity of mine water samples from vanadium mine in South Africa.

Acknowledgments

The authors are grateful to the South African mining industries for allowing the researchers to use their mine water samples and to Princess Ramokolo for assisting with statistical analysis.

Author Contributions

Conceived and designed the experiments: IK. Performed the experiments: IK. Analyzed the data: IK MNBM. Contributed reagents/materials/analysis tools: IK MNBM. Wrote the paper: IK MNBM. Obtained permission for sample collection: MNBM. Obtained permission for use of FT-IR: IK. Statistical analysis: IK.

References

1. Oelofse SHH (2009) Mine water pollution–acid mine decant, effluent and treatment: a consideration of key emerging issues that may impact the state of the environment. In: Krishna CS, editor. Mining: Environment and health concerns 1st edn. India: The Icfai University Press. pp. 84–91.

2. Christensen B, Laake M, Lien T (1996) Treatment of acid mine water by sulphate-reducing bacteria; results from a bench scale experiment. Water Res 30(7): 1617–1624.

3. Adler RA, Claassen M, Godfrey L, Turton AR (2007) Water, mining, and waste: an historical and economic perspective on conflict management in South Africa. Econ Peace Secur J 2(2): 33–44.

4. Zhou Z-J, Yin H-Q, Liu Y, Xie M, Qiu G-Z, et al. (2010) Diversity of microbial community at acid mine drainages from Dachang metals-rich mine, China. T Nonferr Metal Soc 20(6): 1097–1103.

5. Wang J, Yang D, Zhang Y, Shen J, van der Gast C, et al. (2011) Do Patterns of bacterial diversity along salinity gradients differ from those observed for macroorganisms. PLOSOne 6(11): e27597.

6. Johnson DB, Hallberg KB (2003) The microbiology of acidic mine waters. Res Microbiol 154: 466−473.

7. Imarla T, Hector SB, Deane SM, Rawlings DE (2006) Resistance determinants of a highly arsenic-resistant strain of *Leptospirillum ferriphilum* isolated from a commercial biooxidationtank. Appl Environ Microbiol 72: 2247−2253.

8. Tekere M, Lotter A, Olivier J, Jonker N, Venter S (2011) Metagenomic analysis of bacterial diversity of Siloam hot water spring, Limpopo, South Africa. Afr J Biotechnol 10(78): 18005–18012.

9. Rantsiou K, Urso R, Lacumin L, Cantoni C, Cattaneo P, et al. (2005) Culture-dependent and -independent methods to investigate the microbial ecology of Italian fermented sausages. Appl Environ Microbiol 71(4): 1977–1986.

10. Handelsman J (2004) Metagenomics: Application of genomics to uncultured microorganisms. Microbiol Mol Biol Rev 68(4): 669–685.

11. Riesenfeld CS, Schloss PD, Handelsman J (2004) Metagenomics: genomic analysis of microbial communities. Annu Rev Genet 38: 525–552.

12. Oarga A (2009) Life in extreme environments. Revista de Biologia e ciencias da Terra 9(1): 1–10.

13. APHA (2001) *Standard methods for the examination of water and wastewater.* 20th ed. Washington DC: American Public Health Association (APHA).

14. DeSantis TZ, Brodie EL, Moberg JP, Zubieta IX, Piceno YM, et al. (2007) High-density universal 16S rRNA microarray analysis reveals broader diversity than typical clone library when sampling the environment. Microb. Ecol. 53: 371–383.

15. Huu NB, Denner EBM, Ha DTC, Wanner G, Stan-Lotter H (1999) Marinobacter aquaeolei sp. nov., a halophilic bacterium isolated from a Vietnamese oil-producing well. Int J Syst Bacteriol 49: 367–375.

16. Momba MNB, Cloete TE (1996) The relationship of biomass to phosphate uptake by *Acinetobacter junii* activated sludge mixed liquor. Water Res 30(2): 364–370.

17. Roh SW, Quan ZX, Nam YD, Chang HW, Kim KH, et al. (2008) Marinobacter goseongensis sp. nov., from seawater. Int J Syst Evol Microbiol 58(Pt 12): 2866–2870.

18. Kamika I, Momba MNB (2011) Comparing the tolerance limits of selected bacterial and protozoan species to nickel in wastewater systems. Sci Total Environ 410: 172–181.

19. Lyer A, Mody K, Bhavanath J (2004) Biosorption of heavy metals by a marine bacterium. Marine Pollut Bull 50: 340–343.

20. Madoni P, Davoli D, Gorbi G, Vescovi L (1996) Toxic effect of heavy metals on the activated sludge protozoan community. Water Res 30(1): 135–141.

21. Malik A, Jaiswal R (2000) Metal resistance in *Pseudomonas* strains isolated from soil treated with industrial wastewater. World J Microbiol Biotechnol 16: 177–182.

22. Shirdam R, Khanafari A, Tabatabaee A (2006) Cadmium, nickel and vanadium accumulation by three strains of marine bacteria. Iran J Biotechnol 4(3): 180–187.

23. Gray NF (2007) The use of an objective index for the assessment of the contamination of surface water and groundwater by acid mine drainage. Water Environ J 10(5): 332–340.

24. South Africa (1998) *National Water Act No 36 of 1998.* Government Gazette, 398(19182), Aug. 26.

25. Kuang J-L, Huang L-N, Chen L-X, Hua Z-S, Li S-J, et al. (2012) Contemporary environmental variation determines microbial diversity patterns in acid mine drainage. The ISME J. 139: 1–13.

26. Government Gazette (1984) Requirements for the purification of wastewater or effluent. Gazette No. 9225, Regulation, 991.

27. US-EPA (2006) Depleted Uranium Technical Brief. Office of Air and Radiation, Washington, EPA 402-R-06-011.

28. FAO (Food and Agriculture Organisation) (1985) *Water quality for agriculture.* Ayers ORS, Westcot DW. FAO Irrigation and Drainage Paper 29 (Rev 1), FAO, Rome, p. 174.

29. Kamika I, Momba MNB (2013) Assessing the resistance and bioremediation ability of selected bacterial and protozoan species to heavy metals in metal-rich industrial wastewater. MBC Microbiol 13: 28. doi:10.1186/1471-2180-13-28.

30. Aslam Z, Yasir M, Khaliq A, Matsui K, Chung YR (2010) Too much bacteria still unculturable. Crop Environ 1(1): 59–60.

31. Lozupone C, Knight R (2007) Global patterns in bacterial diversity. Proc Natl Acad Sci USA 104: 11436–11440.

32. He Z, Xie X, Xiao S, Li J, Qiu G (2007) Microbial diversity of mine water at Zhong TIaoshan copper mine, China. J Basic Microbiol 47: 485–495.

33. Raji AI, Moller C, Litthauer D, van Heerden E, Piater LA (2008) Bacterial diversity of biofilm samples from deep mines in South Africa. BIOKEMISTRI 20(2): 53–62.

34. Duxbury T (1981) Toxicity of heavy metals to soil bacteria. FEMS Microbiol 11: 217–220.

35. Naz N, Young HK, Ahmed N, Gadd GM (2005) Cadmium accumulation and DNA homology with metal resistance genes in sulfate-reducing bacteria. Appl Environ Microbiol 71(8): 4610–4618.

36. Zucconi L, Ripa C, Alianiello F, Benedetti A, Onofri S (2003) Lead resistance, sorption and accumulation in a *Paecilomyces lilacinus* strain. Biol Fert Soils 37: 17–22.

37. Guo B, Gu J, Ye Y-G, Tang Y-Q, Kida K, et al. (2007) Marinobacter segnicrescens sp. nov., a moderate halophile isolated from benthic sediment of the South China Sea. Int J Syst Evol Microbiol 57: 1970–1974.

38. Amin SA, Green DH, Al Waheed D, Gardes A, Carrano CJ (2012) Iron transport in the genus *Marinobacter*. Biometals 25(1): 135–147.

39. Singer E, Webb EA, Nelson WC, Heidelberg JF, Ivanova N, et al. (2011) Genomic potential of *Marinobacter aquaeolei*, a biogeochemical "Opportunitroph". Appl Environ Microbiol 77(8): 2763–2771.

40. Liao S, Zhou J, Wang H, Chen X, Wang H, et al. (2013) Arsenite oxidation using biogenic manganese oxides produced by deep-sea manganese-oxidizing bacterium *Marinobacter* sp. MnI7-9. Geomicrobiol J 30: 150–159.

41. Lameiras S, Quintelas C, Tavares T (2008) Biosorption of Cr (VI) using a bacterial biofilm supported on granular activated carbon and on zeolite. Bioresource Technology 99: 801–806.

42. Tunali SC, Abuk A, Akar T (2006) Removal of lead and copper ions from aqueous solutions by bacterial strain isolated from soil. Chem. Eng. J. 115: 203–211.

8

Genotoxicity of Heterocyclic PAHs in the Micronucleus Assay with the Fish Liver Cell Line RTL-W1

Markus Brinkmann[1]*, **Henning Blenkle**[1], **Helena Salowsky**[2], **Kerstin Bluhm**[1], **Sabrina Schiwy**[1], **Andreas Tiehm**[2], **Henner Hollert**[1]

1 Department of Ecosystem Analysis, Institute for Environmental Research, RWTH Aachen University, Aachen, Germany, 2 Department of Environmental Biotechnology, Water Technology Center, Karlsruhe, Germany

Abstract

Heterocyclic aromatic hydrocarbons are, together with their un-substituted analogues, widely distributed throughout all environmental compartments. While fate and effects of homocyclic PAHs are well-understood, there are still data gaps concerning the ecotoxicology of heterocyclic PAHs: Only few publications are available investigating these substances using *in vitro* bioassays. Here, we present a study focusing on the identification and quantification of clastogenic and aneugenic effects in the micronucleus assay with the fish liver cell line RTL-W1 that was originally derived from rainbow trout (*Oncorhynchus mykiss*). Real concentrations of the test items after incubation without cells were determined to assess chemical losses due to, e.g., sorption or volatilization, by means of gas chromatography-mass spectrometry. We were able to show genotoxic effects for six compounds that have not been reported in vertebrate systems before. Out of the tested substances, 2,3-dimethylbenzofuran, benzothiophene, quinoline and 6-methylquinoline did not cause substantial induction of micronuclei in the cell line. Acridine caused the highest absolute induction. Carbazole, acridine and dibenzothiophene were the most potent substances compared with 4-nitroquinoline oxide, a well characterized genotoxicant with high potency used as standard. Dibenzofuran was positive in our investigation and tested negative before in a mammalian system. Chemical losses during incubation ranged from 29.3% (acridine) to 91.7% (benzofuran) and may be a confounding factor in studies without chemical analyses, leading to an underestimation of the real potency. The relative potency of the investigated substances was high compared with their un-substituted PAH analogues, only the latter being typically monitored as priority or indicator pollutants. Hetero-PAHs are widely distributed in the environment and even more mobile, e.g. in ground water, than homocyclic PAHs due to the higher water solubility. We conclude that this substance class poses a high risk to water quality and should be included in international monitoring programs.

Editor: Aditya Bhushan Pant, Indian Institute of Toxicology Reserach, India

Funding: The authors acknowledge financial support by a Seed funds project of RWTH Aachen University supported by the German Excellence Initiative. M.B. received a personal stipend from the German National Academic Foundation ("Studienstiftung des deutschen Volkes"). The funders had no role in study design, data collection and analysis, decision to publish, or preparation of the manuscript.

Competing Interests: The authors have declared that no competing interests exist.

* E-mail: markus.brinkmann@bio5.rwth-aachen.de

Introduction

One of the key processes that built the foundation of the organic chemical industry in Germany and many other European countries in the 19th century was the distillation of coal tar-oil, by which common building-blocks for numerous syntheses were derived, e.g. for textile dyes [1]. However, tar-oil and coal tar can contain up to 85% polycyclic aromatic hydrocarbons (PAHs) and 5–13% heterocyclic aromatic hydrocarbons containing nitrogen, oxygen or sulphur (hetero-PAHs) [2,3], while the latter can constitute 40% of the water-soluble fraction. In industrial areas, e.g. gas plants, coke manufacturing or wood preservation sites, long ground water plumes contaminated with hetero-PAHs have been detected [4,5], potentially also endangering drinking water resources [6] and aquatic ecosystems.

While PAHs were already extensively investigated with regard to fate, biodegradation, toxicology and ecotoxicology [7,8,9], only limited knowledge exists for the hetero-PAHs. Many studies focused on few individual compounds and toxicological effects, e.g. mutagenicity of carbazoles [10] or chromosome aberrations

induced by quinolines [11]. Researchers just recently began to conduct comparative studies investigating a range of hetero-PAHs using different mechanism-specific *in vitro* bioassays. The biological effects so far cover, e.g. toxicity to *Daphnia* and green algae, mutagenicity in the Ames assay, embryotoxicity to embryos of the zebrafish (*Danio rerio*) or effects mediated by the aryl hydrocarbon and retinoid receptor, respectively [12,13,14,15,16,17,18].

To further fill the data gap mentioned above, we report on the first comparative study in which the clastogenic and aneugenic effects of heterocyclic PAHs typically found at tar-oil contaminated sites were investigated using the micronucleus assay with the permanent rainbow trout liver cell line RTL-W1 [19]. It is one of the best characterized cell lines derived from fish for mechanism-specific biotests [20] and was extensively used in genotoxicity studies [21,22,23,24]. The *in vitro* micronucleus assay was validated [25] and standardized by an international guideline [26] with the Chinese hamster lung fibroblast cell line V79. Furthermore, it is well-accepted that a substance's potential to induce micronuclei represents an important toxicological effect with potential adverse effects on the population level [21,27,28]. The substances tested were chosen to match the set of compounds identified earlier by the project framework KORA (retention and degradation processes to reduce contaminations in groundwater and soil) [12,29] and comprised indole, 1-benzothiophene, benzofuran, 2-methyl benzofuran, 2,3-dimethyl benzofuran, quinoline, 6-methyl quinoline, carbazole, dibenzothiophene, dibenzofuran, acridine, and xanthene. To account for the dissipation of the compounds from the test medium, changes in chemical concentration during incubation were measured by means of gas chromatography-mass spectrometry (GC-MS) analyses [13].

Materials and Methods

Chemicals

Stock solutions of the used heterocyclic PAHs were prepared in dimethyl sulfoxide (DMSO). Indole (>99%), quinoline (>98%), carbazole (approx. 95%), 6-methylquinoline (>98%), benzothiophene (>98%), dibenzothiophene (>98%) were supplied by abcr (Karlsruhe, Germany). Acridine (>98%) was purchased from Merck (Darmstadt, Germany). Xanthene (99%), benzofuran (>99%), 2-methylbenzofuran (≥96%), 2,3-dimethylbenzofuran (≥97%) and dibenzofuran (approx. 98%) were supplied by Sigma-Aldrich (Deisenhofen, Germany).

Micronucleus Assay with RTL-W1 Cells

The protocols recently published by Rocha et al. [30] for cell culture and the micronucleus assay with RTL-W1 cells were followed, with slight modifications. Cells were originally derived from rainbow trout liver [19] and generously provided by Drs Niels C Bols and Lucy Lee (University of Waterloo, Canada). Cells were cultured at 20°C in Leibovitz L15 medium with L-glutamine (Sigma–Aldrich) containing 9% fetal bovine serum (FBS, Biochrom, Berlin, Germany) and 1% (v/v) penicillin/streptomycin solution (Biochrom) according to Klee et al. [31]. Before use in the micronucleus assay, cells were rinsed twice with phosphate buffered saline (PBS, Sigma-Aldrich) and suspended following trypsinisation [32]. Cells of passage number 83 were used for the experiments.

A volume of 2 ml of the cell suspension at a density of 5–6 10^4 cells/ml was seeded onto ethanol pre-cleaned microscopic glass cover slips in 6-well plates (TPP, Trassadingen, Switzerland) and incubated for 12 h at 20°C (resulting in approx. 6–7 10^3 cells/cm^2) in a cooling incubator (Binder, Tuttlingen, Germany). Subsequently, the medium was aspirated and completely exchanged with dilutions of the investigated hetero-PAHs and the plates incubated for 20 h at 20°C. For each substance, a serial dilution of each stock solution (1:2) comprising five concentrations was tested in duplicate (i.e. on two different slides), while the highest concentration was equal to the NR$_{80}$ of the substance, i.e. the concentration at which 80% viability of RTL-W1 cells was measured in the neutral red retention assay (Table 1). The maximum concentration of DMSO in the test was 1%. The exposure medium was aspired and completely exchanged with fresh L15 medium and the plates incubated for 72 h at 20°C to give cells enough time to divide at least once [33]. Subsequently, cells were fixed for 10 min in a PBS-diluted (1:1 v/v) mixture of methanol and glacial acetic acid (4:1 v/v). Fixation was repeated for 10 min in the undiluted mixture. After air-drying, the cover slips were mounted onto glass slides using DePeX (Serva, Heidelberg, Germany). Acridine orange was used for staining of the slides after fixation [34]. A total number of 2000 cells per slide were analyzed under an epifluorescence microscope (Nikon Instruments, Düsseldorf, Germany) with oil-immersion at 1000× magnification. The scoring criteria of the ISO guideline 21427-2

were used: (a) only cells with intact cellular structure were read, micronuclei shall have (b) the same staining intensity as and (c) a maximum size of about 30% of the main nucleus. Furthermore, cells must be (d) clearly separated from the nucleus [26]. A representative photomicrograph of a cell with micronucleus is shown in Figure 1.

Calculation of EC and REP Values

Induction factors (fold-changes) relative to blanks, i.e. controls without treatment containing only the Leibovitz L15 medium, were calculated for each concentration by dividing the micronucleus rate of the respective concentration level by the mean of the blank replicates. Resulting concentration-response curves for individual substances and the well-characterized standard substance NQO (4-nitroquinoline oxide) were calculated with the software GraphPad Prism 5 (GraphPad, San Diego, CA, USA) using the four-parameter logistic equation model following log-transformation of the concentration values. Since a number of samples did not exceed 50% effect of the maximum NQO concentration, fixed-effect-level based REP (relative potency) values (Equation 1) were calculated according to Brack et al. [35]. Unlike in receptor-mediated assays, these REP values cannot be used for mass-balance analyses (i.e. to answer which portion of a measured effect is caused by which compound classes or single compounds) due to the different possible modes of action and are just intended for reference and comparison among samples.

$$REP = \frac{EC_{25,NQO}(NQO)}{EC_{25,NQO}(chemical)} \quad (1)$$

Effect concentrations (EC$_{25,NQO}$) refer to the concentration of the substance causing 25% of the maximum effect level of NQO. The arbitrary level of 25% effect was chosen since it includes more valid curves than using the EC$_{50}$ but is still within the linear portion of the concentration-response curves. The EC$_{25}$ value for NQO, EC$_{25,NQO}$ (NQO), was derived from the same test repetition of the micronucleus assay.

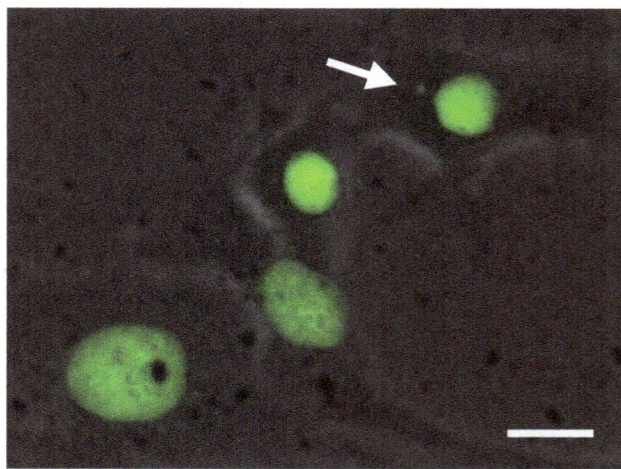

Figure 1. Composite photomicrograph of a micronucleus in RTL-W1 cells directly after cytokinesis (arrow). Nuclei and micronuclei were stained using Acridine Orange dye. The micrograph was captured at 1000× magnification and is a composite of bright-field and epifluorescence microscopy. Scale bar = 5 µm.

Table 1. Corrected and uncorrected REPs relative to NQO, as well as respective corrected and uncorrected EC$_{25}$ values in the micronucleus assay with RTL-W1 cells.

Substance	Maximum concentration[1] (mg L^{-1})	Uncorrected EC$_{25}$ (mg L^{-1})	Uncorrected REP	Corrected EC$_{25}$ (mg L^{-1})	Corrected REP	Genotoxicity in mammalian models	LOQ (µg L^{-1})	Loss (%)
4-nitroquinoline oxide (STD)	0.19	$20.44 \cdot 10^{-3}$	1					
Benzofuran	120.9	41.7	$4.9 \cdot 10^{-4}$	3.5	$5.90 \cdot 10^{-3}$	+ [51]	0.2	91.7
2-Methylbenzofuran	210.0	51.2	$4.0 \cdot 10^{-4}$	8.8	$2.33 \cdot 10^{-3}$	n.a.	0.2	82.9
2,3-Dimethylbenzofuran	110.0	n.d.	n.d.	n.d.	n.d.	n.a.	0.1	51.8
Dibenzofuran	56.9	21.8	$9.4 \cdot 10^{-4}$	5.0	$4.12 \cdot 10^{-3}$	− [52]	0.1	77.2
Benzothiophene	136.1	n.d.	n.d.	n.d.	n.d.	n.a.	0.2	75.0
Dibenzothiophene	40.8	10.8	$1.9 \cdot 10^{-3}$	3.2	$6.45 \cdot 10^{-3}$	n.a.	0.3	70.6
Acridine	40.9	10.3	$2.0 \cdot 10^{-3}$	7.3	$2.82 \cdot 10^{-3}$	+ [60,61]	0.3	29.3
Xanthene	75.5	47.9	$4.3 \cdot 10^{-4}$	8.4	$2.44 \cdot 10^{-3}$	n.a.	0.1	82.5
Carbazole	25.6	7.5	$2.7 \cdot 10^{-3}$	3.5	$5.81 \cdot 10^{-3}$	+ [48]	0.2	53.1
Indole	162.7	55.5	$3.7 \cdot 10^{-4}$	11.6	$1.76 \cdot 10^{-3}$	n.a.	0.1	79.1
Quinoline	226.0	n.d.	n.d.	n.d.	n.d.	+ [11,38,39,40,41,42,43]	0.3	61.7
6-Methylquinoline	528.0	n.d.	n.d.	n.d.	n.d.	− [62]	0.3	41.1

Chemical losses in the micronucleus assay used for calculation of corrected EC$_{25}$ (i.e. by multiplying the residual compound fraction with the EC$_{25}$) and REP values were derived from GC-MS measurements in 6-well microplates without cells.

n.d.: Inactive in assay system, i.e. substances which did not reach 25% induction of the NQO standard; n.a.: not available; STD: standard substance; LOQ: limit of quantification.

[1]Maximun tested concentrations based on cytotoxicity data from Hinger & Brinkmann et al. [13].

Chemical Analysis

To account for the effects of e.g. volatilization or sorption to plastic plates, real concentrations were compared to nominal concentrations comparable to the methodology recently published in Hinger & Brinkmann et al. [13] and Peddinghaus et al. [17]. Here, 6-well microplates were prepared in the same way as for the micronucleus assay without adding cells. Before addition to the plate, and after a 20 h incubation period in the microplate, solutions were stored in a glass vial with PTFE cap. The heterocyclic PAHs were extracted (liquid – liquid extraction) with methyl *tert*-butyl ether (MTBE). For this purpose 45 mL of the diluted sample were spiked with 10 μL internal standard solution in acetone (0.63 μg/μl toluene d_8 and 0.66 μg/μl naphthalene d_8, both obtained from Merck, Darmstadt, Germany) and extracted with 5 mL MTBE. Extraction time was 20 minutes. After phase separation, the extract was dried with sodium sulphate and subsequently analysed using gas chromatography (Agilent technologies GC 6890 N). The GC was equipped with an autosampler (Agilent technologies) and mass selective detector (MSD) (Agilent technologies MS 5973 Network) operated in SIM (Selective Ion Monitoring) mode. For Separation of the substances, a ZB-5 Inferno column (60 m×0.25 mm×0.25 μm) by Phenomenex was used. The concentrations of the internal standard in samples and external standards were equal. The limit of detection was 0.1 to 0.3 μg L^{-1} for the investigated substances. To correct the bioassay data for the measured chemical losses, EC_{25} values were multiplied with the remaining compound fraction and the REPs re-calculated with the corrected EC_{25}.

Statistical Analysis

All spreadsheet calculations were performed using Microsoft Excel^{TM} 2007. Statistical analyses were conducted with Sigma Stat 3.11 (Systat Software, Erkrath, Germany). All tested treatments and levels were tested for statistically significant differences from the blanks, i.e. controls without treatment, by use of one-way ANOVA ($p \leq 0.001$). Dunnett's test was used as the multiple range test to identify significant differences between treatments and blanks. The probability of Type I error (α) was set to $p \leq 0.05$. Values are expressed as mean value ± standard deviation, unless indicated.

Results and Discussion

Clastogenic and Aneugenic Effects of Heterocyclic PAHs

The substances 2,3-dimethylbenzofuran, benzothiophene, quinoline and 6-methylquinoline did not cause significant induction of micronuclei in the permanent fish liver cell line RTL-W1, i.e. did not reach 25% induction of the nitroquinoline oxide (NQO) standard (Table 1). Acridine caused the highest absolute induction (approx. 3-fold compared to blanks). Carbazole, acridine and dibenzothiophene were the most potent substances. To be able to better compare the investigated substances, relative potency (REP) values were calculated, comparing the potency of the reference compound NQO with the potency of the test item. Figure 2 illustrates this data analysis approach. REP values (Figure 3) ranged from $3.7 \cdot 10^{-4}$ (indole) to $2.7 \cdot 10^{-3}$ (carbazole). The three substances with the highest potency were only approx. 500-fold less potent than 4-nitroquinoline oxide. Full concentration-response curves are given for reference (Figure 4). While concentrations of hetero-PAHs in contaminated aquifers are typically in the μg L^{-1} range, values up to the mg L^{-1} range (at which effects in the micronucleus assay were observed) have occasionally been detected at highly contaminated sites [29]. However, care should be taken when comparing *in vitro* results

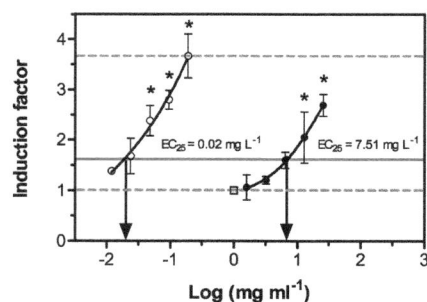

Figure 2. Exemplary concentration-response curve (carbazole) measured in the micronucleus assay with RTL-W1 cells (closed circles). The concentration-response curve for NQO (open circles) and the blanks, i.e. control cells without treatment (open square and lower dashed line), are given for reference. Induction factors are fold-changes relative to blanks. EC_{25} values relative to the maximum induction of NQO (upper dashed line) were calculated to derive fixed-effect-level-based REPs. Concentration values on the x-axis refer to nominal medium concentrations of the substances. Circles represent mean values measured in duplicate experiments, error bars the standard deviation. The red line depicts 25% of the maximum induction caused by the standard NQO. Asterisks denote statistically significant differences compared to blanks (one-way ANOVA with Dunnet's test, $p \leq 0.05$).

with aqueous exposure concentrations, since substances with differing physicochemical properties can be absorbed and accumulated to a varying extent and through different tissues and organs [36].

Eisenträger et al. [12] found that out of the tested substances (similar to the set tested here) only quinoline, 6-methylquinoline and xanthene caused mutagenic effects in the Ames assay with *Salmonella typhimurium*, while these effects were only apparent in treatments with metabolic activation, i.e. when supplemented with rat liver S9. The same effects for the 6-methylated and the parent quinoline were found by Debnath et al. [37]. The latter substance was also shown to cause significant induction of liver micronuclei and chromosome aberrations in rats and the hamster lung fibroblast cell line CHL/IU [11,38,39,40] and is a potent hepatocarcinogen in mice and rats [41,42,43]. Indole was positive in the Ames assay [44]. Acridine was shown to be positive in the SOS chromotest with *Escherichia coli* K12 and a yeast-based reporter-gene assay in which DNA damage induces the expression of green fluorescent protein (GFP) via the RAD54 promoter [45]. It was not mutagenic in the Ames assay [12,46] but produced clastogenic effects detected with the yeast DEL assay without metabolic activation [47]. Carbazole was demonstrated to be clastogenic by Jha et al. [48] but was not mutagenic in the Ames assay [49] or carcinogenic in mice [43]. The O-heterocyclic compound benzofuran showed very interesting effects, although being negative in the Ames assay [12,50]; DNA damage, as measured with the Comet assay, as well as the formation of micronuclei was higher in female specimens, both *in vivo* in rat kidney and *in vitro* using primary rat and human kidney cultures [51]. Dibenzofuran, a substance that was previously reported not to be genotoxic in a mammalian system was identified to be potent genotixicant in the fish system [52].

The following general trends can be deduced: For heterocyclic three-ring PAHs with two benzene rings fused with one five-membered aromatic ring (dibenzofuran, dibenzothiophene, carbazole), the genotoxic potential decreased in the following order of hetero-atoms: nitrogen>sulphur>oxygen. A comparable trend was observed for the three-ring hetero-PAHs with central six-

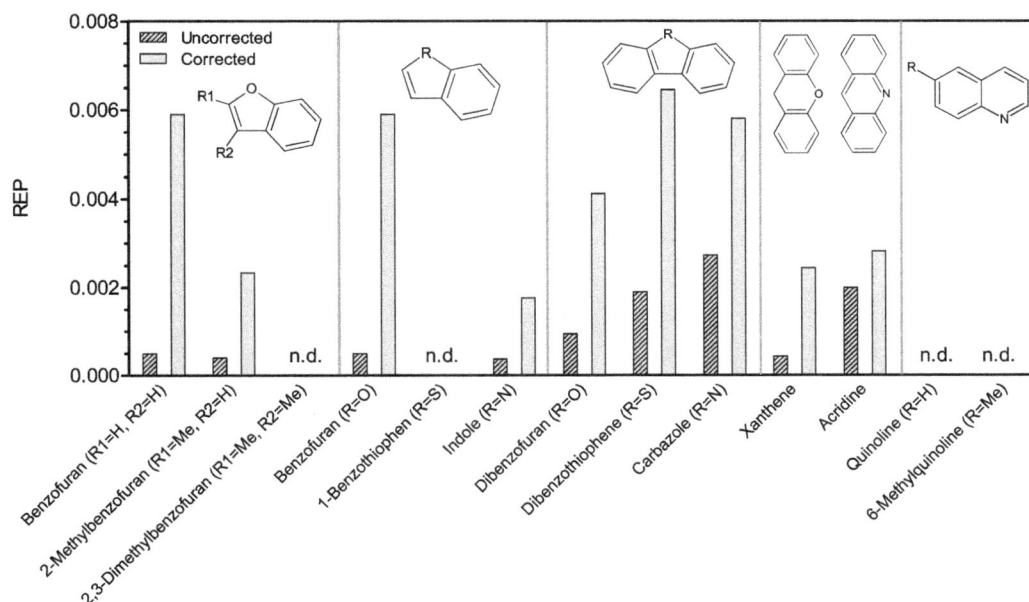

Figure 3. Uncorrected and corrected relative potency (REP) values of hetero-PAHs. Uncorrected values refer to nominal concentrations and corrected to measured concentrations in the exposure medium after incubation without cells. *N.d.: inactive in assay system, i.e. substances which did not reach the stipulated level of 25% induction of the NQO standard.*

membered ring, where the toxicity of acridine (nitrogen hetero-atom) was markedly higher than that of xanthene (oxygen hetero-atom). In accordance with the aforementioned studies, the azaarenes had the highest potency to induce genotoxic effects. Furthermore, for derivates of benzofuran, methylation had a decreasing effect on the genotoxicity (benzofuran >2-methylben-zofuran >2,3-dimethlybenzofuran).

Interestingly, the molar effect concentrations observed in the present study significantly correlated (Pearson's correlation coefficient $r = 0.88$, $p = 0.02$, data not shown) with the embryotoxic potential of hetero-PAHs as observed by Peddinghaus et al. [17]. In this study, acridine and carbazole caused the highest toxicity to embryos of *Danio rerio* and showed the highest potency to induce micronuclei in RTL-W1 cells in the present study. Furthermore, Hinger & Brinkmann et al. [13] have demonstrated that these two substances had the highest relative potency for luciferase induction in the DR-CALUX assay, an assay for aryl hydrocarbon receptor (AhR) agonists, and generally seem to be of high concern. The metabolic pathways of PAHs and hetero-PAHs were lately reviewed by Xue et al. [53] and need to be considered when investigating the genotoxic effects of these substances.

Influence of Substance Losses during Incubation

Chemical losses during incubation ranged from 29.3% (acri-dine) to 91.7% (benzofuran) and may lead to a drastic underestimation of the real potency of the substances, e.g. in mass-balance analyses (Table 1). Here, we attempted to correct for chemical losses from the exposure medium by correcting EC_{25} and REP values by applying analytically-derived correction factors. These correction factors were based on the compound concentration after incubation for 20 h (without cells but under the same conditions) and represent a worst-case scenario in which only the residual concentration after incubation caused the effect. The resulting REP values differed by a factor of 1.5–2 for substances with low losses during the incubation period (acridine and carbazole) compared to the uncorrected REP values. For

benzofuran, the substance with the highest loss (91.7%), the REPs differed by a factor of 12 compared with the uncorrected value (Figure 3). As it has been shown by other studies from the KORA project framework, accounting for the chemical losses is a prerequisite to adequately judge the real toxicological potency [12,13,17]. Furthermore, experimental precautions are warranted when experimenting with these volatile genotoxicants.

Implications for Water Quality

The substances we investigated here showed a relatively high potency to induce micronuclei in the permanent cell line RTL-W1. We were able to show genotoxic effects for six compounds that have not been reported for vertebrate systems before. Dibenzofuran was positive in our investigation and tested negative before in a mammalian system. Homocyclic PAHs are commonly monitored as priority pollutants and are indicator substances for contamination with substances originating from historically contaminated sites, e.g. gas-plants, wood preservation or dyestuff industry. However, heterocyclic aromatic compounds are mark-edly more mobile than homocyclic PAHs, e.g. in ground water, due to higher water solubility. Because ground water wells are an important drinking water resource, we conclude that this substance class poses a high risk to human health and should be included in groundwater monitoring programs, e.g. according to the German LAWA-GFS (threshold values for groundwater) that are based on toxicological and ecotoxicological effect data [54] or the health-related indicator value (HRIV) introduced by the German Federal Environment Agency (UBA) [55]. The interna-tional NORMAN network, which was originally founded by the European Commission to set up a permanent network of reference laboratories and research groups, recently emphasized the importance of also screening for compounds that are not commonly part of monitoring programs [56]. In fluvial environ-ments, heterocyclic PAHs contribute to the overall genotoxicity. In studies applying the concept of effect-directed analysis (EDA) to sediment and suspended particulate matter samples, as well as soil

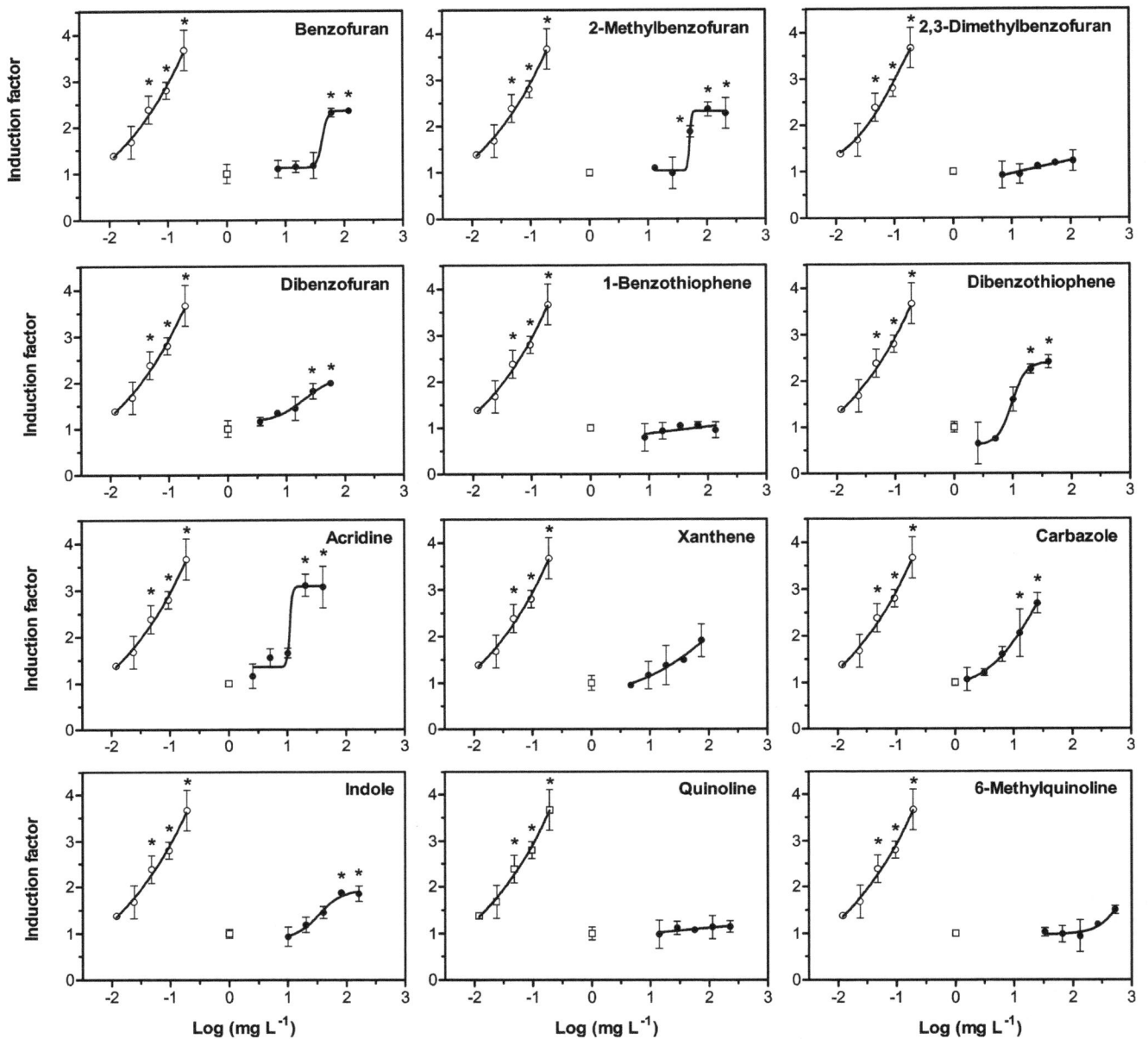

Figure 4. Dose-response curves in the micronucleus assay with RTL-W1 cells for all investigated heterocyclic compounds (closed circles). Standard curves for NQO (open circles) and blanks, i.e. control cells without treatment (open squares, grey line), are given for reference. Induction factors are fold-changes relative to blanks. Concentration values on the x-axis refer to nominal medium concentrations of the substances. Dots represent mean values measured in duplicate experiments, error bars the standard deviation. Asterisks denote statistically significant differences compared to blanks (one-way ANOVA with Dunnet's test, $p \leq 0.05$).

samples from related flood plains, those fractions containing heterocyclic substances showed strong mutagenic or genotoxic effects [57,58]. Re-suspension of contaminated sediments can ultimately lead to an increased bioavailability of such particle-bound pollutants with potentially adverse effects in aquatic biota [59].

Acknowledgments

The authors would like to express their thanks to Drs Niels C. Bols and Lucy Lee (University of Waterloo, Canada) for providing RTL-W1 cells. Furthermore, we want to express our gratitude to Simone Hotz for technical assistance during parts of the experiment.

Author Contributions

Conceived and designed the experiments: MB HH AT KB. Performed the experiments: HB HS. Analyzed the data: MB HB HS. Contributed reagents/materials/analysis tools: HH AT. Wrote the paper: MB KB SS HH AT.

References

1. Johnston W (2008) The discovery of aniline and the origin of the term "aniline dye". Biotech Histochem 83: 83–87.

2. Meyer S, Steinhart H (2000) Effects of heterocyclic PAHs (N, S, O) on the biodegradation of typical tar oil PAHs in a soil/compost mixture. Chemosphere 40: 359–367.

3. Dyreborg S, Arvin E, Broholm K (1997) Biodegradation of NSO-compounds under different redox-conditions. J Contam Hydrol 25: 177–197.

4. Blum P, Sagner A, Tiehm A, Martus P, Wendel T, et al. (2011) Importance of heterocyclic aromatic compounds in monitored natural attenuation for coal tar contaminated aquifers: A review. J Contam Hydr 126: 181–194.

5. Tiehm A, Müller J, Alt S, Jacob H, Schad H, et al. (2008) Development of a groundwater biobarrier for the removal of PAH, BTEX, and heterocyclic hydrocarbons. Water Sci Technol 58: 1349–1355.

6. Reineke A-K, Göen T, Preiss A, Hollender J (2007) Quinoline and Derivatives at a Tar Oil Contaminated Site: Hydroxylated Products as Indicator for Natural Attenuation? Environ Sci Technol 41: 5314–5322.

7. Douben PET, editor (2003) PAHs: An Ecotoxicological Perspective. Chichester-England: John Wiley & Sons Ltd. 36 p.

8. Tiehm A, Schulze S (2003) Intrinsic aromatic hydrocarbon biodegradation for groundwater remediation. Oil Gas Sci Technol 58: 449–462.

9. Schulze S, Tiehm A (2004) Assessment of microbial natural attenuation in groundwater polluted with gasworks residues. Water Sci Technol 50: 347–353.

10. Dutson SM, Booth GM, Schaalje GB, Castle RN, Seegmiller RE (1997) Comparative developmental dermal toxicity and mutagenicity of carbazole and benzo[a]carbazole. Environ Toxicol Chem 16: 2113–2117.

11. Asakura S, Sawada S, Sugihara T, Daimon H, Sagami F (1997) Quinoline-induced chromosome aberrations and sister chromatid exchanges in rat liver. Environ Mol Mutagen 30: 459–467.

12. Eisentraeger A, Brinkmann C, Hollert H, Sagner A, Tiehm A, et al. (2008) Heterocyclic compounds: Toxic effects using algae, daphnids, and the Salmonella/microsome test taking methodical quantitative aspects into account. Environ Toxicol Chem 27: 1590–1596.

13. Hinger G, Brinkmann M, Bluhm K, Sagner A, Takner H, et al. (2011) Some heterocyclic aromatic compounds are Ah receptor agonists in the DR-Calux and the EROD assay with RTL-W1 cells. Environ Sci Pollut Res 18: 1297–1304.

14. Sovadinová I, Bláha L, Janoscaronek J, Hilscherová K, Giesy JP, et al. (2006) Cytotoxicity and aryl hydrocarbon receptor-mediated activity of N-heterocyclic polycyclic aromatic hydrocarbons: Structure-activity relationships. Environ Toxicol Chem 25: 1291–1297.

15. Benisek M, Kubincova P, Blaha L, Hilscherova K (2011) The effects of PAHs and N-PAHs on retinoid signaling and Oct-4 expression in vitro. Toxicol Lett 200: 169–175.

16. Larsson M, Orbe D, Engwall M (2012) Exposure time–dependent effects on the relative potencies and additivity of PAHs in the Ah receptor-based H4IIE-luc bioassay. Environ Toxicol Chem 31: 1149–1157.

17. Peddinghaus S, Brinkmann M, Bluhm K, Sagner A, Hinger G, et al. (2012) Quantitative assessment of the embryotoxic potential of NSO-heterocyclic compounds using zebrafish (Danio rerio). Reprod Toxicol 33: 224–232.

18. Hawliczek A, Nota B, Cenijn P, Kamstra J, Pieterse B, et al. (2012) Developmental toxicity and endocrine disrupting potency of 4-azapyrene, benzo[b]fluorene and retene in the zebrafish Danio rerio. Reprod Toxicol 33: 213–223.

19. Lee LE, Clemons JH, Bechtel DG, Caldwell SJ, Han KB, et al. (1993) Development and characterization of a rainbow trout liver cell line expressing cytochrome P450-dependent monooxygenase activity. Cell Biol and Toxicol 9: 279–294.

20. Hallare A, Seiler T-B, Hollert H (2011) The versatile, changing, and advancing roles of fish in sediment toxicity assessment–a review. J Soils Sediments 11: 141–173.

21. Boettcher M, Grund S, Keiter S, Kosmehl T, Reifferscheid G, et al. (2010) Comparison of in vitro and in situ genotoxicity in the Danube River by means of the comet assay and the micronucleus test. Mut Res 700: 11–17.

22. Boettcher M, Kosmehl T, Braunbeck T (2011) Low-dose effects and biphasic effect profiles: Is trenbolone a genotoxicant? Mut Res 723: 152–157.

23. Kosmehl T, Hallare AV, Braunbeck T, Hollert H (2008) DNA damage induced by genotoxicants in zebrafish (Danio rerio) embryos after contact exposure to freeze-dried sediment and sediment extracts from Laguna Lake (The Philippines) as measured by the comet assay. Mut Res 650: 1–14.

24. Seitz N, Böttcher M, Keiter S, Kosmehl T, Manz W, et al. (2008) A novel statistical approach for the evaluation of comet assay data. Mut Res 652: 38–45.

25. Reifferscheid G, Ziemann C, Fieblinger D, Dill F, Gminski R, et al. (2008) Measurement of genotoxicity in wastewater samples with the in vitro micronucleus test–Results of a round-robin study in the context of standardisation according to ISO. Mut Res 649: 15–27.

26. ISO 21427-2:2006 (2006) Water quality - Evaluation of genotoxicity by measurement of the induction of micronuclei - Part 2: Mixed population method using the cell line V79. International Organization of Standardization.

27. Diekmann M, Hultsch V, Nagel R (2004) On the relevance of genotoxicity for fish populations I: effects of a model genotoxicant on zebrafish (Danio rerio) in a complete life-cycle test. Aquat Toxicol 68: 13–26.

28. Diekmann M, Waldmann P, Schnurstein A, Grummt T, Braunbeck T, et al. (2004) On the relevance of genotoxicity for fish populations II: genotoxic effects in zebrafish (Danio rerio) exposed to 4-nitroquinoline-1-oxide in a complete life-cycle test. Aquat Toxicol 68: 27–37.

29. Blotevogel J, Reineke AK, Hollender J, Held T (2008) Identification of NSO-heterocyclic priority substances for investigating and monitoring creosote-contaminated sites. Grundwasser 13: 147–157.

30. Rocha PS, Luvizotto GL, Kosmehl T, Böttcher M, Storch V, et al. (2009) Sediment genotoxicity in the Tietê River (São Paulo, Brazil): In vitro comet assay versus in situ micronucleus assay studies. Ecotoxicol Environ Saf 72: 1842–1848.

31. Klee N, Gustavsson L, Kosmehl T, Engwall M, Erdinger L, et al. (2004) Toxicity and genotoxicity in an industrial sewage sludge containing nitro- and amino-aromatic compounds during treatment in bioreactors under different oxygen regimes. Environ Sci Pollut Res 11: 313–320.

32. Kosmehl T, Krebs F, Manz W, Erdinger L, Braunbeck T, et al. (2004) Comparative genotoxicity testing of rhine river sediment extracts using the comet assay with permanent fish cell lines (rtg-2 and rtl-w1) and the ames test. J Soils Sediments 4: 84–94.

33. Schnurstein A, Braunbeck T (2001) Tail Moment versus Tail Length–Application of an In Vitro Version of the Comet Assay in Biomonitoring for Genotoxicity in Native Surface Waters Using Primary Hepatocytes and Gill Cells from Zebrafish (Danio rerio). Ecotoxicol Environ Saf 49: 187–196.

34. Jernbro S, Rocha PS, Keiter S, Skutlarek D, Färber H, et al. (2007) Perfluorooctane sulfonate increases the genotoxicity of cyclophosphamide in the micronucleus assay with V79 cells. Further proof of alterations in cell membrane properties caused by PFOS. Environ Sci Pollut Res Int 14: 85–87.

35. Brack W, Segner H, Möder M, Schüürmann G (2000) Fixed-effect-level toxicity equivalents-a suitable parameter for assessing ethoxyresorufin-O-deethylase induction potency in complex environmental samples. Environ Toxicol Chem 19: 2493–2501.

36. Mackay D, Fraser A (2000) Bioaccumulation of persistent organic chemicals: mechanisms and models. Environ Pollut 110: 375–391.

37. Debnath AK, Decompadre RLL, Hansch C (1992) Mutagenicity of quinolines in Salmonella typhimurium TA100 - a QSAR study based on hydrophobicity and molecular-orbital determinants. Mut Res 280: 55–65.

38. Suzuki H, Takasawa H, Kobayashi K, Terashima Y, Shimada Y, et al. (2009) Evaluation of a liver micronucleus assay with 12 chemicals using young rats (II): a study by the Collaborative Study Group for the Micronucleus Test/Japanese Environmental Mutagen Society–Mammalian Mutagenicity Study Group. Mutagenesis 24: 9–16.

39. Suzuki T, Takeshita K, Saeki KI, Kadoi M, Hayashi M, et al. (2007) Clastogenicity of quinoline and monofluorinated quinolines in Chinese hamster lung cells. J Health Sci 53: 325–328.

40. Hakura A, Kadoi M, Suzuki T, Saeki KI (2007) Clastogenicity of quinoline derivatives in the liver micronucleus assay using rats and mice. J Health Sci 53: 470–474.

41. Hirao K, Shinohara Y, Tsuda H, Fukushima S, Takahashi M, et al. (1976) Carcinogenic Activity of Quinoline on Rat Liver. Cancer Res 36: 329–335.

42. La Voie EJ, Dolan S, Little P, Wang CX, Sugie S, et al. (1988) Carcinogenicity of quinoline, 4- and 8-methylquinoline and benzoquinolines in newborn mice and rats. Food Chem Toxicol 26: 625–629.

43. Weyand EH, Defauw J, McQueen CA, Meschter CL, Meegalla SK, et al. (1993) Bioassay of quinoline, 5-fluoroquinoline, carbazole, 9-methylcarbazole and 9-ethylcarbazole in newborn mice. Food Chem Toxicol 31: 707–715.

44. Ochiai M, Wakabayashi K, Sugimura T, Nagao M (1986) Mutagenicities of indole and 30 derivatives after nitrite treatment. Mutat Res 172: 189–197.

45. Bartoš T, Letzsch S, Škarek M, Flegrová Z, Čupr P, et al. (2006) GFP assay as a sensitive eukaryotic screening model to detect toxic and genotoxic activity of azaarenes. Environ Toxicol 21: 343–348.

46. Brown BR, Firth Iii WJ, Yielding LW (1980) Acridine structure correlated with mutagenic activity in salmonella. Mutat Res 72: 373–388.

47. Kirpnick Z, Homiski M, Rubitski E, Repnevskaya M, Howlett N, et al. (2005) Yeast DEL assay detects clastogens. Mutat Res 582: 116–134.

48. Jha AM, Singh AC, Bharti MK (2002) Clastogenicity of carbazole in mouse bone marrow cells in vivo. Mutat Res 521: 11–17.

49. LaVoie EJ, Briggs G, Bedenko V, Hoffmann D (1982) Mutagenicity of substituted carbazoles in Salmonella typhimurium. Mutat Res 101: 141–150.

50. Weill-Thevenet N, Buisson JP, Royer R, Hofnung M (1981) Mutagenic activity of benzofurans and naphthofurans in the Salmonella/microsome assay: 2-nitro-7-methoxynaphtho[2,1-b]furan (R7000), a new highly potent mutagenic agent. Mutat Res 88: 355–362.

51. Robbiano L, Baroni D, Carrozzino R, Mereto E, Brambilla G (2004) DNA damage and micronuclei induced in rat and human kidney cells by six chemicals carcinogenic to the rat kidney. Toxicology 204: 187–195.

52. Galloway SM, Armstrong MJ, Reuben C, Colman S, Brown B, et al. (1987) Chromosome aberrations and sister chromatid exchanges in chinese hamster ovary cells: Evaluations of 108 chemicals. Environ Mol Mutagen 10: 1–175.

53. Xue WL, Warshawsky D (2005) Metabolic activation of polycyclic and heterocyclic aromatic hydrocarbons and DNA damage: A review. Toxicol Appl Pharmacol 206: 73–93.

54. Frank D, Dieter H, Herrmann H, Konietzka R, Moll B, et al. (2010) Ableitung von Geringfügigkeitsschwellenwerten für das Grundwasser: NSO-Heterozyklen. Bund/Länder-Arbeitsgemeinschaft Wasser (LAWA).

55. Grummt T, Kuckelkorn J, Bahlmann A, Baumstark-Khan C, Brack W, et al. (2013) Tox-Box: securing drops of life - an enhanced health-related approach for risk assessment of drinking water in Germany. Environmental Sciences Europe 25: 27.

56. Brack W, Dulio V, Slobodnik J (2012) The NORMAN Network and its activities on emerging environmental substances with a focus on effect-directed analysis of complex environmental contamination. Environmental Sciences Europe 24: 29.

57. Higley E, Grund S, Jones PD, Schulze T, Seiler T-B, et al. (2012) Endocrine disrupting, mutagenic, and teratogenic effects of upper Danube River sediments using effect-directed analysis. Environ Toxicol Chem 31: 1053–1062.

58. Wölz J, Schulze T, Lübcke-von Varel U, Fleig M, Reifferscheid G, et al. (2011) Investigation on soil contamination at recently inundated and non-inundated sites. J Soils Sediments 11: 82–92.

59. Schüttrumpf H, Brinkmann M, Cofalla C, Frings R, Gerbersdorf S, et al. (2011) A new approach to investigate the interactions between sediment transport and ecotoxicological processes during flood events. Environmental Sciences Europe 23: 39.

60. Moir D, Poon R, Yagminas A, Park G, Viau A, et al. (1997) The subchronic toxicity of acridine in the rat. Journal of Environmental Science and Health Part B-Pesticides Food Contaminants and Agricultural Wastes 32: 545–564.

61. Phelps JB, Garriott ML, Hoffman WP (2002) A protocol for the in vitro micronucleus test: II. Contributions to the validation of a protocol suitable for regulatory submissions from an examination of 10 chemicals with different mechanisms of action and different levels of activity. Mut Res 521: 103–112.

62. Wild D, King MT, Gocke E, Eckhard K (1983) Study of artificial flavouring substances for mutagenicity in the Salmonella/microsome, BASC and micronucleus tests. Food Chem Toxicol 21: 707–719.

9

The Added Value of Water Footprint Assessment for National Water Policy: A Case Study for Morocco

Joep F. Schyns*, Arjen Y. Hoekstra

Twente Water Centre, University of Twente, Enschede, The Netherlands

Abstract

A Water Footprint Assessment is carried out for Morocco, mapping the water footprint of different activities at river basin and monthly scale, distinguishing between surface- and groundwater. The paper aims to demonstrate the added value of detailed analysis of the human water footprint within a country and thorough assessment of the virtual water flows leaving and entering a country for formulating national water policy. Green, blue and grey water footprint estimates and virtual water flows are mainly derived from a previous grid-based (5×5 arc minute) global study for the period 1996–2005. These estimates are placed in the context of monthly natural runoff and waste assimilation capacity per river basin derived from Moroccan data sources. The study finds that: (i) evaporation from storage reservoirs is the second largest form of blue water consumption in Morocco, after irrigated crop production; (ii) Morocco's water and land resources are mainly used to produce relatively low-value (in US$/m^3 and US$/ha) crops such as cereals, olives and almonds; (iii) most of the virtual water export from Morocco relates to the export of products with a relatively low economic water productivity (in US$/m^3); (iv) blue water scarcity on a monthly scale is severe in all river basins and pressure on groundwater resources by abstractions and nitrate pollution is considerable in most basins; (v) the estimated potential water savings by partial relocation of crops to basins where they consume less water and by reducing water footprints of crops down to benchmark levels are significant compared to demand reducing and supply increasing measures considered in Morocco's national water strategy.

Editor: Vanesa Magar, Centro de Investigacion Cientifica y Educacion Superior de Ensenada, Mexico

Funding: The research was funded by Deltares (http://www.deltares.nl/en). Karen Meijer (Deltares) had a role in study design. Wil van der Krogt (Deltares) had a role in data collection. The writing stage of the manuscript was funded by the Institute for Innovation and Governance Studies (IGS) of the University of Twente (http://www.utwente.nl/igs/). The funders had no role in data analysis, decision to publish, or preparation of the manuscript.

Competing Interests: The authors have declared that no competing interests exist.

* E-mail: j.f.schyns@utwente.nl

Introduction

Morocco is a semi-arid country in the Mediterranean facing water scarcity and deteriorating water quality. The limited water resources constrain the activities in different sectors of the economy of the country. Agriculture is the largest water consumer and withdrawals for irrigation peak in the dry period of the year, which contributes to low surface runoff and desiccation of streams. Currently, 130 reservoirs are in operation to deal with this mismatch in water demand and natural water supply and to serve for generation of hydroelectricity and flood control [1]. Groundwater resources also play an important role in the socio-economic development of the country, in particular by ensuring the water supply for rural communities [2]. However, a large part of the aquifers is being overexploited and suffer from deteriorating water quality by intrusion of salt water, caused by the overexploitation, and nitrates and pesticides that leach from croplands, caused by excessive use of fertilizers. Surface water downstream of some urban centres is also polluted, due to untreated wastewater discharges.

In 1995, the Moroccan Water Law (no. 10–95) came into force and introduced decentralized integrated water management and rationalisation of water use, including the user-pays and polluter-pays principles. It also dictates the development of national and river basin master plans [3], which are elaborated in accordance with the national water strategy. To cope with water scarcity and

pollution, the national water strategy includes action plans to reduce demand, increase supply and preserve and protect water resources [1]. It also proposes legal and institutional reforms for proper implementation and enforcement of these actions. Demand management focuses on improving the efficiency of irrigation and urban supply networks and pricing of water to rationalise its use. Plans to increase supply include the construction of more dams and a large North-South inter-basin water transfer, protection of existing hydraulic infrastructure, desalinization of sea water and reuse of treated wastewater.

Although the national water strategy considers options to reduce water demand in addition to options to increase supply, it does not include the global dimension of water by considering international virtual water trade, nor does it consider whether water resources are efficiently allocated based on physical and economic water productivities of crops (the main water consumers). Analysis of the water footprint of activities in Morocco and the virtual water trade balance of the country therefore might reveal new insights to alleviate water scarcity.

The concept of water footprint was introduced by Hoekstra [4]; this subsequently led to the development of Water Footprint Assessment as a distinct field of research and application [5,6]. The water footprint is an indicator of freshwater use that looks not only at direct water use of a consumer or producer, but also at the indirect water use. As such, it provides a link between human consumption and human appropriation of freshwater systems. Water Footprint Assessment refers to a variety of methods to quantify and map the water footprint of specific processes, products, producers or consumers, to assess the environmental, social and economic sustainability of water footprints at catchment or river basin level and to formulate and assess the effectiveness of strategies to reduce water footprints in prioritized locations. The water footprint of a product is the volume of freshwater used to produce the product, measured over the full supply chain [6]. Three different components of a water footprint are distinguished: green, blue and grey. The green water footprint is the volume of rainwater evaporated or incorporated into the product. Blue water refers to the volume of surface- or groundwater evaporated, incorporated into the product or returned to another catchment or the sea. The grey water footprint relates to pollution and is defined as the volume of freshwater that is required to assimilate the load of pollutants given natural background concentrations and existing ambient water quality standards [6]. The total freshwater volume consumed or polluted within the territory of a nation as a result of activities within the different sectors of the economy is called the water footprint of national production. International trade of products creates 'virtual water flows' leaving and entering a country. The virtual-water export from a nation refers to the water footprint of the products exported. The virtual-water import into a nation refers to the water footprint of the imported products.

Several authors have assessed the water footprint and virtual water trade balance of nations and regions and state the relevance of the tool for well-informed water policy on the national and river basin level [7–10]. In a case study for a Spanish region, Aldaya *et al.* [10] conclude that water footprint analyses can provide a transparent framework to identify potentially optimal alternatives for efficient water use at the catchment level and that this can be very useful to achieve an efficient allocation of water and economic resources in the region. Chahed *et al.* [8] state that integration of all water resources at the national scale, including the green water used in rain-fed agriculture and as part of the foodstuffs trade balance, is essential in facing the great challenges of food security in arid countries.

The objective of this study is to explore the added value of analysing the water footprint of activities in Morocco and the virtual water flows from and to Morocco in formulating national water policy. The study includes an assessment of the water footprint of activities in Morocco (at the river basin level on a monthly scale) and the virtual water trade balance of the country and, based on this, response options are formulated to reduce the water footprint within Morocco, alleviate water scarcity and

allocate water resources more efficiently. Results and conclusions from the Water Footprint Assessment are compared with the scope of analysis of, and action plans included in Morocco's national water strategy and river basin plans in order to address the added value of Water Footprint Assessment relative to these existing plans.

The water footprint of Morocco has not been assessed previously on the river basin level on a monthly scale. Morocco has been included in a number of global studies, but these studies did not analyse the spatial and temporal variability of the water footprint within the country [11–13]. Furthermore, this study is the first to include specific estimates of the evaporative losses from the irrigation supply network and from storage reservoirs as part of a comprehensive Water Footprint Assessment. Finally, it is new in providing quantitative estimates of the potential water savings by partial relocation of crop production to regions with lower water consumption per ton of crop by means of an optimization and by reducing water footprints of crops down to benchmark levels.

Several insights and response options emerged from the Water Footprint Assessment, which are currently not considered in the national water strategy of Morocco and the country's river basin plans. Therefore, Water Footprint Assessment is considered to have an added value for formulating national water policy in Morocco.

Method and Data

Water Footprint of Morocco's Production

This study follows the terminology and methodology developed by Hoekstra *et al.* [6]. The water footprint of Morocco's production is estimated at river basin level on a monthly scale for the activities included in Table 1. The river basins are chosen such that they coincide with the action zones of Morocco's river basin agencies (Figure 1A). Due to data limitations, the grey water footprint is analysed on an annual scale and the water footprints of grazing and animal water supply are analysed at national and annual level. The study considers the average climate, production and trade conditions over the period 1996–2005. The water footprints of agriculture, industry and households are obtained from Mekonnen and Hoekstra [13,14], who estimated these parameters globally at a 5 by 5 arc minute spatial resolution. The annual blue water footprint estimates for industries and households by Mekonnen and Hoekstra [13] are distributed throughout the year according to the monthly distribution of public water supply obtained from Ministry EMWE (unpublished data 2013). These distributions are available for the basins Loukkos, Sebou,

Table 1. Water footprint estimates included in this study.

Water footprint of	Components	Period	Source
Crop production	Green, blue, grey	1996–2005	[14]
Grazing	Green	1996–2005	[13]
Animal water supply	Blue	1996–2005	[13]
Industrial production	Blue, grey	1996–2005	[13]
Domestic water supply	Blue, grey	1996–2005	[13]
Storage reservoirs	Blue	-	Own elaboration
Irrigation water supply network	Blue	1996–2005	Own elaboration

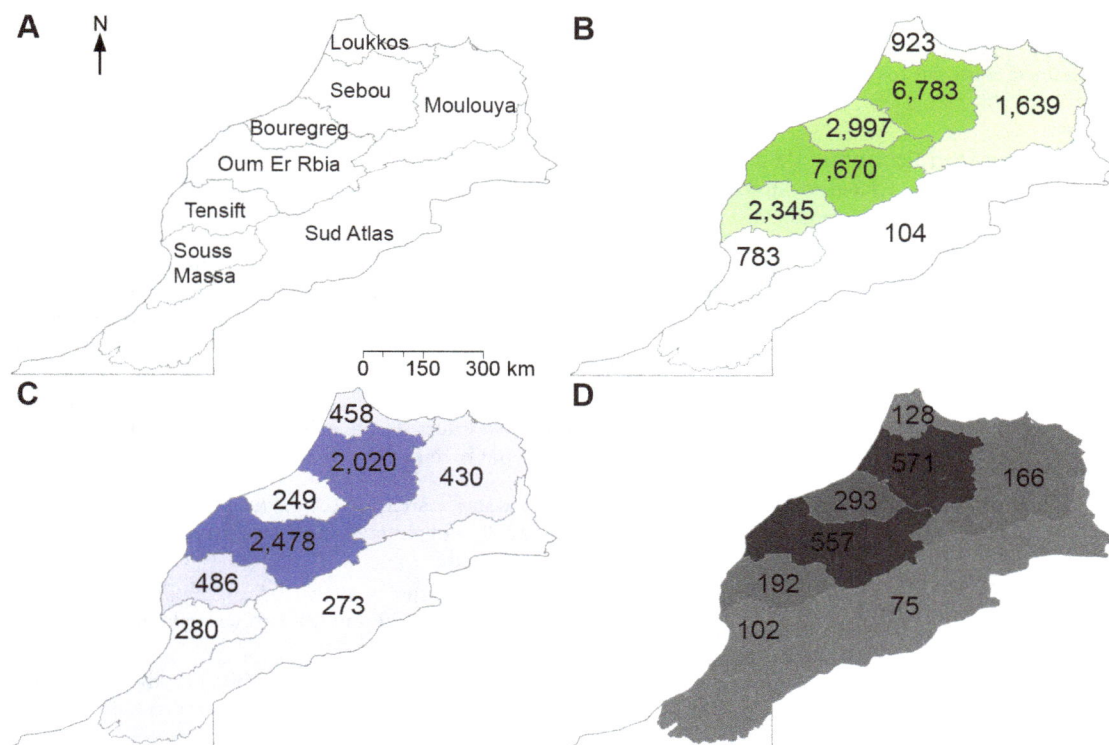

Figure 1. Water footprint of Morocco's production per river basin. Period: 1996–2005. Morocco's river basins (A) and total green (B), blue (C) and grey (D) water footprint of Morocco's production per river basin (in Mm³/yr).

Bouregreg and Oum Er Rbia. For the other basins an average of these distributions is taken.

The monthly water footprint of storage reservoirs (in m³/yr) is calculated as the open water evaporation (in m/yr) times the surface area of storage reservoirs (in m²). Data on open water evaporation from the reservoirs in the basins Loukkos, Sebou, Bouregreg and Oum Er Rbia is obtained from Ministry EMWE (unpublished data 2013) and for the other basins from a model simulation with the global hydrological model PCR-GLOBWB carried out by Sperna Weiland et al. [15]. The surface area of reservoirs at upper storage level is derived from Ministry EMWE (unpublished data 2013) and FAO [16]. Since storage levels vary throughout the year (and over the years), and reservoir areas accordingly, this gives an overestimation of the evaporation from reservoirs. To counteract this overestimation, but due to lack of data on monthly storage level and reservoir area, for all months a fraction of the evaporation at upper storage level (43%) is taken as estimate of the water footprint of storage reservoirs. This fraction represents the average reservoir area as fraction of its area at upper storage level, calculated as the average over the reservoirs in the basins Loukkos, Sebou, Bouregreg and Oum Er Rbia for which data on surface area at different reservoir levels is available from Ministry EMWE (unpublished data 2013).

The water footprint of the irrigation supply network refers to the evaporative loss in the network and is estimated based on a factor K, which is defined as the ratio of the blue water footprint of the irrigation supply network to the blue surface water footprint of crop production at field level (i.e. crop evapotranspiration of irrigation water stemming from surface water). The blue water footprint of crop production at field level is taken from Mekonnen and Hoekstra [14] and the split to surface water is made according

to the fraction of irrigation water withdrawn from surface water (as opposed to groundwater) per river basin based on data from the associated river basin plans. K is calculated as:

$$K = \left[\frac{1}{e_a \times e_c} - \frac{1}{e_a} \right] \times f_E$$

in which e_a represents the field application efficiency, e_c the irrigation canal efficiency and f_E the fraction of losses in the irrigation canal network that evaporates (as opposed to percolates). The irrigation efficiencies e_a and e_c are estimated based on data from a local river basin agency and FAO [17]. The value of f_E is assumed at fifty percent. The resultant K for Morocco's irrigated agriculture as a whole is 15%, i.e. the evaporative loss from the irrigation water supply network represents a volume equal to 15% of the blue surface water footprint of crop production at field level on average.

Water Footprint and Economic Water and Land Productivity of Crops

The water footprint of crops per unit of production (in m³/ton) is calculated by dividing the water footprint per hectare (in m³/ha/yr) by the yield (in ton/ha/yr), for which data are obtained from Mekonnen and Hoekstra [14]. Economic water productivity (in US$/m³) represents the economic value of farm output per unit of water consumed and is calculated as the average producer price for the period 1996–2005 (in US$/ton) obtained from FAO [18] divided by the green plus blue water footprint (in m³/ton). Similarly, economic land productivity (in US$/ha) represents the economic value of farm output per hectare of harvested land and is calculated as the same producer price multiplied by crop yield (in ton/ha), which is also obtained from Mekonnen and Hoekstra [14].

Virtual Water Flows and Associated Economic Value

Green, blue and grey virtual water flows related to Morocco's import and export of agricultural and industrial commodities for the period 1996–2005 are obtained from Mekonnen and Hoekstra [13], who estimated these flows at a global scale based on trade matrices and water footprints of traded products at the locations of origin. The virtual water export that originates from domestic water resources (another part is re-export) is estimated based on the relative share of the virtual water import to the total water budget:

$$V_{e,dom.res.} = \frac{WF_{national}}{V_i + WF_{national}} \times V_e$$

in which $WF_{national}$ is the water footprint within the nation, V_i the virtual water import and V_e the virtual water export.

The average earning per unit of water exported (in US\$/m³) is calculated by dividing the value of export (in US\$/yr) by virtual water export (in m³/yr). Similarly, the cost per unit of virtual water import is calculated by dividing the import value (in US\$/yr) by virtual water import (in m³/yr). The average economic value of import and export for the period 1996–2005 are derived from the Statistics for International Trade Analysis (SITA) database from the International Trade Centre [19].

Water Footprint versus Water Availability and Waste Assimilation Capacity

To assess the environmental sustainability of the water footprint within Morocco, the total blue (surface- plus groundwater) water footprint of production is placed in the context of monthly natural runoff and the groundwater footprint in the context of annual groundwater availability. The water needed to assimilate the nitrogen fertilizers that reach the water systems due to leaching is compared with the waste assimilation capacity of aquifers.

The groundwater footprint is calculated by splitting the blue water footprint of crop production, industrial production and domestic water supply according to the fraction withdrawn from groundwater per river basin based on data from the associated river basin plans. Assuming that none of the water abstracted from groundwater for industrial production and domestic water supply returns (clean) to the groundwater in the same period of time, the groundwater footprints of these purposes are increased to equal water withdrawal (as opposed to consumption) by dividing them by the consumptive fractions assumed by Mekonnen and Hoekstra [13]: 5% for industries and 10% for households.

Long-term average monthly natural runoff (1980–2011) for the river basins of Loukkos, Sebou, Bouregreg and Oum Er Rbia is derived from Ministry EMWE (unpublished data 2013). Natural runoff is estimated as the inflow of reservoirs. It is considered undepleted runoff, since large-scale blue water withdrawals come from the reservoirs. For the other basins, long-term average annual natural runoff is derived from the river basin plans for the respective river basins and subsequently distributed over the months according to intra-annual rainfall patterns [20,21] or monthly natural discharge [22]. Due to lack of data, for the Souss Massa basin the same monthly variation is applied as for the adjacent Tensift basin. Groundwater availability is assessed on river basin scale and defined as the recharge by percolation of

rainwater and from rivers, minus the direct evaporation from aquifers. These data are obtained from the river basin plans and from Laouina [23] for the basin of Souss Massa.

Blue water scarcity is defined as the ratio of the total blue water footprint in a catchment over the blue water availability in that catchment [6]. In this study, this ratio is calculated as the total blue water footprint to monthly natural runoff and as the groundwater footprint to annual groundwater availability. Following Hoekstra et al. [24], blue water scarcity values have been classified into four levels of water scarcity. The classification in this study corresponds with their classification, with the note that the current study does not account for environmental flow requirements in the definition of blue water availability, since they are generally not considered in Morocco's river basin plans and local studies on the level of these requirements are lacking. This is compensated for by using stricter threshold values for the different scarcity levels, so that the resultant scheme is equivalent to that of Hoekstra et al. [24]:

- low blue water scarcity (<0.20): the blue water footprint is lower than 20% of natural runoff; river runoff is unmodified or slightly modified.
- moderate blue water scarcity (0.20–0.30): the blue water footprint is between 20 and 30% of natural runoff; runoff is moderately modified.
- significant blue water scarcity (0.30–0.40): the blue water footprint is between 30 and 40% of natural runoff; runoff is significantly modified.
- severe water scarcity (>0.40): the monthly blue water footprint exceeds 40% of natural runoff, so runoff is seriously modified.

The water pollution level is defined as the total grey water footprint in a catchment divided by the waste assimilation capacity [6]. In other words, it shows the fraction of actual runoff that is required to dilute pollutants in order to meet ambient water quality standards. A water pollution level greater than 1 means that ambient water quality standards are violated. The nitrate-related grey water footprint of crop production as computed in this study is assumed to mostly contribute to groundwater pollution and is therefore compared with the waste assimilation capacity of groundwater. As a measure of the latter, we use the actual groundwater availability, calculated as (natural) groundwater availability minus the groundwater footprint.

Relocation of Crop Production and Reducing Water Footprints of Crops to Benchmark Levels

The potential water savings by changing the pattern of crop production across river basins (which is possible due to spatial differences in crop water use) are quantified by means of an optimization model. The total green plus blue water footprint of twelve main crops in the country (in Mm³/yr) is minimized by changing the spatial pattern of production (in ton/yr) over the river basins under constraints for production demand (in ton/yr) and land availability (in ha/yr). The analysed crops are: almonds, barley, dates, grapes, maize, olives, oranges, sugar beets, sugar cane, mandarins, tomatoes and wheat. Results are compared with a base case, which corresponds with the average green plus blue water footprint of the analysed crops over the period 1996–2005. Land availability is restricted per river basin and taken equal to the average harvested area in the period 1996–2005 obtained from Mekonnen and Hoekstra [14]. Two cases are distinguished: 1) all crops can be relocated; 2) only annual crops (barley, maize, sugar beets, tomatoes and wheat) can be relocated, perennials cannot.

For both cases, the restriction is imposed that the total national production per crop (in ton/yr) should be equal to (or greater than) the total national production of the crop in the base case, which is defined as the average production in the period 1996–2005 obtained from Mekonnen and Hoekstra [14].

Additionally, an assessment is made of the potential water savings by reducing the water footprints of the twelve main crops down to certain benchmark levels. For each basin and crop a benchmark is set in the form of the lowest water consumption (green plus blue) of that crop which is achieved in a comparable river basin in Morocco. In this case, basins are considered comparable when the reference evapotranspiration (ET$_0$ in mm/yr) is in the same order of magnitude (see Table 2). Reference evapotranspiration expresses the evaporating power of the atmosphere at a specific location (and time of the year) and does not consider crop characteristics and soil factors [6]. Differences in soil and development conditions are thus not accounted for.

Results

Water Footprint of Morocco's Production

The total water footprint of Morocco's production in the period 1996–2005 was 38.8 Gm^3/yr (77% green, 18% blue, 5% grey), see Table 3. Crop production is the largest contributor to this water footprint, accounting for 78% of all green water consumed, 83% of all blue water consumed (evaporative losses in irrigation water supply network included) and 66% of the total volume of polluted water. Evaporative losses from storage reservoirs are estimated at 884 Mm^3/yr, which is 13% of the total blue water footprint within Morocco. For most reservoirs, these losses are ultimately linked to irrigated agriculture and in some cases potable water supply.

Largest water footprints (green, blue and grey) are found in the basins Oum Er Rbia and Sebou, the basins containing the main agricultural areas of Morocco (see Figure 1B–D). Together, these two basins account for 63% of the total water footprint of national production. In general, the green water footprint is largest in the rainy period December–May, while the blue water footprint is largest in the period April–September when irrigation water use increases.

In the basins Bouregreg and Loukkos, evaporation from storage reservoirs accounts for 45% and 55% of the total blue water footprint, respectively. Irrigated agriculture is the largest blue water consumer in the other basins, but evaporation from storage reservoirs is also significant in these basins. Main irrigated crops in the Oum Er Rbia basin are maize, wheat, olives and sugar beets,

which together account for 60% of the total irrigation water consumed in the period 1996–2005. In the basin of Sebou, 56% of the blue water footprint of crop production relates to the irrigation of wheat, olives, sugar beets, sugar cane and sunflower seed.

Water Footprint and Economic Water and Land Productivity of Main Crops

In the period 1996–2005, most green water was consumed by the production of wheat, barley and olives (Figure 2). The largest blue water footprints relate to the production of wheat, olives and maize. For wheat, the number one blue water consuming crop, the blue water footprint was largest in the period March–May and peaked in April.

Water consumption of crops (green plus blue, in m^3/ton) varies significantly per river basin due to differences in climatic conditions. In general, water consumption of crops is above country-average in the basins Oum Er Rbia and Tensift and below country-average in the northern basins Bouregreg, Sebou, Loukkos and Moulouya (Figure 3). In the basins Sud Atlas and Souss Massa the picture is not so clear, with some crops having above and others below country-average water footprints (in m^3/ton).

The five crops that consumed the most green plus blue water in the period 1996–2005 are the crops with the lowest economic water productivity, ranging from 0.08 US\$/m^3 for wheat to only 0.02 US\$/m^3 for almonds (Figure 2). Production of tomatoes yielded 22 times more value per drop than production of wheat. The same five crops also have the lowest economic land productivity, ranging from 375 US\$/ha for olives to 112 US\$/ha for almonds (Figure 4). The highest value per hectare cultivated was obtained by production of tomatoes.

Virtual Water Trade Balance of Morocco

Morocco's virtual water trade balance for the period 1996–2005 is shown in Figure 5. Virtual water import exceeds virtual water export, which makes Morocco a net virtual water importer. Only 31% of the virtual water export originates from Morocco's water resources, the other 69% is related to re-export of imported virtual water. By import of products instead of producing them domestically, Morocco saved 27.8 Gm^3/yr (75% green, 21% blue and 4% grey) of domestic water in the period 1996–2005, equivalent to 72% of the water footprint within Morocco.

The value of the total virtual water imported in the period 1996–2005 was 12.4 billion US\$/yr. Import of industrial products accounted for 83%, import of crop products for 16% and import

Table 2. Comparison of river basins based on reference evapotranspiration (ET$_0$ in mm/yr, period: 1961–1990).

No.	River basin	ET$_0$ (mm/yr)	Considered comparable with no.
1	Sud Atlas	1,652	–
2	Souss Massa	1,450	3
3	Moulouya	1,409	2
4	Tensift	1,389	5
5	Oum Er Rbia	1,387	4
6	Sebou	1,266	7,8
7	Bouregreg	1,239	6,8
8	Loukkos	1,212	6,7

Source: ET$_0$ from FAO [31].

Table 3. Water footprint of Morocco's production in the period 1996–2005 (in Mm^3/yr).

Water footprint of	Green	Blue	Grey	Total
Crop production[a]	23,245	5,097	1,378	29,719
Grazing[a]	6,663	-	-	6,663
Animal water supply[a]	-	151	-	151
Industrial production[a]	-	18	69	88
Domestic water supply[b]	-	125	640	765
Storage reservoirs[b]	-	884	-	884
Irrigation water supply network[b]	-	549	-	549
Total water footprint	29,908	6,824	2,087	38,819

Source: [a] [13], [b] Own elaboration.

of animal products for 1%. The average cost of imported commodities per unit of virtual water imported was 0.98 US$/m^3. The value of the total virtual water exported in this period was 7.1 billion US$/yr (industrial products: 51%, crop products: 48%, animal products: 1%). The average earning of exported commodities per unit of virtual water exported was 1.66 US$/m^3.

The total volume of Morocco's water virtually exported out of the country (i.e. excluding re-export) in the period 1996–2005 is estimated at 1,333 Mm^3/yr. This means that about 4% of the water used in Morocco's agricultural and industrial sector is used for making export products. The remainder is used to produce products that are consumed by the inhabitants of Morocco. Virtual export of blue water from Morocco's resources was 435 Mm^3/yr, which is to equivalent 3.4% of long-term average natural runoff (13 Gm^3/yr).

Most of the virtual water export from Morocco's resources returns relatively little foreign currency per unit of virtual water exported. Export of crop products had the largest share in the virtual water export from Morocco's water resources (1,305 Mm^3/yr), returning 0.87 US$/m^3 on average. Specific crop products associated with large virtual water export from Moroccan origin are olives, oranges, wheat, sugar beets and mandarins. Out of these products, only export of mandarins (122 Mm^3/yr) returned a

value (1.37 US$/m^3) larger than the average for crop products (0.87 US$/m^3). On the other hand, virtual water export related to Moroccan tomatoes (24 Mm^3/yr) yielded 7.13 US$/m^3.

Water Footprint versus Water Availability and Waste Assimilation Capacity

Blue water scarcity manifests itself in specific months of the year (Figure 6; Table 4). The average monthly water scarcity indicates severe water scarcity, more severe than annual (total) water scarcity values suggest. In all basins, the total blue water footprint exceeds natural runoff during a significant period of the year. In the months June, July and August, severe water scarcity occurs in all river basins. Crops with a large blue water footprint in July are: sugar beets in Oum Er Rbia and Sebou; grapes in the basins of Sud Atlas, Souss Massa and Oum Er Rbia; dates in Oum Er Rbia and Sebou; sunflower seed in the Sebou basin; maize in the basin of Oum Er Rbia. Demand for potable water peaks in the months June, July and August due to tourism and evaporation from storage reservoirs is large in these months due to the strong evaporative power of the atmosphere. Annual runoff in the Oum Er Rbia basin is almost completely consumed (inter-basin water transfers not yet considered), which raises the question whether it

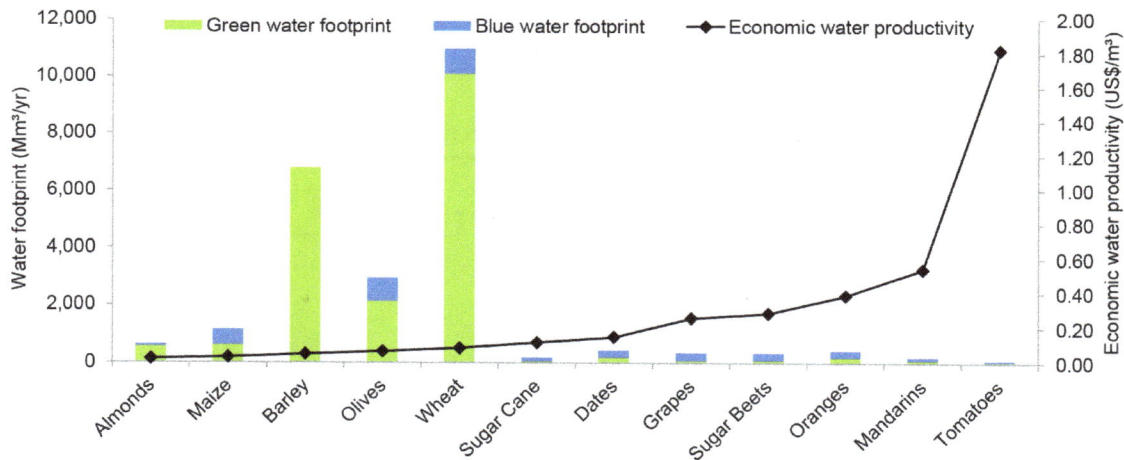

Figure 2. Economic water productivity and green and blue water footprint of main crops in Morocco. Period: 1996–2005. Source: Water footprint from Mekonnen and Hoekstra [14], producer prices from FAO [18].

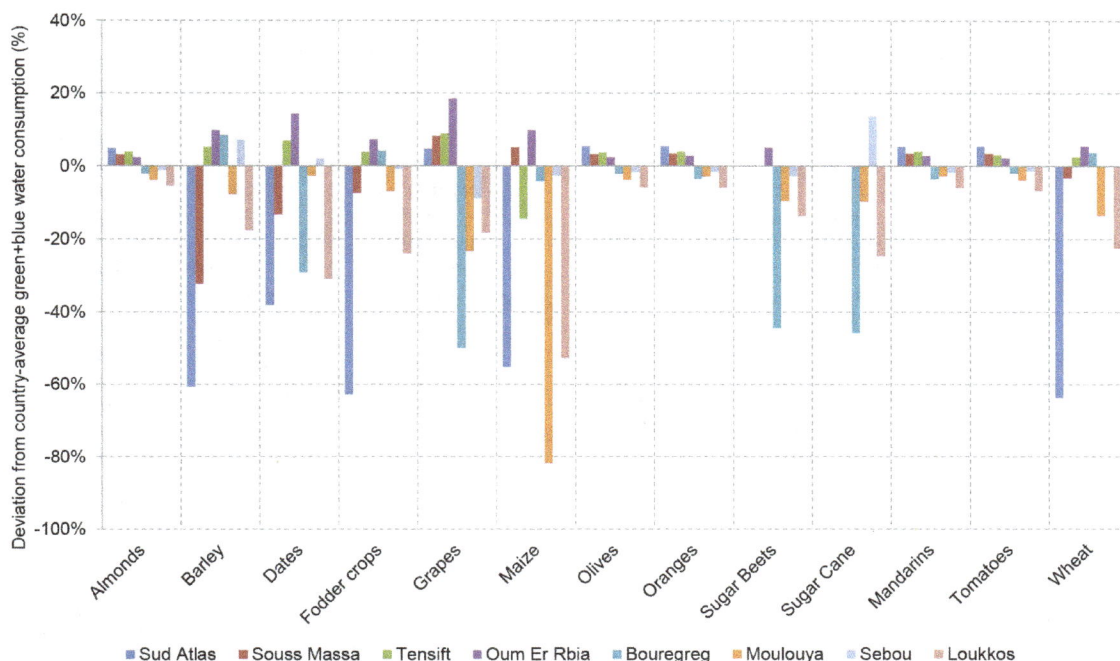

Figure 3. Variation in green plus blue water consumption (in m³/ton) across river basins. Period: 1996–2005.

is wise to export water out of this basin to the basins of Bouregreg and Tensift as is common practice.

The total groundwater footprint in Morocco constitutes about half of the country's groundwater availability (Table 5). Groundwater stress is severe in all river basins, except for the basins of Loukkos and Sud Atlas. In the Bouregreg basin, the annual groundwater footprint exceeds annual groundwater availability. As confirmed in the 2012 river basin plan for this basin, most of the aquifers in this basin are indeed overexploited, especially the main aquifers of Berrechid and Chaouia côtière.

In the Bouregreg basin there is no waste assimilation capacity of the groundwater left (because the blue groundwater footprint exceeds groundwater availability), which results in an infinite

water pollution level (Table 6). In the basins of Tensift and Oum Er Rbia, waste assimilation capacity of the groundwater is also exceeded, even by 43 times the natural groundwater availability in the Tensift basin. These findings correspond with figures reported in the river basin plans for these three basins, which indicate severely high nitrate concentrations in the groundwater (at some measurement stations exceeding the maximum permissible limit in drinking water), mainly caused by diffuse nitrate pollution by the irrational use of nitrogen fertilizers, but in the case of the Sahel-Doukkala aquifer in the Oum Er Rbia basin also by the infiltration of untreated domestic wastewater.

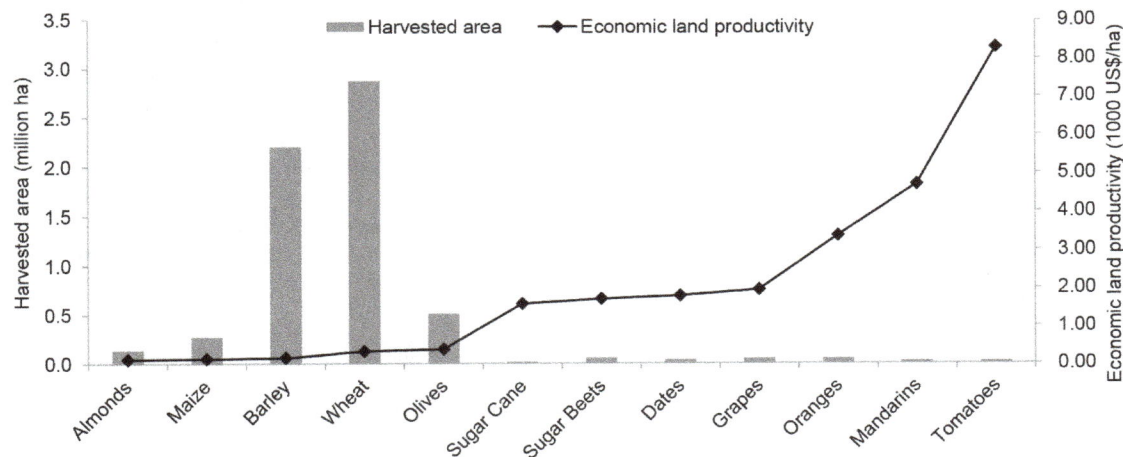

Figure 4. Economic land productivity and harvested area of main crops in Morocco. Period: 1996–2005. Source: Harvested area and yield from Mekonnen and Hoekstra [14], producer prices from FAO [18].

Morocco
Use of domestic resources in agricultural and industrial sector = 37 Gm³/yr
Natural runoff = 13 Gm³/yr

Main import partners:

USA (1,496 Mm³/yr)
wheat = 684
maize = 396
soybeans = 327

France (1,307 Mm³/yr)
wheat = 558
seed cotton = 358

Argentina (1,194 Mm³/yr)
soybeans = 532
maize = 334
wheat = 240

Brazil (1,140 Mm³/yr)
sugar cane = 493
soybeans = 465

Canada (862 Mm³/yr)
wheat = 759

Russia (611 Mm³/yr)
wheat = 371

China (544 Mm³/yr)
tea = 461

Ukraine (529 Mm³/yr)
wheat = 289

Main products (in Mm³/yr):
wheat = 3,657
seed cotton = 2,391
soybeans = 1,742
maize = 876
coffee, green = 749
sugar cane = 630
barley = 491
tea = 467
sunflower seed = 332
industrial products = 540
animal products = 152

Virtual water import = 12,643 Mm³/yr
■80% green ■9% blue ■11% grey

Main products (in Mm³/yr):
olives = 292
oranges = 181
wheat = 180
sugar beets = 129
mandarins = 122
industrial products = 28

Virtual water export = 1,333 Mm³/yr
■60% green ■33% blue ■8% grey

Virtual water re-export = 2,974 Mm³/yr
■80% green ■9% blue ■11% grey

Main products (in Mm³/yr):
seed cotton = 2,641
industrial products = 175

Main export partners:

Libya (210 Mm³/yr)
wheat = 192

Italy (239 Mm³/yr)
olives = 142

Russia (77 Mm³/yr)
oranges = 46
mandarins = 28

Spain (618 Mm³/yr)
olives = 66
seed cotton = 507

UK (565 Mm³/yr)
sugar beets = 42
oranges = 22
seed cotton = 485

France (1,056 Mm³/yr)
seed cotton = 922

Germany (356 Mm³/yr)
seed cotton = 313

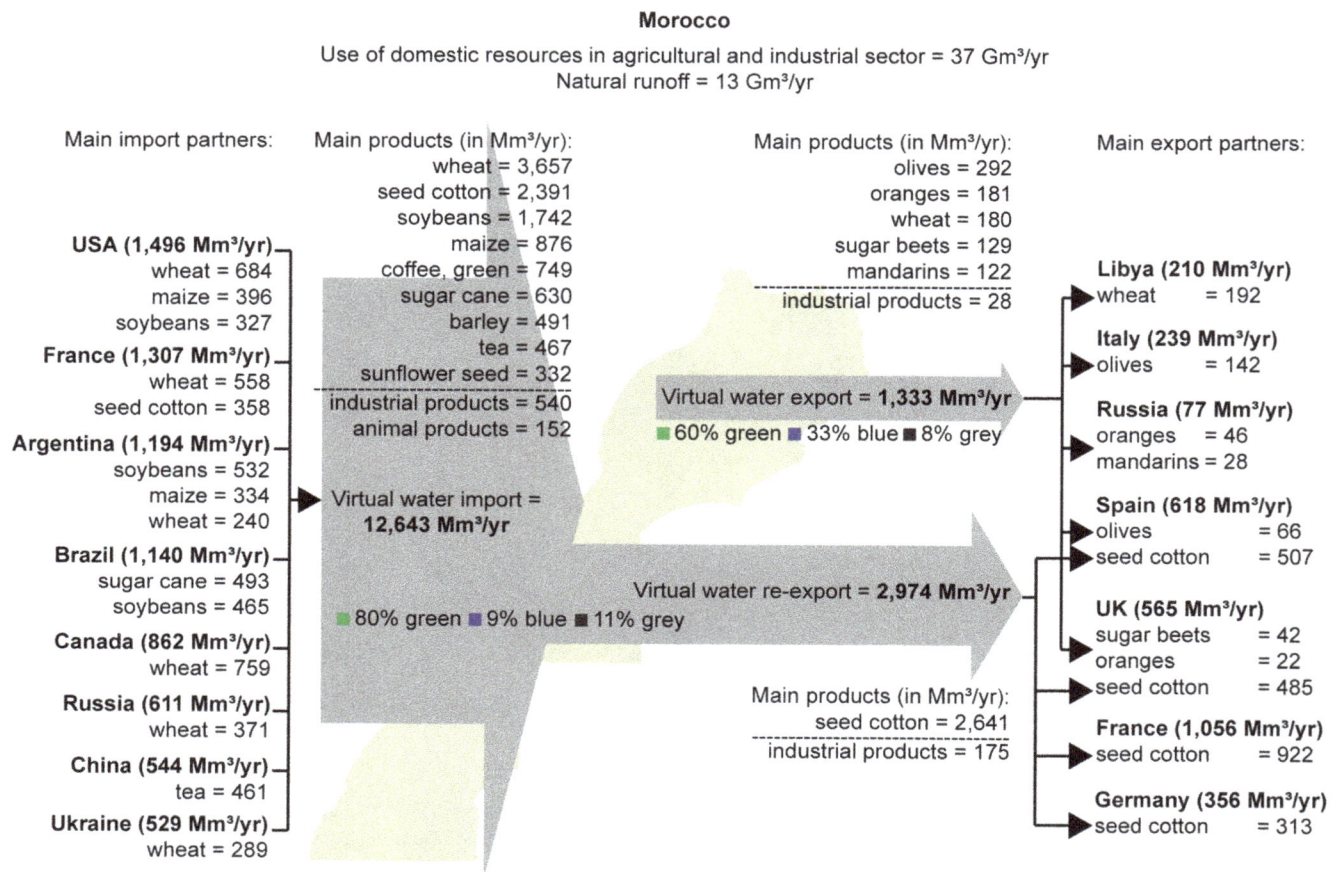

Figure 5. Morocco's virtual water trade balance related to trade in agricultural and industrial commodities. Period: 1996–2005. Source: Virtual water import and (total) virtual water export from Mekonnen and Hoekstra [13].

Reducing the Water Footprint of Crop Production in Morocco

The regional differences in crop water use (Figure 3) provide an opportunity for reduction of the water footprint of crop production in Morocco. Potential water savings (green plus blue) are in the order of 1.9 and 1.2 billion m³ per year when all crops (case A) and when only annual crops (case B) are relocated over the river basins, respectively (Table 7). Blue water savings are 1,276 Mm³/yr in case A and 697 Mm³/yr in case B. These are significant savings when put in the context of Morocco's national water strategy, which includes actions plans to mobilize 1.7 billion m³/yr by 2030 through the construction of 60 large and 1000 small local dams and an additional 0.8 billion m³/yr with the North-South inter-basin water transfer [1].

Largest potential water savings can be obtained by partial relocation of the production of maize and wheat (Table 7), particularly by moving maize production from the Oum Er Rbia basin to the Moulouya basin and wheat production from the Bouregreg basin to the basin of Sebou. Partial relocation of crop production in case A results in decreased water footprints (green plus blue) in all basins, except for the basin of Bouregreg where the water footprint increases (Table 8). In case B, the water footprints in the basins Bouregreg, Sebou and Loukkos increase, particularly due to increased wheat production in these basins, while the water footprints in the other basins decrease. Precipitation in the basins

of Sebou and Loukkos is generally larger than in other parts of Morocco [1].

Reducing the water footprints of crops to benchmark levels leads to a potential green plus blue water saving of 2,768 Mm³/yr, a reduction of 11% (Table 9). Fifty-two per cent of this saving is related to reduced water footprints (i.e. improved water productivities) in the Sebou basin alone. Largest potential water savings are associated with reducing the water footprints of cereals, especially wheat. Blue water savings are estimated at 422 Mm³/yr and are largest in the basins of Sebou and Oum Er Rbia.

Added Value of Water Footprint Assessment for Morocco's Water Policy

Several insights and response options emerged from the Water Footprint Assessment, which are currently not considered in the national water strategy of Morocco and the country's river basin plans. They include:

(i)   New insights in the water balance of Morocco and the country's main river basins:

● The evaporative losses from storage reservoirs account for a significant part of the blue water footprint within Morocco. This sheds fresh light on the national water strategy that

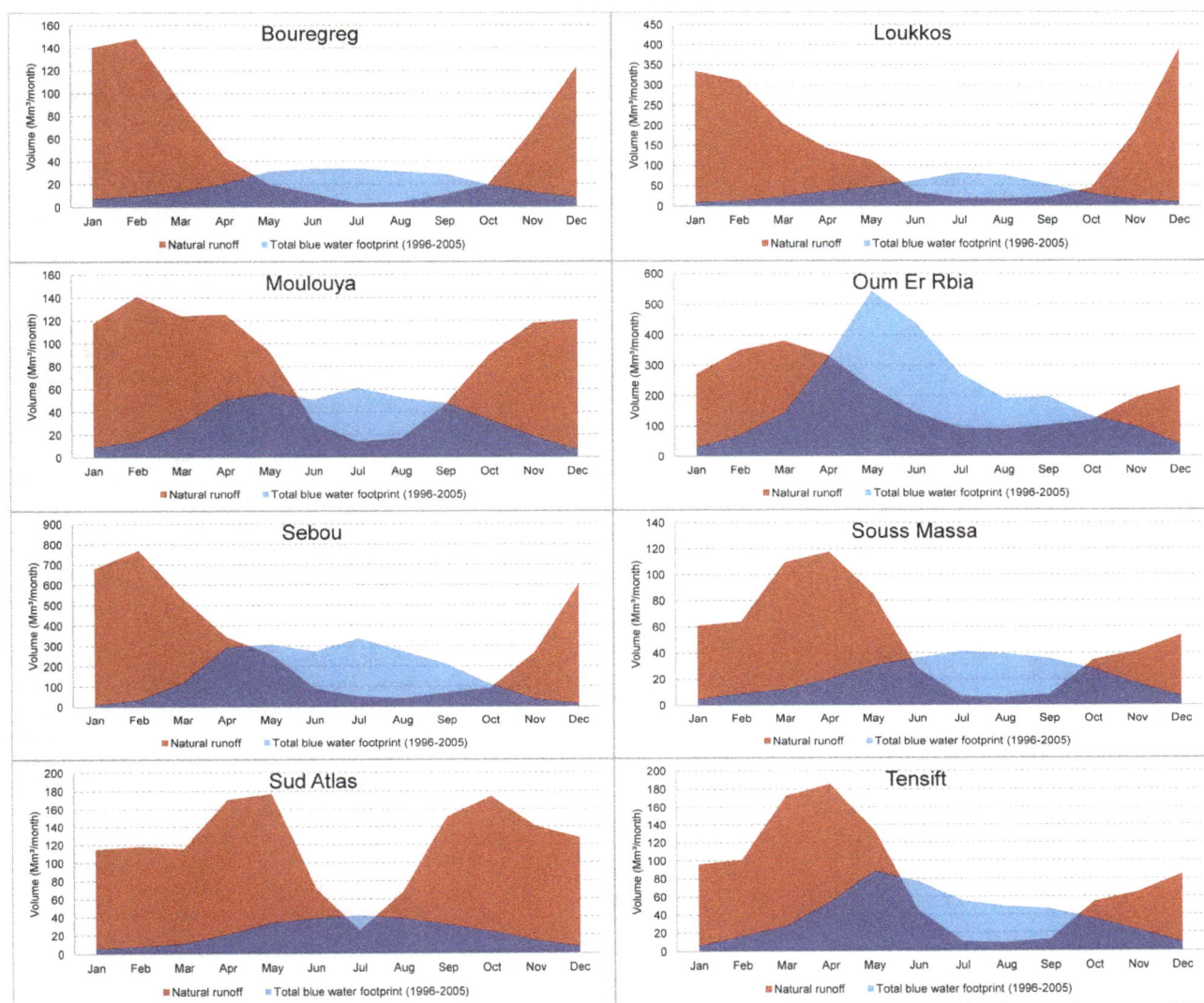

Figure 6. Total blue water footprint and natural runoff per river basin. Period of blue water footprint: 1996–2005. Natural runoff is estimated as the long-term average inflow of reservoirs. It is considered undepleted runoff, since large-scale blue water withdrawals come from the reservoirs. The estimates can be considered conservative, because net precipitation in areas downstream of reservoirs is not included.

proposes to build another 60 large and 1000 small dams by 2030.

- Blue water scarcity on a monthly scale is severe and hidden by annual analysis of demand versus supply, which is the common scale of analysis in Morocco's river basin plans.

(ii) New insights in how economically efficient water and land resources are used:

- Analysis of the economic value of crop products per unit of water and land used in the period 1996–2005 indicate that agricultural policy may be better brought in line with water policy by reconsidering which crops to grow.
- It is shown that the export policy in this period was not optimal from a water-economics point of view, which raises the question whether the foreign income generated by export covers the direct and indirect costs of mobilization and (over) exploitation of Morocco's water resources. This might not be

the case considering the costs of the construction and maintenance of the large dams and intra- and inter-basin water transfers in the country and the costs associated with the negative externalities of water (over) consumption, such as the salt-intrusion in Morocco's coastal aquifers.

(iii) New response options to reduce the water footprint of crop production:

- Analysis of the water footprint of the main crops in Morocco and its variation across the river basins offers new ways of looking at reducing water consumption in the agricultural sector. The estimated potential water savings by partial relocation of crops to basins where they consume less water and by reducing water footprints of crops down to benchmark levels are significant compared to demand reducing and supply increasing measures considered in the national water strategy of Morocco.

Water Quality and Pollution

Table 4. Blue water scarcity per river basin.

River basin	Jan	Feb	Mar	Apr	May	Jun	Jul	Aug	Sep	Oct	Nov	Dec	Tot	Avg
Bouregreg	0.05	0.06	0.14	0.47	1.57	2.89	11.3	7.30	2.78	1.01	0.19	0.06	0.37	2.32
Loukkos	0.03	0.04	0.12	0.25	0.42	1.85	4.04	4.11	2.49	0.69	0.08	0.02	0.25	1.18
Moulouya	0.07	0.10	0.23	0.40	0.62	1.65	4.41	3.09	1.03	0.37	0.16	0.05	0.41	1.02
Oum Er Rbia	0.11	0.20	0.38	0.98	2.42	3.08	2.91	2.14	1.93	1.10	0.51	0.16	0.98	1.33
Sebou	0.02	0.04	0.22	0.86	1.19	3.01	6.66	6.72	3.05	1.21	0.14	0.02	0.53	1.93
Souss Massa	0.07	0.14	0.11	0.17	0.36	1.28	6.35	6.82	4.45	0.81	0.40	0.12	0.46	1.76
Sud Atlas	0.05	0.07	0.09	0.12	0.19	0.54	1.67	0.56	0.21	0.14	0.10	0.06	0.19	0.32
Tensift	0.06	0.16	0.16	0.29	0.66	1.72	5.39	5.40	3.66	0.64	0.34	0.11	0.50	1.55
Total	0.05	0.09	0.22	0.56	1.03	2.23	4.15	2.98	1.55	0.66	0.22	0.06	0.52	1.15

Blue water scarcity is defined as the ratio of the total blue water footprint in a catchment over the natural runoff in that catchment. Classification: low blue water scarcity (<0.20); moderate blue water scarcity (0.20–0.30); significant blue water scarcity (0.30–0.40); severe water scarcity (>0.40).

Discussion

Morocco's water footprint is mostly green (77%). This underlines the importance of green water resources, also (or especially) in semi-arid countries with a high dependency on blue water, and is in line with other studies showing the dominance of the green over the blue water flow in Africa (and most of the world) [25,26]. The relevance of the green water footprint should not be underestimated. Although rain is free and evaporation happens anyway, green water that is used for one purpose cannot be used for another purpose [27].

Storage reservoir evaporation accounts for a significant share (13%) in the blue water footprint in Morocco. The need for seasonal storage of water is evident given the large mismatch in natural runoff and water demand (Figure 6). However, the large evaporation from reservoirs shows that these should be seen as water consumers, besides their role in water supply. This water footprint can ultimately be linked to the end-purpose of the reservoir, which for most cases in Morocco is primarily serving irrigated agriculture. Therefore, to reduce the need for seasonal storage and hence the water footprint of storage reservoirs, it would be worthwhile to take the timing of crop water demands with respect to natural water availability into account in deciding which crops or crop varieties to grow. Furthermore, local alternatives to the large surface water reservoirs are groundwater dams, which enhance underground water storage in alluvial aquifers and thereby loose less water by evaporation [28].

Our analysis shows that from a strictly water-economics point of view it would be worthwhile to reconsider which crops to grow in Morocco (due to the low value in US$/m^3 and US$/ha for some crops compared to others). In practice, the choice of which crops to produce is part of the national strategy regarding food security and of course closely linked to the demand for crops (national and global). Nevertheless, we consider it useful and important to analyse economic water and land productivities (as done in this study) in addition to these considerations. Especially for water-short countries as Morocco it is relevant to evaluate the economic efficiency of water allocation. This also relates to the question whether the foreign income generated by export products, which have a footprint on national resources, outweighs the direct and indirect costs associated with the resource use.

Uncertainties and Limitations

The water footprint of crop production is largely influenced by the input data used and assumptions made by Mekonnen and Hoekstra [14] and can easily contain an uncertainty of ±20% [14,29,30]. The calculated economic water and land productivities of crops are, apart from the water footprints and yields, dependent on the producer prices. Variations in these prices largely influence the economic water and land productivity of crops. The water footprints of industrial production and domestic water supply are very sensitive to the consumptive fractions applied.

Although figures on water availability are based on data from the river basin plans and the Ministry EMWE (unpublished data 2013), the way they are estimated exactly is often unclear and so is the uncertainty in them. Since natural runoff is estimated as the inflow of reservoirs (thus excluding small-scale local abstractions upstream) and net precipitation in areas downstream of reservoirs is not included, the estimates of natural runoff can be considered conservative.

In general, the river basin plans indicate larger pressure on groundwater resources than suggested in this study. This might be caused by the fact that the river basin plans include more recent withdrawals and because the unit of analysis in this study (river

Table 5. Blue water scarcity related to groundwater.

River basin	Groundwater footprint (Mm³/yr)	Groundwater availability (1996–2005) (Mm³/yr)	Blue water scarcity (−)	Level of water scarcity
Bouregreg	106	66	1.60	Severe
Tensift	259	262	0.99	Severe
Oum Er Rbia	510	667	0.77	Severe
Souss Massa	219	349	0.63	Severe
Sebou	689	1,502	0.46	Severe
Moulouya	144	351	0.41	Severe
Loukkos	93	377	0.25	Moderate
Sud Atlas	137	697	0.20	Moderate
Total	2,159	4,347		

Basins are sorted top-down from highest to lowest scarcity.

basin agency action zone) is larger than the unit used in the river basin plans (individual aquifers), whereby in this study overexploitation of one aquifer might be masked by low exploitation of another. Also local groundwater pollution according to the river basin plans is sometimes worse than the water pollution level estimated here. This could be explained by the fact that the water quality measurements recorded in the basin plans are partly more recent and are measured at specific points, whereas this study considered homogeneous distribution of nitrates in the groundwater.

Given the uncertainties and limitations of the study, the presented water footprint estimates and water scarcity values should be interpreted with care. Nevertheless, the order of magnitude of the estimates in this study gives a good indication to which activities and crops Morocco's water resources are allocated, in which months and basins the water footprints are relatively large or small and where and when this leads to highest water scarcity.

Uncertainties in the estimated potential savings by relocation of crop production and reducing the water footprints of crops to benchmark levels are closely linked to the uncertainties in the estimates of the water footprint of crop production and the results should be interpreted carefully. However, the order of magnitude of the estimated savings gives a rough indication of the potential of these measures. When considering relocation of crop production it is necessary to assess how the green and blue water footprints of crops manifest themselves on a monthly scale. This study looked at annual water savings, but the associated relocation of crops might well aggravate monthly water scarcity in some river basins. Furthermore, the feasibility and desirability of relocation of crop production are of course largely determined by social and economic factors which are not considered in this study.

Conclusion

The study finds that: (i) evaporation from storage reservoirs is the second largest form of blue water consumption in Morocco, after irrigated crop production; (ii) Morocco's water and land resources are mainly used to produce relatively low-value (in US$/m³ and US$/ha) crops such as cereals, olives and almonds; (iii) most of the virtual water export from Morocco relates to the export of products with a relatively low economic water productivity (in US$/m³); (iv) blue water scarcity on a monthly scale is severe in all river basins and pressure on groundwater

Table 6. Water pollution level related to nitrate-nitrogen in groundwater.

River basin	Grey water footprint of crop production (1996–2005) (Mm³/yr)	Actual groundwater availability/Waste assimilation capacity (Mm³/yr)	Water pollution level (−)	Waste assimilation capacity exceeded?
Bouregreg	148	0	∞	Yes
Tensift	129	3	43.2	Yes
Oum Er Rbia	435	157	2.78	Yes
Sebou	428	813	0.53	No
Moulouya	99	207	0.48	No
Souss Massa	51	130	0.39	No
Loukkos	63	284	0.22	No
Sud Atlas	25	560	0.04	No
Total	1,378	2,188	0.63	No

Basins are sorted top-down from highest to lowest pollution level.

Table 7. Potential water savings by partial relocation of crop production per crop.

	Base case green plus blue water footprint (Mm³/yr)	Partial relocation considered for all crops*		Partial relocation considered for annual crops only**	
		Saving (green+blue) (Mm³/yr)	Relative saving (%)	Saving (green+blue) (Mm³/yr)	Relative saving (%)
Almonds	641	14	2%	0	0%
Barley	6,787	−116	−2%	−202	−3%
Dates	449	131	29%	0	0%
Grapes	367	183	50%	0	0%
Maize	1,148	939	82%	939	82%
Olives	2,951	58	2%	0	0%
Oranges	440	15	3%	0	0%
Sugar Beets	353	157	44%	157	44%
Sugar Cane	200	91	46%	0	0%
Mandarins	209	7	3%	0	0%
Tomatoes	99	2	2%	2	2%
Wheat	10,981	413	4%	278	3%
Total	24,625	1,896	8%	1,174	5%

*All analysed crops are: almonds, barley, dates, grapes, maize, olives, oranges, sugar beets, sugar cane, mandarins, tomatoes and wheat.
**Annual crops are: barley, maize, sugar beets, tomatoes and wheat.

resources by abstractions and nitrate pollution is considerable in most basins; (v) the estimated potential water savings by partial relocation of crops to basins where they consume less water and by reducing water footprints of crops down to benchmark levels are significant compared to demand reducing and supply increasing measures considered in Morocco's national water strategy.

On the basis of these new insights and response options it is concluded that Water Footprint Assessment has an added value for national water policy in Morocco. Water Footprint Assessment forces to look at end-users and -purposes of freshwater, which is key in determining efficient and equitable water allocation within the boundaries of what is environmentally sustainable, both on the river basin and on the national level. This is especially relevant for water-scarce countries such as Morocco. Furthermore, considering the green and grey components of a water footprint provides new perspectives on blue water scarcity, because pressure on blue water resources might be reduced by more efficient use of green water and by less pollution.

Table 8. Potential water savings by partial relocation of crop production per river basin.

	Base case green plus blue water footprint (Mm³/yr)	Partial relocation considered for all crops		Partial relocation considered for annual crops only**	
		Saving (green+blue) (Mm³/yr)	Relative saving (%)	Saving (green+blue) (Mm³/yr)	Relative saving (%)
Sud Atlas	306	189	62%	12	4%
Souss Massa	903	175	19%	14	2%
Tensift	2,525	388	15%	124	5%
Oum Er Rbia	8,498	1,229	14%	821	10%
Bouregreg	2,813	−994	−35%	−95	−3%
Moulouya	1,737	605	35%	412	24%
Sebou	6,905	154	2%	−95	−1%
Loukkos	939	151	16%	−19	−2%
Total	24,625	1,896	8%	1,174	5%

*All analysed crops are: almonds, barley, dates, grapes, maize, olives, oranges, sugar beets, sugar cane, mandarins, tomatoes and wheat.
**Annual crops are: barley, maize, sugar beets, tomatoes and wheat.

Table 9. Potential water savings by benchmarking water productivities of main crops* (in Mm^3/yr).

	Sud Atlas	Souss Massa	Tensift	Oum Er Rbia	Bouregreg	Moulouya	Sebou	Loukkos	Total
Almonds	0	2	1	0	3	0	8	0	14
Barley	0	0	0	100	158	222	238	0	717
Dates	0	0	0	10	0	4	48	0	63
Grapes	0	20	0	5	0	0	18	4	48
Maize	0	13	0	175	32	0	33	0	254
Olives	0	9	4	0	10	0	35	0	59
Oranges	0	1	1	0	1	0	6	0	9
Sugar Beets	0	0	0	0	0	0	70	4	73
Sugar Cane	0	0	0	0	0	0	79	10	89
Mandarins	0	1	0	0	0	0	3	0	4
Tomatoes	0	0	0	0	1	0	1	0	3
Wheat	0	14	0	102	417	0	904	0	1,436
Total (gn+bl)	0	60	6	392	623	226	1,444	18	2,768
Total (blue)**	0	23	2	113	11	2	258	12	422
Total (blue) (% of natural runoff)	0%	4%	0%	4%	2%	0%	7%	1%	3%

*Analysed crops are: almonds, barley, dates, grapes, maize, olives, oranges, sugar beets, sugar cane, mandarins, tomatoes and wheat.
**Assuming that the green/blue water ratio remains the same for all basins and crops.

Acknowledgments

We like to thank Mesfin Mekonnen (University of Twente, Enschede, Netherlands), Karen Meijer and Wil van der Krogt (Deltares, Delft, Netherlands), Abdelkader Larabi (Mohammadia School of Engineers, Rabat, Morocco) and Siham Laraichi (Ministry of Energy, Mining, Water and Environment, Rabat, Morocco) for their feedback during the various stages of research.

Author Contributions

Conceived and designed the experiments: JFS AYH. Performed the experiments: JFS. Analyzed the data: JFS AYH. Wrote the paper: JFS AYH.

References

1. Ministry EMWE (2011) Stratégie Nationale de l'Eau. Available: www.minenv.gov.ma/PDFs/EAU/strategie_eau.pdf, Department of Water, Ministry of Energy, Mining, Water and Environment, Rabat, Morocco. Accessed 2013 January 21.
2. Ministry EMWE (2012) Les eaux souterraines, http://www.water.gov.ma/index.cfm?gen=true&id=12&ID_PAGE=42, Department of Water, Ministry of Energy, Mining, Water and Environment, Rabat, Morocco.Accessed 2012 December 8.
3. Official State Gazette (1995) Royal Decree no. 1-95-154 promulgating Law no. 10-95 on water. Available: http://ocid.nacse.org/rewab/docs/Royal_Decree_No_1-95-154_Promulgating_Law_on_Water_EN.pdf. Accessed 2013 January 3.
4. Hoekstra AY, ed. (2003) Virtual water trade: Proceedings of the International Expert Meeting on Virtual Water Trade, Delft, The Netherlands, 12–13 December 2002. Value of Water Research Report Series No. 12, UNESCO-IHE, Delft, The Netherlands. Available: www.waterfootprint.org/Reports/Report12.pdf. Accessed 2013 July 5.
5. Hoekstra AY, Chapagain AK (2008) Globalization of water: Sharing the planet's freshwater resources. Oxford: Blackwell Publishing.
6. Hoekstra AY, Chapagain AK, Aldaya MM, Mekonnen MM (2011) The water footprint assessment manual: Setting the global standard. London: Earthscan.
7. Aldaya MM, Garrido A, Llamas MR, Varelo-Ortega C, Novo P, et al. (2010) Water footprint and virtual water trade in Spain. In: Garrido A, Llamas MR, eds. Water policy in Spain. Leiden: CRC Press. 49–59.
8. Chahed J, Besbes M, Hamdane A (2011) Alleviating water scarcity by optimizing "Green Virtual-Water": the case of Tunisia. In: Hoekstra AY, Aldaya MM, Avril B, eds. Proceedings of the ESF Strategic Workshop on accounting for water scarcity and pollution in the rules of international trade, Amsterdam, 25–26 November 2010. Value of Water Research Report Series No. 54, UNESCO-IHE. 99–113. Available: http://www.waterfootprint.org/Report54-Proceedings-ESF-Workshop-Water-Trade.pdf. Accessed 2013 February 4.
9. Hoekstra AY, Mekonnen MM (2012) The water footprint of humanity. Proc Natl Acad Sci U S A 109(9): 3232−3237.
10. Aldaya MM, Martinez-Santos P, Llamas MR (2010) Incorporating the water footprint and virtual water into policy: Reflections from the Mancha Occidental Region, Spain. Water Resources Management 24(5): 941–958.
11. Hoekstra AY, Chapagain AK (2007) Water footprints of nations: water use by people as a function of their consumption pattern. Water Resources Management 21(1): 35–48.
12. Hoekstra AY, Chapagain AK (2007) The water footprints of Morocco and the Netherlands: Global water use as a result of domestic consumption of agricultural commodities. Ecol Econ 64(1): 143–151.
13. Mekonnen MM, Hoekstra AY (2011) National water footprint accounts: the green, blue and grey water footprint of production and consumption. Value of Water Research Report Series No.50, UNESCO-IHE, Delft, The Netherlands. Available: http://www.waterfootprint.org/Reports/Report50-NationalWaterFootprints-Vol1.pdf. Accessed 2012 November 23.
14. Mekonnen MM, Hoekstra AY (2010) The green, blue and grey water footprint of crops and derived crop products. Value of Water Research Report Series No.47, UNESCO-IHE, Delft, The Netherlands. Available: http://www.waterfootprint.org/Reports/Report47-WaterFootprintCrops-Vol1.pdf. Accessed 2012 November 23.
15. Sperna Weiland FC, van Beek LPH, Kwadijk JCJ, Bierkens MFP (2010) The ability of a GCM-forced hydrological model to reproduce global discharge variability. Hydrol Earth Syst Sci 14: 1595–1621. doi:10.5194/hess-14-1595-2010.
16. FAO (2013) AQUASTAT online database. Geo-referenced database of African dams. Food and Agriculture Organization, Rome, Italy. Available: http://www.fao.org/nr/water/aquastat/dams/region/D_Africa.xlsx. Accessed 2013 March 28.
17. FAO (2013) AQUASTAT online database. Country Fact Sheet: Morocco. Food and Agriculture Organization, Rome, Italy. Available: www.fao.org/nr/aquastat/. Accessed 2013 February 22.
18. FAO (2013) FAOSTAT online database. Food and Agriculture Organization, Rome, Italy. Available: http://faostat.fao.org. Accessed 2013 July 6.
19. ITC (2007) SITA version 1996–2005 in SITC [DVD-ROM], International Trade Centre, Geneva.
20. Riad S (2003) Typologie et analyse hydrologique des eaux superficielles à partir de quelques bassins versants représentatifs du Maroc. Unpublished thesis. Available: http://ori-nuxeo.univ-lille1.fr/nuxeo/site/esupversions/e5d351a6-ce4c-4b64-b891-84d85f3d8f02. Accessed 2013 June 19.
21. Tekken V, Kropp JP (2012) Climate-Driven or Human-Induced: Indicating Severe Water Scarcity in the Moulouya River Basin (Morocco). Water 4: 959–982. doi:10.3390/w4040959.
22. JICA MATEE, ABHT (2007) Etude du plan de gestion intégrée des ressources en eau dans la plaine du Haouz, Royaume du Maroc, Rapport intermédiaire. Available: http://eau-tensift.net/fileadmin/user_files/pdf/etudes/JICA_ETUDE_PLAN_DE_GESTION_INTEGREE_RE_HAOUZ.pdf. Accessed 2013 August 12.
23. Laouina A (2001) Compétition irrigation/eau potable en région de stress hydrique: le cas de la région d'Agadir (Maroc). In: Camarda D, Grassini L, ed. Interdependency between agriculture and urbanization: Conflicts on sustainable use of soil and water. Bari: CIHEAM, 2001, 17–31. Available: ressources.ciheam.org/om/pdf/a44/02001585.pdf. Accessed 2013 June 20.
24. Hoekstra AY, Mekonnen MM, Chapagain AK, Mathews RE, Richter BD (2012) Global monthly water scarcity: Blue water footprints versus blue water availability. PLoS ONE 7(2): e32688.
25. Rockstrom J, Falkenmark M, Karlberg L, Hoff H, Rost S, et al. (2009) Future water availability for global food production: The potential of green water for increasing resilience to global change. Water Resour Res 45: W00A12.
26. Schuol J, Abbaspour KC, Yang H, Srinivasan R, Zehnder AJB (2008) Modeling blue and green water availability in Africa. Water Resour Res 44: W07406.
27. Hoekstra AY (2013) The Water Footprint of Modern Consumer Society. London: Routledge.
28. Al-Taiee TM (2012) Groundwater Dams: a Promise Option for Sustainable Development of Water Resources in Arid and Semi-Arid Regions. In: UNESCO. Proceedings of the Second International Conference on integrated water resources management and challenges of the sustainable development, Agadir, 24–26 March 2010. IHP-VII Series on Groundwater No. 4, UNESCO. 35–41.
29. Hoff H, Falkenmark M, Gerten D, Gordon L, Karlberg L, et al. (2010) Greening the global water system. J Hydrol 383(3–4): 177–186.
30. Mekonnen MM, Hoekstra AY (2010) A global and high-resolution assessment of the green, blue and grey water footprint of wheat. Hydrol Earth Syst Sci 14(7): 1259–1276.
31. FAO (2013) Global map of monthly reference evapotranspiration-10 arc minutes. GeoNetwork: grid database. Food and Agriculture Organization, Rome, Italy. Available: www.fao.org/geonetwork/srv/en/resources.get?id=7416&fname=ref_evap_fao_10min.zip&access=private. Accessed 2013 March 19.

10

Bioremediation of a Complex Industrial Effluent by Biosorbents Derived from Freshwater Macroalgae

Joel T. Kidgell[1], Rocky de Nys[1], Yi Hu[2], Nicholas A. Paul[1], David A. Roberts[1]*

1 MACRO - The Centre for Macroalgal Resources and Biotechnology, and School of Marine and Tropical Biology, James Cook University, Townsville, Queensland, Australia,
2 Advanced Analytical Centre, James Cook University, Townsville, Queensland, Australia

Abstract

Biosorption with macroalgae is a promising technology for the bioremediation of industrial effluents. However, the vast majority of research has been conducted on simple mock effluents with little data available on the performance of biosorbents in complex effluents. Here we evaluate the efficacy of dried biomass, biochar, and Fe-treated biomass and biochar to remediate 21 elements from a real-world industrial effluent from a coal-fired power station. The biosorbents were produced from the freshwater macroalga *Oedogonium* sp. (Chlorophyta) that is native to the industrial site from which the effluent was sourced, and which has been intensively cultivated to provide a feed stock for biosorbents. The effect of pH and exposure time on sorption was also assessed. These biosorbents showed specificity for different suites of elements, primarily differentiated by ionic charge. Overall, biochar and Fe-biochar were more successful biosorbents than their biomass counterparts. Fe-biochar adsorbed metalloids (As, Mo, and Se) at rates independent of effluent pH, while untreated biochar removed metals (Al, Cd, Ni and Zn) at rates dependent on pH. This study demonstrates that the biomass of *Oedogonium* is an effective substrate for the production of biosorbents to remediate both metals and metalloids from a complex industrial effluent.

Editor: Tilmann Harder, University of New South Wales, Australia

Funding: This research is part of the MBD Energy Research and Development program for Biological Carbon Capture and Storage with the co-operation of Stanwell Energy Corporation. This project is supported by the Advanced Manufacturing Cooperative Research Centre (AMCRC), funded through the Australian Government's Cooperative Research Centre Scheme, and the Australian Renewable Energy Agency (ARENA). Stanwell Energy Corporation provided effluents for the study. The funders had no role in study design, data collection and analysis, decision to publish, or preparation of the manuscript.

Competing Interests: We have the following interests. This research is part of the MBD Energy Research and Development program for Biological Carbon Capture and Storage with the co-operation of Stanwell Energy Corporation. Stanwell Energy Corporation provided effluents for the study. The data contained within this study form a component of Australian Provisional Patent AU2013902101: " Biosorbent and methods of use", James Cook University. There are no further patents, products in development or marketed products to declare.

* E-mail: david.roberts1@jcu.edu.au

Introduction

Mining, mineral processing and energy generation produce large quantities of contaminated effluent. For example, coal-fired power stations produce complex effluents containing dissolved elements from the flushing of ash from the flue and furnace [1]. The resulting effluent contains elements at concentrations of potential environmental concern, such as Al, As, B, Cd, Mo, Se, Sr, V, and Zn, and extensive treatment is required before the effluent can be discharged [1,2]. As the cost and operational conditions of treatment options can be prohibitive [1,3], the effluent is often retained in large storages known as Ash Dams (AD). However, despite the apparent confinement of these water bodies, AD remain a significant source of toxic elements to local organisms [4]. Consequently, there is a need for a cost effective, sustainable and comprehensive approach to the remediation of complex industrial effluents.

Biosorption with biomass is an alternative to existing waste water treatment technologies with promising results at the laboratory scale [5]. Biosorption exploits the ability of dead or denatured biomass, such as dried macroalgae, to passively bind ions from aqueous solutions [6,7]. Dried macroalgae are particularly effective biosorbents due to the high abundance of functional groups which have a strong affinity for dissolved cationic metals despite also having relatively high concentrations of these same metals in the biomass [8]. Many functional groups can be involved in biosorption and this can vary according to taxonomic groupings. For example, in brown algae the carboxylic groups of alginates are typically dominant in biosorption processes, while some freshwater green algae, such as *Oedogonium*, have cellulosic cell walls that resemble those of higher plants [9,10]. These functional groups can passively bind dissolved metals through various processes, including passive electrostatic attraction, ion exchange with "light" metal ions (Ca^{2+}, Na^+, K^+ and Mg^{2+}), or complexation processes [10].

Macroalgae (and other biosorbents such as activated carbon) only have an affinity for dissolved cations and are relatively ineffective at treating oxyanions, such as selenate (SeO_4^{2-}) that are common constituents of effluents [11]. However, dried macroalgae can be manipulated to improve its affinity for specific contaminants. Biomass can be converted to carbon-rich biochar through slow pyrolysis, resulting in a product with similar properties to activated carbon [12]. Additionally, biomass and biochar can be pre-treated with an iron (Fe) solution to improve the adsorption of anionic metalloids, including SeO_4^{2-} [13]. Deposition of Fe onto the surface of either dried biomass or biochar provides a positive charge, promoting the formation of inner-sphere complexes between oxyanionic metalloids and Fe-treated biosorbents, where there would otherwise be no natural affinity for sorption [14–16].

Despite the promise of macroalgae as a biosorbent, industrial application has been limited. One key factor limiting the application of biosorption is the lack of a sustainable and sufficient source of biomass [6,17]. Wild harvests of biomass to support biosorption are simply not sustainable when one considers the volumes required [6], and commercially cultivated seaweeds have existing applications in other markets [17]. However, in recent work we have shown that native species of macroalgae can be cultured to provide sustainable biomass for bioremediation [17,18]. This cultivated biomass represents a sustainable source of biomass, but little research has considered the efficacy of cultivated biomass in biosorption applications.

An additional limitation of existing biosorption research is that it has focused on simple synthetic effluents. These studies often focus on the kinetics of sorption and the mechanisms of uptake of select elements under idealized conditions. In contrast, real-world industrial effluents are complex, involving multiple interacting and competing contaminants that occur in a variety of speciation and oxidation states, which are influenced by environmental conditions [2,8]. Biosorption research that has been conducted in multi-element systems has shown that non-target elements can interfere with [19,20] or competitively exclude [21] biosorption of target elements. Consequently, in multi-element systems the capacity of a biosorbent for individual elements typically decreases in comparison to results obtained in idealized single-element effluents [22]. Macroalgal biosorbents have not yet been proven to be an effective means of treating complex effluents with multiple co-existing contaminants [7] and it is rare for studies to consider systems with more than three elements [8]. In fact, very little is known about the performance of biosorbents of any type in multi-elemental systems, or the effects that physical parameters such as pH and exposure time have in these scenarios.

Here we address key constraints to the industrial application of algal-based biosorption by assessing the efficacy of a macroalgal biosorbent for use in a real-world complex industrial effluent. We collect a native isolate of the cosmopolitan freshwater macroalgal genus *Oedogonium* (Link ex Hurn, 1900) from the AD of a coal-fired power station and cultivate it in intensive production systems as a means of providing sustainable biomass for biosorption. Specifically, we test *Oedogonium* dried biomass, derived biochar, Fe-treated biomass and Fe-treated biochar as biosorbents for 21 metals and metalloids in an effluent taken from coal-fired power production under a range of pH conditions and exposure times. These results will establish the potential of biosorption for the remediation of complex industrial effluents using purposely cultivated biomass.

Materials and Methods

Industrial effluent

This study targeted Ash Dam Water (ADW) from Tarong coal-fired power station in south-east Queensland, Australia (26.76°S, 151.92°E). Tarong is one of Queensland's largest power stations with a generation capacity of 1400 MW, and a 46,000 ML AD storing contaminated waste water. ADW was sourced directly from the AD and transported to James Cook University (JCU), Townsville in 1000 L Intermediate Bulk Containers (IBCs) in November 2012. The ADW was then stored at ambient temperature in 12,000 L storage tanks until use. The effluent was collected and transported to JCU by Stanwell Energy Corporation.

Algal biosorbent production & preparation

Oedogonium sp. (Genbank: KF606974) [23] hereafter *Oedogonium*, was used as the source biomass for the production of biosorbents (see below). *Oedogonium* is a native filamentous, freshwater green alga in the Tarong AD [18]. *Oedogonium* samples were initially collected from the Tarong AD in October 2012 but could not be identified to species using taxonomic keys based on morphological characteristics [24]. The species was therefore assessed using molecular techniques, arguably the most accurate means to identify cryptic species, and this isolate has been assigned the Genbank accession number KC606974 with no current matches for this species in the database [23].

After collection from Tarong AD, *Oedogonium* was cultivated in Manutec f/2 algal growth media in 2500 L tanks during the austral summer months (January – March) in the aquaculture facility at JCU (19.33°S, 146.76°E). Prior to experiments, 2 kg of algae was harvested from the tanks and oven dried to a constant mass at 60°C for 48 hours (h). Subsequently, 1 kg of the dried *Oedogonium* biomass was converted into biochar by slow pyrolysis under conditions previously developed for macroalgae [12]. Briefly, *Oedogonium* was suspended within a muffle furnace (Labec CEMLS-1200) and continuously purged with N_2 (BOC) gas at 4.0 L min^{-1} while being heated to a hold temperature of 450°C for 1 h. Additionally, a sub-sample of both the dried biomass and biochar were also treated with a 5% Fe solution, prepared by diluting $FeCl_3$ (Sigma Aldrich 45% w/v) in deionized (DI) water (Millipore Direct-Q3), to become Fe-loaded biosorbents. Dried biomass and biochar were exposed to separate Fe solutions at a density of 25 g L^{-1} for 24 h on a shaker plate (100 rpm) at 20°C, then filtered from the solutions and rinsed three times with DI water at a rate of 20 ml g^{-1}, then dried at 60°C for 48 h.

Biosorption experiments

Biosorption experiments were conducted to quantify the rate and composition of metal and metalloid adsorption from ADW by *Oedogonium* dried biomass, biochar, Fe-treated dried biomass and Fe-treated biochar at three alternate initial pH levels (2.5, 4, 7.1 see below). Filtered (0.45 μm sterile Starstedt syringe filters) samples of ADW were analyzed prior to experimental treatments to serve as a benchmark for initial conditions.

Two solutions of pH-manipulated ADW were produced with 1 M HCl (pH 2.5 and 4, Sigma Aldrich TraceSelect Ultra), while a third remained at the native pH of the ADW, 7.07±0.01. The experiment was fully factorial in design, with independent samples being destructively sampled at each time point. Each of the treatments consisted of a plastic beaker with 60 ml of ADW and 0.6 g of biomass, biochar or the Fe-treated derivatives (10 g L^{-1} of biosorbent). The beakers were shaken (100 rpm at 20°C) in incubator shaker cabinets. At the end of the allocated exposure time (0:15, 0:30, 1:00, 4:00, 24:00, or 168:00 hours) the samples were removed from the cabinets and filtered with 75 μm nylon filter paper. The solution was then filtered to 0.45 μm using a glass fiber filter and syringe, then analysed as described below. Samples containing no biosorbent were processed in the same manner to serve as negative controls to quantify losses of elements to experimental glassware and filtration. The experiment was replicated three times. All plastic and glassware was acid washed in a 5% HNO_3 (Sigma Aldrich) bath for 48 h, then rinsed in DI water prior to use.

Elemental analysis

The concentrations of Al, As, Ba, Cd, Co, Cr, Cu, Fe, Mn, Mo, Ni, Pb, Se, Sr, V and Zn were measured with a Bruker 820-MS Inductively Coupled Plasma Mass Spectrometer (ICP-MS), and

Ca, K, Mg and Na with a Varian Liberty series II Inductively Coupled Plasma Optical Emissions Spectrometer (ICP-OES). An external calibration strategy was used for both instruments, where a standard solution of 0.45 μm filtered ADW was used as the vector to calculate the concentration of elements. Collisional Reaction Interface (CRI) was used for As (H_2) and V (He), while ^{82}Se isotope was used for Se quantification, to eliminate polyatomic interferences for these elements. A 1% HCl solution was spiked with 1 ppb As, Se and V and measured three times for quality control; recovery between 98.5 and 110% indicated no significant interferences. All analyses were conducted at the Advanced Analytical Centre at JCU, Townsville.

Data analysis

Multivariate patterns in biosorption were visualized using Principal Components Analysis (PCA) from a correlation matrix with some elements log-transformed to create a normal distribution [25]. Univariate analysis took the form of three-way fully-factorial Analysis of Variance (ANOVA), with the factors biosorbent (fixed), pH (fixed) and time (random). Data were examined for normality and homogeneity of variance using normal-probability plot of raw residuals and predicted-residual scatter plot, and were transformed as necessary to meet assumptions [26]. Both the PCA and ANOVA test were conducted in Statistica for Windows (Ver. 10, C. Statsoft Inc. 1984–2011).

Results

Characteristics of ADW

Twelve (Al, As, B, Cd, Cr, Cu, Pb, Mn, Mo, Ni, Se and Zn) of the 21 elements measured in the ADW have trigger levels established by the Australian and New Zealand Environmental Conservation Council (ANZECC) [27]. Of these twelve elements, eleven (Al, As, B, Cd, Cr, Cu, Pb, Mo, Ni, Se and Zn) were in excess of the trigger values (Table 1). Given that these elements have quantifiable remediation goals they are the focus of the following results section.

Biochar, biomass and ANZECC metals

Biochar was the most effective biosorbent, removing a broad suite of metals (Mn, Al, Cr, Cu, Cd, Ni, Pb and Zn). from solution (Figures 1 and 2). The PCA shows that effluent treated by biochar clustered along the positive PC1 axis, being characterized by lower concentrations of metals than the effluents treated by the remaining biosorbents (Figure S1). There was, however, a significant effect of pH on the biosorption of most metals by biochar ("Biosorbent x pH" Table S1). At high initial pH (4.0 & 7.1) the raw biochar rapidly adsorbed metals from solution but at low initial pH (2.5), leached metals into solution (Figure 1a).

Of the metals included in the ANZECC guidelines, Al, Ni and Zn were most effectively removed from solution by biochar, and each of these had a pH-dependent response in the rate of biosorption (Figure 2a, c, e; Table S1). All three of these metals were reduced below their respective ANZECC trigger levels by biochar at high initial pH, in the case of Al and Zn within 30 min (Table 1; Figure 2a, c, e). When pH was initially low (2.5), the concentration of Al and Zn increased in solution in the first 15 min, then over the next four hours adsorbed onto the biochar to finally reach levels below the limits of detection (Figure 2a, e). While Cd was also adsorbed by biochar at varying rates under different initial pH conditions (2.5, 4.0 and 7.1), it was not reduced to below the trigger level (Figure 2g; Table 1).

The response of metals to biomass varied greatly and, as with biochar, often in a pH-mediated fashion (Figure 1b). Overall, there was an increase in element concentrations in ADW treated with biomass (Figure 1b) which is supported by effluent treated by biomass being broadly distributed around the centroid in the PCA, demonstrating it was a relatively ineffective biosorbent (Figure S1). Al was the only metal reduced below its respective trigger level when exposed to biomass, and this only occurred at high initial pH (4 and 7.1) (Figure 2b). In contrast, at an initial pH of 2.5 the concentration of Al increased substantially and continuously for the entire exposure duration (Figure 2b). Again, Zn displayed a similar pattern (Figure 2f). When exposed to biomass, Ni and Cd both displayed a similar pattern of initial decrease in concentration at high pH (4.0 and 7.1) followed by no significant change for the remaining duration of exposure, however, in low initial pH (2.5) both Ni and Cd did not differ from the initial concentration (Figure 2d, h; Table S1). Mn displayed substantial pH-mediated leaching when exposed to both biochar and biomass (Figure. S3g, h).

Fe-biochar, Fe-biomass and ANZECC metalloids

Fe-biochar was an effective adsorbent of As, Mo and Se, and initial pH had no impact on the rate or extent of adsorption of these elements (Figure 1c, Table S1). The net concentration of all ANZECC oxyanionic metalloids (As, B, Mo, Se) decreased by 2700 μg L^{-1} (30%) within the first 15 min of exposure to Fe-biochar when the initial pH was 4 or 7.1 (Figure 1c). This can be visualized in the PCA, in which ADW treated with Fe-biochar clusters along the positive PC2 axis in the PCA, demonstrating the effluent treated with Fe-biochar tended to have lower concentrations of As, Se and Mo than the remaining treatments (Figure S1).

The concentrations of As, Mo and Se all dropped significantly lower with Fe-biochar than for any other biosorbent. As and Mo were adsorbed by Fe-biochar to below their respective trigger levels for all initial pH conditions (Table 1; Figure 3a, c). Se was substantially reduced within the first 15 mins by Fe-biochar but not to the point of the AZNECC trigger level (Figure 3e; Table 1). As and Mo followed the characteristic pattern of rapid initial adsorption within the first 15 mins and continued decline at a slower rate for the remaining exposure (Figure 3a, c). Initial concentrations of B were 20 times in excess of the trigger level and despite a drop in concentration of approximately 20% when exposed to Fe-biochar at an initial pH of 4 (Table 1), the concentration of B did not approach the trigger level for any of the treatments (Figure 3g, h).

Fe-biomass behaved in a similar manner to Fe-biochar, albeit not as successfully. There was an initial reduction of 1700 μg L^{-1} (18%) of ANZECC metalloids in the first 15 min of exposure to Fe-biomass (Figure 1c). Interestingly, untreated biomass at low pH (2.5) showed a similar total effectiveness with metalloids as Fe-biochar, with an initial decrease in metalloid concentration of 2400 μg L^{-1} (28%) in the first 15 min (Figure. S2d).Mo and Se were slightly reduced in ADW when exposed to Fe-biomass (Figure 3d, f; Table S1), however, these were the only elements to do so.

Discussion

This study demonstrates that the biomass of *Oedogonium* is an effective substrate for the production of biosorbents to remediate both metals and metalloids from a complex industrial effluent. Conversion of biomass to biochar through slow pyrolysis, and Fe-treatment of this biochar, produces biosorbents that effectively bind metals and metalloids respectively. The affinity of each

Table 1. The ANZECC trigger level and initial concentration for each element investigated, in addition to the lowest final concentration and the biosorbent, time and pH conditions responsible.

Element	ANZECC Trigger ($\mu g\ L^{-1}$)	Initial Concentration [$\mu g\ L^{-1}$ ($\pm$ SE)]		Final Concentration [$\mu g\ L^{-1}$ ($\pm$ SE)]		Best Biosorbent	Fastest Time	Best pH
Aluminium	**55**	**144**	**(35)**	**35**	**(28)**	**Biochar**	**0:30**	**≥4**
Arsenic	**13**	**43**	**(5.5)**	**9**	**(1.0)**	**Fe-Biochar**	**24**	**NA**
Boron	370	7475	(893)	5767	(462)	Fe-Biochar	0:15	4
Cadmium	0.2	2.3	(0.2)	0.9	(0.2)	Biochar	0:30	≥4
Chromium	1	5.2	(3.3)	2.6	(2.1)	Biochar	0:15	NA
Copper	**1.4**	**1.9**	**(0.9)**	**1.2**	**(0.9)**	**Biochar**	**0:30**	**NA**
Lead	3.4	0.3	(0.1)	0.03	(<0.01)	Biochar	0:30	8
Manganese	1900	3	(0.8)	-		-	-	-
Molybdenum	**34**	**1437**	**(127)**	**28**	**(5.9)**	**Fe-Biochar**	**168**	**NA**
Nickel	**11**	**53**	**(7.3)**	**11**	**(2.7)**	**Biochar**	**24**	**≥4**
Selenium	11	82	(3.9)	21	(2.6)	Fe-Biochar	0:30	NA
Selenium	11	82	(3.9)	13	(1.8)	Biomass	168	4
Zinc	**8**	**64**	**(11)**	**5**	**(2.4)**	**Biochar**	**0:30**	**≥4**
Barium	-	108	(2.3)	90	(4.2)	Biochar	168	NA
Calcium	-	330500	(1528)	293000	(3464)	Biochar	168	≥4
Cobalt	-	0.6	(0.1)	0.3	(0.1)	Biochar	0:30	≥4
Iron	-	1372	(360)	671	(211)	Biochar	1	NA
Magnesium	-	93700	(302)	-		-	-	-
Potassium	-	30022	(11416)	-		-	-	-
Sodium	-	446000	(2363)	396667	(14170)	Biochar	4	≥4
Strontium	-	1648	(275)	-		-	-	-
Vanadium	-	1098	(102)	149	(31)	Biomass	168	≥4

Elements which were reduced below ANZECC trigger level are in bold.
"NA" indicates that the lowest concentration was not significantly different between pH conditions.
"-" indicates element did not change or only increased in concentration.

biosorbent for different constituents of an extremely complex waste water effluent is clearly demonstrated where biochar binds metals from solution at a rate that is affected by pH, while Fe-biochar consistently binds metalloids from solution in a manner that is unaffected by pH. Our results therefore highlight the complexities of biosorption that are only apparent in experiments conducted on real-world industrial effluents. No single biosorbent was effective at holistically treating the range of elements in the complex ADW and so biosorption strategies for real-world effluents may require multiple stages of treatment.

The greatest change in metal and metalloid concentration within the ADW occurred in the first hour of exposure. Rapid initial sorption of metals and metalloids is commonly reported [28–33]. The effect of pH was pronounced for the untreated biosorbents, biomass and biochar, with biosorption patterns at low initial pH (2.5) often differing to those at higher initial pH (4.0 and 7.1). The effect of pH for Fe-biomass and Fe-biochar was, however, negligible. The pH-independent sorption of metalloids by Fe-treated biosorbents may be due to the formation of inner-sphere complexes [14], which are largely unaffected by ionic strength and act without the restrictions of electrostatic attraction, allowing bond formation irrespective of net biosorbent charge [34].

The suites of elements targeted by the most successful biosorbents, Fe-biochar and biochar, were distinct and complementary. Fe-biochar removed the oxyanionic metalloids As, Mo

and Se, with As and Mo being reduced to below their respective ANZECC trigger levels, which is particularly notable for Mo as the concentration was initially 40 times in excess of the trigger level. While the ability of Iron Based Sorbents (IBS) to remediate anions has been established [11,12,13,35], we have shown that the remediation of these metalloids in a complex effluent comes at the expense of substantial leaching of metals back into solution. Conversely, biochar was able to remove a suite of metals from solution and did not leach any ANZECC elements into solution at high pH. The ability to simultaneously remove multiple metals from solution makes biochar a very successful biosorbent in a multi-elemental context, and offers the potential to combine biochar and Fe-biochar in sequential treatment strategies to sequester both metals and metalloids from complex effluents.

For biochar, there was significant variation in metal sorption with half of the ANZECC metals (Al, Cu, Ni & Zn) being remediated to below their respective trigger levels, while metals such as Mn and K leached off the biochar and into solution. Ionic affinity for biosorbents is not fully understood, however, the ionic radius and electronegativity of a metal may have a significant effect [36–38]. For example, cation adsorption onto freeze-dried fungus *Rhizopus arrhizus* may be related to the electronegativity and ionic radius of each ion, otherwise known as the Covalent Index, which suggests Mn^{2+} has a very low sorption affinity [39,40]. Consequently, it is possible that during this study ion exchange is

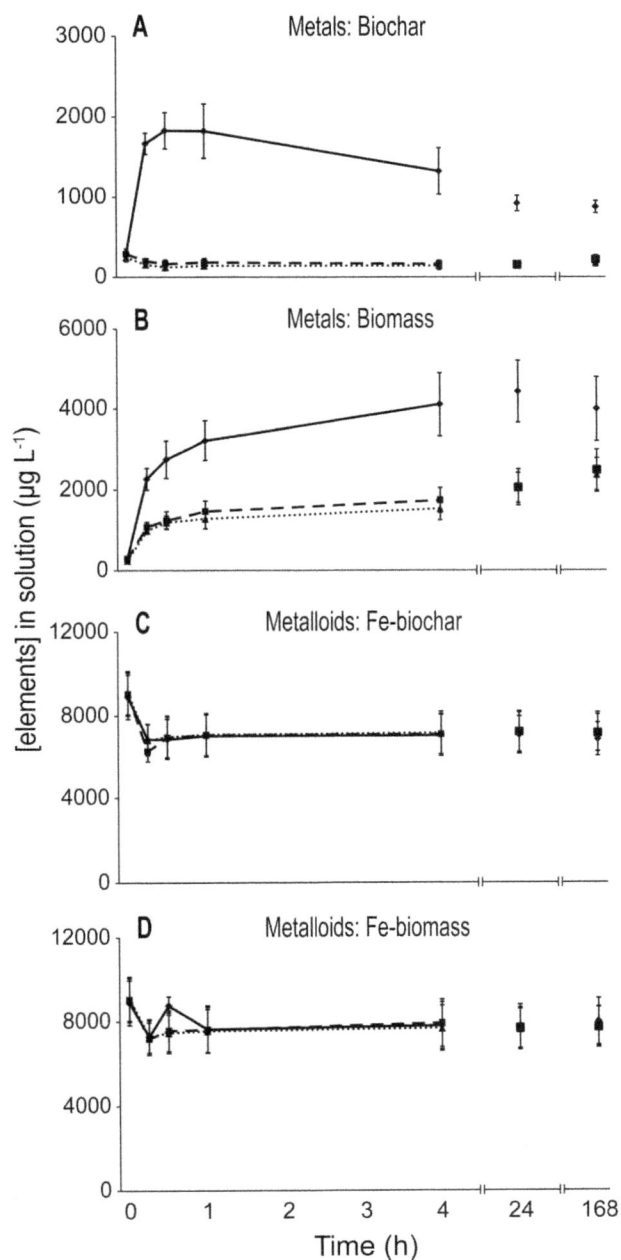

Figure 1. Total respective biosorption of ANZECC metals by biochar and biomass, and metalloids by Fe-treated biosorbents. Biosorption of metals (Al, Cd, Cr, Cu, Pb, Mn, Ni, and Zn) by (a) biochar and (b) biomass and biosorption of metalloids (As, B, Mo and Se) by (c) Fe-biochar and (d) Fe-biomass. Initial pH of 2.5, 4 and unmanipulated (7.1) are shown by solid, dashed and dotted lines, respectively. Error bars show standard errors

occurring involving the release of Mn, with its relatively low affinity and high abundance on *Oedogonium*, in exchange for metals of higher affinity such as Zn, Pb or Cu [36].

Metals and metalloids behaved differently when exposed to biomass and biochar under low initial pH conditions. When exposed to biomass and biochar at an initial pH of 2.5, several metals (Al, Cd, Mn, Ni, Zn) had higher concentrations in solution than the higher pH (4 & 7.1) treatments. As described earlier, the biomass was initially sourced from Tarong Ash Dam and cultivated in f/2 media. Consequently, the resulting biomass contained elements from f/2 media that are required for growth and some of these elements leached when the dried biomass was returned to water at low pH (particularly Cu, Mo, Mn, Zn and Fe). Our finding that some elements leached from biomass at low pH further highlights the importance of measuring a broad suite of analytes in biosorption experiments to uncover unexpected interactions between target and non-target elements. Conversely, metalloids (Se, Mo and V) had lower concentrations at low initial pH. There are several possible explanations for this pH-mediated response. Firstly, the increased metal concentration is a result of increased availability of free-ions at lower pH [41,42]. Second, the metals could be competitively excluded from the biosorption sites by the increased number of protons at lower pH [43,44]. Third, the lowering of the pH below the isoelectric point of the biosorbent resulted in a net charge reversal and therefore enhanced the adsorption of metalloids while limiting the adsorption of metals [45–47]. In reality the pH-dependent adsorption of ions onto biosorbents is probably due to a combination of factors [6,7]. Overall however, lowering the pH to 2.5 in this study had no benefit to the removal of ions as elements were most successfully removed from the effluent at an un-manipulated initial pH of 7.1.

Interestingly, when the biomass was converted to biochar, the metal leaching at low pH was reduced by more than 50%. While the behavior of complex feed stocks during slow pyrolysis is relatively poorly understood, it is known that biochar produced from element-rich biomass typically has a lower exchangeable fraction of metals than the feed stock. Some elements that are constituents of biomass are volatile and may not report to the biochar fraction during slow pyrolysis. Furthermore, converting biomass to biochar changes the speciation of bound metals, rendering them less liable to dissociation [48]. This is clearly supported by the significantly lower leaching of metals from biochar at low pH in our study. Clearly, therefore, biomass cultivated using f/2 media – or any similar growth media – can be considered an appropriate feedstock for biosorption despite containing elements that are also targets for bioremediation, and this biomass is most effective when converted to biochar and used at an unmodified pH.

In an overall sense, there is a developing dichotomy in the study of biosorption of metals and metalloids. The majority of research to date has focused on the kinetics and mechanisms of biosorption in synthetic effluents, which are in essence abstract and simplified conditions. While these studies are important in understanding the processes involved in biosorption, they lack the authenticity of complex effluents in which biosorption is to be applied [5–7]. Our results clearly demonstrate that macroalgae are a versatile feedstock for biosorbents, as *Oedogonium* biomass was able to be converted to biochar, Fe-biomass and Fe-biochar, each of which displayed differential affinity for metals and metalloids. Determination of whether the metal leaching that occurred for Fe-biochar (during the removal of the problematic metalloids) is an acceptable outcome or if the metals could be remediated using another process or biosorbent such as a sequential treatment, requires further investigation. A sequential approach in which alternative macroalgal-derived biosorbents are used in sequence on the same effluent solution, each targeting a specific suite of elements in the effluent, may result in a more comprehensive treatment. Regardless, our results highlight the critical importance of research that evaluates biosorbent performance in industrial effluents to fully understand the potential and

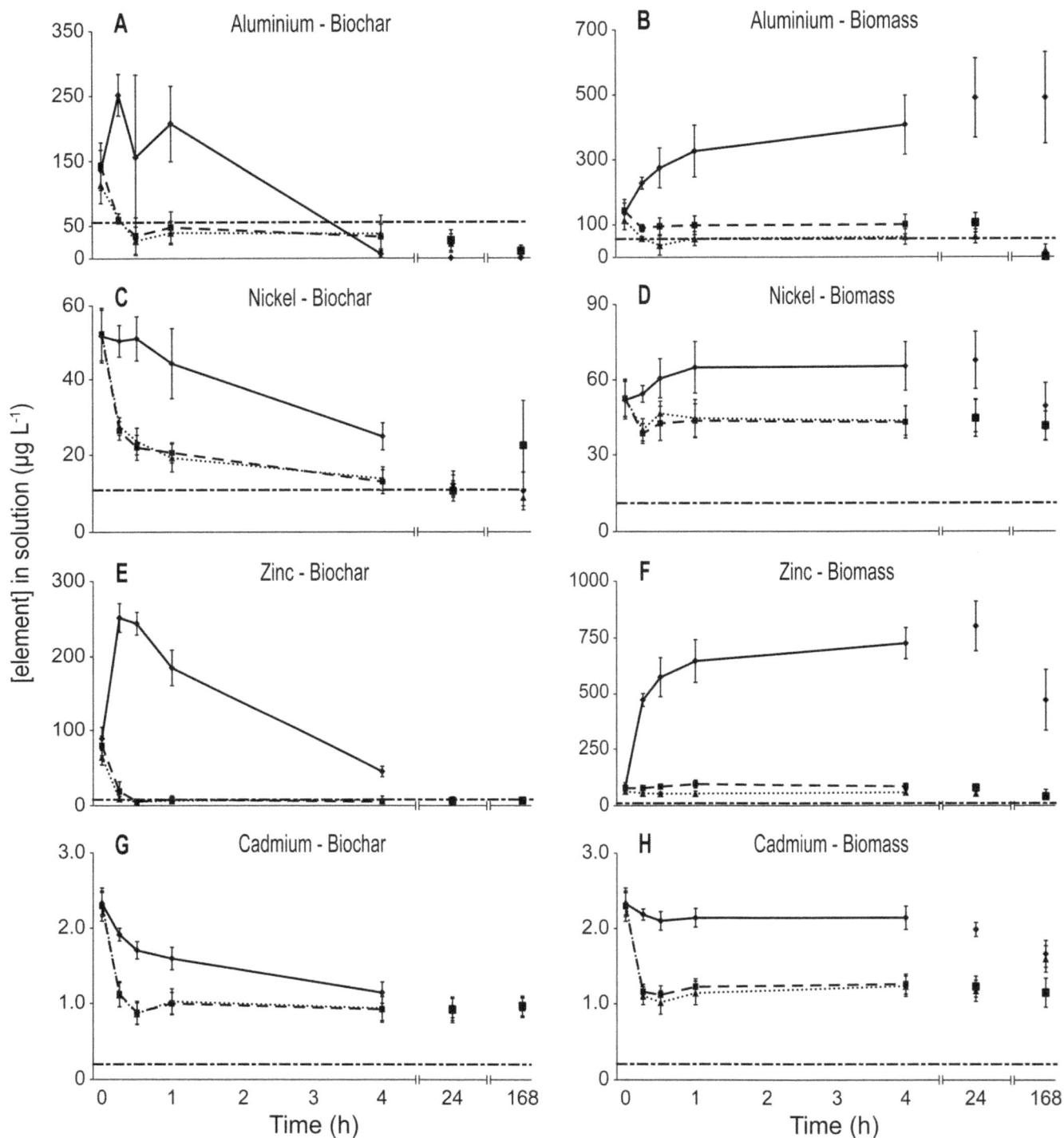

Figure 2. Biosorption of ANZECC metals (Al, Ni, Cd, and Zn) when exposed to biochar and biomass. Initial pH of 2.5, 4 and un-manipulated (7.1) are shown by solid, dashed and dotted lines, respectively. Error bars show standard errors. Horizontal dashed line indicates the respective ANZECC trigger concentration for each element.

viability of algal-based biosorption as a water treatment technology.

Conclusions

In conclusion we have not only demonstrated that the macroalga *Oedogonium* is an effective biosorbent in a complex industrial effluent, but we have done so with a macroalga that can be produced on-site at industrial facilities [18]. The biomass used in this study was cultivated at large scale in f/2 media to provide a rapidly growing source of biomass for waste water treatment in industry. To achieve this rapid growth, some elements must be added as part of any standard algal growth media, but these

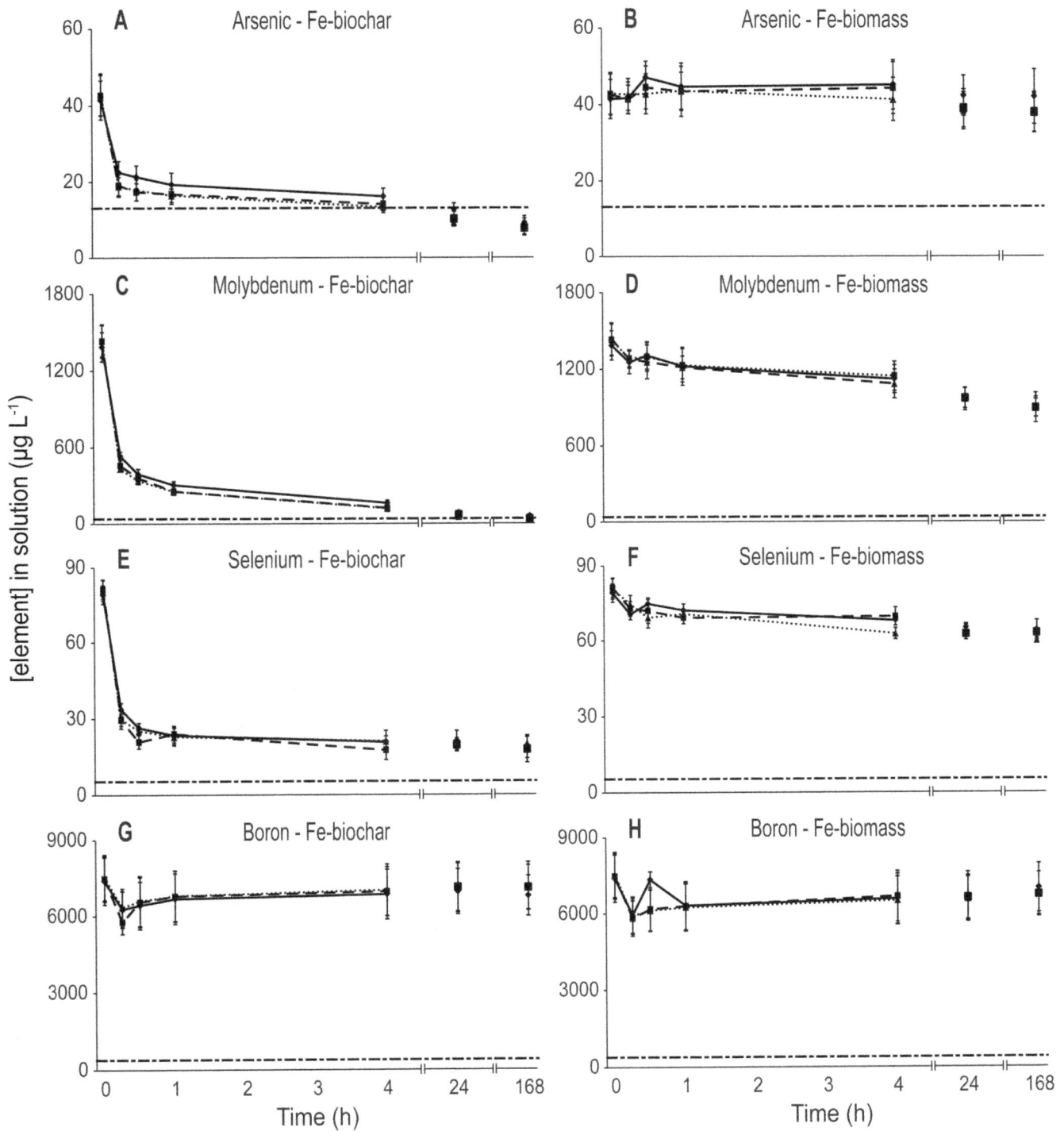

Figure 3. Biosorption of ANZECC metalloids (As, B, Mo and Se) when exposed to Fe-biochar and Fe-biomass. Initial pH of 2.5, 4 and unmanipulated (7.1) are shown by solid, dashed and dotted lines, respectively. Error bars show standard errors. Horizontal dashed line indicates the respective ANZECC trigger concentration for each element.

elements are minor components of the biomass relative to the positive effects of biosorption. The intensive cultivation of the biomass delivers the productivities required to support scaled biosorption which circumvents a critical barrier to application of biosorption. Furthermore, the on-site production of a native macroalga negates one of the most problematic components in the

use of algal-based biosorbents, the source and transport of the biomass [6]. This offers a new paradigm in sustainable waste water treatment, where biomass for bioremediation is produced on-site at industrial facilities while delivering carbon capture. Through this strategic integration of industries, algal-based biosorption will have much greater prospects for industrial application.

Supporting Information

Figure S1 Principal Components Analysis of solution concentration for 12 ANZECC elements. (A) PCA and (B) factor loadings for 12 elements include all biosorbent, time periods (excluding t0) and pH conditions, grouped by biosorbent. Vectors (factor loadings) indicate the direction and magnitude of correlation between a specific element and the biosorbent which resulted in the lowest respective concentration.

Figure S2 The total respective biosorption of metals (Al, Cd, Cr, Cu, Pb, Mn, Ni, and Zn) by (a) Fe-biomass and (b) Fe-biochar and total respective biosorption of metalloids (As, B, Mo and Se) by (c) biochar and (d) biomass. Initial pH of 2.5, 4 and un-manipulated (7.1) are shown by solid, dashed and dotted lines, respectively. Error bars show standard errors.

Figure S3 The biosorption of ANZECC metals (Pb, Cr, Cu, and Mn) when exposed to biochar and biomass. Initial pH of 2.5, 4 and un-manipulated (7.1) are shown by solid, dashed and dotted lines, respectively. Error bars show standard errors. Horizontal

dashed line indicates the respective ANZECC trigger concentration for each element.

Table S1 Three factor factorial Analysis of Variance tests run on each of the 12 ANZECC elements. Factorial analysis of variance tests were run on elemental concentration under the factors of Biosorbent, pH (Fixed) and Time (Random). Type III sum of squares was used. All tests met the assumption of homogeneity of variance, normality of residuals and independence. Transformation of the data were required for some elements, the transformation applied is listed next to the title. Factors in bold indicate significance under alpha of 0.05.

Acknowledgments

We thank Stanwell Energy Corporation for providing effluents for the experimental studies, and Charlotte Johansson, Thomas Mannering, Michael Bird and Tony Forsyth for their invaluable input and assistance.

Author Contributions

Conceived and designed the experiments: JK RN NP DR. Performed the experiments: JK. Analyzed the data: JK NP DR. Contributed reagents/materials/analysis tools: RN YH. Wrote the paper: JK RN NP DR. Chemical analyses: YH.

References

1. Volesky B (2001) Detoxification of metal-bearing effluents: Biosorption for the next century. Hydrometallurgy 59: 203–216.
2. Jankowski J, Ward CR, French D, Groves S (2006) Mobility of trace elements from selected Australian fly ashes and its potential impact on aquatic ecosystems. Fuel 85: 243–256.
3. Frankenberger WT Jr, Amrhein C, Fan TWM, Flaschi D, Glater J, et al. (2004) Advanced treatment technologies in the remediation of seleniferous drainage waters and sediments. Irrigation Drainage Syst 18: 19–42.
4. Oman J, Dejanovic B, Tuma M (2002) Solutions to the problem of waste deposition at a coal-fired power plant. Waste Manage 22: 617–623.
5. Mehta SK, Gaur JP (2005) Use of algae for removing heavy metal ions from wastewater: Progress and prospects. Crit Rev Biotechnol 25: 113–152.
6. Volesky B (2007) Biosorption and me. Water Res 41: 4017–4029.
7. Gadd GM (2009) Biosorption: Critical review of scientific rationale, environmental importance and significance for pollution treatment. J Chem Technol Biotechnol 84: 13–28.
8. Volesky B, Holan ZR (1995) Biosorption of heavy metals. Biotechnol Prog 11: 235–250.
9. Domozych DS, Ciancia M, Fangel JU, Mikkelsen MD, Ulvskov P, et al. (2012). The cell walls of green algae: a journey through evolution and diversity. Front Plant Sci 3: 1–7.
10. Davis TA, Volesky B, Mucci A (2003). A review of heavy metal biosorption by brown algae. Water Res 37: 4311–4330.
11. Latva S, Peraniemi S, Ahlgren M (2003) Study of metal-loaded activated charcoals for the separation and determination of selenium species by energy dispersive X-ray fluorescence analysis. Anal Chim Acta 478: 229–235.
12. Bird MI, Wurster CM, Silva PHD, Bass AM, de Nys R (2011) Algal biochar: Production and properties. Bioresour Technol 102: 1886–1891.
13. Roberts DA, Paul NA, de Nys R (2013) Biosorbents and methods of use. Provisional patent 27326AU1-DJH/MAR. Available: http://www.ipaustralia.com.au/applicant/james-cook-university/patents/AU2013902101/. Accessed 21 March 2014.
14. Manceau A, Charlet L (1994) The mechanism of selenate adsorption on goethite and hydrous ferric oxide. J Colloid Interface Sci 168: 87–93.
15. Lalvani SB (2004) Selemium removal from agricultural drainage water: Lab scale studies. Sacramento: Department of Water Resources. 86 p.
16. Yang T, Chen M-L, Liu L-H, Wang J-H, Dasgupta PK (2012) Iron(III) Modification of Bacillus subtilis membranes provides record sorption capacity for arsenic and endows unusual selectivity for As(V). Environ Sci Technol 46: 2251–2256.
17. Saunders RJ, Paul NA, Hu Y, de Nys R (2012) Sustainable sources of biomass for bioremediation of heavy metals in waste water derived from coal-fired power generation. PLoS ONE 7: e36470.
18. Roberts D, de Nys R, Paul N (2013) The effect of CO2 on algal growth in industrial waste water for bioenergy and bioremediation applications. PLoS ONE 8: e81631.
19. Lee HS, Suh JH, Kim B, Yoon T (2004). Effect of aluminium in two-metal biosorption by an algal biosorbent. Miner Eng 17: 487–493.
20. Mehta SK, Tripathi BN, Gaur JP (2000). Influence of pH, temperature, culture age and cations on adsorption and uptake of Ni by Chlorella vulgaris. Eur J Protistol 36 443–450.
21. Figueira MM, Volesky B, Azarian K, Ciminelli VST (2000). Biosorption column performance with a metal mixture. Environ Sci Technol 34: 4320–4326.
22. Figueira MM, Volesky B, Ciminelli VST (1997). Assessment of interference in biosorption of a heavy metal. Biotechnol Bioeng 54: 344–350.
23. Lawton RJ, de Nys R, Skinner S, Paul NA (2014) Isolation and selection of Oedogonium species and strains for biomass applications. PLoS ONE 9: e90223.
24. Entwistle TJ, Skinner S, Lewis SH, Foard HJ (2007). Algae of Australia: Batrachospermales, Thoreales, Oedogoniales and Zygnemaceae. Canberra: CSIRO Publishing. 191 p.
25. Jolliffe IT (2002) Principal Component Analysis. New York: Springer-Verlag. 487 p.
26. Quinn G, Keough M (2002) Experimental design and data analysis for biologists. Cambridge: Cambridge University Press. 556 p.
27. ANZECC (2000) Australian and New Zealand guidelines for fresh and marine water quality. Sydney: Australian Water Association. 314 p.
28. Vieira DM, da Costa ACC, Henriques CA, Cardoso VL, de Franca FP (2007) Biosorption of lead by the brown seaweed Sargassum filipendula - Batch and continuous pilot studies. Electron J Biotechnol 10: 368–375.
29. Pavasant P, Apiratikul R, Sungkhum V, Suthiparinyanont P, Wattanachira S, et al. (2006) Biosorption of Cu^{2+}, Cd^{2+}, Pb^{2+}, and Zn^{2+} using dried marine green macroalga Caulerpa lentillifera. Bioresour Technol 97: 2321–2329.
30. Karthikeyan S, Balasubramanian R, Iyer CS (2007) Evaluation of the marine algae Ulva fasciata and Sargassum sp. for the biosorption of Cu(II) from aqueous solutions. Bioresour Technol 98: 452–455.
31. Esmaeili A, Ghasemi S, Sohrabipour J (2010) Biosorption of copper from wastewater by activated carbon preparation from alga Sargassum sp. Nat Prod Res 24: 341–348.
32. Schneegurt MA, Jain JC, Menicucci JA, Brown SA, Kemner KM, et al. (2001) Biomass byproducts for the remediation of wastewaters contaminated with toxic metals. Environ Sci Technol 35: 3786–3791.
33. Sheng PX, Ting YP, Chen JP, Hong L (2004) Sorption of lead, copper, cadmium, zinc, and nickel by marine algal biomass: Characterization of biosorptive capacity and investigation of mechanisms. J Colloid Interface Sci 275: 131–141.
34. Stumm W, Morgan J (1996) Aquatic chemistry: Chemical equilibria and rates in natural waters. New York: Wiley. 1040 p.
35. Zhang N, Lin LS, Gang DC (2008) Adsorptive selenite removal from water using iron-coated GAC adsorbents. Water Res 42: 3809–3816.
36. Can C, Jianlong W (2007) Correlating metal ionic characteristics with biosorption capacity using QSAR model. Chemosphere 69: 1610–1616.
37. Zamil SS, Ahmad S, Choi MH, Park JY, Yoon SC (2009) Correlating metal ionic characteristics with biosorption capacity of Staphylococcus saprophyticus BMSZ711 using QICAR model. Bioresour Technol 100: 1895–1902.

38. Kogej A, Pavko A (2001) Comparison of *Rhizopus nigricans* in a pelleted growth form with some other types of waste microbial biomass as biosorbents for metal ions. World J Microbiol Biotechnol 17: 677–685.

39. Brady JM, Tobin JM (1995) Binding of hard and soft metal-ions to *Rhizopus arrhizus* biomass. Enzyme Microb Technol 17: 791–796.

40. Tobin JM, Cooper DG, Neufeld RJ (1984) Uptake of metal ions by *Rhizopus arrhizus* biomass. Appl Environ Microbiol 47: 821–824.

41. Sigg L, Xue H (1994) Metal speciation: Concepts, analysis and effects. In: Bidoglio G, Stumm W, editors. Chemistry of aquatic systems: Local and global perspectives. Brussels: ECSE EAEC.pp. 153–181.

42. Esposito A, Pagnanelli F, Veglio F (2002) pH-related equilibria models for biosorption in single metal systems. Chem Eng Sci 57: 307–313.

43. Peterson HG, Healey FP, Wagemann R (1984) Metal toxicity to algae - A highly ph dependent phenomenon. Can J Fish Aquat Sci 41: 974–979.

44. Pirszel J, Pawlik B, Skowronski T (1995) Cation-exchange capacity of algae and cyanobacteria: A parameter of their metal sorption abilities. J Ind Microbiol 14: 319–322.

45. Crist RH, Martin JR, Carr D, Watson JR, Clarke HJ, et al. (1994) Interaction of metals and protons with algae.4. Ion-exchange vs adsorption models and a reassessment of scatchard plots - ion-exchange rates and equilibria compared with calcium alginate. Environ Sci Technol 28: 1859–1866.

46. Garnham GW, Avery SV, Codd GA, Gadd GM (1994) Interactions of microalgae and cyanobacteria with toxic metals and radionuclides: Physiology and environmental implications. In: Dyer KR, Orth RJ, editors. Changes in fluxes in estuaries - Implications from science to management. Fredensborg: Olsen and Olsen. pp. 289–293.

47. Schijf J, Ebling AM (2010) Investigation of the ionic strength dependence of *Ulva lactuca* acid functional group pKas by manual alkalimetric titrations. Environ Sci Technol 44: 1644–1649.

48. Farrell M, Ragnott G, Krull E (2013). Difficulties in using soil-based methods to assess plant availability of potenitally toxic elements in biochars and their feedstocks. J Hazard Mat 250–251: 29–36.

11

The Duration of Gastrointestinal and Joint Symptoms after a Large Waterborne Outbreak of Gastroenteritis in Finland in 2007 – A Questionnaire-Based 15-Month Follow-Up Study

Janne Laine[1,2*¶], Jukka Lumio[3], Salla Toikkanen[2], Mikko J. Virtanen[2], Terhi Uotila[1], Markku Korpela[1,3], Eila Kujansuu[4¤], Markku Kuusi[2]

1 Tampere University Hospital, Department of Internal Medicine, Tampere, Finland, 2 National Institute of Health and Welfare, Epidemiologic Surveillance and Response Unit, Helsinki, Finland, 3 University of Tampere, School of Medicine, Tampere, Finland, 4 Nokia Health Centre, Nokia, Finland

Abstract

An extensive drinking water-associated gastroenteritis outbreak took place in the town of Nokia in Southern Finland in 2007. 53% of the exposed came down with gastroenteritis and 7% had arthritis-like symptoms (joint swelling, redness, warmth or pain in movement) according to a population-based questionnaire study at 8 weeks after the incident. *Campylobacter* and norovirus were the main pathogens. A follow-up questionnaire study was carried out 15 months after the outbreak to evaluate the duration of gastrointestinal and joint symptoms. 323 residents of the original contaminated area were included. The response rate was 53%. Participants were inquired about having gastroenteritis during the outbreak and the duration of symptoms. Of those with gastroenteritis, 43% reported loose stools and abdominal pain or distension after the acute disease. The prevalence of symptoms declined promptly during the first 3 months but at 15 months, 11% reported continuing symptoms. 32% of the respondents with gastroenteritis reported subsequent arthritis-like symptoms. The disappearance of arthritis-like symptoms was more gradual and they levelled off only after 5 months. 19% showed symptoms at 15 months. Prolonged gastrointestinal symptoms correlated to prolonged arthritis-like symptoms. High proportion of respondents continued to have arthritis-like symptoms at 15 months after the epidemic. The gastrointestinal symptoms, instead, had declined to a low level.

Editor: Oliver Schildgen, Kliniken der Stadt Köln gGmbH, Germany

Funding: This study was financially supported by the Competitive Research Funding of the Tampere University Hospital (Grants 9E129 and 9M124). The funders had no role in study design, data collection and analysis, decision to publish, or preparation of the manuscript.

Competing Interests: The authors have declared that no competing interests exist.

* E-mail: janne.laine@pshp.fi

¤ Current address: Department of Social Services and Health Care, Tampere, Finland

¶ The Pirkanmaa Waterborne Outbreak Study Group: Jaakko Antonen[1,3], Pekka Collin[1,3], Tina Katto[1], Anna-Leena Kuusela [1,3], Sami Mustajoki[1], Jukka Mustonen[1,3], Heikki Oksa[1,3], Petri Ruutu[2] and Sirpa Räsänen[1]

Introduction

Although acute gastroenteritis is a common disease, little is known about its consequences after passing the acute phase of the disease. As nearly everyone meets episodes of acute gastroenteritis occasionally, the possible connection with later health problems is easily missed. Food- and waterborne epidemics give special opportunity to study this connection prospectively as a cohort of people is infected in a relatively short period, often with identified pathogens. Epidemics large enough to conduct a study are, however, uncommon in developed countries with a good level of investigation capabilities.

Reactive arthritis (ReA) and milder forms of joint complaints are well known and fairly common acute complications of bacterial gastroenteritis [1]. There is also growing evidence of increased incidence of irritable bowel syndrome (IBS) after bacterial [2,3] and possibly viral [4] gastroenteritis. The most intensively studied waterborne outbreak is the Walkerton epidemic in Ontario, Canada. According to the Walkerton Health Study, increased risk of reactive arthritis, irritable bowel syndrome, pregnancy-related hypertension, hypertension, kidney disease, and even cardiovascular events has been observed after the epidemic [5,6,7,8,9,10].

A large drinking-waterborne epidemic occurred in a Finnish town Nokia in November–December 2007. Shortly after the incident, a comprehensive outbreak investigation of short- and long-term health effects was initiated. Details on the epidemiology, microbial findings, and early arthritic complaints have been published previously [11,12,13,14,15]. We performed a questionnaire study at 15 months after the exposure to contaminated water to observe the duration of symptoms and the remaining health complaints at 15 months after the incident. Here, we present data on the persistence of gastrointestinal and joint symptoms after the outbreak.

Methods

Setting

The town of Nokia is located in Southern Finland and has a population of 30 000. At the end of November 2007, maintenance work was carried out in the town's wastewater plant. During the work, a valve connecting the wastewater plant's effluent line and household water distribution line was opened and it was accidentally left open for two days. Approximately 450 m^3 of plant's effluent water contaminated the drinking water of 9 500 residents. This resulted in a large gastroenteritis outbreak in the contaminated area of the town. Some excess morbidity was also detected in the uncontaminated area [11]. The epidemic peaked four days after the incident, and most gastroenteritis cases occurred within two weeks.

Seven pathogens were found from the patients' stools. Six of those were also detected from tap water or water-distribution network samples. *Campylobacter* was the most common pathogen found in 27% (N = 148) of stool samples. In addition, norovirus was also considered as a major pathogen [16]. Non-typhoidal *Salmonellae* and *Giardiae* [13] were detected in less number of samples. In a study focusing on children with gastroenteritis, other viruses and infections with mixed pathogens were also observed [12].

According to a population-based questionnaire survey performed eight weeks after the incident 53% of the population in the contaminated area reported falling ill with gastroenteritis and 6.7% reported having arthritis-like joint symptoms [15]. In spite of active encouragement to refer new cases of probable arthritis to a rheumatologist at the local university hospital, only 21 confirmed cases of ReA were detected [14].

Questionnaire studies

Two consecutive questionnaires were mailed 13 months apart. The first was mailed eight weeks after the incident to detect the immediate morbidity [15]. The second, a follow-up study, was performed fifteen months after the outbreak in order to study the persistence of gastrointestinal and joint symptoms.

Details of the first questionnaire at 8 weeks were presented in previous articles [11,15]. In brief, two target populations were defined from Nokia; those residing in the part of town with contaminated water supply ("contaminated group") and those residing in the uncontaminated area ("uncontaminated group"). A third group ("control group") was chosen among the citizens of another municipality in the same district. Approximately, 1000 participants were randomly picked from the national population registry for each group (Table 1). All the ages were included and the groups were matched for age and gender. Only one participant per household was allowed.

The second questionnaire at 15 months ("follow-up") was mailed to persons who had participated in the first survey and given their consent for being contacted again. Therefore, the groups became smaller (Table 1.). No reminder was mailed to those who did not respond. Participants were asked whether they had gastroenteritis during the outbreak period (from November 28 to December 31 2007). They were also asked for how long (weeks or months) abdominal or joint symptoms persisted after the acute gastroenteritis.

Gastrointestinal symptoms, such as the presence of loose stools, constipation, nausea, abdominal pain, and abdominal distension were recorded. After analyzing these symptoms separately, a combination of gastrointestinal symptoms was created to observe IBS-like symptoms, such as loose stools and abdominal pain or distension. This combination is close to the Manning 3 and/or Rome I IBS-criteria [17].

Of joint symptoms, joint pain, pain in joint movement and the presence of swelling, redness, or warmth of the joint were recorded. The condition was classified as arthritis-like if pain in joint movement or any of the symptoms of joint swelling, redness, or warmth was present as defined in our previous study of early joint symptoms [15].

The persistence of gastrointestinal and arthritis-like symptoms was then analyzed as the remaining prevalence of a symptom in the course of time among those who had declared as having had gastroenteritis during the epidemic.

Use of microbiological data

Data of the stool specimens sent for microbiological analysis were obtained from the database of the Fimlab laboratories. This institution is a district-wide and public-run clinical laboratory serving the entire public health-care sector in the district. From year 2005 to 2009, the quantity, dates, and microbiological findings among the residents of Nokia and the control municipality were collected for comparison.

Statistical methods

Completed forms were stored in a database and analysed in the National Institute of Health and Welfare. Estimation of the effect of selection was done in two steps. First, the probability to be included in the study (i.e., those who gave permission to be contacted again) within all the initial study subjects was estimated by using univariate logistic regression. Secondly, the probability of returning the questionnaire within the included subjects was estimated in a similar manner. Confidence intervals for the proportion of subjects continuing to have symptoms at each time point were calculated by using ordinary bootstrap. All analyses were done with R version 2.15.2.

Ethical considerations

When answering the first questionnaire, participants were asked whether they would accept a follow-up contact. An informed consent was attached in the second questionnaire. The follow-up study was approved by the ethical committee of the Tampere University Hospital.

Results

Response rates and selection

Answer on the first questionnaire was received from 808 (79%) persons of the contaminated group, 717 (73%) of the uncontaminated group and 598 (60%) of the control group (Table 1). Taking all groups together, 667 (31.4%) of those responding declined the request for further contact. Response rates to the follow-up questionnaire were 53% (323/615), 46% (230/498), and 54% (186/343), respectively.

The odds ratios (ORs) for probability to be included in the follow-up questionnaire (i.e., giving permission to be contacted again) were 1.2 (95% CI 1.0–1.5) for the female gender, 1.7 (1.3–2.1) for those who had reported gastroenteritis and 1.7 (1.3–2.2) for those reporting joint symptoms in the first questionnaire. For these features the ORs for responding to the follow-up questionnaire were 1.1 (0.9–1.3), 1.0 (0.8–1.2) and 1.03 (0.8–1.3), respectively.

Of those participating in the follow-up study, 54% (174/323) in the contaminated group, 13% (31/230) in the uncontaminated group, and 3% (6/186) in the control group reported that they had gastroenteritis during the outbreak.

Table 1. Evolution of the study groups in the follow-up study.

	Nokia, contaminated	Nokia, uncontaminated	Control population
Population 2007	9 538	20 478	27 259
Size of the study groups in the first study	1 021	979	1 000
Responded to the first study	808	717	598
Sample size in the follow-up study (i.e., those who gave permission to be contacted again)	615	498	343
Responded to the follow-up study	323	230	186
Gastroenteritis during the epidemic, according to the follow-up study	174		

Study groups were based on the original population samples that were used in the first survey. The study groups became step-wise smaller because of lack of response to the first survey, denying further contact, and lack of response to the follow-up survey.

Gastrointestinal symptoms

Of the 174 persons in the contaminated group who had gastroenteritis, 53.6% (93) reported continuing loose stools, 47.9% (83) abdominal pain, and 32.7% (57) abdominal distension after the acute disease. 42.7% (74) experienced loose stools and abdominal pain or distension in the beginning of the follow-up (Figure 1). After a rapid decrease in the proportion of symptomatic persons in one month, the decline started to level off. At 15 months, 10.9% (19) of the respondents continued to have loose stools and abdominal pain or distension. This figure was 11.5% for adults 16 years of age or older and 8.9% for children.

Arthritis-like symptoms

Among the 174 participants in the contaminated group who had experienced gastroenteritis, 31.8% (55/174) reported arthritis-like symptoms after acute gastroenteritis. The proportion of persons with arthritis-like symptoms declined more slowly over time than the corresponding proportion with gastrointestinal symptoms (Figure 1). At 15 months, 19.0% (33/174) reported as continuing to have arthritis-like symptoms. The persistence of gastrointestinal symptoms (loose stools and abdominal pain or distension) predicted the presence of arthritis-like symptoms at 3 and 15 months (Table 2).

Use of microbiological tests

In Nokia, a sharp peak in the number of microbiological stool specimens was observed during the outbreak. Thereafter, the number of specimens diminished gradually and reached the pre-outbreak level in six months. In Nokia, the incidence of positive findings was high during the outbreak and for two months thereafter, but no difference was observed compared to the control municipality after one month (Figure 2).

Discussion

The faecal contamination of household water and the subsequent large epidemic of gastroenteritis was an exceptional accident [11]. The contamination was extensive as over half of the exposed population became ill. A remarkable proportion of the population suffered also from joint symptoms within the following eight weeks [15]. Two questionnaire surveys done thirteen months apart, based on representative population samples, focused on short- and long-term health effects of the epidemic. The 15-month follow-up questionnaire study reported here describes the duration of gastrointestinal and arthritis-like symptoms after the epidemic.

53.6% of those having had gastroenteritis during the epidemic reported loose stools and abdominal pain after the acute disease and 42.7% loose stools and abdominal pain or distension. Although the prevalence decreased rapidly within five months, 10.9% was still having symptoms resembling IBS (loose stools and abdominal pain or distension) at 15 months.

Irritable bowel syndrome (IBS) is a condition characterized by altered bowel function and abdominal pain or discomfort. The studies on the prevalence of symptoms compatible with IBS in the general population have given widely variable figures depending on the definition used. For example, in a population-based Finnish study, the prevalence of IBS ranged from 5.1% to 16.2%, depending on diagnostic criteria [17]. According to a recent systematic review of worldwide IBS-studies, the pooled prevalence of IBS was 11.2% among adult subjects [18].

IBS occurring after an episode of infective gastroenteritis has been referred as post-infectious irritable bowel syndrome (PI-IBS). Generally about 10% of IBS-patients regard that their symptoms began after an infective gastroenteritis [3]. According to a meta-analysis of eight studies published until December 2005, the median prevalence of IBS was 9.8% in the infective gastroenteritis groups and 1.2% in the control groups [19]. The observed frequency of PI-IBS is likely to depend on the pathogen(s) involved. In Walkerton, Canada, the prevalence of IBS was 27.5% among adult subjects who had had gastroenteritis during the epidemic while the prevalence among those without gastroenteritis was 10.1% at 2–3 years after the outbreak [9]. Diarrhoea was found to be a more prominent feature in gastroenteritis-associated IBS than in the case of sporadic IBS. The prognosis of PI-IBS is thought to be better than IBS without infectious onset [3]; however, the condition can still take more than 8 years to resolve as was observed by the Walkerton group [3,20]. If PI-IBS occurs after viral gastroenteritis, it is short lived and rarely lasts for 6 months or longer [4,21].

Commonly used diagnostic criteria for IBS were not utilized in this study. However, abdominal pain or discomfort or distension and loose stools are the principal parts of these criteria. Analyzing these symptoms in such a combination makes it reasonably close to Manning 3 and/or Rome I criteria for IBS. The high prevalence of gastrointestinal symptoms within the first five months after the epidemic in our study may therefore indicate PI-IBS. However, at the end of the follow-up, the prevalence of these symptoms was 11.5% among adults ≥16 years of age. In the population-based study among adults in Finland, the prevalence of IBS was 9.7% according to the Manning 3 criteria [17]. Therefore, the

Loose stools and abdominal pain or distension

Arthritis-like symptoms

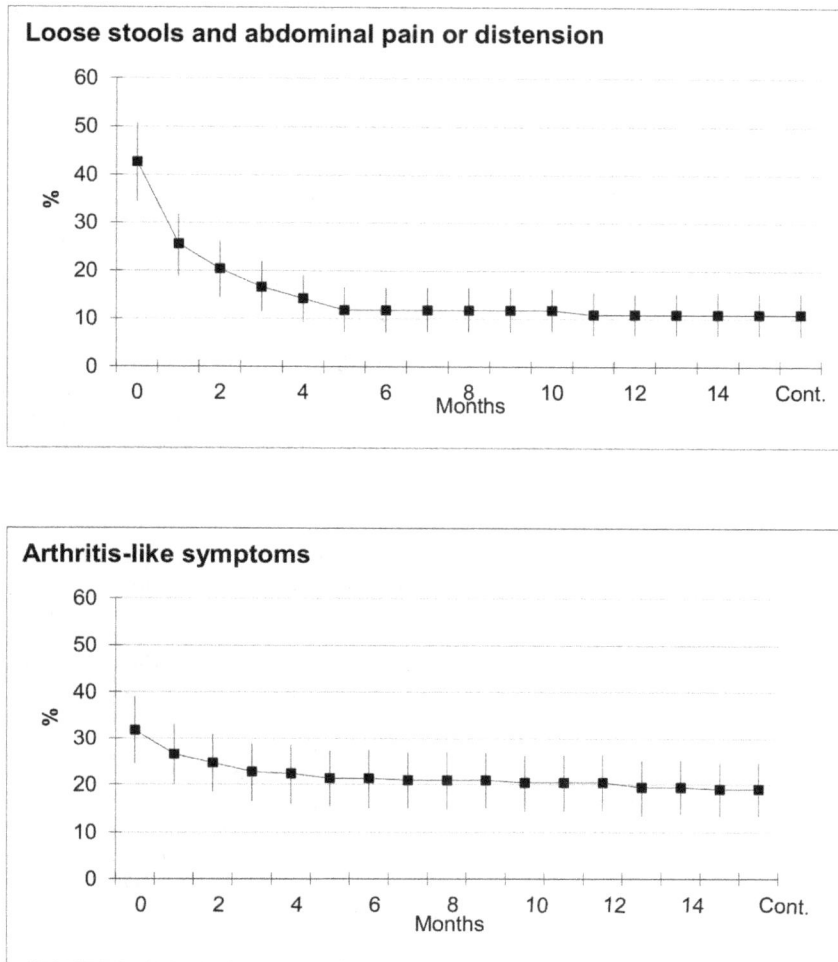

Figure 1. The prevalence of gastrointestinal and arthritis-like symptoms in the contaminated group within fifteen months from the outbreak. The proportions were counted as dividing the number of subjects with symptoms by the number of subjects having fallen ill with gastroenteritis at the time of the outbreak (N = 174). The curves represent the proportion of subjects with symptoms and the vertical lines are the 95% confident intervals.

remaining prevalence of gastrointestinal symptoms in the present study may be close to the natural occurrence of these symptoms.

Other reasons for long-lasting symptoms than PI-IBS must be considered in an epidemic where more than half of the exposed residents fell ill with gastroenteritis caused by mixed pathogens. There may have been prolonged circulation of the pathogens in the population leading to secondary infections. Bacterial stool cultures were extensively examined during the following months in

Table 2. The prevalence of arthritis-like symptoms among participants with and without gastrointestinal symptoms (loose stools and abdominal pain or distension) at 3 and 15 months after the water contamination.

	At 3 months	At 15 months
With gastrointestinal symptoms	54.3% (19/35)	52.2% (12/23)
Without gastrointestinal symptoms	10.9% (6/55)	10.4% (7/67)
p-value	<0.0001	<0.0001

the affected town. No excess frequency of positive findings could be noticed after the peak of the epidemic (Figure 2). This evidence does not support the prolonged circulation of bacterial pathogens in the population. Viral pathogens (especially, norovirus), are highly contagious and could well have started to spread horizontally after the outbreak. Such a possibility could not be excluded reliably because samples for viral diagnostics were not routinely taken during the study period. Giardiasis often causes long-lasting gastrointestinal symptoms. This epidemic was the first domestic Giardia-cluster in Finland [13]. The whole population was informed about the possibility of giardiasis and individuals were called for parasitological investigation if they suffered from prolonged symptoms. In spite of that, only 55 cases were found out of the 872 persons tested, a number that is probably too low to solely explain the magnitude of prolonged gastrointestinal symptoms.

As the data relies on self-reported symptoms, psychological factors like anger, anxiety or distress may have influenced the participants' expressions. The connection of psychological factors on the duration of IBS has been observed in epidemics caused by drinking water [20]. The public anger and anxiety in the town of Nokia was considerable for several months after the epidemic.

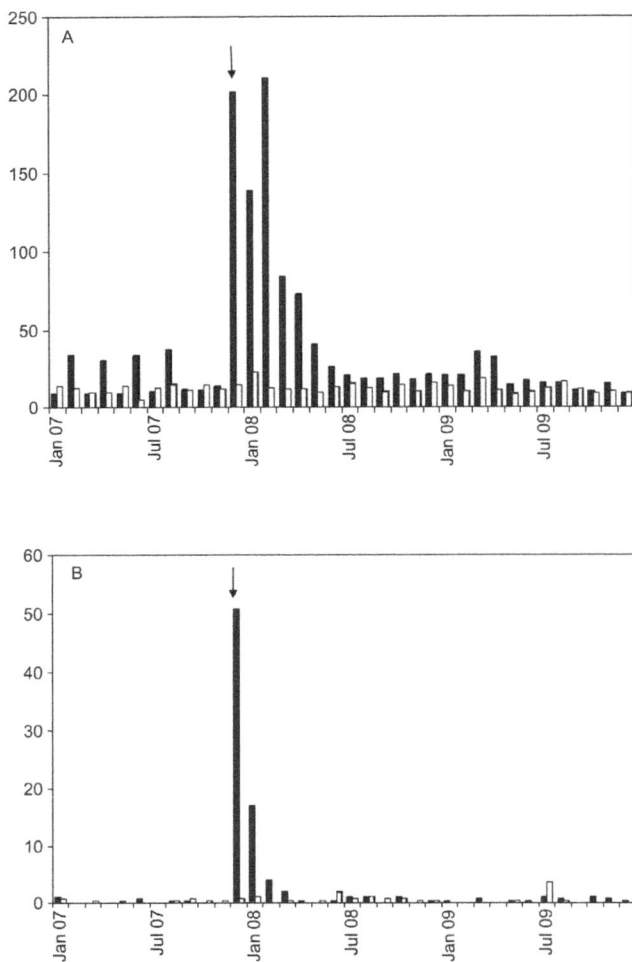

Figure 2. a–b. Monthly incidence (/10 000 inhabitants) of microbiological stool specimens (Fig. 2a) and positive findings (Fig. 2b). Black bars represent the town of Nokia and white bars the control municipality. The arrows indicate the outbreak month, December 2007. Only findings of *Campylobacter* and *Giardia* from January 2007 to December 2009 were counted.

Psychological influences of the incident on health experience will be investigated in another project and reported later.

Previously, we have reported a frequency of 6.7% of arthritis-like symptoms among the participants of the contaminated group within eight weeks from the incident [15]. According to the follow-up study, 31.8% of those who reported having had gastroenteritis during the epidemic announced subsequent arthritis-like symptoms. The explanation for the difference lies in the fact that joint symptoms in the follow-up study were asked for only from those with gastroenteritis and through a selection bias towards those with joint symptoms in the first study.

The arthritis-like symptoms were fairly longstanding. About a third of the symptoms were resolved within five months. Thereafter, the decline was slow and 19% of the respondents were still experiencing arthritis-like symptoms 15 months after the epidemic. In a previously published population-based study of verified ReA cases related to this outbreak, 21 cases of ReA were observed [14], mostly relatively mild cases. According to a follow-up study (one year) of this cohort, over a half of all the patients were still taking analgesics because of ReA symptoms and 33%

was on antirheumatic medication [22]. This proportion of residual symptoms among verified ReA-cases is quite well in line with the findings of the present study.

ReA is a well-known complication of bacterial gastroenteritis [1]. *Giardia* has also been associated rarely with ReA [23]. To our knowledge, there are hardly any observations possibly linking rotavirus or norovirus to ReA. Considering this and the microbiological findings, we regard that campylobacter infections probably have been the major trigger of joint symptoms in this epidemic.

There are few population-based studies on the duration of self-reported joint symptoms following gastroenteritis epidemics. The previous literature deals with those fulfilling the criteria for ReA, either according to self-reporting or clinical judgement. In studies done in Finland, the average duration of ReA symptoms has been 3–5 months [1]. Symptoms persisting over six months have often been considered as a sign of chronic disorder. Development to the chronic condition has been reported in 12–16% of the cases [1]. The proportion of subjects with arthritis-like symptoms (19%), after 15 months in this study, is therefore, reasonably well in line with previous observations, especially, since this was a questionnaire study with no clinical verification of the symptoms.

The persistence of gastrointestinal symptoms correlated with the presence of arthritis-like symptoms (Table 2.). As PI-IBS and ReA are inflammatory disorders, pathogenesis of both gastrointestinal and arthritis-like symptoms may have common immunological mechanisms, although the genetic predisposition for these two conditions seems to be different [24].

There are some limitations of this study. First, although the initial sample for the first questionnaire was carefully randomized, there were three subsequent steps in which a selection bias could have taken place. They are responding to the first questionnaire, giving permission to be re-contacted, and finally responding to the follow-up survey (Table 1). The response rates of the first survey were good and the authors concluded that the survey was representative. For the follow-up survey, some selection in sample formation took place, but not any more in responding to the follow-up survey. Those with joint symptoms in the original cohort were more prone (OR 1.7, 95% CI 1.3–2.2) to participate in the 15-month follow-up. As the aim of the present study was to observe the duration of symptoms, selection towards participants who experienced symptoms during the outbreak does not seriously hamper the conclusions. Secondly, prolonged symptoms were asked only from those who fell ill with gastroenteritis during the outbreak. As one could assume, there were hardly any such cases in the control group and this made meaningful comparisons impossible. Thirdly, as the follow-up survey was done 15 months after the exposure, recalling the symptoms by those surveyed may have been difficult, thereby, creating a recall-bias of some degree. Fourthly, the outbreak and the faults causing it were widely discussed in the public media and several claims for compensation were made. Therefore, it is possible that some subjects with a disease that may have been potentially connected to the water contamination were prone to overweigh their symptoms.

The contamination of drinking water and the subsequent epidemic in the town of Nokia was a potentially dangerous situation. Large amounts of pathogens were distributed through household water and thousands of people fell ill. Given these circumstances, consequences could have been much more serious. In addition, more virulent bacterial strains, such as *Escherichia coli* O157:H7 were avoided. Despite the relatively favourable outcome of this epidemic, however, there were cases of people who suffered from consequences, especially, joint symptoms up to one year or more. In addition, individual patients have suffered from serious, prolonged, or permanent damages caused by joint destruction [22], IBS, or chronic fatigue syndrome.

In conclusion, prolonged gastrointestinal and joint symptoms were common among those who fell ill with gastroenteritis during the epidemic. The frequency of gastrointestinal symptoms declined during the follow-up period of 15 months. However, over half of the subjects with arthritis-like symptoms after the outbreak were still symptomatic at the end of the follow-up.

Acknowledgments

We thank Anneli Keinonen for providing the register data of microbiological specimens.

Author Contributions

Conceived and designed the experiments: JLaine JLumio MJV MKuusi. Performed the experiments: JLaine JLumio MJV MKuusi. Analyzed the data: JLaine JLumio ST MJV TU MKorpela EK MKuusi. Wrote the paper: JLaine JLumio ST MJV TU MKKorpela EK MKuusi.

References

1. Leirisalo-Repo M (2005) Reactive arthritis. Scand J Rheumatol 34: 251–259.
2. DuPont AW (2008) Postinfectious irritable bowel syndrome. Clin Infect Dis 46: 594–599.
3. Spiller R, Garsed K (2009) Postinfectious irritable bowel syndrome. Gastroenterology 136: 1979–1988.
4. Marshall JK, Thabane M, Borgaonkar MR, James C (2007) Postinfectious irritable bowel syndrome after a food-borne outbreak of acute gastroenteritis attributed to a viral pathogen. Clin Gastroenterol Hepatol 5: 457–460.
5. Clark WF, Sontrop JM, Macnab JJ, Salvadori M, Moist L, et al. (2010) Long term risk for hypertension, renal impairment, and cardiovascular disease after gastroenteritis from drinking water contaminated with Escherichia coli O157:H7: a prospective cohort study. Bmj 341: c6020.
6. Garg AX, Moist L, Matsell D, Thiessen-Philbrook HR, Haynes RB, et al. (2005) Risk of hypertension and reduced kidney function after acute gastroenteritis from bacteria-contaminated drinking water. Cmaj 173: 261–268.
7. Garg AX, Pope JE, Thiessen-Philbrook H, Clark WF, Ouimet J (2008) Arthritis risk after acute bacterial gastroenteritis. Rheumatology (Oxford) 47: 200–204.
8. Marshall JK (2009) Post-infectious irritable bowel syndrome following water contamination. Kidney Int Suppl: S42–43.
9. Marshall JK, Thabane M, Garg AX, Clark WF, Salvadori M, et al. (2006) Incidence and epidemiology of irritable bowel syndrome after a large waterborne outbreak of bacterial dysentery. Gastroenterology 131: 445–450; quiz 660.
10. Moist LM, Sontrop JM, Garg AX, Clark WF, Suri RS, et al. (2009) Risk of pregnancy-related hypertension within five years of exposure to bacteria-contaminated drinking water. Kidney Int Suppl: S47–49.
11. Laine J, Huovinen E, Virtanen MJ, Snellman M, Lumio J, et al. (2011) An extensive gastroenteritis outbreak after drinking-water contamination by sewage effluent, Finland. Epidemiol Infect: 1–9.
12. Rasanen S, Lappalainen S, Kaikkonen S, Hamalainen M, Salminen M, et al. (2010) Mixed viral infections causing acute gastroenteritis in children in a waterborne outbreak. Epidemiol Infect 138: 1227–1234.
13. Rimhanen-Finne R, Hanninen ML, Vuento R, Laine J, Jokiranta TS, et al. (2010) Contaminated water caused the first outbreak of giardiasis in Finland, 2007: a descriptive study. Scand J Infect Dis 42: 613–619.
14. Uotila T, Antonen J, Laine J, Kujansuu E, Haapala AM, et al. (2011) Reactive arthritis in a population exposed to an extensive waterborne gastroenteritis outbreak after sewage contamination in Pirkanmaa, Finland. Scand J Rheumatol 40: 358–362.
15. Laine J, Uotila T, Antonen J, Korpela M, Kujansuu E, et al. (2012) Joint symptoms after a large waterborne gastroenteritis outbreak–a controlled, population-based questionnaire study. Rheumatology (Oxford) 51: 513–518.
16. Maunula L, Kllemola P, Kauppinen A, Söderberg K, Nguyen T, et al. (2009) Enteric viruses in a large waterborne outbreak of acute gastroenteritis in Finland. Food Environ Virol 1: 31–36.
17. Hillila MT, Farkkila MA (2004) Prevalence of irritable bowel syndrome according to different diagnostic criteria in a non-selected adult population. Aliment Pharmacol Ther 20: 339–345.
18. Lovell RM, Ford AC (2012) Global Prevalence of, and Risk Factors for, Irritable Bowel Syndrome: a Meta-analysis. Clin Gastroenterol Hepatol.
19. Halvorson HA, Schlett CD, Riddle MS (2006) Postinfectious irritable bowel syndrome–a meta-analysis. Am J Gastroenterol 101: 1894–1899; quiz 1942.
20. Marshall JK, Thabane M, Garg AX, Clark WF, Moayyedi P, et al. (2010) Eight year prognosis of postinfectious irritable bowel syndrome following waterborne bacterial dysentery. Gut 59: 605–611.
21. Porter CK, Faix DJ, Shiau D, Espiritu J, Espinosa BJ, et al. (2012) Postinfectious gastrointestinal disorders following norovirus outbreaks. Clin Infect Dis 55: 915–922.
22. Uotila TM, Antonen JA, Paakkala AS, Mustonen JT, Korpela MM (2013) Outcome of reactive arthritis after an extensive Finnish waterborne gastroenteritis outbreak: a 1-year prospective follow-up study. Clin Rheumatol.
23. Morris D, Inman RD (2012) Reactive arthritis: developments and challenges in diagnosis and treatment. Curr Rheumatol Rep 14: 390–394.
24. Nielsen H, Steffensen R, Ejlertsen T (2012) Risk and prognosis of campylobacteriosis in relation to polymorphisms of host inflammatory cytokine genes. Scand J Immunol 75: 449–454.

12

Sources of Heavy Metals in Surface Sediments and an Ecological Risk Assessment from Two Adjacent Plateau Reservoirs

Binbin Wu[1], Guoqiang Wang[1]*, Jin Wu[1], Qing Fu[2], Changming Liu[1]

1 College of Water Sciences, Beijing Normal University, Key Laboratory of Water and Sediment Sciences, Ministry of Education, Beijing, China, **2** Chinese Research Academy of Environmental Sciences, Beijing, China

Abstract

The concentrations of heavy metals (mercury (Hg), cadmium (Cd), lead (Pb), chromium (Cr), copper (Cu) and arsenic (As)) in surface water and sediments were investigated in two adjacent drinking water reservoirs (Hongfeng and Baihua Reservoirs) on the Yunnan-Guizhou Plateau in Southwest China. Possible pollution sources were identified by spatial and statistical analyses. For both reservoirs, Cd was most likely from industrial activities, and As was from lithogenic sources. For the Hongfeng Reservoir, Pb, Cr and Cu might have originated from mixed sources (traffic pollution and residual effect of former industrial practices), and the sources of Hg included the inflows, which were different for the North (industrial activities) and South (lithogenic origin) Lakes, and atmospheric deposition resulting from coal combustion. For the Baihua Reservoir, the Hg, Cr and Cu were primarily derived from industrial activities, and the Pb originated from traffic pollution. The Hg in the Baihua Reservoir might also have been associated with coal combustion pollution. An analysis of ecological risk using sediment quality guidelines showed that there were moderate toxicological risks for sediment-dwelling organisms in both reservoirs, mainly from Hg and Cr. Ecological risk analysis using the Hakanson index suggested that there was a potential moderate to very high ecological risk to humans from fish in both reservoirs, mainly because of elevated levels of Hg and Cd. The upstream Hongfeng Reservoir acts as a buffer, but remains an important source of Cd, Cu and Pb and a moderately important source of Cr, for the downstream Baihua Reservoir. This study provides a replicable method for assessing aquatic ecosystem health in adjacent plateau reservoirs.

Editor: Jonathan H. Freedman, NIEHS/NIH, United States of America

Funding: This research was supported by Beijing Higher Education Young Elite Teacher Project (Grant No. YETP0275), the Program for New Century Excellent Talents in University (Grant No. NCET-12-0058) and the Fundamental Research Funds for the Central Universities (Grant No. 2012LZD10). The funders had no role in study design, data collection and analysis, decision to publish, or preparation of the manuscript.

Competing Interests: The authors have declared that no competing interests exist.

* Email: wanggq@bnu.edu.cn

Introduction

There is worldwide concern about heavy metal contamination because of the environmental persistence of these elements, biogeochemical recycling and the ecological risks that metals present [1,2]. Large numbers of anthropogenically generated heavy metals from urban areas, agricultural areas and industrial sites are discharged into aquatic environments where they are transported in the water column, accumulated in sediment, and biomagnified through the food chain [3], resulting in significant ecological risk to benthic organisms, fish and humans [4]. Sediments are the main sink for heavy metals in aquatic environments [5], and sediment quality has been recognized as an important indicator of water pollution [6]. However, heavy metals are not permanently bound to sediments [7], and they may be released into the water column when the environmental conditions change (e.g., temperature and pH) or when sediments undergo other physical or biological disturbances [8]. Furthermore, reservoir construction generally leads to an increase in residence time, resulting in high accumulations of heavy metals in sediments. Consequently, it is important to analyze sediments from

reservoirs for heavy metals to support environmental management, particularly for sediments from drinking water reservoirs.

Understanding the sources of pollutants in aquatic sediments is important for pollution control. Statistical approaches, such as Pearson correlation analysis, principal components analysis (PCA), and cluster analysis, are considered to be effective tools for uncovering pollution sources and have been used successfully in many studies of heavy metal pollution in sediments [1,2,3,7,9,10,11]. Risk assessments of the environmental pollution are also critical for sediment analysis. The ecological risk of heavy metals in sediments differs for different receptors (e.g., sediment-dwelling organisms, fish or humans). The thresholds in sediment quality guidelines (SQGs) have been used to evaluate the potential adverse effects of heavy metals on sediment-dwelling organisms in freshwater systems [12,13,14,15]. However, few SQGs have been developed to assess the adverse effects of heavy metals in sediment on higher trophic levels (fish or other wildlife) [16,17]. The potential ecological risk index proposed by Hakanson [18] is based on heavy metal concentrations in sediment, and it is the simplest and most popular method for assessing the human health risk from fish consumption.

The rapid growth of urbanization and industrial development has resulted in increasing heavy metal pollution in the aquatic sediment of the Yunnan-Guizhou Plateau [11,19]. Several cascade hydropower stations have been built along the region's large rivers (e.g., Wujiang, Jinshajiang and Nanpanjiang) since the 1950s, and stations are still being built for electricity production today, leading to a continuous series of reservoirs along the rivers [20]. Previous studies have evaluated the carbon (C) cycle [21,22] and the mercury (Hg) balance [20,23] in adjacent plateau reservoirs, and have demonstrated how upstream reservoirs influence downstream reservoirs. However, little research has been conducted on other heavy metals in the adjacent reservoirs on this plateau. In addition, several decades after their construction, the functions of these reservoirs were changed, so they now supply drinking water to the human population, which has grown rapidly due to economic growth. Currently, these reservoirs are the main drinking water sources on the Yunnan-Guizhou Plateau. Although some pollution sources were closed or moved when the reservoir functions were changed, the residue of previous pollutants still remains in reservoir sediments. Furthermore, the Yunnan-Guizhou Plateau is famous for its karst landforms, and the hydrogen carbonate (HCO_3^-) concentration and pH are both high in the aquatic environment [24,25,26]. The alkaline environment favors heavy metal accumulation in sediment [27], while the karst landform promotes interactions between groundwater and surface water through fractures (sinkholes, conduits and caves) or carbonate bedrock [26,28,29]. These interactions can complicate heavy metal transport and increase the ecological risk of secondary pollution. Therefore, it is important and necessary to investigate the heavy metal pollution and to assess the associated pollution sources and ecological risks from reservoir sediments on the Yunnan-Guizhou Plateau, particularly for drinking water reservoirs. The objectives of this paper are to (1) identify the pollution sources of heavy metals in the sediment from two adjacent drinking water reservoirs on the Yunnan-Guizhou Plateau, (2) estimate the associated ecological risk by considering different receptors, and (3) discuss the influence of upstream reservoirs (as buffers or sources of heavy metals) on downstream reservoirs.

Materials and Methods

Study Areas

The Hongfeng and Baihua Reservoirs are two adjacent reservoirs on the Yunnan-Guizhou plateau, just northwest of Guiyang City, the capital of Guizhou Province, Southwest China (Fig. 1). These two reservoirs were constructed on the main channel of the Maotiao River, a branch of the Wujiang River in the Yangtze River Basin, in 1958 and 1960, respectively. The Maotiao River was one of the first rivers to be used for cascade hydropower in China. The Hongfeng is the first reservoir and the Baihua is the second of seven cascade hydropower stations along the Maotiao River. The Hongfeng Reservoir covers a water surface area of 57.2 km², while the Baihua Reservoir covers an area of 14.5 km². Both reservoirs are very deep, with each having a maximum depth of approximately 45 m. The Hongfeng Reservoir consists of the North and South Lakes (which have different flow directions, Fig. 1), and has five main inflows, two into the North Lake and three into the South Lake. The Maotiao River is the only outlet of the Hongfeng Reservoir, and also serves as the major inlet of the Baihua Reservoir. The Baihua Reservoir has eight additional minor inflows and one outlet. For their first 30 years, the reservoirs were mainly used for electricity generation and flood control. During this period, as industry, agriculture, tourism and fishery production were established and developed in

the basin, the water quality in both reservoirs declined. However, because of an increasing demand for water through the 1990s, the two reservoirs were designated as drinking water sources for Guiyang City. The major function of both reservoirs was changed to drinking water supply in 2000, at which point the government strengthened their environmental protection. The pollution sources have gradually decreased, but the sediments may still hold residue from earlier pollution.

Heavy metal pollution is one of the most prominent environmental problems in the study reservoirs, mainly owing to intense anthropogenic activities. Figure 2 shows the distribution of main point sources in Hongfeng and Baihua Reservoirs basin from the first China pollution source census in 2008 (provided by the Guiyang Research Academy of Environmental Sciences). There are many mining, smelting, mechanical manufacture, chemical and other industries (e.g., building material, food and pharmaceutical factories) in the catchment area of the Hongfeng Reservoir (1596 km²), which are major sources of heavy metals (Fig. 2). There is also a large coal-fired power plant (300 MW) situated on the southeast bank of the Hongfeng Reservoir [30], which is the main source of atmospheric deposition, especially for Hg and other heavy metals associated with coal combustion. The catchment area of the Baihua Reservoir is 1895 km², but pollutants are first transported into the Hongfeng Reservoir, which may serve as a buffer for the Baihua Reservoir. The Baihua Reservoir receives direct inputs from an area of only 299 km². Even so, there are many intense point sources in this small area, including various heavy and light industries (Fig. 2), which have resulted in serious heavy metal pollution in the Baihua Reservoir. In particular, the Baihua Reservoir is noted for its Hg contamination from the Guizhou Organic Chemical Plant (GOCP) [31], which is located in the upper reaches of the Baihua Reservoir and downstream of the Hongfeng Reservoir. The GOCP used Hg-based technology to produce acetaldehyde and discharged Hg-laden wastewater to the Baihua Reservoir via the Dongmenqiao River until 1997. The pollution caused by the GOCP persists to the present day.

Sampling and Analysis

Two field surveys were conducted in December 2010 and April 2012. Water and sediment samples were collected from 26 sites in the Hongfeng and Baihua Reservoirs (Fig. 1 and Table 1). The field studies were permitted by the Administration of Hongfeng, Baihua and Aha Reservoirs, and did not involve endangered or protected species. There were 13 sites in each reservoir, comprising inlets of main tributaries (sites 1–5 in the Hongfeng Reservoir and sites 14–22 in the Baihua Reservoir) and representative sites within both reservoirs (sites 6–13 in the Hongfeng Reservoir and sites 23–26 in the Baihua Reservoir). Surface water samples were collected in acid-washed polyethylene sample bottles and were acidified with 1:1 nitric acid: deionized water. Water samples were stored at 4°C immediately upon returning from the field. The upper 0–10 cm of sediment was collected, placed into pre-cleaned polyethylene bags, and taken to the laboratory. All sediment samples were freeze-dried and passed through a 2 mm nylon sieve to discard the coarse debris. A pestle and mortar was then used to grind the sieved sediments until all particles were fine enough to pass through a 0.147 mm nylon sieve. Sediment samples were digested in a microwave digestion system with a HNO_3-HF-$HClO_4$-HCl acid mixture solution before analysis for total heavy metal content. All water samples and the solutions of the digested sediment samples were analyzed by inductively coupled plasma atomic emission spectroscopy for Cr, Cu and Pb. Cd concentrations were determined by graphite

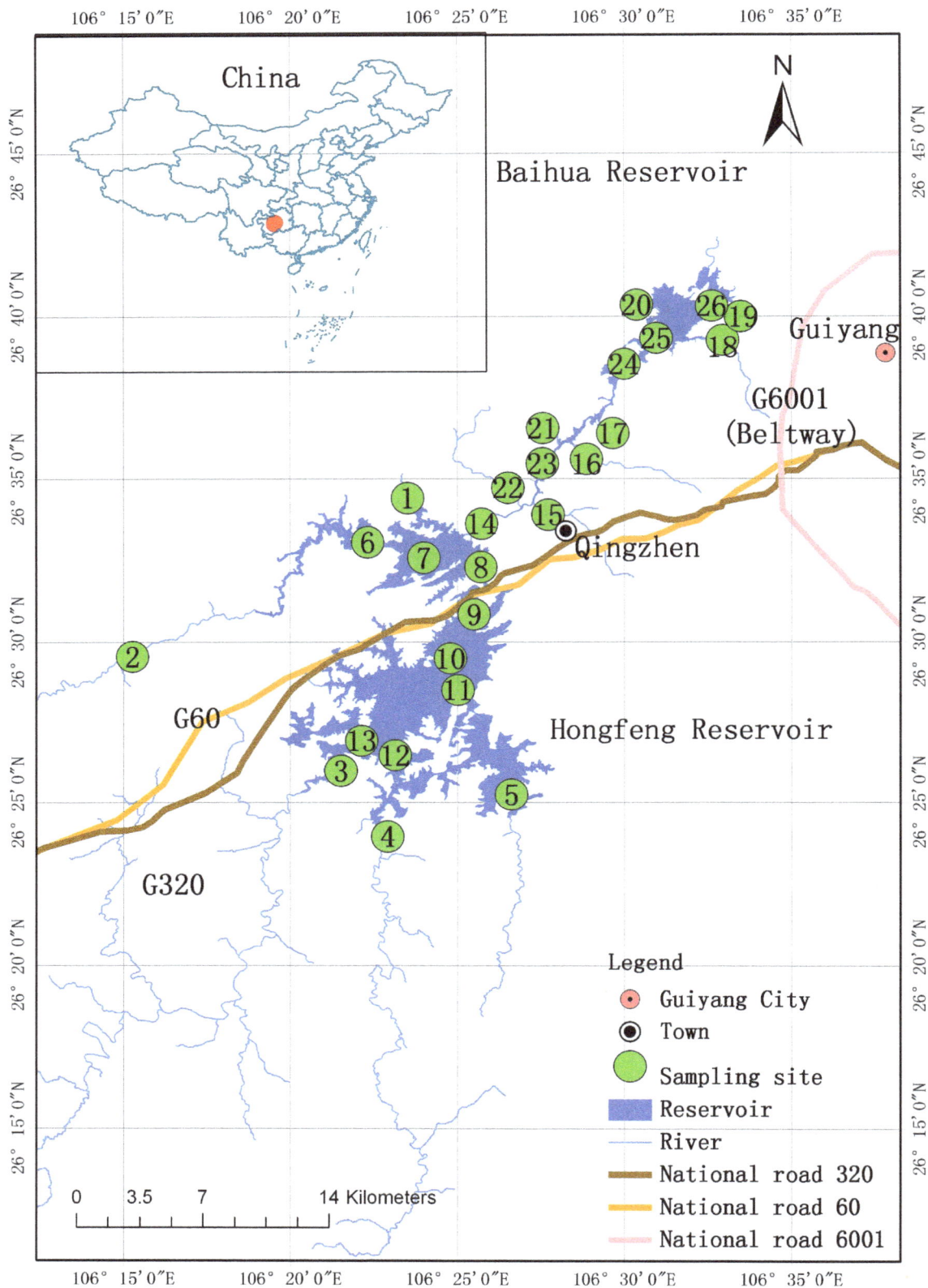

Figure 1. Map of Hongfeng and Baihua Reservoirs on the Yunnan-Guizhou Plateau, Southwest China.

furnace atomic absorption spectrophotometry. As and Hg were measured using atomic fluorescence spectrometry. Quality assur-ance and quality control of the analyses processes were assessed by duplicates, method blanks and standard reference materials [11].

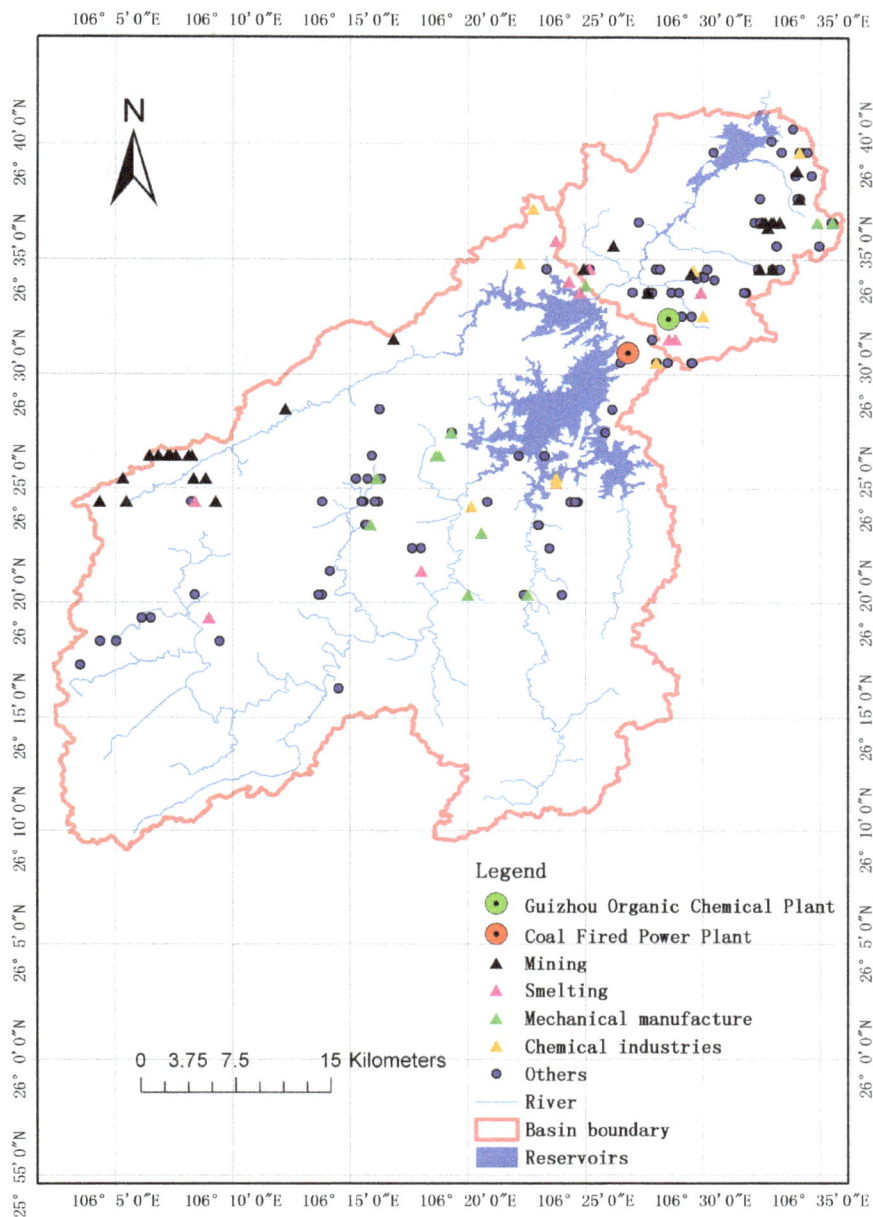

Figure 2. Distribution of main point sources of pollution in the Hongfeng and Baihua Reservoir basins.

Spatial and Statistical Analyses

Spatial and statistical analyses were performed by using Arc GIS 9.3 and SPSS 17.0 for Windows (SPSS Inc., Chicago, IL) to investigate the heavy metal pollution sources separately for the Hongfeng and Baihua Reservoirs. A one-way ANOVA was performed on heavy metal concentrations in sediment to determine whether the differences between the two field surveys were significant. Pearson correlation analysis was used to determine the relationships between the heavy metals in sediment. To obtain more reliable information about the relationships between the heavy metals, a PCA with Varimax normalized rotation was performed separately for the Hongfeng and Baihua Reservoirs. The PCA calculated eigenvectors to determine the common pollution sources, and components with eigenvalues greater than 1 were considered to be relevant [32]. Components

with factor loadings above 0.75, between 0.5 and 0.75, and between 0.3 and 0.5 were considered to be strong, moderate and weak, respectively [33]. Boxplot is a convenient way to depict the full range of data and compare the distributions among different datasets. In this study, the boxplot was used to compare the heavy metal concentrations at site 14, the outlet of the Hongfeng Reservoir and inlet of the Baihua Reservoir, with concentrations at other sites in the tributaries and at sites within the reservoirs, in order to discuss the influence of the upstream Hongfeng Reservoir on the downstream Baihua Reservoir.

Ecological Risk Assessment

We used two methods to assess the ecological risk of the heavy metals in surface sediments to benthic organisms and humans. First, we used the consensus-based SQGs for freshwater ecosys-

Table 1. Locations of sampling sites in the Hongfeng and Baihua Reservoirs on the Yunnan-Guizhou Plateau, Southwest China.

Reservoir	Site	Longitude	Latitude	Description	
Hongfeng Reservoir	1	106°23′30.64″	26°34′24.52″	At main tributaries	Maibao River
	2	106°15′3.58″	26°29′22.80″		Maiweng River
	3	106°21′30.89″	26°25′58.34″		Yangchang River
	4	106°22′54.88″	26°23′56.79″		Maxian River
	5	106°26′42.01″	26°25′4.80″		Houliu River
	6	106°22′11.10″	26°33′44.07″	Within reservoir	Taipingdi
	7	106°24′2.05″	26°32′34.80″		Center of the North Lake
	8	106°23′17.43″	26°32′27.13″		Junction of the North and South Lakes
	9	106°26′07.06″	26°30′58.03″		Houwu
	10	106°24′39.31″	26°29′19.07		Center of the South Lake
	11	106°25′04.17″	26°28′33.89″		Jiangjundong
	12	106°23′01.81″	26°26′26.73″		Yangjiajun
	13	106°22′01.22″	26°26′37.57″		Sanjiazhai
Baihua Reservoir	14	106°25′43.88″	26°33′37.65″	At main tributaries	Outlet of Hongfeng Reservoir
	15	106°27′19.60″	26°34′7.06″		Dongmenqiao River
	16	106°28′51.88″	26°35′36.16″		Maicheng River
	17	106°29′39.21″	26°36′23.43″		Dianzishanggou River
	18	106°32′56.81″	26°39′15.61″		Maixi River
	19	106°33′16.38″	26°40′6.00″		Banpochanggou River
	20	106°30′22.33″	26°40′21.27″		Maolizhaigou River
	21	106°27′33.58″	26°36′32.72″		Xiaohekou River
	22	106°26′47.46″	26°34′17.11″		Changchong River
	23	106°27′25.20″	26°35′27.60″	Within reservoir	Huaqiao
	24	106°30′4.42″	26°38′31.42″		Xuantiandong
	25	106°30′50.23″	26°39′1.37″		Tangerpo
	26	106°32′48.39″	26°40′18.16″		Chafan

tems that were proposed by MacDonald et al. [15], which included a threshold effect concentration (TEC) and a probable effect concentration (PEC). TECs are the concentrations below which adverse effects are not expected on sediment-dwelling organisms, while PECs are concentrations above which adverse effects are expected to occur frequently [34,35]. The mean PEC

Table 2. Ecological risk assessment criteria for the sediment quality guidelines (SQGs) and Hakanson index.

Method	C or E_r^i	Potential ecological risk for single heavy metal	m-PEC-Q or RI	Ecological risk for all Heavy metals
SQGs	$C<TEC$	Low	m-PEC-Q<0.1	Low (<14%)[a]
	$TEC <C<PEC$	Moderate	0.1< m-PEC-Q <1.0	Moderate (15–29%)[a]
	$C>PEC$	High	1.0<m-PEC-Q<5.0	Considerable (33–58%)[a]
			m-PEC-Q >5.0	Very high (75–81%)[a]
Hakanson index	$E_r^i<40$	Low	$RI<95$	Low
	$40<E_r^i<80$	Moderate	$95<RI<190$	Moderate
	$80<E_r^i<160$	Considerable	$190<RI<380$	Considerable
	$160<E_r^i<320$	High	$RI>380$	Very high
	$E_r^i>320$	Very high		

C: concentration of heavy metal in surface sediment.
TEC: threshold effect level; PEC: probable effect level [15].
m-PEC-Q: mean PEC quotient; [a]incidence of toxicity [36].

Table 3. Summary statistics for heavy metal concentrations in surface sediments from the Hongfeng and Baihua Reservoirs.

		Hg	Cd	Pb	Cr	Cu	As	
HongfengReservoir	Mean (Min, Max)	0.32(0.08, 1.03)	0.28(0.01, 0.85)	28.41(0.10, 89.20)	86.91(34.10, 141.00)	43.50(15.70, 93.60)	15.33(0.12, 45.54)	This study (2010.12)
	S.D.	0.26	0.36	27.55	29.65	25.09	14.92	
	CV(%)	82.86	130.78	96.97	34.12	57.67	97.31	
	Mean (Min, Max)	0.27(0.04, 0.56)	0.53(0.31, 1.37)	30.22(1.21, 89.20)	82.78(41.00, 141.00)	45.46(23.20, 73.80)	23.31(17.31, 29.61)	This study (2012.4)
	S.D.	0.17	0.27	25.63	23.45	17.72	3.98	
	CV(%)	62.33	51.27	84.78	28.32	38.99	17.07	
	2007.3	0.99		34.31		89.11	49.90	Huang et al. [40]
	2008.8				120.16	69.84		Zeng et al. [41]
	2008.10	0.66	0.77	35.91	87.98	91.85	29.74	Liu et al. [42]
	2009.5	0.46	0.65		118.00	88.00	40.80	He et al. [27]
Baihua Reservoir	Mean (Min, Max)	0.68(-[a], 2.20)	0.58(0.01, 1.00)	27.28(0.10, 51.90)	76.24(30.80, 143.00)	36.71(0.36, 65.90)	29.95(7.95, 48.19)	This study (2010.12)
	S.D.	0.73	0.37	16.89	31.50	19.77	14.80	
	CV(%)	108.24	64.15	61.90	41.31	53.86	49.42	
	Mean (Min, Max)	0.45(0.01, 1.25)	0.61(0.23, 1.00)	27.84(0.10, 51.90)	76.38(30.12, 143.10)	43.16(9.38, 73.55)	26.23(7.95, 34.75)	This study (2012.4)
	S.D.	0.48	0.28	18.15	31.45	21.49	7.30	
	CV(%)	107.21	46.13	65.19	41.17	49.79	27.82	
	2007	18.90	0.88	16.05	59.75	74.97	53.34	Huang [39]
	2010.5		0.95	38.90	66.00	67.50		Tian et al. [43]
Natural Background Value		0.08-0.15	0.08-0.12	18.50-23.90	73.90-94.60	27.30-36.70	27.00-50.00	NEPA [44]

All concentrations are in mg/kg dry weight. [a]: not detected. S.D.: standard deviation; CV: coefficients of variation.

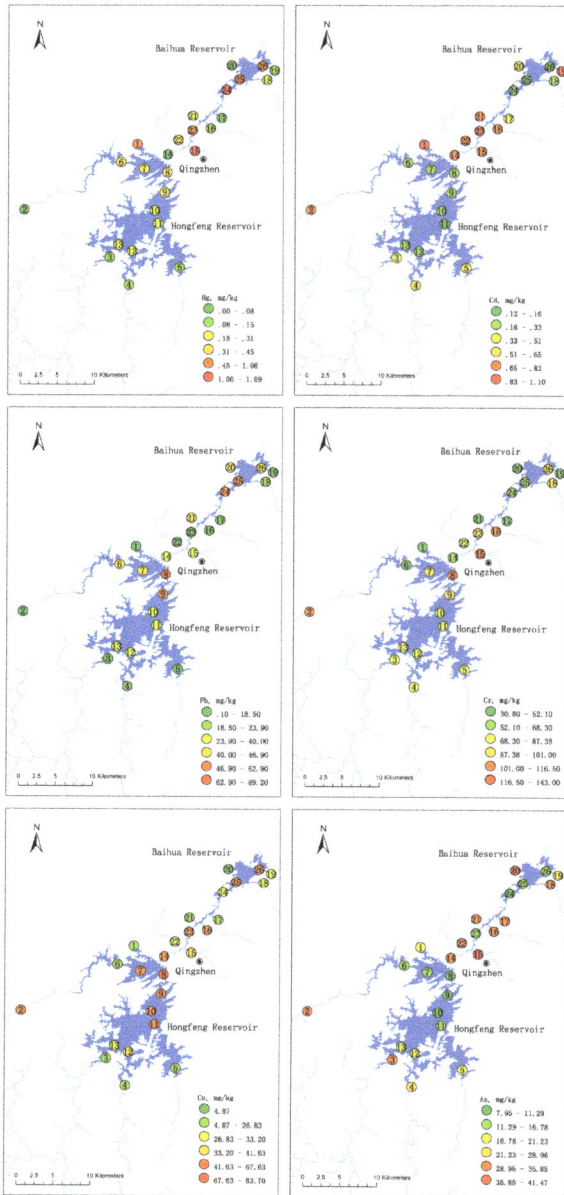

Figure 3. Heavy metal concentrations in surface sediments from Hongfeng and Baihua Reservoirs.

quotient (*m-PEC-Q*) [36] was also calculated for each sediment sample to assess the biological significance of the contaminant mixtures as follows:

$$m-PEC-Q = \frac{\sum_{i=1}^{n}(C_i/PEC_i)}{n} \qquad (1)$$

where C_i is the sediment concentration of compound i, PEC_i is the PEC for compound i and n is the number of compounds i. Four ranges of the mean PEC quotient were developed by Long et al. [36] for ranking samples in terms of toxicity incidence (Table 2).

We also used the Hakanson index, which reflects the risk to human health from fish consumption. This index is based on the assumption that the sensitivity of the aquatic system depends on its productivity [3,18]. The potential ecological risk index (*RI*) was introduced to evaluate heavy metal pollution in sediments by considering the toxicity of heavy metals and the environmental response. The *RI* is calculated as follows:

$$RI = \sum E_r^i \qquad (2)$$

$$E_r^i = T_r^i C_f^i \qquad (3)$$

$$C_f^i = C_0^i / C_n^i \qquad (4)$$

where *RI* is the total potential ecological risk index for multiple metals, E_r^i is the potential ecological risk index for a single metal, and T_r^i is the toxic-response factor for a given metal, considering both toxicity and the sensitivity. C_f^i is the contamination factor, C_0^i is the metal concentration in the sediment and C_n^i is a reference value for metals. In this study, because both reservoirs are moderately eutrophic [37], T_r^i was described as Hg (40) > Cd (30) > As (10) > Cu = Pb (5) > Cr (2), based on the assumption that the bioproduction index was 5 [18]. C_n^i was defined as the upper limit of the natural background value for a given metal in the study area (Table 3). Four ranges of the risk factor *RI* were suggested by Hakanson, based on eight metals (polychlorinated biphenyls (PCBs), Hg, Cd, As, Pb, Cu, Cr, and zinc (Zn)). PCBs and Zn were not considered in this study. Based on the different contributions of these elements to the ecological risk index *RI*, the adjusted evaluation criteria for *RI* based on the six metals in this study are listed in Table 2.

Table 4. Correlations between heavy metals in surface sediments from the Hongfeng and Baihua Reservoirs.

	Hg	Cd	Pb	Cr	Cu	As
Hg	**1.000**	0.090	0.321	−0.234	−0.038	−0.343
Cd	−0.422*	**1.000**	−0.540**	−0.168	−0.237	0.625**
Pb	0.511**	−0.506**	**1.000**	0.438*	0.595**	−0.524**
Cr	0.495*	0.013	−0.086	**1.000**	0.785**	0.046
Cu	0.241	−0.013	0.006	0.509**	**1.000**	−0.183
As	−0.414*	0.476*	−0.314	0.067	−0.333	**1.000**

Hongfeng Reservoir in the upper right corner (blod); Baihua Reservoir in the lower left corner.
Levels of significance: *p<0.05; **p<0.01.

Table 5. Principal Component Analysis (PCA) for heavy metals in surface sediments from the Hongfeng and Baihua Reservoirs.

Heavy metal	Hongfeng Reservoir			Baihua Reservoir	
	F1	F2	F3	F1	F2
Hg			0.973	−0.669	0.523
Cd		0.924		0.821	
Pb	0.605	−0.575	0.378	−0.800	
Cr	0.929				0.902
Cu	0.930				0.799
As		0.851	−0.334	0.705	
Variance (%)	35.29	32.24	21.35	37.84	29.32
Cumulative (%)	35.29	67.53	88.88	37.84	67.16

Factor loadings smaller than 0.3 have been removed.
Extraction method: PCA, Rotation method: Varimax with Kaiser normalization.

Results and Discussion

Heavy Metal Concentrations in Water Samples and Surface Sediments

Of the 6 heavy metals, only Hg and As were detected in the water samples in December 2010, while Hg, Cd, Cr (VI) and As were detected in April 2012. Their concentrations were similar in the Hongfeng and Baihua Reservoirs, with Cd, Cr (VI), and As concentrations lower than Class I as defined in the Chinese Environmental Quality Standards for Surface Water (GB3838-2002, <0.001 mg/L for Cd, <0.01 mg/L for Cr (VI), and < 0.05 mg/L for As) and Hg concentrations ranging from Class I (GB3838-2002, <0.00005 mg/L) to Class IV (GB3838-2002, 0.0001–0.001 mg/L) among different sites. The low heavy metal concentrations in water were primarily due to the accumulation in sediments because the alkaline environment in both reservoirs provides ideal conditions for adsorption and precipitation [27]. Moreover, the sediment accumulation rate in both reservoirs was quite high [38], contributing to the removal of heavy metals from the water column. A prior one-way ANOVA analysis was conducted to examine the variation in heavy metal concentrations in sediment between the two field surveys. None of the heavy metals in Hongfeng and Baihua Reservoirs displayed significant variation in means ($p>0.05$), although Cd in Hongfeng Reservoir and As in both reservoirs showed significant changes in their variances ($p<0.05$). The significant differences in variance for Cd and As are mainly because of their higher concentrations in the sites within the reservoirs and the reduced spatial heterogeneity in the second field sampling comparing to the first one. However, the general spatial patterns for Cd and As (with higher concentrations at sites in the tributaries than at sites within the reservoirs) were still similar between the two field surveys. Those results indicate that the pollution sources for the metals were relatively stable between the two surveys. The concentrations of heavy metals in surface sediments of both reservoirs from the two field surveys are summarized in Table 3. Heavy metal concentrations in sediment were much higher than those in water. In general, the mean Hg, Cd and As sediment concentrations in the Baihua Reservoir were higher than those in the Hongfeng Reservoir, while Pb, Cr and Cu were higher in the Hongfeng Reservoir. Comparison with the results of previous studies [27,39,40,41,42,43] shows that most of the heavy metal concentrations in both reservoirs have decreased, though by differing amounts (Table 3). In particular, the Hg concentrations in the Baihua Reservoir have decreased signifi-

cantly, indicating that measures taken in recent years have been effective and resulted in improvements. During the two field surveys, the mean concentrations of Hg, Cd, Pb and Cu, and the maximum concentrations of Cr in both reservoirs exceeded the upper limit of the natural background values for the study area [44], indicating anthropogenic sources. However, the concentrations of As (including the minimum and maximum values) in both reservoirs were well within the range of the natural background values [44], implying no significant anthropogenic impact and primarily lithogenic sources.

Heavy Metal Pollution Sources

To develop control strategies for environmental pollution, it is very important to identify its source. Spatial and statistical analyses were performed to identify the possible pollution sources for heavy metals in the Hongfeng and Baihua Reservoirs. The average concentrations of heavy metals in the sediments from the two field surveys were used to study the spatial distributions, while all data in the sediments from the two field surveys were used in a Pearson correlation analysis and PCA. The spatial distribution patterns of Hg, Cd, Pb, Cr, Cu and As in surface sediments of both reservoirs are shown in Figure 3. The Pearson correlation coefficients and the results of the PCA for the investigated metals are shown in Table 4 and Table 5, respectively. All of the results were generally consistent with each other.

Specifically, the PCA yielded three significant components for Hongfeng Reservoir and two significant components for Baihua Reservoir, accounting for 88.88% and 67.16% of the cumulative variance, respectively (Table 5). For the Hongfeng Reservoir, the first component (F1), explaining 35.29% of the total variance, had strong positive loadings of Cr and Cu, and moderate positive loading of Pb. Those three heavy metals exhibited similar spatial distributions in the Hongfeng Reservoir, with unexpectedly higher concentrations at reservoir sites than at tributary sites. In particular, site 8 (at the junction of the North and South Lakes) showed the highest concentrations for all of the three heavy metals, and site 9 (near Houwu) also showed relatively high concentrations (Fig. 3). In addition, those three heavy metals were highly correlated (Table 4, $p<0.01$ for Cr-Cu and Cu-Pb, $p<0.05$ for Cr-Pb), indicating their similar origins or comparable chemical properties [45]. This phenomenon might be caused by two possible reasons. Firstly, Bai et al. [46] found that traffic pollution was responsible for the high heavy metal concentrations (including comparable Cr, Cu and Pb concentrations with our study) along

Table 6. Results of ecological risk assessments for single heavy metal from two methods for the Hongfeng and Baihua Reservoirs.

	Hg	Cd	Pb	Cr	Cu	As
TEC	0.18	0.99	35.8	43.4	31.6	9.79
PEC	1.06	4.98	128	111	149	33
Hongfeng Reservoir						
% samples which exceeded TEC	69.23	7.69	38.46	92.31	61.54	84.62
% samples which exceeded PEC	0.00	0.00	0.00	15.38	0.00	15.38
% samples with E_r^i<40	30.77	7.69	100.00	100.00	100.00	100.00
% samples with 40<E_r^i<80	23.08	53.85	0.00	0.00	0.00	0.00
% samples with 80<E_r^i<160	38.46	23.08	0.00	0.00	0.00	0.00
% samples with 160<E_r^i<320	7.69	15.38	0.00	0.00	0.00	0.00
% samples with E_r^i>320	0.00	0.00	0.00	0.00	0.00	0.00
Baihua Reservoir						
% samples which exceeded TEC	53.85	7.69	38.46	84.62	46.15	92.31
% samples which exceeded PEC	30.77	0.00	0.00	7.69	0.00	38.46
% samples with E_r^i<40	38.46	23.08	100.00	100.00	100.00	100.00
% samples with 40<E_r^i<80	15.38	0.00	0.00	0.00	0.00	0.00
% samples with 80<E_r^i<160	7.69	15.38	0.00	0.00	0.00	0.00
% samples with 160<E_r^i<320	15.38	61.54	0.00	0.00	0.00	0.00
% samples with E_r^i>320	23.08	0.00	0.00	0.00	0.00	0.00

All concentrations are in mg/kg dry weight.

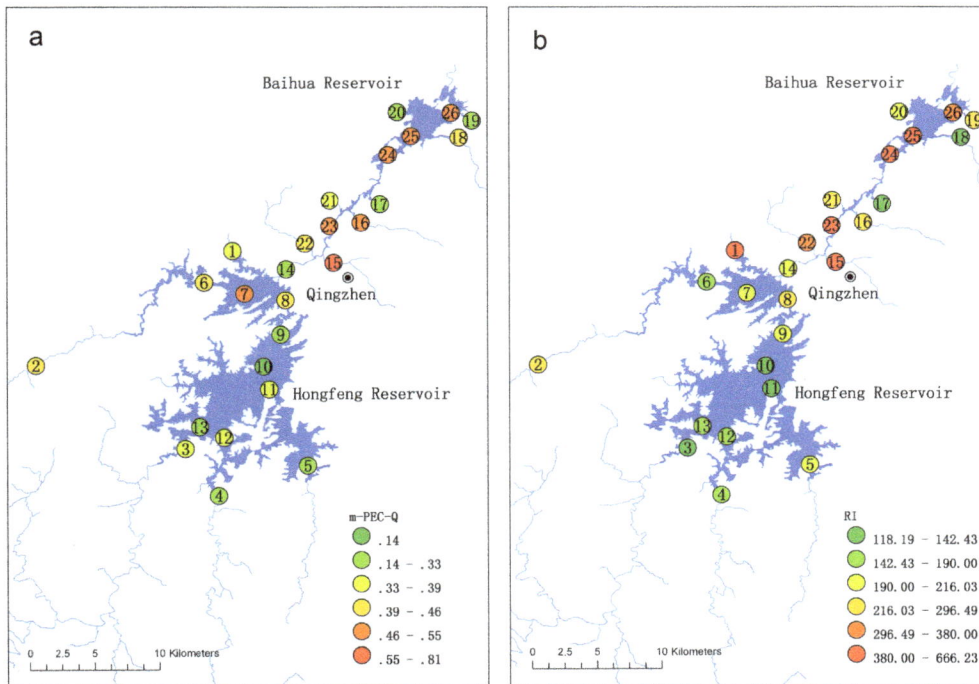

Figure 4. Mean PEC quotient (a) and potential ecological risk indexes (b) of heavy metals in sediments.

the roadside of National Road 320 in the Yunnan province (adjacent to Guizhou Province), and Zhu et al. [47] also found that road dust samples were severely polluted by Cr, Cu and Pb in another metal smelting/processing industrial city in Guizhou Province. In this study, the National Road 60 and the National Road 320 pass close to the junction of the South and North Lakes of the Hongfeng Reservoir (Fig. 1), which suggests that the traffic emissions, through atmospheric deposition and road runoff, could result in heavier pollution in sites near the roadway (site 8 and 9). Secondly, the high concentrations of Cr, Cu and Pb at reservoir sites are likely to be related to the residual effect from former industrial activities (e.g., mining, smelting, mechanical manufacture and chemical industry). The metals are more likely retained in the sediment of sites within the reservoirs rather than sites in the tributaries because heavy metal accumulation in lake sediments is generally higher than that in rivers [3]. Additionally, the complex hydrodynamic conditions at the junction of the South and North Lakes may affect the heavy metal distributions in sediment, which requires further research. Therefore, the first component (F1) might reflect mixed sources from traffic pollution and the residual effect of former industrial influence. The second component (F2), explaining 32.24% of the total variance, was dominated by Cd and As. Similar spatial patterns were observed for Cd and As, with higher concentrations at tributary sites than at sites within the reservoirs, indicating that they mainly come from the inflows (Fig. 3). As expected, significantly positive correlations were found between Cd and As (Table 4, $p<0.01$). However, Cd showed apparent anthropogenic origin, with most sites exceeding its natural background values, while As levels suggested natural origins, with all sites well within the natural background values (Table 3). Cd is closely related to industrial activities, such as smelting, electroplating and plastics production in the upstream areas. Hence, F2 may reflect the pollution through inflows from both industrial activities and natural weathering and erosion. The

third component (F3) had strong positive loading on Hg and a weak positive loading on Pb, accounting for 21.35% of the total variance. The highest Hg concentration in Hongfeng Reservoir was found at site 1 in the tributary of Maibao River, which has several smelting and chemical industries in its upstream (Fig. 2). Meanwhile, for both Hg and Pb, the North Lake were more polluted than the South Lake, and three tributaries in the South Lake showed concentrations well within the natural background values. Feng et al. [30] found that runoff due to soil erosion was the main source of Hg in sediment in the South Lake of the Hongfeng Reservoir. Thus, F3 may reflect the pollution from inflows from industrial activities in the North Lake and lithogenic origin in the South Lake. In addition, He [48] found that atmospheric deposition from coal combustion was also an important source of Hg in the Hongfeng Reservoir, which was not clearly distinguished by the PCA.

For Baihua Reservoir, the first component (F1), explaining 37.84% of total variance, showed strong positive loadings on Cd and As. Similar spatial distributions (with higher concentrations at tributary sites than at sites within the reservoirs) (Fig. 3) and positive correlations were also found between Cd and As (Table 4, $p<0.05$). As discussed above, F1 in Baihua Reservoir might be similar to F2 in Hongfeng Reservoir, including the pollution through inflows from both industrial activities and natural origin. The second component (F2) had a moderate positive loading of Hg and strong positive loadings on Cr and Cu. The highest Hg and Cr concentration in the Baihua Reservoir was found at site 15 in the Dongmengqiao tributary, which received wastewater from the GOCP and many other industries (Fig. 2). The highest Cu concentration in the Baihua Reservoir was at site 16 in the tributary of the Maicheng River, which has several industries (especially mining) in its upstream (Fig. 2). Strong associations were found between Cr and Cu ($p<0.01$) and between Cr and Hg ($p<0.05$) (Table 4). F2 obviously represented industrial activity

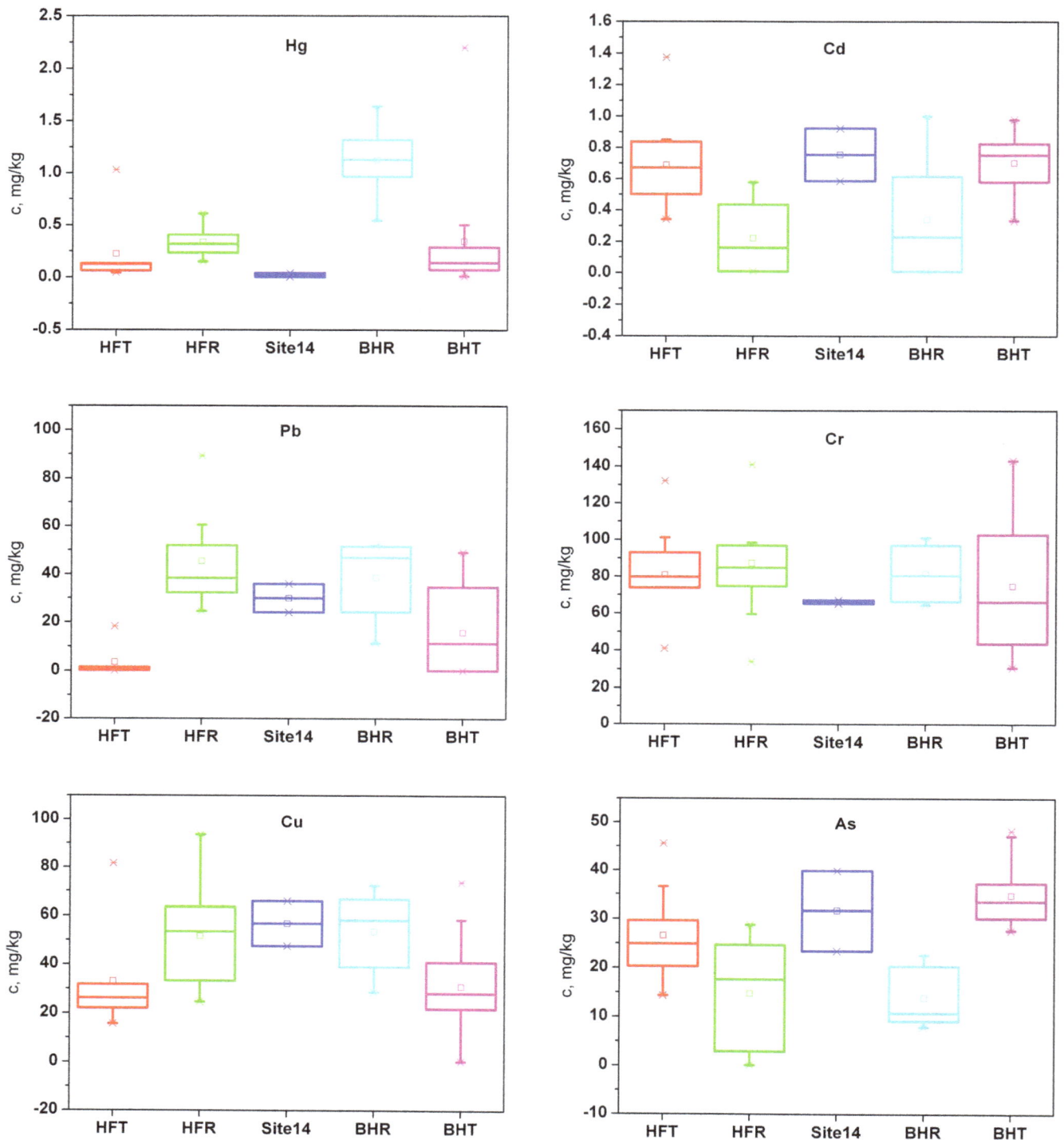

Figure 5. Comparison of heavy metal concentrations in sediments. (HFT: sites at inlets of main tributaries in the Hongfeng Reservoir, namely sites 1–5; HFR: representative sites within Hongfeng Reservoir, namely sites 6–13; BHT: sites at inlets of main tributaries in the Baihua Reservoir (except site 14), namely sites 15–22; BHR: representative sites within the Baihua Reservoir, namely sites 23–26).

upstream. On the other hand, the PCA failed to identify Pb sources in Baihua Reservoir, but only showed its strong negative associations with F1, indicating that there may be significant sources other than F1 and F2 for Pb. The spatial pattern of Pb in the Baihua Reservoir also showed much higher concentrations at sites within the reservoir than at sites in the tributaries (Fig. 3).

Although no main road crosses Baihua Reservoir directly, it is close to the urban district of Guiyang City, which has several roads and high traffic density (Fig. 1 only shows the main beltway and there are many other crisscrossed roads inside the beltway). In addition, Pb was added to gasoline in China until June 2000 [49]. Hence, Pb may originate from traffic pollutants deposited atmospherically. Moreover, highly positive correlations were found between Pb and Hg ($p<0.01$), implying that Hg might also come from atmospheric deposition (coal combustion) in addition to from the factors associated with F2.

Ecological Risk Assessment of Heavy Metals in Surface Sediments

Because there was no significant change in the mean heavy metal concentrations in sediment between the two field surveys, the average concentrations were used to study the ecological risk of single heavy metals and the combined ecological effects of six heavy metals for the two study reservoirs using both the SQG and Hakanson index methods (Table 6). The SQG method revealed that the metal concentrations were within the TEC and PEC ranges for Hg, Cd, Pb and Cu at 69.23%, 7.69%, 38.46% and 61.54% of the sites in the Hongfeng Reservoir, and for Cd, Pb and Cu at 7.69%, 38.46% and 46.15% of sites in the Baihua Reservoir. The heavy metals at the remaining sites in the corresponding reservoirs fell below the TEC. Cr and As exceeded the PEC at 15.38% of sites in the Hongfeng Reservoir and Hg, Cr and As exceeded the PEC at 30.77%, 7.69% and 38.46%, respectively, of sites in the Baihua Reservoir. Previous studies have shown that ecological risk assessments should consider the regional background values and that exceeding the SQG values does not always lead to adverse ecological effects [50]. Therefore, the As concentrations within background ranges in both reservoirs should be excluded. The Cr in the Hongfeng Reservoir and both Hg and Cr in the Baihua Reservoir may pose significant ecological risks for sediment-dwelling organisms, and they deserve special attention. Heavy metals within the TEC–PEC (viz., Hg in Hongfeng Reservoir and Cd, Pb and Cu in both reservoirs) are also a cause for concern because this uncertain area may be considered to be moderately polluted [10]. The toxicity, derived from mean PEC quotients, that results from the mixture of the six heavy metals at each sampling site in both reservoirs is shown in Figure 4a. Overall, mean PEC quotients for samples in the Baihua Reservoir (range 0.28–0.81) were slightly higher than those in the Hongfeng Reservoir (range 0.14–0.55). However, the mean PEC quotients for all of the samples in both reservoirs were well within the range of 0.1 to 1.0, indicating moderate toxicological risks for sediment-dwelling organisms, with a toxicity incidence of between 15 and 29% in the study areas (Table 2).

The Hakanson method expresses the threat to humans from fish consumption. The results from this index were quite different from those for the SQG method (Table 6). Both Hg and Cd posed high potential ecological risks at 7.69% and 15.38% of sites, considerable risks at 38.46% and 23.08% of sites and moderate risks at 23.08% and 53.85% of sites, respectively, in the Hongfeng Reservoir. The risks were higher in the Baihua Reservoir, in which there was a very high potential ecological risk from Hg at 23.08% of sites, high risks from Hg and Cd at 15.38% and 61.54% of sites, considerable risks from Hg and Cd at 7.69% and 15.38% of sites, and a moderate risk from Hg at 15.38% of sites. However, the other heavy metals (Pb, Cr, Cu and As) posed little potential ecological risks for all sites in both reservoirs, with E_r^i values lower than 40. The high concentrations and toxic-response factors of Hg and Cd in both reservoirs contribute to their posing higher ecological risks than the other metals we examined. RI illustrates the potential ecological risk from heavy metal mixtures, and RI at all sites in both reservoirs were higher than 95 (Fig. 4b). Site 1 showed the highest potential ecological risk ($RI = 484.25$) in the Hongfeng Reservoir, at a level that should cause concern because it poses a very high risk. Sites 2, 5 and 7–9 exhibited considerable ecological risks, while other sites showed moderate ecological risks

in the Hongfeng Reservoir. The combined ecological risk was more severe in the Baihua Reservoir. The RI at site 15 (Dongmenqiao tributary) and 23–25 (within the reservoir) were much higher than 380, which indicates a very high potential ecological risk. All of the other sites exhibited considerable ecological risks (except for the moderate ecological risks at site 17 and 18). Therefore, there are moderate to very high potential ecological risks from heavy metal mixtures in the sediments of both reservoirs. In addition, the contribution of the monomial potential ecological risk to RI for the six heavy metals in both reservoirs decreased in the following order: Hg $\approx$ Cd > As > Cu > Pb > Cr, with the greatest ecological risk from Hg and Cd.

Overall, the ecological risks from either a single heavy metal or from mixed heavy metals were different for the two receptors (viz., sediment-dwelling organisms and human beings through fish consumption) in both prior contaminants and risk level. However, hot spots with higher ecological risks were similar even though two different methods were used, and they were mainly located in the North Lake and the Houwu area of the Hongfeng Reservoir and in the key tributaries and at all of the sites in the Baihua Reservoir. Therefore, the need for industrial wastewater and mining tailings treatment in upstream watersheds of both reservoirs should be highlighted, especially for the tributaries in the North Lake of the Hongfeng Reservoir and in the key tributaries of the Baihua Reservoir. Additionally, given that the lakes are sources for drinking water, continuous monitoring should be increasingly implemented in areas near their inflows. Finally, there is uncertainty in both the SQG and Hakanson index methods because the SQGs were developed in North America and the toxic-response factor in the Hakanson method is not very sophisticated. Therefore, further on-site or laboratory toxicological experiments should be carried out to ascertain the actual adverse effects on sediment-dwelling organisms and different fish species [10,50], as well as to determine the impacts on human health from consuming fish from the study area.

Influence of the Hongfeng Reservoir on the Baihua Reservoir

Reservoir construction generally leads to an increase in residence time and a decrease in suspended solids and turbidity. For an alkaline reservoir on the Yunnan-Guizhou Plateau such as the Hongfeng Reservoir, heavy metals tend to be adsorbed to suspended solids, and sediments then settle on the lake bed, resulting in fewer heavy metals in water. Hence, in heavily polluted areas, reservoirs may serve as a sink for pollutants and a buffer for downstream receiving areas. In this study, the catchment area upstream of the Hongfeng Reservoir occupies 84% of the Baihua Reservoir catchment area, and the heavy metal concentrations at site 14, the outlet of the Hongfeng Reservoir and inlet of the Baihua Reservoir, reflect the buffering effect of the Hongfeng Reservoir according to our two field surveys. The metals (Hg, Cd, Cr (VI) and As) in the surface water at site 14 had lower concentrations than they did at the inflow tributary sites of the Hongfeng Reservoir (results not shown). The sediment concentration of Hg was much lower at site 14 than at most sites at the tributaries and within the Hongfeng Reservoir (Fig. 5). He [48] also found that the Hongfeng Reservoir functioned as a net sink for Hg and that it intercepted a large amount of Hg before it was conveyed to the Baihua Reservoir. The sediment concentrations of Cd and As were lower at site 14 than the maximum concentration at the tributaries of Hongfeng Reservoir, although they were generally higher there than at sites within Hongfeng Reservoir (Fig. 5). Due to the different pollution sources (some indirect rather than through inflows), higher concentrations of Pb, Cr and Cu were found at the reservoir sites than at the tributary sites, and the sediment concentrations of Pb, Cr and Cu were generally lower at site 14 than at sites within the Hongfeng Reservoir (Fig. 5). On the other hand, the outflow of the Hongfeng Reservoir has accounted for an average of 70% of the total inflow of the Baihua Reservoir over the last 6 years (2005–2010, provided by the Administration of Hongfeng, Baihua and Aha Reservoirs), implying that the Hongfeng Reservoir may also serve as an important source for total metals in the Baihua Reservoir. For example, the concentrations of the metals (Hg, Cd, Cr (VI) and As) detected in water samples at site 14 fell in the mid-range of the concentrations in the other tributaries of the Baihua Reservoir (results not shown). The concentrations of all of the heavy metals in sediments at site 14 were generally within the concentration ranges of heavy metals in other tributaries of the Baihua Reservoir, with Hg at the low end, Cr in the medium range, and Cd, Pb, Cu and As at the high end (Fig. 5). However, the sediment concentrations of Cd and As at site 14 were generally higher than at sites within the Hongfeng Reservoir. This pattern was not found in the other heavy metals, resulting in uncertainty when considering whether the relatively high concentrations of Cd and As at site 14 came from the Hongfeng Reservoir or not (Fig. 5). Because Cd was mainly from industrial activities and no emission sources exist near site 14 (as determined by the field investigation), the high concentrations at site 14 might result from the pollution of site 1 (Fig. 3) because our sampling sites within the Hongfeng Reservoir did not cover the area near the outlet. In terms of the spatial distribution and the lithogenic source of As, the high concentrations at site 14 might be affected by other factors, such as soil type and land use of the nearby banks. Therefore, for heavy metal concentrations in sediment, the Hongfeng Reservoir might be an important source of Cd, Cu and Pb, and a moderately important source of Cr, but might not be an important source of Hg and As for the Baihua Reservoir. The results also indicate that the Hongfeng Reservoir is not always the most important source for total metals in the Baihua Reservoir, and other tributaries contribute large quantities of pollutants to the Baihua Reservoir, some of which even exceed the levels in the Hongfeng Reservoir. It should also be noted that our field surveys only indirectly reflect the potential long-term impacts from Hongfeng Reservoir on heavy metals in the sediment of the Baihua Reservoir, and continuous monitoring of inflows and outflows of both reservoirs are needed for the specific contribution of the upstream reservoir to the downstream reservoir in future studies.

Other factors contribute to the adverse effects of the Hongfeng Reservoir on the Baihua Reservoir. The Hongfeng Reservoir is a deep reservoir with thermal stratification from May to November [37], and its release water is mainly from the hypolimnion, which has a lower DO concentration, higher CO_2 concentration and lower pH than surface water during those months [21,37]. The water chemistry in the hypolimnion favors the release of heavy metals from the sediments and changes the speciation and toxicity of heavy metals. He et al. [37] found that the low DO and pH in hypolimnion accelerated Hg methylation at Houwu (near site 9) and enhanced the release of methylmercury from sediments at Daba (near the outlet) in the Hongfeng Reservoir in summer. He et al. [37] also concluded that the Hongfeng Reservoir was a net source of methylmercury for the Baihua Reservoir. In addition,

the release of hypolimnetic water has a cooling effect in the summer and a warming effect in winter, which may have a significant influence on the temperature downstream, and thus may indirectly influence the heavy metal distribution. Therefore, the outflow of the Hongfeng Reservoir may pose serious risks to ecosystems in the Baihua Reservoir. Further research is needed to help understand the influence of heavy metals and their chemical forms as they are transported downstream from reservoirs.

Conclusions

This study of heavy metal (Hg, Cd, Pb, Cr, Cu and As) concentrations in surface water and sediments from two adjacent drinking water reservoirs (the Hongfeng and Baihua Reservoirs) on the Yunnan-Guizhou Plateau, Southwest China, showed that surface water was polluted by Hg, and sediments were polluted by Hg, Cd, Pb, Cr and Cu. In both reservoirs, Cd and As mainly came from industrial activities and lithogenic source through inflows, respectively. The Pb, Cr and Cu in Hongfeng Reservoir may have arisen from a mixture of sources (traffic pollution and residual effect of former industrial influence), and they were present at higher concentrations at the junction of the North and South Lakes. Hg sources in the Hongfeng Reservoir might include the sources that contribute Hg through inflows, which were different for the North (industrial activities) and South Lakes (lithogenic origin), and atmospheric deposition resulting from coal combustion. For the Baihua Reservoir, Hg, Cr and Cu were primarily derived from upstream industrial activities, and the Pb originated from traffic pollution. Additionally, the Hg in Baihua Reservoir might have come from atmospheric deposition (coal combustion). Ecological risk was assessed using the SQGs and the Hakanson potential ecological risk index. There were moderate toxicological risks for sediment-dwelling organisms (with the main risks from Hg and Cr) and moderate to very high potential ecological risks for humans from fish consumption (with the main risk coming from Hg and Cd) in both reservoirs. Overall, the risks were higher in the Baihua Reservoir. Improved treatment of industrial wastewater and mining tailings in upstream watersheds would alleviate the pollution and ecological risk in both reservoirs, especially for tributaries of the North Lake of the Hongfeng Reservoir and the key tributaries of the Baihua Reservoir. Ecological restoration could be considered to counteract the residual effects from previous pollution; however, more research is needed in this area. In terms of heavy metal concentrations, the Hongfeng Reservoir acts as a buffer, but it is still an important source of Cd, Cu and Pb and a moderately important source of Cr for the Baihua Reservoir. The Hongfeng Reservoir also had adverse effects on the Baihua Reservoir and merits further research. These findings provide useful information about sediment quality in adjacent reservoirs on the Yunnan-Guizhou Plateau.

Author Contributions

Conceived and designed the experiments: GQW CML. Performed the experiments: BBW GQW JW QF. Analyzed the data: BBW JW. Contributed reagents/materials/analysis tools: BBW JW QF. Wrote the paper: BBW GQW.

References

1. Liu WX, Li XD, Shen ZG, Wang DC, Wai OWH, et al. (2003) Multivariate statistical study of heavy metal enrichment in sediments of the Pearl River Estuary. Environ Pollut 121: 377–388.

2. Chabukdhara M, Nema AK (2012) Assessment of heavy metal contamination in Hindon River sediments: A chemometric and geochemical approach. Chemosphere 87: 945–953.

3. Yi YJ, Yang ZF, Zhang SH (2011) Ecological risk assessment of heavy metals in sediment and human health risk assessment of heavy metals in fishes in the middle and lower reaches of the Yangtze River Basin. Environ Pollut 159: 2575–2585.

4. Uluturhan E, Kucuksezgin F (2007) Heavy metal contaminants in Red Pandora (Pagellus erythrinus) tissues from the Eastern Aegean Sea, Turkey. Water Res 41: 1185–1192.

5. Singh KP, Mohan D, Singh VK, Malik A (2005) Studies on distribution and fractionation of heavy metals in Gomti river sediments-a tributary of the Ganges. J Hydrol 312: 14–27.

6. Larsen B, Jensen A (1989) Evaluation of the sensitivity of sediment monitoring stationary in pollution monitoring. Mar Pollut Bull 20: 556–560.

7. Li XD, Wai OWH, Li YS, Coles BJ, Ramsey MH, et al. (2000) Heavy metal distribution in sediment profiles of the Pearl River estuary, South China. Appl Geochem 15: 567–581.

8. Agarwal A, Singh RD, Mishra SK, Bhunya PK (2005) ANN-based sediment yield river basin models for Vamsadhara (India). Water SA 31: 95–100.

9. Loska R, Wiechula D (2003) Application of principal component analysis for the estimation of source of heavy metal contamination in surface sediments from the Rybnik Reservoir. Chemoshere 51: 723–733.

10. Larrose A, Coynel A, Schafer J, Blanc G, Masse L, et al. (2010) Assessing the current state of the Gironde Estuary by mapping priority contaminant distribution and risk potential in surface sediment. Appl Geochem 25: 1912–1923.

11. Bai JH, Cui BS, Chen B, Zhang KJ, Deng W, et al. (2011) Spatial distribution and ecological risk assessment of heavy metals in surface sediments from a typical plateau lake wetland, China. Ecol Model 222: 301–306.

12. Persaud D, Jaagumagi R, Hayton A (1993) Guidelines for the protection and management of aquatic sediment quality in Ontario. Water Resources Branch. Ontario Ministry of the Environment, Toronto, 27.

13. Smith SL, MacDonald DD, Keenleyside KA, Ingersoll CG, Field J (1996) A preliminary evaluation of sediment quality assessment values for freshwater ecosystems. J Great Lakes Res 22: 624–638.

14. Ingersoll CG, Haverland PS, Brunson EL, Canfield TJ, Dwyer FJ, et al. (1996) Calculation and evaluation of sediment effect concentrations for the amphipod Hyalella azteca and the midge Chironomus riparius. J Great Lakes Res 22: 602–623.

15. Macdonald DD, Ingersoll CG, Berger TA (2000). Development and evaluation of consensus-based sediment quality guidelines for freshwater ecosystems. Arch Environ Contam Toxical 39: 20–31.

16. Word JQ, Albrecht BB, Anghera ML, Baudo R, Bay MS, et al. (2002). Predictive ability of sediment quality guidelines. In: Wenning RJ, Batley GE, Ingersoll CG, Moore DW, editors. Use of sediment quality guidelines and related tools for the assessment of contaminated sediments. Pensacola (FL): SETAC. p 121–162.

17. Bhavsar SP, Gewurtz SB, Helm PA, Labencki TL, Marvin CH, et al. (2010) Estimating sediment quality thresholds to prevent restrictions on fish consumption: Application to polychlorinated biphenyls and dioxins–furans in the Canadian Great Lakes. Integr Environ Assess Manag 6: 641–652.

18. Hakanson L (1980) An ecological risk index for aquatic pollution control: A sedimentological approach. Water Res 14: 975–1001.

19. Liu Y, Guo HC, Yu YJ, Huang K, Wang Z (2007) Sediment chemistry and the variation of three altiplano lakes to recent anthropogenic impacts in southwestern China. Water SA 33: 305–310.

20. Feng XB, Jiang HM, Qiu GL, Yan HY, Li GH, et al. (2009a) Mercury mass balance study in Wujiangdu and Dongfeng Reservoirs, Guizhou, China. Environ Pollut 157: 2594–2603.

21. Wang FS, Wang BL, Liu CQ, Wang YC, Guan J, et al. (2011a) Carbon dioxide emission from surface water in cascade reservoirs-river system on the Maotiao River, southwest of China. Atmos Environ 45: 3827–3834.

22. Wang FS, Liu CQ, Wang BL, Liu XL, Li GR, et al. (2011b) Disrupting the riverine DIC cycling by series hydropower exploitation in Karstic area. Appl Geochem 26: S375–S378.

23. Feng XB, Jiang HM, Qiu GL, Yan HY, Li GH, et al. (2009b) Geochemical processes of mercury in Wujiangdu and Dongfeng reservoirs, Guizhou, China. Environ Pollut 157: 2970–2984.

24. Han GL, Liu CQ (2004) Water geochemistry controlled by carbonate dissolution: a study of the river waters draining karst-dominated terrain, Guizhou Province, China. Chem Geo 204: 1–21.

25. Wang J, Liu CQ, Wu Y (2005) Effect of heavy metals on the activity of external carbonic anhydrase of microalga chlamydomonas reinhardtii and microalgae from karst lakes. Bull Environ Contam Toxicol 74: 227–233.

26. Lang YC, Liu CQ, Zhao ZQ, Li SL, Han GL (2006) Geochemistry of surface and ground water in Guiyang, China: Water/rock interaction and pollution in a karst hydrological system. Appl Geochem 21: 887–903.

27. He SL, Li CJ, Pan ZP, Luo MX, Meng W, et al. (2012) Geochemistry and environmental quality assessment of Hongfeng Lake sediments, Guiyang. Geophysical and Geochemical Exploration 36: 273–297 (in Chinese).

28. Wang Y, Luo TMZ (2001) Geostatistical and geochemical analysis of surface water leakage into groundwater on a regional scale: a case study in the Liulin karst system, northwestern China. J Hydrol 246: 223–234.

29. Sophocleous M (2002) Interactions between groundwater and surface water: the state of the science. Hydrogeol J 10: 52–67.

30. Feng XB, Foucher D, Hintelmann H, Yan HY, He TR, et al. (2010) Tracing mercury contamination sources in sediments using mercury isotope compositions. Environ Sci Technol 44: 3363–3368.

31. Yan HY, Feng XB, Shang LH, Qiu GL, Dai QJ, et al. (2008) The variations of mercury in sediment profiles from a historically mercury-contaminated reservoir, Guizhou province, China. Sci Total Environ 407: 497–506.

32. Kaiser HF (1960) The application of electronic computers to factor analysis. Educ Psychol Measure 20: 141–151.

33. Liu CW, Lin KH, Kuo YM (2003) Application of factor analysis in the assessment of groundwater quality in a Blackfoot disease area in Taiwan. Sci Total Environ 313: 77–89.

34. Macdonald DD, Carr RS, Calder FD, Long ER, Ingersoll CG (1996) Development and evaluation of sediment quality guidelines for Florida coastal waters. Ecotoxicology 5: 253–278.

35. Swartz RC (1999) Consensus sediment quality guidelines for PAH mixtures. Environ Toxicol Chem 18: 780–787.

36. Long ER, Ingersoll CG, Macdonald DD (2006) Calculation and uses of mean sediment quality guideline quotients: a critical review. Environ Sci Technol 40: 1726–1736.

37. He TR, Feng XB, Guo YN, Qiu GL, Li ZG, et al. (2008) The impact of eutrophication on the biogeochemical cycling of mercury species in a reservoir: A case study from Hongfeng Reservoir, Guizhou, China. Environ Pollut 154: 56–67.

38. Bai ZG, Wan GJ, Liu TS, Huang RG (2002) A comparative study on accumulation characteristics of ^7Be and ^{137}Cs in sediments of Lake Erhai and Lake Hongfeng, China, Geochimica 31: 113–118 (in Chinese).

39. Huang XF (2008) Studies on characteristics of pollution in sediments from Baihua Lake. Guiyang (in Chinese).

40. Huang XF, Qin FX, Hu JW, Li CX (2008) Characteristic and ecological risk of heavy metal polltuion in sediments from Hongfeng Lake. Res Environ Sci 21: 18–23 (in Chinese).

41. Zeng Y, Zhang W, Chen JA, Zhu ZJ (2010) Analysis of heavy metal pollution in the sediment of the inflow-lake rivers of the Hongfeng Lake, Earth Environ 38: 470–475 (in Chinese).

42. Liu F, Hu JW, Wu D, Qin FX, Li CX, et al. (2011) Speciation characteristics and risk assessment of heavy metals in sediments from Hongfeng Lake, Guizhou Province. Environmental Chemistry 30: 440–446 (in Chinese).

43. Tian LF, Hu JW, Luo GL, Ma JJ, Huang XF, et al. (2012) Ecological risk and stability of heavy metals in sediments from Lake Baihua in Guizhou Province. Acta Scientiae Circumstantiae 32: 885–894 (in Chinese).

44. NEPA: National Environmental Protection Agency (Presently known as MEP; Ministry of Environmental Protection) (1994) The Atlas of Soil Environmental Background Value in the People's Republic of China. China Environmental Science Press.

45. Hakanson L, Jasson M (1983) Principles of Lake Sedimentology. Springer Verlag, Berlin.

46. Bai JH, Cui BS, Wang QG, Gao HF, Ding QY (2009) Assessment of heavy metal contamination of roadside soils in Southwest China. Stoch Environ Res Risk Assess 23: 341–347.

47. Zhu ZM, Li ZG, Bi XY, Han ZX, Yu GH (2013) Response of magnetic properties to heavy metal pollution in dust from three industrial cities in China. J Hazard Mater 246–247: 189–198.

48. He TR (2007) Biogeochemical cycling of mercury in Hongfeng Reservior, Guizhou, China. Guiyang (in Chineses).

49. SEPA: State Environmental Protection Administration (Presently known as MEP; Ministry of Environmental Protection) (2000) Report on the State of the Environment in China. Available: http://english.mep.gov.cn/SOE/soechina2000/english/atmospheric/atmospheric_e.htm.

50. Farkas A, Claudio E, Vigano L (2007) Assessment of the environmental significance of heavy metal pollution in surficial sediments of the River Po. Chemosphere 68: 761–768.

13

Removal Efficiency of Radioactive Cesium and Iodine Ions by a Flow-Type Apparatus Designed for Electrochemically Reduced Water Production

Takeki Hamasaki, Noboru Nakamichi, Kiichiro Teruya, Sanetaka Shirahata*

Department of Bioscience and Biotechnology, Faculty of Agriculture, Kyushu University, Higashi-ku, Fukuoka, Japan

Abstract

The Fukushima Daiichi Nuclear Power Plant accident on March 11, 2011 attracted people's attention, with anxiety over possible radiation hazards. Immediate and long-term concerns are around protection from external and internal exposure by the liberated radionuclides. In particular, residents living in the affected regions are most concerned about ingesting contaminated foodstuffs, including drinking water. Efficient removal of radionuclides from rainwater and drinking water has been reported using several pot-type filtration devices. A currently used flow-type test apparatus is expected to simultaneously provide radionuclide elimination prior to ingestion and protection from internal exposure by accidental ingestion of radionuclides through the use of a micro-carbon carboxymethyl cartridge unit and an electrochemically reduced water production unit, respectively. However, the removability of radionuclides from contaminated tap water has not been tested to date. Thus, the current research was undertaken to assess the capability of the apparatus to remove radionuclides from artificially contaminated tap water. The results presented here demonstrate that the apparatus can reduce radioactivity levels to below the detection limit in applied tap water containing either 300 Bq/kg of ^{137}Cs or 150 Bq/kg of ^{125}I. The apparatus had a removal efficiency of over 90% for all concentration ranges of radio–cesium and –iodine tested. The results showing efficient radionuclide removability, together with previous studies on molecular hydrogen and platinum nanoparticles as reactive oxygen species scavengers, strongly suggest that the test apparatus has the potential to offer maximum safety against radionuclide-contaminated foodstuffs, including drinking water.

Editor: Vishal Shah, Dowling College, United States of America

Funding: All experiments were performed using Kyushu university's finance (Trust Accounts No. JAKF650803). The funder had no role in study design, data collection and analysis, decision to publish, or preparation of the manuscript.

Competing Interests: The authors have declared that no competing interests exist.

* Email: sirahata@grt.kyushu-u.ac.jp

Introduction

The Great East Japan Earthquake of magnitude 9 struck the northeastern coast of Japan on March 11, 2011. The earthquake caused a catastrophic tsunami, with the wave height of nearly 40.5 m, which caused failures in the nuclear reactor cooling system in the Fukushima Daiichi Nuclear Power Plant (FDNPP) [1,2]. Soon after, these failures triggered hydrogen explosions in the nuclear reactors, discharging radioactive steam and liberating various radionuclides into the air over several days [2,3]. Following the incident, natural factors such as wind flow, air streams, and rainfall caused dispersion and precipitation of various levels of radionuclides on land surfaces and vegetation in the Tohoku and Kanto regions [4–8]. Radionuclides were also detected in Fukuoka, 1,000 km away from the FDNPP [9], indicating the wide spread of the radioactive plume over Japan. Urgent action to cope with the situation involves decontamination of terrestrial and aquatic radioactivity sources, including drinking water. Incineration of contaminated materials such as plants, wood bark, garbage, and house wreckage is one choice for disposition, although it leaves cesium-enriched ash. An entire system for safe incineration, removal of ash radioactivity and safe disposal has been reported, with promising results [10]. Numerous conventional methods using ion exchange, various membrane processes, coagulation and co-precipitation and other technologies for eliminating radionuclides from radioactive wastewaters have been reported to be effective [11,12]. Numerous approaches have been shown to remove radionuclides from contaminated water, including a mixture of activated carbon and/or zeolite-based media [13–15], co-precipitation with zinc hexacyanoferrate (II) followed by precipitation [16], sorption of radionuclides with biomaterials such as diatomite [17], Prussian blue immobilized diatomite or alginate/calcium beads or magnetic nanoparticles [18–20], arca shell [21], sulphuric acid-modified persimmon waste [22], nickel (II) hexacyanoferrate (III) functionalized walnut shell [23], mesoporous silica monoliths conjugated with dibenzo-18-crown-6 ether [24], and cobalt ferrocyanide impregnated anion exchange beads [25]. Additionally, a layered chalcogenide with a CdI_2 crystal structure for adsorbing several cations has been explored [26].

Although these technologies are encouraging for removal of various levels of radionuclides and further improvements are expected to arise in the future, securing safe drinking water is also of prime importance. Rainwater samples collected in Fukushima in early April, 2011 have been reported to contain ^{131}I (1470 ± 26.5 Bq/L), ^{134}Cs (100 ± 25.3 Bq/L) and ^{137}Cs (129 ± 9.47 Bq/L) [27]. The fallout contaminates surface waters, including lakes and rivers, which are the main sources for preparing tap water to supply the residents in these regions. As a result, drinking water prepared from several water purification plants was reported to be contaminated. Subsidiary methods to reinforce conventional water purification systems have been reported to eliminate radioactivity from contaminated water sources. The efficacies of the coagulation-flocculation-sedimentation method in water purification plants, with removal efficiencies of 17% and 56% for ^{131}I and [134]Cs, respectively [28,29], and radionuclide absorption by algal strains for environmental remediation [30,31] have been assessed. Another significant point to consider is the contamination of drinking water via distribution system such as pipes, storage tanks, water pumps and heaters, which may be persistent contaminating sources. A recent review concluded that cesium appears to be removed by flushing water pipes with a low pH solution containing sodium or magnesium as ion competitors [32]. However, further assessment will be required before applying this approach to the vast areas of regional contamination. Approximately one month later, the radioactivity levels had decreased to below the limit values in the water purification plants [33]. Whereas even after 2 years, total Cs radioactivities above the limit values are reported in some foodstuffs, such as Chinese mushrooms, rice, soybean, adzuki-bean and several fish obtained from the areas surrounding the FDNPP [34,35]. Moreover, low levels of radioactive Cs species are still detected in the drinking water of many cities around FDNPP [36]. These results imply that the fallout still remains on land surfaces and nearby mountain areas and that rainfall wash down is a highly probable contaminant of tap water sources [7,8]. Precautions to avoid consumption of such foodstuffs, including drinking water, have been taken by measuring radioactivity levels prior to distribution. Nevertheless, following the accident, the concentrations of ^{131}I in the tap water distributed by these purification plants were 210 Bq/L in Tokyo, 189 Bq/L in Ibaraki, and 220 Bq/L in Chiba, all of which exceeded the upper limit of ^{131}I concentration set as 100 Bq/L for infants under 1 year of age by the Ministry of Health, Labour and Welfare, 1947 [3,37]. Therefore, it is highly desirable to have terminal security systems that can achieve the removal of even lower levels of radioactive contaminants in tap water because, for example, radiocesium accumulates in the body. However, only limited studies examining removability of radionuclides from household water purifiers are available to date. Several domestic pot-type water purifiers have been suggested as a possible final security treatment to eliminate contaminated radionuclides in tap water [27,38]. Although most of these pot-type water purifiers are efficacious, with varying degrees of radionuclide removal from contaminated water, they are useless against the biological effects exerted by unconscious ingestion of radionuclides via drinking water and/or foodstuffs.

Ionizing radiation emitted by ingested radionuclides causes water radiolysis by acting on the water molecules, which comprise approximately 80% of body weight [39]. Water radiolysis yields a variety of reactive oxygen species (ROS) including hydrogen peroxide (H_2O_2), the hydroxyl radical ($^{\bullet}OH$), superoxide anion radicals ($^{\bullet}O_2^{-}$), and other molecular species [40]. These free radicals cause extensive oxidative damage to biologically critical macromolecules such as DNA, RNA, proteins and lipids [41–45]. Such damage eventually induces cellular apoptosis or carcinogenic transformation [46,47]. Therefore, an ideal apparatus should have the potential to provide both the elimination of radionuclides prior to ingestion and protection from detrimental ROS effects generated by the accidentally and/or unconsciously internalized radionuclides.

Considering these requirements, an apparatus designed to produce electrochemically reduced water (ERW) could be thought to fulfill such demands because it contains two functional units; an electrolysis unit for molecular hydrogen enrichment, and a micro-carbon carboxymethyl (CM) cartridge unit for removing various impurities. ERW produced from tap water by this apparatus contains as much as 0.587 ppm dissolved hydrogen (Table 1, [48]). Dissolved molecular hydrogen has been shown to exert a radioprotective effect in both *in vitro* and *in vivo* studies [49,53]. These compelling results strongly support the suggestion that molecular hydrogen dissolved in ERW could function as a radioprotective agent in the body. Moreover, ERW was shown to contain platinum nanoparticles (Pt NPs) at up to 2.5 ppb as an ROS scavenger, liberated from Pt-electrodes during electrolysis [39,54].

As for the second requirement, a micro-carbon CM cartridge unit composed of a nonwoven-fabric filter, several types of activated carbon and an ion-exchange material was present in the current test apparatus to remove particulate matters, microorganisms and 13 designated impurities [55]. However, this micro-carbon CM cartridge has not been assessed for its ability to remove radionuclides from contaminated tap water. Therefore, the present research was aimed at evaluating whether the test apparatus as a whole is capable of removing radionuclides from contaminated tap water.

Materials and Methods

Chemicals
Cesium chloride (CsCl) and potassium iodide (KI) were purchased from Wako Pure Chemical Industries (Osaka, Japan).

Radioisotopes
[137]CsCl [0.0021 MBq/g] and Na[125]I [12.950 TBq/g] were purchased from Japan Radioisotope Association (JRIA, Tokyo, Japan). We used ^{125}I because Kyushu University Radioisotope Center has an approval to use this radionuclide. Tap water distributed by the Fukuoka City Waterworks Bureau, Fukuoka, Japan was used in all experiments except ultrapure water (Milli Q water, Merck Millipore, Tokyo, Japan) for the preparation of standard solutions for inductively coupled plasma-mass spectrometry (ICP-MS) analysis.

Electrochemically reduced water (ERW)-producing apparatus
A water flow-type apparatus, Trim Ion NEO, was provided by Nihon Trim Co. Ltd., Osaka, Japan as the test apparatus. This test apparatus is composed of two units, a micro-carbon CM cartridge unit (Fig. 1B) and an electrolysis unit (Fig. 1C). Tap water flows into the cartridge unit, where tap water passes through the nonwoven-fabric filter to remove macroparticles, and pre-cleaned water flows into mixed layers of activated charcoal powders and cationic ion-exchange material to remove most of the impurities, including dissolved lead and 13 other elements that must be removed. The remaining contaminants, such as microorganisms and iron rust particles larger than 0.1 μm in size, are also eliminated by the cartridge (Fig. 1B). The micro-carbon CM cartridge unit is certified to withstand filtration of at least 12 tons of tap water per year or 35 liters per day for 1 year. In the present study, we used a new cartridge unit for each experiment. Purified tap water flows into the electrolysis unit, which is composed of five platinum (Pt)-coated electrode plates, separated by semi-permeable membranes and the water is electrolyzed while passing through the gaps between the electrodes (Fig. 1C). Platinum-

Table 1. Characteristics of the sample waters.

| | Tap Water | Filtered Water | ERW | | | |
			Lv 1	Lv 2	Lv 3	Lv 4
pH	7.6±0.0	7.6±0.0	8.0±0.0	8.5±0.0	9.1±0.0	9.4±0.1
ORP (mV)	555.3±15.5	550.0±20.1	140.0±5.0	110.0±7.5	−673.3±2.5	−688.0±9.5
EC (ms/m)	49.3±0.1	49.5±0.1	49.7±0.1	49.7±0.1	49.0±0.2	48.1±0.2
DH (ppb, µg/l)	N.D.	N.D.	70.0±19.3	163.3±18.0	321.7±47.5	587.0±44.6
DO (ppm, g/l)	7.5±0.0	7.5±0.0	7.5±0.1	7.1±0.1	6.6±0.2	6.1±0.3

Filtered water: tap water was passed through the micro carbon cartridge without electrolysis. Lv 1: electrochemically reduced water (ERW) generated by electrolyzing the filtered water at level 1 with constant electric current at 50 volts (V) upper limit voltage and a flow rate of 1.8–2.0 l/min. Likewise, other ERWs were produced using identical conditions, except selecting the Lv 2 to Lv 4 switch. ORP: oxidation-reduction potential. EC: electrical conductivity. DH: dissolved hydrogen. DO: dissolved oxygen. Measurements were conducted at ambient temperatures. N.D.: Not Detected.

coated titanium electrodes are certified for at least 1,400 hours use without a marked deterioration with respect to the efficacy of water electrolysis, suggesting that the loss of a small amount of Pt nanoparticles from the surface of the electrode will not significantly affect the electrolysis efficacy of the device used here. Electrolyzed tap water near the cathode typically exhibits a high pH, low dissolved oxygen, high negative redox potential and a high concentration of dissolved hydrogen (0.4–0.9 ppm) (Table 1, [48]). Water produced in this manner, with the above characteristics, is designated as ERW. The test apparatus is designed to produce five types of water; four types of ERW (Levels 1–4) electrolyzed with a constant electric current for each level (0.8 to 4.2 A) at a maximum of 50 volts and one type of filtered water without electrolysis (Table 1). ERW is produced near the cathode, as indicated by the thick right-facing arrows in Fig. 1c, and positively charged radioactive Cs ions will be attracted to the cathode side during electrolysis, resulting in an increased concentration of Cs^+ ions in ERW, dependent upon the current intensity. Conversely, negatively charged I ions will be attracted to the anode side, resulting in a decreased concentration of I ions in ERW. The electrolysis currents were increased in the order of levels 1 to 4, where Level 4 represents the strongest current, reflecting the highest dissolved hydrogen (DH) and the lowest oxidation-reduction potential (ORP) (Table 1). When the radio-activity of ERW at level 4 is measured as being lower than the background level, then one can conclude that the radioactivity of ERW at levels 1 to 3 is lower than the background level. ERW at levels 1 to 3 is usually used for drinking and at level 4 is used for cooking. We have included Table 1 to aid the readers understanding of the four types of ERW.

Preparation of non-radioactive sample water (CsCl, KI)

Tap water was used as a control. CsCl solutions of 20 liters each with concentrations of 20 and 2,000 ppb were prepared using tap water. Likewise, KI solutions with concentrations of 100 and 4,000 ppb were prepared. These solutions are designated as sample waters. The test system was arranged by placing an adjustable speed pump between the sample waters and the test apparatus to mimic tap water pressure, connected to the inlet of the test apparatus, as shown in Fig. 1A. The water flow rate was set to 1.8–2.0 L/min by adjusting the pump speed throughout the entire experiment. In the experiment, 1–2 liters of tap water was used to wash and equilibrate the system each time the sample concentrations were changed. Fifteen milliliters of filtered, ERW and relevant control waters were collected for ICP-MS analysis. The

removal efficiency was calculated according to a previously described equation [38], shown in Tables 2 and 3.

ICP-MS analysis of Cs and I elements in ERWs

Sample waters were passed through the apparatus, and collected filtered waters were quantitated using ICP-MS (Agilent 7500c, Agilent Technologies Co. Ltd., Santa Clara, CA, USA) in the Radioisotope Center at Kyushu University.

Preparation of radioactive sample water ([137]CsCl and Na[125]I)

Stock solution of [137]CsCl was diluted with 20 liters of tap water to prepare concentrations of 15,000, 3,000, 300, and 30 Bq/Kg. Likewise, Na[125]I stock solution was diluted with 20 liters of tap water to prepare concentrations of 15,000, 1,500, and 150 Bq/Kg. All other experimental conditions, such as water flow rate, system equilibration, the electrolysis conditions of the apparatus were carried out as closely as possible to those used for the non-radioisotope experiments, except that 10 ml of each of the sample waters were collected for radioactivity counting.

Radioactivity counting of ^{137}Cs and ^{125}I in sample waters

Radioactive sample waters were passed through the apparatus, and collected waters were quantitated using a gamma counter (AccuFLEX γ ARC-7001, Hitachi Aloka Medical, Ltd., Tokyo, Japan) in the Center of Advanced Instrumental Analysis at Kyushu University. To evaluate the effect of the electrolysis step on radionuclide removal, filtered waters were electrolyzed by a constant current (4.2 A) at level 4 and radioactivities of ERW were quantitated as above.

Statistical analysis

All experiments were performed in triplicate. Data are expressed as means ± SD for each experiment.

Results

Analysis of Cs and I elements in the filtered water

Prior to radioisotope experiments, CsCl and KI solutions were prepared as described in the Materials and Methods section and their removability was tested. The background Cs concentration in tap water was similar to that for the filtered water (Fig. 2A, column 0 ppb). When 20 and 2,000 ppb CsCl solutions were used, the measured values of the filtered water indicate that the test apparatus had a higher removability (87.4%) for the 20 ppb CsCl

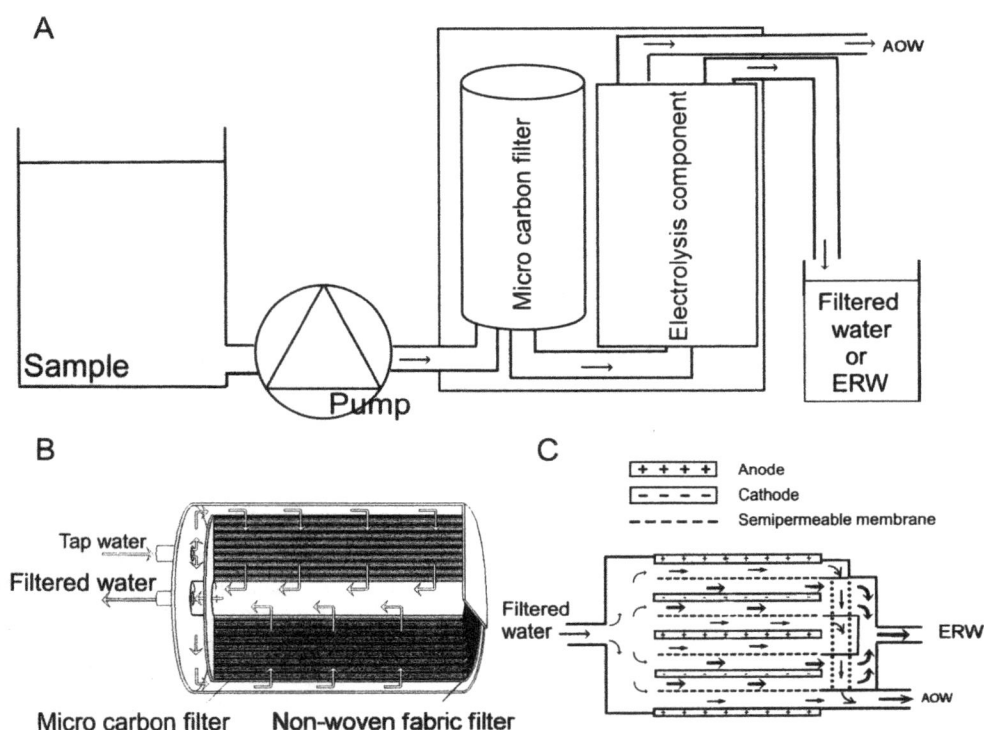

Figure 1. Schematic of the flow-type electrolysis apparatus. The test apparatus is composed of two units, a micro-carbon CM cartridge (B) and an electrolysis unit (C). The overall water flow and equipment set up is shown in (A). Sample water is connected to an adjustable speed pump to maintain a flow rate of 1.8–2.0 l/min and expelled to the inlet of the electrolysis unit (A). Tap water passes through the nonwoven-fabric filter, the mixed layers of activated charcoal powders and cationic ion-exchange material to make filtered water (B). Filtered water then flows into the electrolysis unit composed of platinum-coated 5 electrode plates separated by semi-permeable membranes (C). Filtered water will be electrolyzed at levels 1, 2, 3 and 4 at a maximum of 50 volts while passing through the gaps between the electrodes.

solution than for the 2,000 ppb CsCl solution (58.2%) (Fig. 2A, Table 2). Similar experiments using KI solutions were carried out and the results are shown in Fig. 2B. The background I concentrations in tap water and that for the filtered water were similar (Fig. 2B, column 0 ppb). Removal efficiency after filtration for 100 ppb and 4,000 ppb KI solutions were 91.7% and 84.6%, respectively (Table 3). These results demonstrate that the micro-carbon CM cartridge is capable of removing Cs and I ions at all concentration ranges tested (Fig. 2).

Removal efficiency of ^{137}CsCl and Na^{125}I in the filtered water

Because the test apparatus removed Cs and I ions efficiently, assays were extended to examine the removability of ^{137}CsCl and Na^{125}I. The natural background counts in tap water and filtered water were below the detection limit of the gamma counter (Fig. 3A and B, column 0). Tap water containing 30 (0.0067 ppb as Cs ions), 300 (0.0642 ppb as Cs ions), 3,000 (0.636 ppb as Cs ions) and 15,000 (3.16 ppb as Cs ions) Bq/kg of ^{137}CsCl as controls showed the expected radioactive counts (Fig. 3A and 3B, white bar

Table 2. Removal efficiencies (%) for Cs ion and ^{137}Cs.

Measured (Loaded) amounts		Removal efficiency (%)
as Cs ion (ppb)	as ^{137}Cs (Bq/kg)	
1976.47 (2000)	0	58.2
20.55 (20)	0	87.4
#3.1600	16212.0 (15000)	96.9
#0.6360	3262.0 (3000)	96.9
#0.0642	329.0 (300)	99.2*
#0.0067	34.9 (30)	92.5*

Removal efficiency (%) = (1−[A]/[B])×100 according to [38]. [A], [B]: concentrations of Cs and ^{137}Cs after and before filtration. Each solution was filtered only, without electrolysis. *: [A] values were below the detection limit. *: [A] values used to calculate removal efficiency were below the detection limit. #: equivalent ppb values calculated from the radioactivities loaded. Values within parentheses were prepared and loaded amounts or radioactivities of cesium.

Table 3. Removal efficiencies (%) for I and ^{125}I ions.

Measured (Loaded) amounts		Removal efficiency (%)
as I ion (ppb)	as ^{125}I (Bq/kg)	
3891.0 (4000)	0	84.6
130.0 (100)	0	91.7
#0.0000197	14993.0 (15000)	99.4
#0.00000351	1788.0 (1500)	99.3
#0.000000196	146.3 (150)	99.5*

Removal efficiency (%) = (1−[A]/[B])×100 according to [38]. [A], [B]: concentrations of I and ^{125}I solutions after and before filtration. Each solution was filtered only, without electrolysis. *: [A] values used to calculate removal efficiency were below the detection limit. #: equivalent ppb values calculated from the radioactivities loaded. Values within parentheses were prepared and loaded amounts or radioactivities of iodine.

at each concentration) with a high correlation coefficient (Fig. 3C, R^2 = 0.999). Control waters were then passed through the micro-carbon CM cartridge and the filtrate radioactivities were measured (Fig. 3A and B). It was found that the radioactivities of the filtered water for ^{137}CsCl were reduced significantly (Fig. 3A and 3B) and removal efficiency was 96.9%, even after loading 15,000 Bq/kg of ^{137}CsCl (Table 2).

To evaluate Na^{125}I removability, we prepared Na^{125}I containing sample waters as described above. The natural background count in tap water and filtered water exhibited values below the detection limit (Fig. 4A and B, column 0). Tap water containing 150 (0.000196 ppt as I ions), 1,500 (0.00351 ppt as I ions) and 15,000 (0.0197 ppt as I ions) Bq/kg of Na^{125}I as controls showed expected radioactive counts (Fig. 4A and 4B, white bar at each concentration) with a high correlation coefficient (Fig. 4C, R^2 = 0.999). Radioactive control tap waters were passed through the micro-carbon CM cartridge, reducing the filtrate radioactivities significantly (Fig. 4A and B), with a removal efficiency of over 99% (Table 3). Thus, the micro-carbon CM cartridge was demonstrated to efficiently remove radioactivities up to 15,000 Bq/kg of Na^{125}I.

Effect of electrolysis on the removal efficiency of ^{137}Cs and ^{125}I

In parallel with the preceding experiments, we evaluated the effects of the electrolysis step in terms of efficiencies for ^{137}Cs and ^{125}I removal from the filtered radioactive water. Filtered water was electrolyzed at the highest current level of 4. In this experiment, we selected 300 Bq/kg of ^{137}Cs water, which loaded 30 times more radioactivity than the upper limit value of 10 Bq/kg for drinking water set by the government [56]. Under these conditions, the radioactivity in ERW remained below the detection limit (Fig. 5A). Similarly, we evaluated the removability of ^{125}I by the highest electrolysis level of 4. In this case, we selected 150 Bq/kg of ^{125}I, which is a loading of 1.5 times more radioactivity than the upper limit of 100 Bq/L of ^{131}I concentration for infants under 1 year of age set by the Ministry of Health, Labour and Welfare, 1947 [37]. The radioactive iodine level in ERW remained below the detection limit (Fig. 5B). Therefore, the results indicate that the cartridge substantially removed ^{137}Cs and ^{125}I from tap water prior to the electrolysis step, thereby assuring undetectable levels of radioactivity in ERW produced at the highest current level of 4, which has the highest attraction for ^{137}Cs$^+$, and thus the results hold true for ERWs produced with the current levels 1 to 3.

Discussion

The FDNPP accident liberated various radionuclides, including ^{131}I, ^{132}I, ^{134}Cs, and ^{137}Cs [57]. Amongst these radionuclides, ^{131}I can enter the body through inhalation and by ingesting contaminated foodstuffs including drinking water, which then rapidly concentrates in the thyroid gland, where β-radiation exposure takes place. As its half-life is 8 days, radioactivity levels are expected to be reduced substantially over several months. Therefore, an obvious precaution is not to ingest ^{131}I-contaminated or doubtful foodstuffs including drinking water. Water supply law in Japan limits the lowest chlorine concentration in tap water outlet at 0.1 mg/L [58]. Dissolved ^{131}I is reported to form various species in tap water such as the radioactive iodide ion (^{131}I$^-$), hypoiodous acid (HO^{131}I), the iodate ion (^{131}IO$_3^-$), iodine molecules (^{131}IO$_2$) and organic ^{131}I. ^{131}I$^-$ reacts with chlorine and is transformed mainly into HOI at neutral pH. HOI is further transformed into IO$_3^-$ by reacting with chlorine [29], and as a result, almost all iodine is converted to the iodate ion (IO$_3^-$) in tap water due to the oxidation by chlorine [59]. It is reported that ^{131}I$^-$ removal is increased by water containing 0.1–0.5 mg/L chlorine, with lower concentrations of powdered activated charcoal [29]. However, granular and powdered activated carbons were reported to remove ^{131}I at about 30–40% efficiency. Additionally, it has been reported that ^{125}I$^-$ and ^{125}I$_3^-$ were prepared from ^{125}I and used to test the removability of these species by a granular type charcoal, which resulted in a small amount of adsorption [60]. These results may partly explain the inefficient removability by activated charcoal reported by others, through selective adsorption of iodate and iodine [38,60,61]. Activated carbon was shown to remove iodide (I$^-$) more efficiently than iodate (IO$_3^-$) [27]. Therefore, it appears that combinations between the types of activated carbon/charcoal and iodine species affect overall removability. In the present experiments, we used tap water distributed by the Waterworks Bureau of the City of Fukuoka, expected to contain at least 0.1 mg/L chlorine. Thus, ^{125}I is mostly, if not completely, converted to iodate ions (IO$_3^-$) by chlorine in the tap water. In the present results, KI and ^{125}I were efficiently removed from tap waters by the micro-carbon CM cartridge, suggesting that iodide and iodate ions were removed. The micro-carbon CM cartridge is composed of a nonwoven-fabric filter and activated carbons consisting of a coconut shell activated carbon powder, a coconut shell activated carbon conjugated with a silver compound for antimicrobial effect, and an amorphous titanosilicate-based inorganic compound (BASF Co, Germany) molded with a fibrous binder for shaping. This cartridge was used in the present test apparatus to remove

A

B

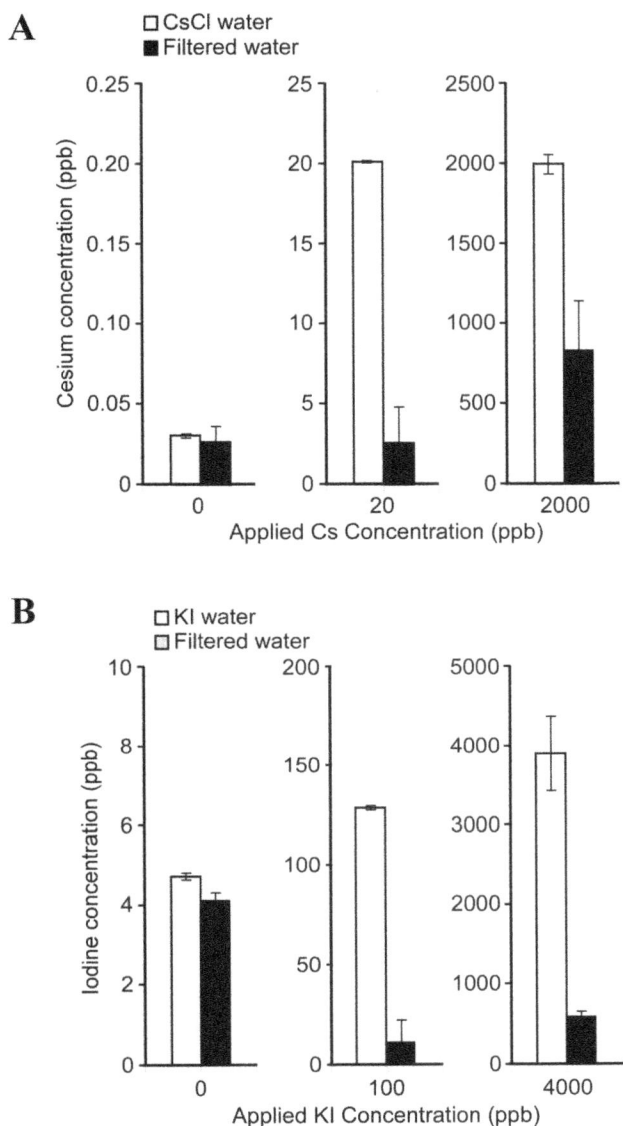

Figure 2. Measurement of Cs and I elements in filtered waters. CsCl solutions at concentrations of 0, 20 and 2,000 ppb were passed through the test apparatus. Collected filtered waters were used to measure Cs concentration by ICP-MS (A). KI solutions at concentrations 0, 100 and 4,000 ppb were passed through the test apparatus. Collected filtered waters as in (A) were used to measure I concentration by ICP-MS (B). White bar: Tap water, gray bar: Filtered water. Experiments were carried out in triplicate.

particulate matters, microorganisms, and for qualified removability of 13 designated impurities, tested according to the standard method set by JIS S 3201, 2004 (Domestic Water Purifier Quality Test) [55]. It is worth mentioning that the test apparatus effectively removed I and ^{125}I (applicable to Cs^+ and ^{137}Cs), even though water was supplied to the apparatus through a pump simulating tap water outlet pressure to attain 1.8–2.0 L/min flow rate, which markedly reduced the contact time of water with the activated carbon surfaces and ion-exchangers compared with those in pot-type water purifiers. It has been reported that the above-mentioned molded activated carbons can replace the hollow fiber membrane filter that is commonly used in other water purifiers to eliminate materials larger than 0.1 μm in size [55]. Incidentally,

hollow fiber membranes do not contribute to the elimination of iodate (IO_3^-) ions because their radius is 0.326 nm, even when their radius is increased several fold in water [61]. Additionally, the ineffectiveness of removing ^{131}I by boiling tap water has been reported [3].

Cesium is an alkaline earth metal that exists as a monovalent cation form (Cs^+) in water and in soils [27]. We found that Cs^+ could be efficiently removed by the micro-carbon CM cartridge tested here. The mechanism for the removal of Cs^+ remains to be investigated. The Cs^+ removal efficiency by the apparatus was 87.4% at 20 ppb, which is comparable to that of several pot-type water purifiers that have efficiencies of around 90% for tap water containing 40–50 μg/l (ppb) cesium chloride [38]. A removal efficiency of 58.2% for Cs^+ appears to be low at the highest concentration (1976.5 ppb) loading. This lower removal efficiency could be explained by the amount of Cs^+ getting close to system over loading because this amount is 625.3 times more Cs^+ ion loading than the 3.16 ppb Cs^+ ion calculated from the highest radioactivity (16,212 Bq/kg) loading where 96.9% removability was attained (Table 2). Therefore, the apparatus could remove ^{137}Cs with above 96% efficiency for less than a 3.16 ppb CsCl loading and the removal efficiency is higher than that reported for two commercialized pot-type water purifiers, composed of activated charcoal and an ion exchanger, or activated charcoal, ceramics and a hollow fiber membrane, with 84.2–91.5% efficiencies for rain water samples [27]. Another set of experiments using commercialized four pot-type purifiers made of materials similar to those above assessed iodine and cesium removability, with efficiencies of approximately 85% and 75–90%, respectively [38]. Others also tested Cs removability using a spongiform adsorbent made of Prussian blue caged within the diatomite cavities and carbon nanotubes, by contacting for 10 hours with low levels of ^{137}Cs, yielding a 99.93% removal efficiency [18]. The present test apparatus showed a removal efficiency of over 96% for Cs and I, which is competitive with or better than previously reported removal efficiencies ranging from 75% to 99.93%. It is emphasized here that the advantages of the test apparatus are that it has long been used for domestic use, is easy to operate, provides a sufficient amount of purified water instantaneously (max. 5 l/min.) and offers an established system for proper disposal and/or recycling of used cartridges. Following the FDNPP accident, tap water contamination monitoring revealed that the maximum of the sum of ^{134}Cs and ^{137}Cs was 180.5 Bq/kg on March 2011 in Tamura, Fukushima Prefecture. It was also reported that a sum of ^{134}Cs and ^{137}Cs less than 32 Bq/kg was sporadically detected in tap water during 22 days of monitoring after the accident [29]. The water purification plants take precautions not to distribute contaminated water through constant monitoring to meet the latest upper limit value, set by the government as 10 Bq/kg for drinking water, effective from April 1, 2012 [56]. In reality, the detection of greater than 10 Bq/kg radioactivities in tap water in general public is most likely to be the result of accidental and sporadic contamination events. In any case, the test apparatus was demonstrated to decontaminate radiocesium levels to below the detection limit, even when tap water was contaminated by up to 300 Bq/kg radiocesium. When loading 300 Bq/kg of ^{137}Cs to the cartridge, the removability obtained was a conditional value of 99.2% and leaving the remaining radioactivity to be below the upper limit value of 10 Bq/kg set by the Government. This indicates that the filtered water right before entering into the electrolysis unit still contains a trace amount of ^{137}Cs and the following electrolysis step may produce ^{137}Cs enriched ERW. The test apparatus is a powerful electrolysis device yet finely tuned to produce various levels of dissolved hydrogen electric current

A

B

C

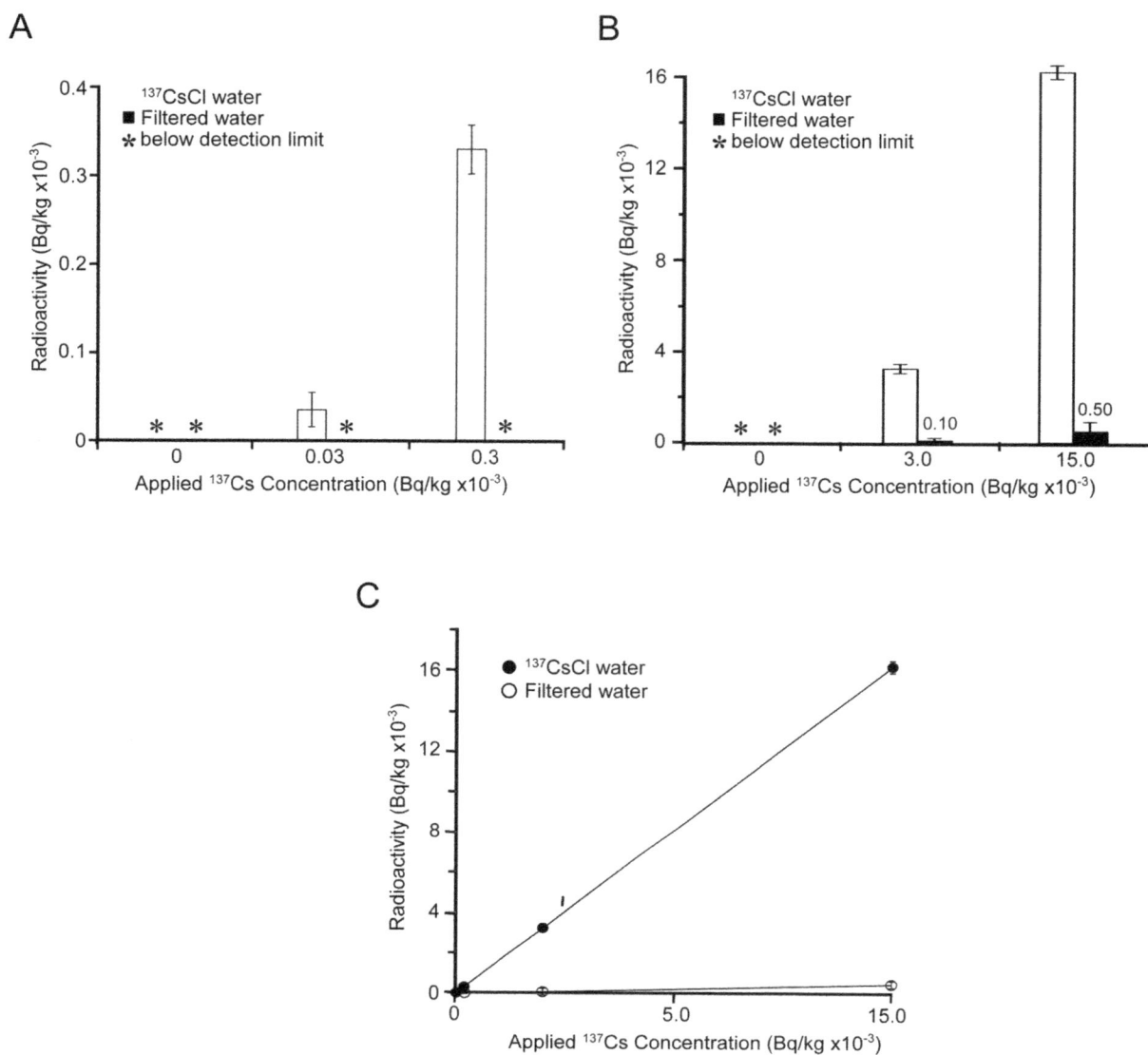

Figure 3. Measurement of ^{137}Cs in sample waters. ^{137}CsCl solutions at concentrations of 0, 0.03, 0.3, 3.0 and 15.0 KBq/kg were passed through the test apparatus. Collected filtered waters were used to measure ^{137}Cs counts by an AccuFLEX γ ARC-7001 gamma counter (A and B). White bar: ^{137}CsCl solutions before filtration, gray bar: ^{137}CsCl solutions after filtration. Radioactivities before and after filtration were evaluated by linear-regression analysis (C). ●: ^{137}CsCl solutions before filtration, ○: ^{137}CsCl solutions after filtration. Experiments were carried out in triplicate.

dependently (Table 1). Under these conditions, we could not definitely exclude a slight possibility that the electrolysis step contributes to ^{137}Cs enrichment in ERW. Thus, the only way to clarify such uncertainty was to conduct the experiments as shown in Fig. 5. Moreover, we judged that it is not sufficient enough by just showing the removability of the cartridge filter unit alone and extrapolating the results for evaluating the entire flow-type system. To this end, we decided to measure the radioactivity in ERW, which allows evaluating cartridge unit and electrolysis unit simultaneously. Therefore, the evaluation of the filtering unit in combination with the electrolysis unit as the complete flow-type system was necessary. Our concern for negatively charged I ion was less intense compared to Cs$^+$ ion due to higher removability by the cartridge unit and attracted to the anode side. Nevertheless, we confirmed the removability of the flow-type system experimentally to provide the data set with Cs$^+$ data.

It is commonly regarded that tap water prepared from lakes and rivers contains varying amounts of organic and inorganic materials. In the present study, we considered these to have a significant impact on removability by the test system because such materials are most likely to compete with the very small amounts of ^{137}Cs and ^{125}I ions present. Only experiments using low levels of radionuclides will answer the question of whether such interactions between the constituents and added radionuclides may affect removability by this apparatus. Another reason to use lower levels of radionuclides is that even a small amount of ^{137}Cs dissolved in water is difficult to remove [11,20] and accumulates in the body, causing prolonged exposure. Moreover, the fact is that low levels of radioactive Cs species currently contaminate drinking water in many cities around FDNPP [36]. This may be partly attributed to the limited removability of solubilized cesium by the conventional coagulation-sedimentation process [11,29]. It is therefore extremely important, for the residents of affected regions,

A

B

C

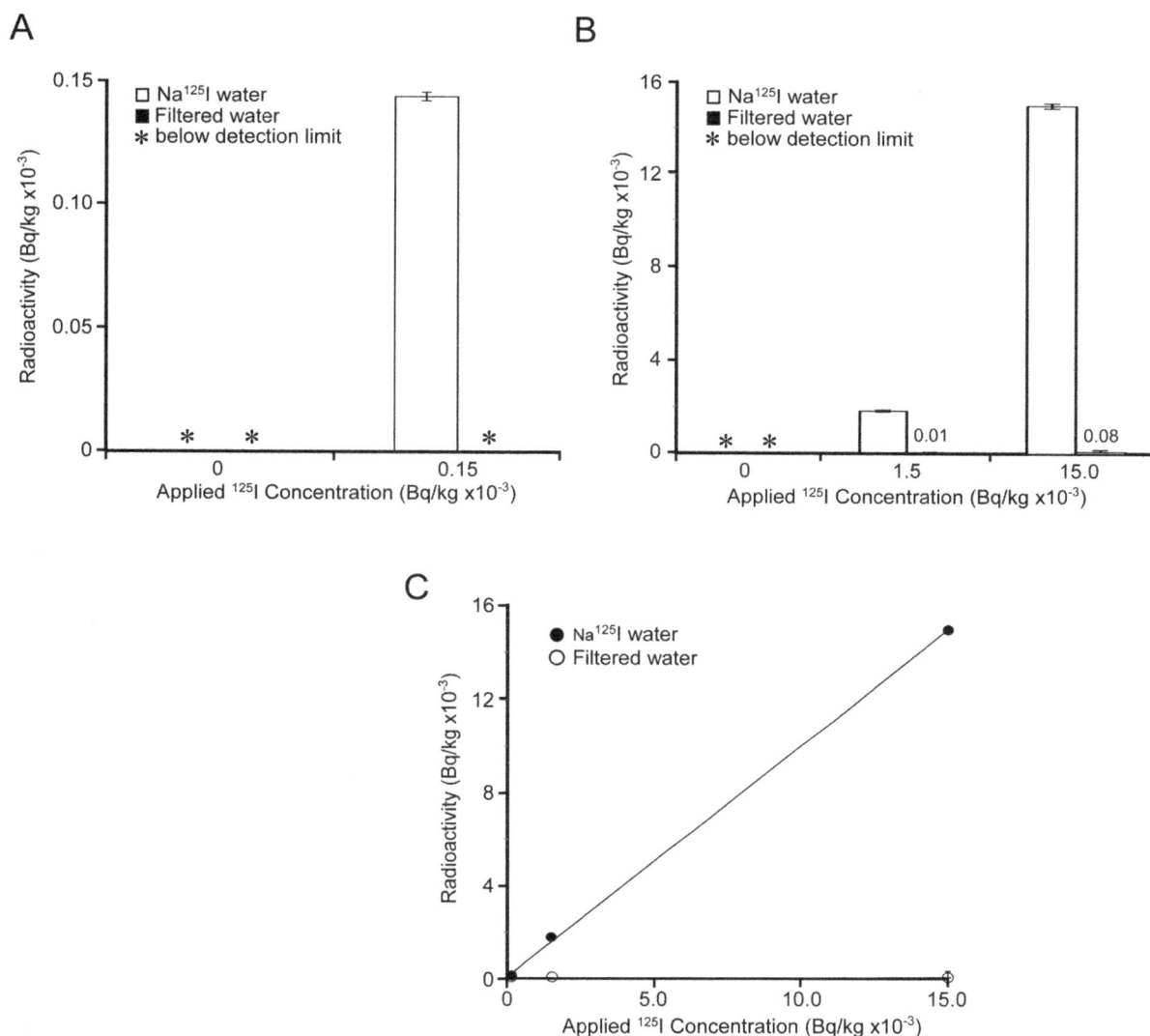

Figure 4. Measurement of ^{125}I elements in sample waters. Na^{125}I solutions at concentrations of 0, 0.15, 1.5 and 15.0 KBq/kg were passed through the test apparatus. Collected filtered waters were used to measure ^{125}I counts by an AccuFLEX γ ARC-7001 gamma counter (A and B). White bar: Na^{125}I solutions before filtration, gray bar: Na^{125}I solutions after filtration. Radioactivities before and after filtration were evaluated by linear-regression analysis (C). ●: Na^{125}I solutions before filtration, ○: Na^{125}I solutions after filtration. Experiments were carried out in triplicate.

to find a way to remove even small amounts of nuclear contaminants from drinking water.

Another concern related to radiocesium is its longer half-life and a characteristic of ready transfer to the human diet through plants [62]. Precipitated Cs$^+$ binds to clay minerals rather tightly [27], and depth distribution studies reveal that approximately 80% of total radiocesium is retained in the upper 2.0 cm of tested soil samples [63]. Another study estimated that ^{137}Cs could reach a depth of only 18 cm after 300 yr [37]. These characteristics of surface area retention of radiocesium in addition to its long physical half-life (^{134}Cs, $T_{1/2} = 2.06$ yr; ^{137}Cs, $T_{1/2} = 30.17$ yr) could be a secondary contamination source for vegetation via roots. Uptake of radiocesium from the root is thought to occur via the potassium transport system and is distributed rapidly within the plants [62]. Indeed, many agricultural products are reported to be contaminated by radiocesium and are their marketing is restricted [64,65]. Ingestion of radiocesium-contaminated food-stuffs will expose the gastrointestinal tract and be absorbed into

tissues and organs in the body. Gastrointestinal, reproductive and hematopoietic systems are sensitive to ionizing radiation due to their high turnover rate [57,66]. As an example, the degeneration of small intestinal mucosa cells is caused by free radicals produced from the interactions of radiation energy with intracellular water molecules [66]. Water radiolysis generates a variety of ROS that cause extensive oxidative damages to biologically critical macro-molecules, leading to cell death [40,42–45]. Therefore, providing a method to counter radiation hazards caused by accidentally ingested radioactive waters and foodstuffs will be a great contribution to human health.

ERW is regarded as beneficial to health because of its ROS scavenging ability [39]. ERW produced from tap water by this apparatus could contain as much as 0.587 ppm of dissolved hydrogen (Table 1, [48]). This hydrogen concentration in ERW is relatively high for a flow-type electrolysis apparatus when compared with the concentration of 1.6 ppm hydrogen in 100% hydrogen-saturated water [53]. Such dissolved molecular hydro-

A

B

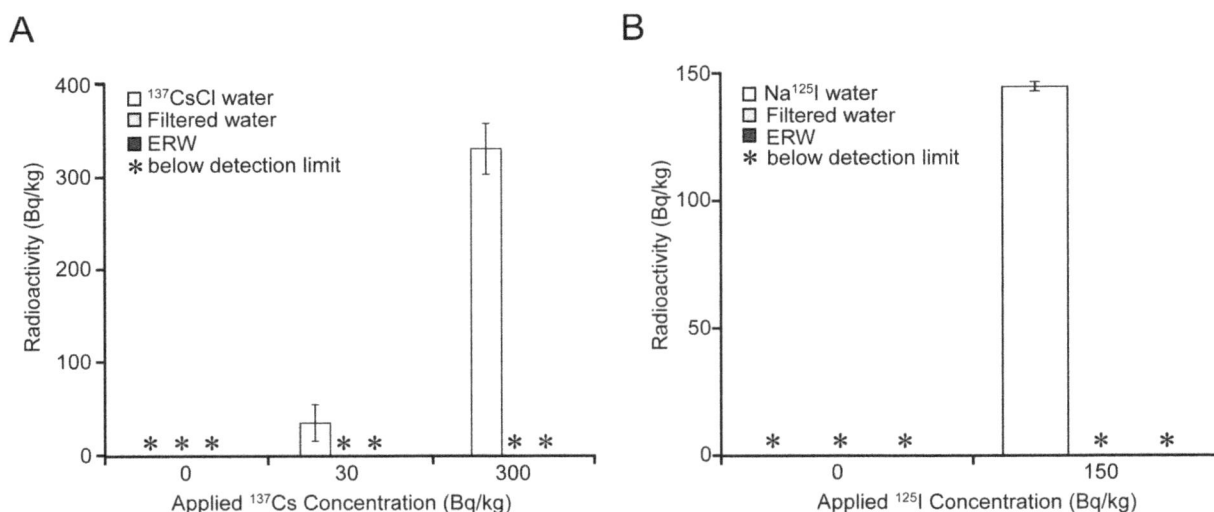

Figure 5. Effects of electrolysis on filtered radioactive sample waters. $^{137}CsCl$ solutions of 30 and 300 Bq/kg were passed through the test apparatus and filtered waters were collected for measurement. Then, filtered water was passed through the electrolysis unit at the highest electrolysis level of 4 and ERW was collected for measurement. Collected waters were used to measure ^{137}Cs counts by an AccuFLEX γ ARC-7001 gamma counter (A and B). Using the same protocol, filtered water and ERW were collected for 150 Bq/kg of Na^{125}I solution. White bar: $^{137}CsCl$ or Na^{125}I solutions, gray bar: Filtered $^{137}CsCl$ or Na^{125}I solutions; black bar: ERWs of filtered $^{137}CsCl$ or Na^{125}I solutions. Experiments were carried out in triplicate.

gen has been shown to exert radioprotective effects in both *in vitro* and *in vivo* studies [49–53]. Molecular hydrogen in ERW prepared from tap water suppressed neuroinflammation in mice [48], and extended the life span of *C. elegans* [54]. Additionally, molecular hydrogen was demonstrated to act as a neuroprotective agent and ROS scavenger [67]. Moreover, ERW produced from an electrolysis unit incorporating Pt-electrodes has been shown to contain 0.1–0.25 ppb Pt nanoparticles [39,54,68]. Pt nanoparticles exhibit protective effects that are attributed to their suppressing ROS production caused by UV-light-induced epidermal inflammation [69]. Synthetic Pt nanoparticles have been shown to scavenge ROS in cultured HeLa cells [70], to induce expression of antioxidant enzyme genes in rat skeletal muscle L6 cells [71], and to act as an SOD/catalase mimetic agent in human lymphoma cells [72]. Model ERW prepared from NaCl, KCl or NaOH solutions has been shown to exert beneficial effects such as anti-diabetic, anti-cancer, and life-span extension of nematodes because of its ROS scavenging ability in numerous *in vitro* and *in vivo* studies [73–78]. Therefore, molecular hydrogen and Pt nanoparticles dissolved in ERW could synergistically contribute to protect gastrointestinal damage caused by ingested radioactive foodstuffs. Furthermore, to maximize protective efficacy against radiation-induced gastrointestinal damage, the consumption of various supplemental foods such as naringin [42], probiotics

[57,66], Kefir [79], melatonin [80] and curcumin [81] are reported to be beneficial.

In conclusion, we demonstrated that radio-cesium and -iodine are efficiently removed by an apparatus containing a micro-carbon CM cartridge filter, prior to ingestion. We also suggest that the ERW produced by the test apparatus will provide maximum protection against accidentally and/or unconsciously ingested radionuclides because it contains dissolved hydrogen and Pt nanoparticles. Therefore, the test apparatus is considered to be a potential alternative tool to minimize radiation hazards caused by contaminated foodstuffs.

Acknowledgments

The authors thank Nihon Trim Co. Ltd. for providing the Trim Ion NEO apparatus and an adjustable flow rate pump. The authors are also grateful to Ms. Yuri Fujimoto and Chika Kubota for their technical assistance.

Author Contributions

Conceived and designed the experiments: TH NN KT SS. Performed the experiments: TH NN SS. Analyzed the data: TH NN KT SS. Contributed reagents/materials/analysis tools: TH NN SS. Wrote the paper: TH NN SS.

References

1. Hamada N, Ogino H (2012) Food safety regulations: what we learned from the Fukushima nuclear accident. J Environ Radioact 111: 83–99.
2. Hamada N, Ogino H, Fujimichi Y (2012) Safety regulations of food and water implemented in the first year following the Fukushima nuclear accident. J Radiat Res 53: 641–671.
3. Tagami K, Uchida S (2011) Can we remove iodine-131 from tap water in Japan by boiling? – Experimental testing in response to the Fukushima Daiichi Nuclear Power Plant accident. Chemosphere 84: 1282–1284.
4. Amano H, Akiyama M, Chunlei B, Kawamura T, Kishimoto T, et al (2012) Radiation measurements in the Chiba Metropolitan Area and radiological aspects of fallout from the Fukushima Dai-ichi Nuclear Power Plants accident. J Environ Radioact 111: 42–52.

5. Koizumi A, Niisoe T, Harada KH, Fujii Y, Adachi A, et al (2013) ^{137}Cs trapped by biomass within 20 km of the Fukushima Daiichi Nuclear Power Plant. Environ Sci Technol 47: 9612–9618.
6. Thakur P, Ballard S, Nelson R (2013) An overview of Fukushima radionuclides measured in the northern hemisphere. Sci Total Environ 458–460: 577–613.
7. Murakami M, Ohte N, Suzuki T, Ishii N, Igarashi Y, et al (2014) Biological proliferation of cesium-137 through the detrital food chain in a forest ecosystem in Japan. Sci Rep 4: 3599.
8. Nakanishi T, Matsunaga T, Koarashi J, Atarashi-Andoh M (2014) ^{137}Cs vertical migration in a deciduous forest soil following the Fukushima Dai-ichi Nuclear Power Plant accident. J Environ Radioact 128: 9–14.
9. Momoshima N, Sugihara S, Ichikawa R, Yokoyama H (2012) Atmospheric radionuclides transported to Fukuoka, Japan remote from the Fukushima Dai-

ichi nuclear power complex following the nuclear accident. J Environ Radioact 111: 28–32.

10. Parajuli D, Tanaka H, Hakuta Y, Minami K, Fukuda S, et al (2013) Dealing with the Aftermath of Fukushima Daiichi Nuclear Accident: Decontamination of Radioactive Cesium Enriched Ash. Environ Sci Technol 47: 3800–3806.

11. Liu X, Chen G-R, Lee D-J, Kawamoto T, Tanaka H, et al (2014) Adsorption removal of cesium from drinking waters: A mini review on use of biosorbents and other adsorbents. Bioresour Technol Available: http://dx.doi.org/10.1016/j.biortech. Last accessed 2014.01.012.

12. Rana D, Matsuura T, Kassim MA, Ismail AF (2013) Radioactive decontamination of water by membrane processes – A review. Desalination 321: 77–92.

13. Song K-C, Lee HK, Moon H, Lee KJ (1997) Simultaneous removal of the radiotoxic nuclides Cs^{137} and I^{129} from aqueous solution. Sep Purif Technol 12: 215–227.

14. El-Kamash AM (2008) Evaluation of zeolite A for the sorptive removal of Cs^+ and Sr^{2+} ions from aqueous solutions using batch and fixed bed column operations. J Hazard Mater 151: 432–445.

15. Borai EH, Harjula R, Malinen L, Paajanen A (2009) Efficient removal of cesium from low-level radioactive liquid waste using natural and impregnated zeolite minerals J Hazard Mater 172: 416–422.

16. Shakir K, Sohsah M, Soliman M (2007) Removal of cesium from aqueous solutions and radioactive waste simulants by coprecipitate flotation. Sep Purif Technol 54: 373–381.

17. Osmanlioglu AE (2007) Natural diatomite process for removal of radioactivity from liquid waste. Appl Radiat Isot 65: 17–20.

18. Hu B, Fugetsu B, Yu H, Abe Y (2012) Prussian blue caged in spongiform adsorbents using diatomite and carbon nanotubes for elimination of cesium. J Hazard Mater 217–218: 85–91.

19. Vipin AK, Hu B, Fugetsu B (2013) Prussian blue caged in alginate/calcium beads as adsorbents for removal of cesium ions from contaminated water. J Hazard Mater 258–259: 93–101.

20. Thammawong C, Opaprakasit P, Tangboriboonrat P, Sreearunothai P (2013) Prussian blue-coated magnetic nanoparticles for removal of cesium from contaminated environment. J Nanopart Res 15: 1689.

21. Dahiya S, Tripathi RM, Hegde AG (2008) Biosorption of heavy metals and radionuclide from aqueous solutions by pre-treated arca shell biomass. J Hazard Mater 150: 376–386.

22. Pangeni B, Paudyal H, Inoue K, Ohto K, Kawakita H, et al (2014) Preparation of natural cation exchanger from persimmon waste and its application for the removal of cesium from water. Chem Eng J 242: 109–116.

23. Ding D, Lei Z, Yang Y, Feng C, Zhang Z (2014) Selective removal of cesium from aqueous solutions with nickel (II)hexacyanoferrate (III) functionalized agricultural residue–walnut shell J Hazard Mater 270: 187–195.

24. Awual MR, Suzuki S, Taguchi T, Shiwaku H, Okamoto Y, et al (2014) Radioactive cesium removal from nuclear wastewater by novel inorganic and conjugate adsorbents. Chem Eng J 242: 127–135.

25. Valsala TP, Roy SC, Shah JG, Gabriela J, Raj K, et al. (2009) Removal of radioactive caesium from low level radioactive waste (LLW) streams using cobalt ferrocyanide impregnated organic anion exchanger. J Hazard Mater 166: 1148–1153.

26. Sengupta P, Dudwadkar NL, Vishwanadh B, Pulhani V, Rao Rekha, et al (2014) Uptake of hazardous radionuclides within layered chalcogenide for environmental protection. J Hazard Mater 266: 94–101.

27. Higaki S, Hirota M (2012) Decontamination Efficiencies of Pot-Type Water Purifiers for ^{131}I, ^{134}Cs and ^{137}Cs in Rainwater Contaminated during Fukushima Daiichi Nuclear Disaster. PLoS ONE 7(5): e37184.

28. Goossens R, Delville A, Genot J, Halleux R, Masschelein WJ (1989) Removal of the typical isotopes of the Chernobyl fall-out by conventional water treatment. War Res 23: 693–697.

29. Kosaka K, Asami M, Kobashigawa N, Ohkubo K, Terada H, et al (2012) Removal of radioactive iodine and cesium in water purification processes after an explosion at a nuclear power plant due to the Great East Japan Earthquake. Water res 46: 4397–4404.

30. Shimura H, Itoh K, Sugiyama A, Ichijo S, Ichijo M, et al (2012) Absorption of Radionuclides from the Fukushima Nuclear Accident by a Novel Algal Strain. PLoS ONE 7(9): e44200.

31. Fukuda S, Iwamoto K, Atsumi M, Yokoyama A, Nakayama T, et al (2014) Global searches for microalgae and aquatic plants that can eliminate radioactive cesium, iodine and strontium from the radio-polluted aquatic environment: a bioremediation strategy. J Plant Res 127: 79–89.

32. Szabo J, Minamyer S (2014) Decontamination of radiological agents from drinking water infrastructure: A literature review and summary. Environ Int Available: http://dx.doi.org/10.1016/j.envint.2014.01.020.

33. Ministry of Health, Labour and Welfare, Japan. Information on the Great East Japan Earthquake-Water supply. Available: www.mhlw.go.jp/english/topics/2011eq/index. html. Accessed Jul. 18, 2013.

34. Ministry of Health, Labour and Welfare, Japan. Measurement results of radionuclides in foodstuffs (No. 522) Available: http://www.mhlw.go.jp/stf/houdou/2r9852000002oo2l-att/2r9852000002oo6v.pdf. Accessed Jul. 18, 2013.

35. Mizuno T, Kubo H (2013) Overview of active cesium contamination of fresh water fish in Fukushima and Eastern Japan. Sci Rep 3: 1742.

36. Nuclear Regulation Authority (NRA). Monitoring information of environmental radioactivity level: Readings of radioactivity level in drinking water by prefecture October–December, 2013, Accessed Mar. 13, 2014.

37. Ohta T, Mahara Y, Kubota T, Fukutani S, Fujiwara K, et al (2012) Prediction of groundwater contamination with ^{137}Cs and ^{131}I from the Fukushima nuclear accident in the Kanto district. J Environ Radioact 111: 38–41.

38. Sato I, Kudo H, Tsuda S (2011) Removal efficiency of water purifier and adsorbent for iodine, cesium, strontium, barium and zirconium in drinking water. J Toxicol Sci 36(6): 829–834.

39. Shirahata S, Hamasaki T, Teruya K (2012) Advanced research on the health benefit of reduced water. Trends Food Sci Technol 23: 124–131.

40. Ewing D, Jones SR (1987) Superoxide Removal and Radiation Protection in Bacteria. Arch Biochem Biophys 254(1): 53–62.

41. Ward JF (1988) DNA damage produced by ionizing radiation in mammalian cells: identities, mechanisms of formation, and reparability. Prog Nucleic Acid Res Mol Biol 35: 95–125.

42. Jagetia GC, Reddy TK (2005) Modulation of radiation-induced alteration in the antioxidant status of mice by naringin. Life Sci 77: 780–794.

43. Nunomura A, Honda K, Takeda A, Hirai K, Zhu X, et al (2006) Oxidative damage to RNA in neurodegenerative diseases. J Biomed Biotechnol 2006: Article ID 82323: 1–6.

44. Tanaka M, Chock PB, Stadtman ER (2007) Oxidized messenger RNA induces translation errors. Proc Natl Acad Sci USA 104(1): 66–71.

45. Radak Z, Zhao Z, Goto S, Koltai E (2011) Age-associated neurodegeneration and oxidative damage to lipids, proteins and DNA. Mol Aspects Med 32: 305–315.

46. Cerutti PA (1985) Prooxidant states and tumor promotion. Science 227: 375–381.

47. Gobbel GT, Bellinzona M, Vogt AR, Gupta N, Fike John R, et al (1998) Response of postmitotic neurons to X-irradiation: implications for the role of DNA damage in neuronal apoptosis. J Neurosci 18(1): 147–155.

48. Spulber S, Edoff K, Hong L, Morisawa S, Shirahata S, et al (2012) Molecular hydrogen reduces LPS-induced neuroinflammation and promotes recovery from sickness behaviour in mice. PLoS ONE 7(7): e42078.

49. Qian L, Cao F, Cui J, Wang Y, Huang Y, et al (2010) The potential cardioprotective effects of hydrogen in irradiated mice. J Radiat Res 51: 741–747.

50. Qian L, Cao F, Cui J, Huang Y, Zhou X, et al (2010) Radioprotective effect of hydrogen in cultured cells and mice. Free Radic Res 44(3): 275–282.

51. Terasaki Y, Ohsawa I, Terasaki M, Takahashi M, Kunugi S, et al (2011) Hydrogen therapy attenuates irradiation-induced lung damage by reducing oxidative stress. Am J Physiol Lung Cell Mol Physiol 301: L415–L426.

52. Chuai Y, Gao F, Li B, Zhao L, Qian L, et al (2012) Hydrogen-rich saline attenuates radiation-induced male germ cell loss in mice through reducing hydroxyl radicals. Biochem J 442: 49–56.

53. Ohno K, Ito M, Ichihara M, Ito M (2012) Molecular hydrogen as an emerging therapeutic medical gas for neurodegenerative and other diseases. Oxid Med Cell Longev 2012: Article ID 353152.

54. Yan H, Tian H, Kinjo T, Hamasaki T, Tomimatsu K, et al (2010) Extension of the lifespan of Caenorhabditis elegans by the use of electrolyzed reduced water. Biosci Biotechnol Biochem 74(10): 2011–2015.

55. Yoshinobu H, Arita S, Kawasaki S (2012) Molded activated charcoal and water purifier involving SAME. Patent application number: US20120132578.

56. Ministry of Health, Labour and Welfare, Japan. Available: http://www.mhlw.go.jp/shinsai_jo uhou/dl/leaflet_120329.pdf. Accessed Jul 18, 2013.

57. Christodouleas JP, Forrest RD, Ainsley CG, Tochner Z, Hahn SM, et al (2011) Short-term and long-term health risks of nuclear-power-plant accidents. N Engl J Med 364: 2334–41.

58. Ministry of Health, Labour and Welfare, Japan. Available: http://www.mhlw.go.jp/shingi/2002/10/s1007-5c.html. Accessed Jul 18, 2013.

59. Kametani K, Matsumura T, Naito M (1992) Separation of iodide and iodate by anion exchange resin and determination of their ions in surface water (In Japanese). Bunseki Kagaku 41: 337–341.

60. Watari K, Imai K, Ohmomo Y, Muramatsu Y, Nishimura Y, et al (1988) Simultaneous adsorption of Cs-137 and I-131 from water and milk on "metal ferrocyanide-anion exchange resin". J Nucl Sci Technol 25(5): 495–499.

61. Kamei D, Kuno T, Sato S, Nitta K, Akiba T (2012) Impact of the Fukushima Daiichi Nuclear Power Plant accident on hemodialysis facilities: An evaluation of radioactive contaminants in water used for hemodialysis. Ther Apher Dial1 6(1): 87–90.

62. Zhu Y-G, Smolders E (2000) Plant uptake of radiocaesium: a review of mechanisms, regulation and application. J Exp Bot 51(351): 1635–1645.

63. Kato H, Onda Y, Teramage M (2012) Depth distribution of ^{137}Cs, ^{134}Cs, and ^{131}I in soil profile after Fukushima Dai-ichi Nuclear Power Plant accident. J Environ Radioact 111: 59–64.

64. Ministry of Health, Labour and Welfare (2013–595): Available: http://www.mhlw.go.jp/stf/houdou/2r9852000002wvi2.html.Accessed Jul. 18, 2013.

65. Ministry of Agriculture, Forestry and Fisheries: Available: http://www.maff.go.jp/j/kanbo/joho/saigai/s_chosa/hinmoku_kekka.html.Accessed Jul. 18, 2013.

66. Spyropoulos BG, Misiakos EP, Fotiadis C, Stoidis CN (2011) Antioxidant properties of probiotics and their protective effects in the pathogenesis of radiation-induced enteritis and colitis. Dig Dis Sci 56: 285–294.

67. Ohsawa I, Ishikawa M, Takahashi K, Watanabe M, Nishimaki K, et al (2007) Hydrogen acts as a therapeutic antioxidant by selectively reducing cytotoxic oxygen radicals. Nat Med 13(6): 688–694.

68. Yan H, Kinjo T, Tian H, Hamasaki T, Teruya K, et al (2011) Mechanism of the lifespan extension of *Caenorhabditis elegans* by electrolyzed reduced water Participation of Pt nanoparticles. Biosci Biotechnol Biochem 75(7): 1295–1299.

69. Yoshihisa Y, Honda A, Zhao Q-L, Makino T, Abe R, et al (2010) Protective effects of platinum nanoparticles against UV-light-induced epidermal inflammation. Exp Dermatol 19: 1000–1006.

70. Hamasaki T, Kashiwagi T, Imada T, Nakamichi N, Aramaki S, et al (2008) Kinetic analysis of superoxide anion radical-scavenging and hydroxyl radical-scavenging activities of platinum nanoparticles. Langmuir 24: 7354–7364.

71. Nakanishi H, Hamasaki T, Kinjo T, Yan Hanxu, Nakamichi N, et al (2013) Low concentration platinum nanoparticles effectively scavenge reactive oxygen species in rat skeletal L6 cells. Nano Biomed Eng 5(2): 76–85.

72. Yoshihisa Y, Zhao Q-L, Hassan MA, Wei Z-L, Furuichi M, et al (2011) SOD/catalase mimetic platinum nanoparticles inhibit heat-induced apoptosis in human lymphoma U937 and HH cells. Free Radic Res 45(3): 326–335.

73. Li Y, Hamasaki T, Nakamichi N, Kashiwagi T, Komatsu T, et al (2011) Suppressive effects of electrolyzed reduced water on alloxan-induced apoptosis and type 1 diabetes mellitus. Cytotechnology 63: 119–131.

74. Kim M-J, Kim HK (2006) Anti-diabetic effects of electrolyzed reduced water in streptozotocin-induced and genetic diabetic mice. Life Sci 79: 2288–2292.

75. Li Y, Nishimura T, Teruya K, Maki T, Komatsu T, et al (2002) Protective mechanism of reduced water against alloxan-induced pancreatic β-cell damage: Scavenging effect against reactive oxygen species. Cytotechnology 40: 139–149.

76. Ye J, Li Y, Hamasaki T, Nakamichi N, Komatsu T, et al (2008) Inhibitory effect of electrolyzed reduced water on tumor angiogenesis. Biol Pharm Bull 31(1): 19–26.

77. Yan H, Kashiwaki T, Hamasaki T, Kinjo T, Teruya K, et al (2011) The neuroprotective effects of electrolyzed reduced water and its model water containing molecular hydrogen and Pt nanoparticles. BMC Proc 5 (Suppl 8): 69–70.

78. Kinjo T, Ye J, Yan H, Hamasaki T, Nakanishi H, et al (2012) Suppressive effects of electrochemically reduced water on matrix metalloproteinase-2 activities and in vitro invasion of human fibrosarcoma HT1080 cells. Cytotechnology 64: 357–371.

79. Teruya K, Myojin-Maekawa Y, Shimamoto F, Watanabe H, Nakamichi N, et al (2013) Protective effects of the fermented milk kefir on X-ray irradiation-induced intestinal damage in B6C3F1 mice. Biol Pharm Bull 36(3): 352–359.

80. Vijayalaxmi, Reiter RJ, Tan D-X, Herman TS, Thomas CR (2004) Melatonin as a radioprotective agent: A review. Int J Radiat Oncol Biol Phys 59(3): 639–653.

81. Akpolat M, Kanter M, Uzal MC (2009) Protective effects of curcumin against gamma radiation-induced ileal mucosal damage. Arch Toxicol 83: 609–617.

14

Development of Composite Indices to Measure the Adoption of Pro-Environmental Behaviours across Canadian Provinces

Magalie Canuel[1]*, Belkacem Abdous[2,3], Diane Bélanger[2,4], Pierre Gosselin[1,2,4]

1 Institut national de santé publique du Québec (INSPQ), Québec City, Canada, 2 Centre de recherche du Centre hospitalier universitaire de Québec, Québec City, Canada, 3 Département de médecine sociale et préventive de l'Université Laval, Québec City, Canada, 4 Institut national de la recherche scientifique, Centre Eau Terre Environnement, Québec City, Canada

Abstract

Objective: The adoption of pro-environmental behaviours reduces anthropogenic environmental impacts and subsequent human health effects. This study developed composite indices measuring adoption of pro-environmental behaviours at the household level in Canada.

Methods: The 2007 Households and the Environment Survey conducted by Statistics Canada collected data on Canadian environmental behaviours at households' level. A subset of 55 retained questions from this survey was analyzed by Multiple Correspondence Analysis (MCA) to develop the index. Weights attributed by MCA were used to compute scores for each Canadian province as well as for socio-demographic strata. Scores were classified into four categories reflecting different levels of adoption of pro-environmental behaviours.

Results: Two indices were finally created: one based on 23 questions related to behaviours done inside the dwelling and a second based on 16 questions measuring behaviours done outside of the dwelling. British Columbia, Quebec, Prince-Edward-Island and Nova-Scotia appeared in one of the two top categories of adoption of pro-environmental behaviours for both indices. Alberta, Saskatchewan, Manitoba and Newfoundland-and-Labrador were classified in one of the two last categories of pro-environmental behaviours adoption for both indices. Households with a higher income, educational attainment, or greater number of persons adopted more indoor pro-environmental behaviours, while on the outdoor index, they adopted fewer such behaviours. Households with low-income fared better on the adoption of outdoors pro-environmental behaviours.

Conclusion: MCA was successfully applied in creating Indoor and Outdoor composite Indices of pro-environmental behaviours. The Indices cover a good range of environmental themes and the analysis could be applied to similar surveys worldwide (as baseline weights) enabling temporal trend comparison for recurring themes. Much more than voluntary measures, the study shows that existing regulations, dwelling type, households composition and income as well as climate are the major factors determining pro-environmental behaviours.

Editor: Judi Hewitt, University of Waikato (National Institute of Water and Atmospheric Research), New Zealand

Funding: This study was funded by the Green Fund for Action 21 of the 2006-2012 Climate Change Action Plan of the Quebec government. The funders had no role in study design, data collection and analysis, decision to publish, or preparation of the manuscript.

Competing Interests: The authors have declared that no competing interests exist.

* Email: magalie.canuel@inspq.qc.ca

Introduction

A significant source of pollution to our natural environment comes from domestic activities and behaviours. For example household-generated waste in Canada accounts for around a third of total waste and household energy use and municipal water consumption for 17% and 57%, respectively [1-3]. Also, 46% of greenhouse gas emissions (GHG), which contribute to climate change, come from direct and indirect household emissions [4]. The impacts of such household pollution can be important.

Municipal waste can impact the environment in various ways including soil and water contamination from leachate in landfills disposal and the production of greenhouse gas emissions (GHG) and air pollution, either from landfills or the incineration process. When solid waste are recycled or composted instead of being landfilled or incinerated, the demand for energy and new-resources can be reduced significantly [3].

The production of energy can impact the environment in various ways, depending on the technology. In Canada, energy production and consumption accounts for around 80% of all GHG emission [5]. A household can reduce its emission of GHG by reducing electric power use. For instance high energy efficiency electronic devices or cleaner energy sources will generate less pollution and GHG.

Water shortages are happening worldwide and one way to limit their occurrence is through water conservation behaviours. In most homes, more than 60% of water use comes from toilet flushing, showers and baths, making water-saving devices like low-flow shower head an efficient way of reducing water consumption. In summer, water use can increase by 50% for yard activities such as watering the lawn. There are behaviours that households can implement to decrease their water consumption in summer time like using sprinklers with a timer or adopting the use of a rain barrel [6].

It thus becomes clear that addressing sustainability concerns has to take into account not only industry or agriculture, but also household behaviours, their impacts on ecosystems and ultimately on human health. Monitoring trends of household behaviours can inform policy and research agendas on the development of incentives or other mechanisms such as information campaigns to reduce domestic pollution and facilitate adaptative measures to minimize related health risks. The adoption of several pro-environmental behaviours, i.e. actions that contribute to the preservation of the environment, should be encouraged to significantly reduce the anthropic impact on the environment.

In Canada, the Households and the Environment Survey (HES) was designed to measure household behaviours with respect to the environment. The HES is a periodic survey conducted by Statistics Canada, the federal government statistical agency, and administered across Canadian provinces. The survey covers 12 broad themes including energy use and heating, water use, transportation decisions, motor vehicle use, recycling and composting (Figure 1) [7]. While this survey provides various estimations of up to 83 Canadian practices (Figure 1) as well as some information on their socio-demographic characteristics, survey reports are limited to analyses of simple cross-tabulation frequencies for some of the 83 separate behaviours [7-13].

It is difficult to follow up on such a wide array of relevant behaviours and their trends over time, unless they are summarized in some way. A composite index is a tool which can be useful to that purpose as it incorporates several aspects of an issue and allow for monitoring across several themes simultaneously, thus facilitating the measurement of trends [14]. While other environmental indices exist, such as the environmental sustainability index [15], to our knowledge no index currently exists to reflect trends of pro-environmental behaviours at the household level in Canada.

This study thus sets out to develop a composite index that summarizes pro-environmental behaviours at the household level across Canadian provinces based on the HES (2007) given the periodicity and geographical coverage of the survey. Pro-environmental behaviours are defined as actions that contribute to the preservation of the environment and can have a positive impact on the health of the population. This study will serve as baseline of the trend of the composite index over time, given the periodicity and geographical coverage of the survey.

Materials and Methods

Ethics statement

This research did not require the approval of an ethics review board as we used an existing and anonymized database made available to universities by Statistics Canada. Statistics Canada obtained consent previous to survey administration. No new data was collected for this study.

Survey

The Households and the Environment Survey (HES) is conducted by Statistics Canada. It was designed to address the needs of the Canadian Environmental Sustainability Indicators project. The project reports on air quality, water quality and greenhouse gas emissions in Canada using indicators to identify areas of importance to Canadians and monitor progress [16].

The survey aimed Canadian households with at least one person aged 18 year or older. The HES covers all 10 of the provinces and excludes the 3 northern territories, Indian reserves and members of the Canadian Armed Forces. The survey was first conducted in 1991 and since 2005 has been carried out biennially. In the present study, the 2007 HES database was used in its Public Use Microdata Files format (PUMF) [16]. As a sub-sample of the dwellings that were part of the Canadian community health survey (CCHS), the sampling allocation for the HES followed that of the CCHS closely. The CCHS used a multistage stratified cluster design in which the dwelling is the final sampling unit. Three sampling frames were used to select the sample of households: 50% of the sample came from an area frame, 49% came from a list frame of telephone numbers and 1% came from a Random Digit Dialing sampling frame [16].

From the 40 584 households selected in the 2007 CCHS, a sub-sample of 29 957 households were selected for the HES. Of those, 21 690 households responded to the survey resulting in an overall response rate of 72%. The survey is representative of 12 932 350 households, corresponding to 97% of all Canadian households [16]. The questionnaire was administered to the 21 690 households by telephone interview spread over a 6-month period, from October 2007 to February 2008.

Questionnaire

The person with the best knowledge of environmental household practices was asked to respond on behalf of the household. The main questionnaire covered 12 themes and included 121 questions (figure 1) [16]. Among the questions, 83 measured behaviours and 7 measured socio-demographic characteristics. The other 31 questions covered knowledge, reasons for not adopting the behaviour, or served to specify some characteristics (e.g. of a good) or to filter for the next question.

Database

The PUMF was used for the analysis and unlike the master file, applies privacy measures to protect personal information [16]. In the PUMF, data were mostly coded as categorical variables. Three different labels (don't know, not stated, and refusal) were used to classify households who did not participate despite eligibility or to protect the anonymity of the household. A 'valid skip' label was used when the provision of a response was not appropriate. For example, a household who answered 'no' to the question for car ownership was allocated a 'valid skip' label for subsequent questions on the characteristics of the car.

Sampling weights

Sampling weights were applied to ensure that any derived composite index is representative of the study population. They were used when proportions and averages were estimated and to weight the relative frequencies of the Burt matrix in the MCA (see Statistical analysis below).

Variables selection

This study focuses on everyday pro-environmental behaviours, defined as actions that contribute to the preservation of the

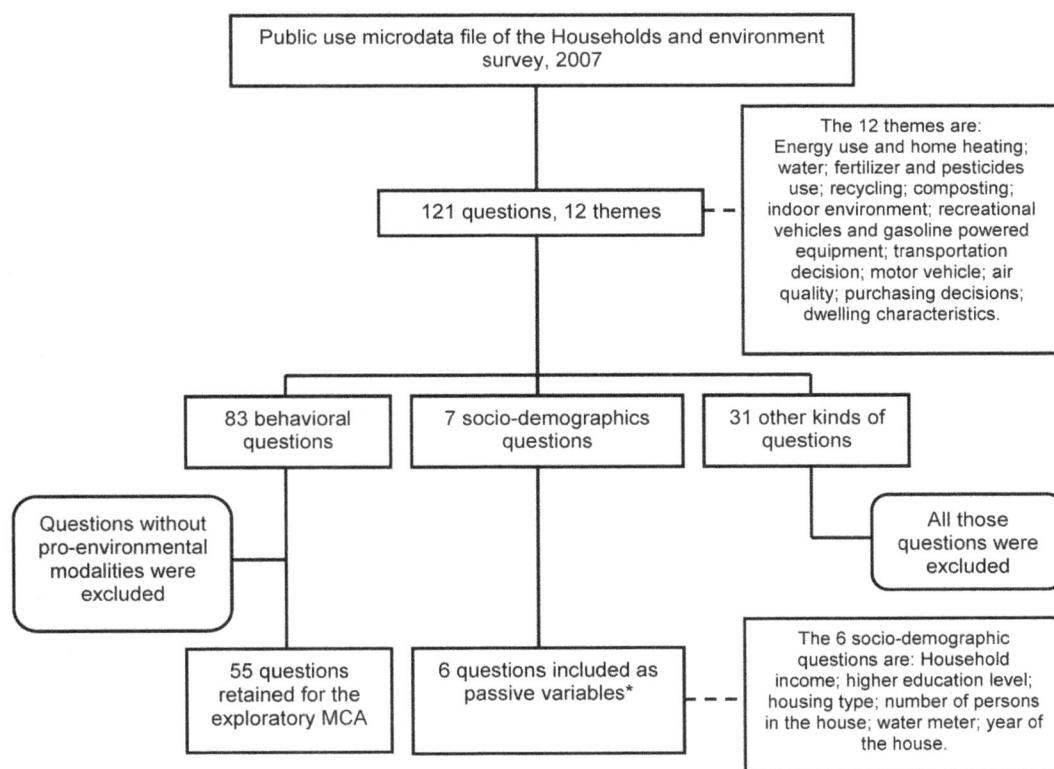

Figure 1. Number and type of questions selected to develop the composite index. Legend: *The composite index was also created for the 7th socio-demographic variable, the census metropolitan area (n = 33), but is not presented in this article.

environment and can have a positive impact on the health of the population. For example, air pollutants can be reduced when households adopt behaviours that decrease their energy consumption such as the use of energy-efficient appliances or when they use more sustainable transport options such as public or active transport.

Based on the above definition, a panel of four environmental health experts applied progressive development consensus after iterations, based on a nominal group technique [17] to evaluate HES variables for exclusion. These were either: variables not measuring a behaviour or questions with no clearly pro-environmental response option. Socio-demographic variables were kept as passive variables with zero mass and no influence on the analysis. They support and complement the interpretation of the map representation of the active variables [18].

Statistical analysis

Given that the data was mostly categorical, the indices in this paper were developed by multiple correspondence analysis (MCA) [18]. Several authors have used Multiple Correspondence Analysis (MCA) as a weighting method for the construction of a composite index [19–23]. MCA is a data reduction procedure for categorical variables (nominal or ordinal) as much as Principal Components Analysis is for quantitative variables [18]. It enables the exploration of associations within a set of variables by transforming the whole data set into dummy variables to form an indicator matrix or upon construction of a matrix from all two-way cross-tabulations among the variables (Burt matrix). This transformed data is treated as a cloud in a space equipped with the classical Chi-square distance. This distance is used in the assessment of

homogeneity and variance (inertia) of rows or columns of the indicator or Burt matrix. The most crucial step of MCA is its use of singular value decomposition and weighted least squares techniques to find low-dimensional best fitting subspaces with minimal inertia and information loss [18].

MCA was conducted using the 'ca' package of the R statistical software [24]. First, the HES database was converted to a Burt matrix taking into consideration the sampling weights. A Burt matrix is a square symmetric categories-by-categories matrix formed from all two-way contingency tables of pairs of variables [18].

Then, an exploratory MCA was performed to project data onto maps where potential outliers were identified and excluded from subsequent analyses. MCA was then applied again to determine the most relevant factorial axes that would serve to build the composite index. There are no universal rules for the determination of the number of dimensions to retain in MCA. However, since the first factorial axis captures the most important part of the total inertia, it plays a central role in the computation of a composite index.

As recommended by Asselin [23], we sought questions having the property of First Axis Ordering Consistency (FAOC). To this end, we projected all the questions on the first axis and tried to identify those having an ordinal structure consistent with respect to this axis, i.e. all questions with pro-environmental responses improving from left to right (or conversely).

The computation of the index score was performed as follows: first, the score of any household was obtained by taking the average of its category-weights generated by the MCA. Then for each province we took the average over all household scores as the value of its composite index. The sampling weight was used in this final step. Coordinates were missing for excluded responses.

The 10 average provincial scores were grouped into categories reflecting different levels of adoption of pro-environmental behaviours. First we applied a cluster analysis and then we used a dendrogram plot using SAS version 9.2 (SAS Institute, Cary, NC) to determine such groups. The categories limits generated for the provincial index were used as reference categories for indices on other socio-demographic variables.

Finally, others indices based on various socio-demographic variables were constructed (Figure 1). Household scores were calculated by taking the average of its category-weights generated by the MCA. Then the index score of the socio-demographic category (e.g. household with annual income less than $40,000) is set as the average of the corresponding household scores.

Results

Multiple Correspondence Analysis

Of the 121 questions in the survey, 55 were kept by the Expert Panel for use in the MCA. These represented 285 response possibilities. On the MCA map projection there was a clear opposition between the missing data (don't know, refusal, not stated) located far from the map center and the other responses which gathered close to the center (Figure S1). Excluding the missing data rebalanced the model (179 remaining responses) (Figure 2). However, since pro-environmental behaviours were spread over both sides of the first axis, we failed to find any meaning to this first dimension.

We then screened the projected responses to identify questions following an ordinal structure, (i.e all pro-environmental responses of a question have negative coordinates on the first factorial axis (or conversely)). Twenty-three such questions with pro-environmental responses deteriorating from left to right on the first axis (group A), and 16 questions with opposite ordinal structure (group B) were identified. The remaining 16 (of 55) questions were excluded from the analysis because their responses were not sufficiently discriminating (i.e. the pro- and anti-environmental responses were on the same side of the axis or they were grouped close together on the map). As well the majority of these questions (10/16) had at least two responses with a contribution of zero to the first axis (Table S1).

These exploration steps led us to consider two separate composite indices. Group A included 96 responses but after excluding missing data, 52 responses were used in the MCA. The majority of excluded responses had frequencies lower than 2.0% and two responses had frequencies of 4.6% and 4.7%. After exclusion of missing data, some responses still looked like extreme values on the map (Figure 3). They were kept in the analysis as they are 2 of the 3 responses for all questions concerning recycling. Excluding these responses would have resulted in the exclusion of all recycling questions. Responses used in this analysis had a frequency of 7.5% or higher, except for two responses with frequencies of 2.5% and 3.5% (responses on recycling).

For group A, the first dimension explained 32.6% of the inertia while the second explained 16.1% (Table 1). Given that the first factorial axis plays a central role in the construction of this composite index, only the first dimension was selected to construct the index. This group respects the FAOC as pro-environmental responses are located on the left of the first axis as opposed to others responses deteriorating to the right (Figure 3). Also, we noted that the retained questions were associated with five themes of the survey: energy use and home heating, water, recycling, composting and, purchasing decisions. All 23 questions assessed

behaviours practiced inside the dwelling and thus the first axis measures these behaviours. Twelve of 15 responses contributing the most to the first factorial axis concerned recycling (Table S2).

The second group of 16 questions (group B) consisted of 86 responses, 41 of which were missing values. The 45 remaining responses used for the MCA had frequencies of 7.0% or higher while excluded responses had frequencies lower than 3.5%. For group B, the first dimension explained 62.1% of the inertia while the second explained only 12.4% (Table 2). Again, the first dimension was selected for the construction of the index and pro-environmental responses were located on the right of the first axis with other responses deteriorating from right to left (Figure 4). The 16 questions cover five themes of the survey: water, fertilizer and pesticide use, recreational vehicles and gasoline powered equipment, transport decisions and air quality, all behaviours being practiced outdoors. Of note, 9 of the 15 responses contributing the most to the first factorial axis concern households with no lawn or garden (i.e. the application of fertilizers or pesticides, yard waste and watering of the lawn or the garden) (Table S3).

Because two distinct behavioural categories resulted from the MCA, two composite indices were created instead of one. The first index (group A) is named the 'Indoor Index' and the second one (group B) the 'Outdoor Index'. Questions included for each index are presented in supporting information, Table S4 and Table S5.

Composite indices by province

The map representations of the final coordinates generated by the MCA are shown in Figure 3 and Figure 4. Coordinates and other results of the MCA are available in supporting information, Table S2 and Table S3. The coordinates of the first dimension were used to construct each of the two composite indices. Coordinates are missing for responses that have been excluded. Only 0.9% and 0.4% of coordinates are missing for the indoor and outdoor indices, respectively.

For the Indoor Index, the households belonging to a province with negative coordinates tend to adopt more pro-environmental behaviours than those of a province with positive coordinates. In contrast, for the Outdoor Index provinces with positive coordinates adopt more outdoor pro-environmental behaviours than those with negative coordinates.

The cluster analysis and dendogram plot resulted in the classification of each province into one of four categories reflecting different levels of adoption of pro-environmental behaviours: 1) adopting the most; 2) adopting slightly fewer; 3) adopting much fewer and; 4) adopting the fewest. The provincial coordinates and the categories generated from the cluster analysis are shown in Table 3 and Table 4. Maps of the Canadian provinces with their categories of pro-environmental behaviours are shown in Figure 5 and Figure 6.

None of the 10 provinces were classified in both indices as adopting the most pro-environmental behaviours. For the Indoor Index, Ontario (ON), Prince Edward Island (PEI) and Nova Scotia (NS) rated in the top category, British Columbia (BC) and Québec (QC) in the next, the three Prairie provinces and New Brunswick (NB) in the third and Newfoundland and Labrador (NL) in "adopting the fewest" category (Figure 5). For the Outdoor Index, QC scored in the top category with BC, NS, NB and PEI following in second, and Manitoba (MN), ON and NL in third, followed by Alberta (AB) and Saskatchewan (SK) in the bottom category (Figure 6).

Four provinces (BC, QC, NS and PEI) were classified in the top two categories for both indices while four provinces (AB, SK, MN and NL) were classified for both indices, in the two lower categories.

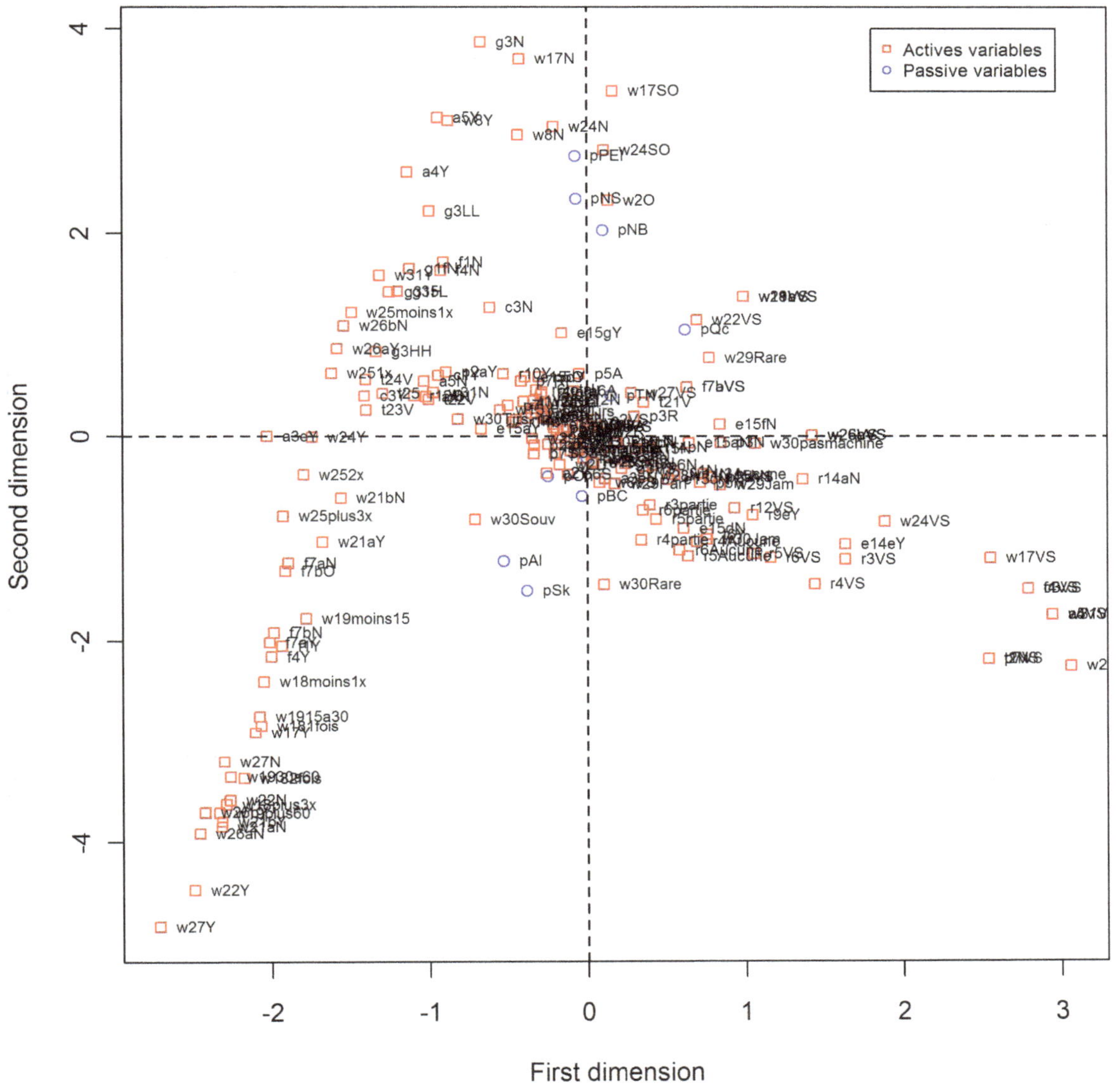

Figure 2. Map representation of the MCA results on the 55 questions without extreme responses.

Composite indices by socio-demographic variables

The coordinates and the classification for the six comparison variables are shown in Table 5. For household income, educational attainment and number of persons in the household, there were oppositions in the classification of the responses. Households with a higher income, or higher educational attainment, or greater number of persons adopted more indoor pro-environmental behaviours, while those with a lower household income, educational attainment, or number of people, adopted more outdoor such behaviours. As well, households with water meters tended to adopt more indoor pro-environmental behaviours than those without, but for outdoors behaviours, the opposite applied – not having a water meter was associated with better adoption of pro-environmental behaviours. And finally, the dwelling's year of construction did not influence the adoption of pro-environmental behaviours as there was no trend on either index (Table 5).

Discussion

This study sought to develop a composite index which measures the overall adoption of pro-environmental behaviours among Canadian households. MCA, our main analytical technique, was used to aggregate survey data and to provide weights to the responses in the construction of the index. Our approach is similar to other studies in different fields [19–23]. This was followed by a cluster analysis to classify the provinces, as well as an exploration of relationships with socio-demographic factors.

The MCA generated two indices based on 39 of the 55 behavioural questions, an Indoor Index and an Outdoor Index, each reflecting environmental behaviours for 5 of the 12 survey themes. Retaining both indices allowed for better representation of the survey; together they cover 9 themes out of 12 (water use is in both) whereas one single index would have covered only 5, excluding important environmental themes such as fertilizer and pesticide use. As well, because the provincial classifications were different for each index and varied as well in the classification by

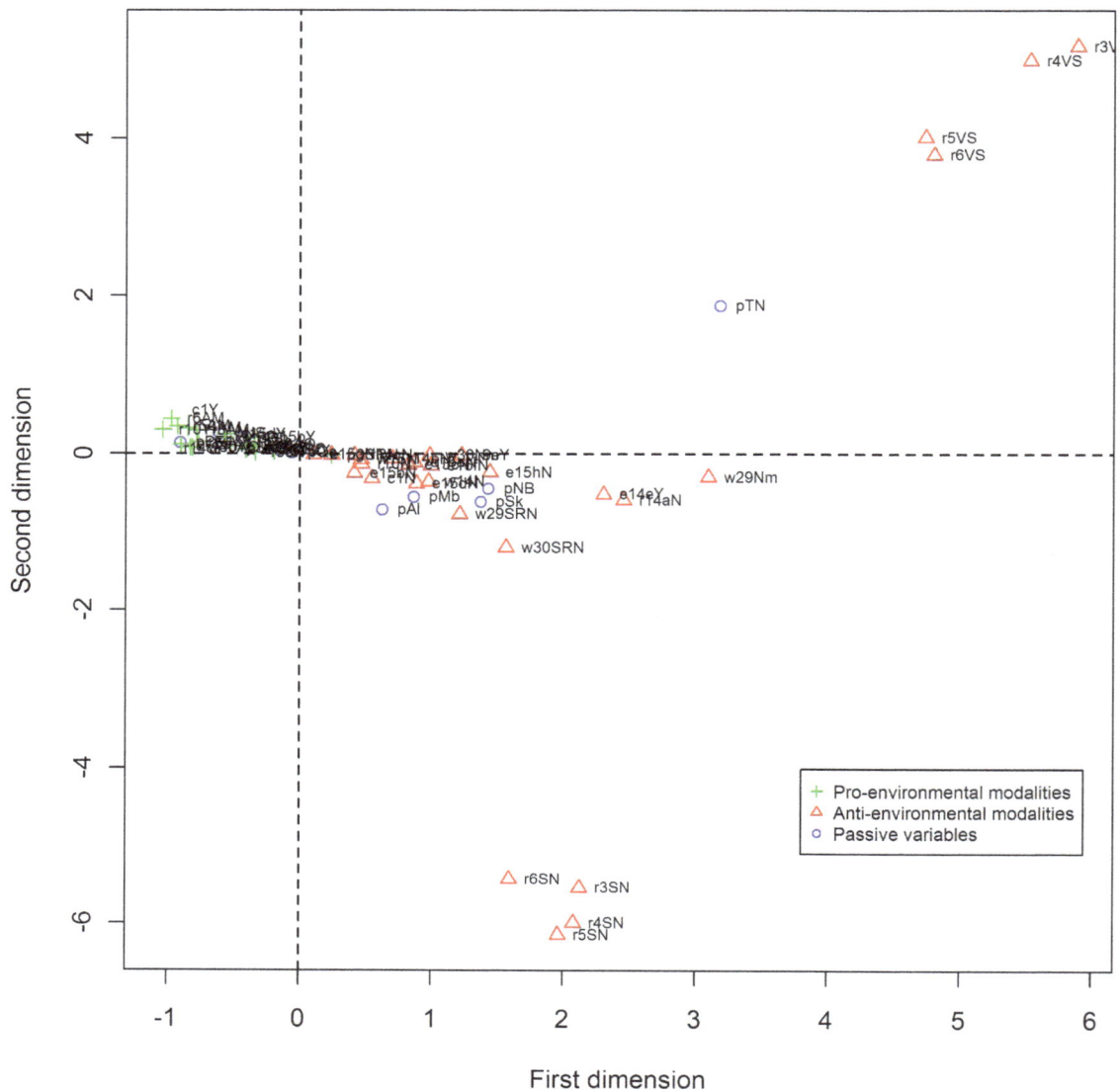

Figure 3. Map representation of the MCA results on the 23 questions of the group A (Indoor).

socio-demographics factors for each index (e.g., household income) it was deemed justifiable to keep both indices.

Most (19/23) questions included in the Indoor Index were asked to all households with the exception of questions on recycling where only those households with access to a program were asked to respond. For the Outdoor Index, most questions (11/16) concerned watering of the lawn or the garden, and the use of fertilizers or pesticides. These (11) questions were answered only by households having a yard. However, even if households living in an apartment did not have to answer these questions, they were still recorded in the Index as households adopting pro-environmental behaviours (i.e., most valid skips were classified as pro-environmental responses and some as anti-environmental ones).

The Indoor Index

One likely explanation for PEI and NS being classified in the top category for the Indoor Index is that nearly 100% of their households recycle and the proportion that compost is substantially above the Canadian average, as reported by Statistics

Canada. In these two provinces, households are obligated by law to recycle and compost [25]. Moreover, questions regarding recycling contributed the most to the Indoor Index.

Recycling and composting are also common in ON but its good ranking is also related to the proportion of households that adopt water conservation behaviours (i.e. use water-efficient shower heads and toilets, run dishwasher and washing machine only when full) [7]. Provinces have been slowly adopting a provincial plumbing code requiring that new buildings use water-saving fixtures, with the exception of NL [26;27]. ON however was the first to adopt such a code in 1996 [26;28], and saw an increase in new residential construction from 1996 to 2002 [29], likely contributing to the higher proportion of households practicing water conservation behaviours [28]. This is an example where building codes may be effective in beneficially influencing the passive uptake of pro-environmental practices.

QC's good classification in the Indoor Index is in part due to its proportion of households adopting recycling behaviours being higher than the Canadian average. There were four questions on

Table 1. Explained inertia by each dimension for group A: Indoor Index, 2007.

Dimension	Inertia	Inertia (%)	cumulative Inertia (%)	*scree plot*
1	0,0270	32,6	32,6	*************************
2	0,0133	16,1	48,8	************
3	0,0078	9,4	58,2	*******
4	0,0032	3,8	62,0	***
5	0,0030	3,6	65,6	***
6	0,0026	3,2	68,8	**
7	0,0024	2,9	71,7	**
8	0,0021	2,5	74,2	**
9	0,0019	2,3	76,5	**
10	0,0018	2,1	78,7	**
11	0,0017	2,0	80,7	**
12	0,0016	1,9	82,6	*
13	0,0015	1,8	84,4	*
14	0,0014	1,7	86,1	*
15	0,0014	1,7	87,7	*
16	0,0013	1,6	89,3	*
17	0,0012	1,5	90,8	*
18	0,0012	1,4	92,2	*
19	0,0011	1,3	93,5	*
20	0,0011	1,3	94,8	*
21	0,0010	1,2	96,1	*
52	0	0,0	100,0	

recycling which contributed significantly to the first dimension, thus contributing to QC's classification. Despite QC having the lowest proportion of households that compost [9] or participate in alternative recycling activities such as donations of furniture and clothing, QC's classification was only slightly affected as these behaviours had only moderate or low contributions to the Index.

In AB, MN, SK and NB, the proportion of households that adopted indoor pro-environmental behaviours is below the Canadian average (data not shown), explaining their lower classification in the Indoor Index. NL had only a few variables above the Canadian average and had most often the lowest proportion of all provinces. For example, the proportion is below the average for all four questions on water conservation and for all questions on recycling. In this province, there is no provincial plumbing code requiring the use of water-saving fixtures in new buildings [26;27]. Also, the proportion of households with access to a recycling program is only 71% [25].

The Outdoor Index

Results for the Outdoor Index show a pattern with respect to Coastal proximity, with coastal provinces, with the exception of NL, rating in the two higher categories, and the continental provinces in the two lower categories with the two lowest rated provinces situated in the Prairies. The climate of the Prairies grasslands is characterized by hot summers combined with low precipitation and periodic drought. The climatic region of the Maritimes however is the one with the greatest annual precipitation [30-32], a pattern which is likely reflected in the frequency of watering lawn and or garden. Although watering of the lawn or garden is around the Canadian average in NL, its inhabitants own more recreational vehicles, use more gas and burn more yard waste on the property (data not shown) which may explain its lower classification than the other coastal provinces.

Also, there was an important difference in the proportion of households that used fertilizers and pesticides and QC had, by far, the lowest proportion. QC was the first province to adopt a provincial law in 2006 prohibiting the sale of pesticides for cosmetic purposes [7;33]. The Prairies on the other hand had the highest proportions of households that used pesticides or fertilizers in 2007 [7;10]. Subsequently, other jurisdictions have adopted similar laws begging the question of whether their classifications in the Outdoor Index will change over time.

It should also be noted that QC and BC have the highest proportion of households living in an apartment [10]. Given that most households living in an apartment do not have a backyard, they do not water neither lawn nor garden, nor do they use pesticides outdoors. Hence, they passively adopt pro-environmental behaviours and are considered as such by the MCA. In fact, these responses, recorded as 'valid skip', had the highest contribution to the Index, likely contributing to the higher classification for BC and QC on the Outdoor Index. Such passive behaviours or external factors were not excluded from the Index as they significantly contribute to the preservation of environmental resources.

Indices for socio-demographic variables

For most socio-demographic variables, there were oppositions in the classification of the modalities, which means that it is not the

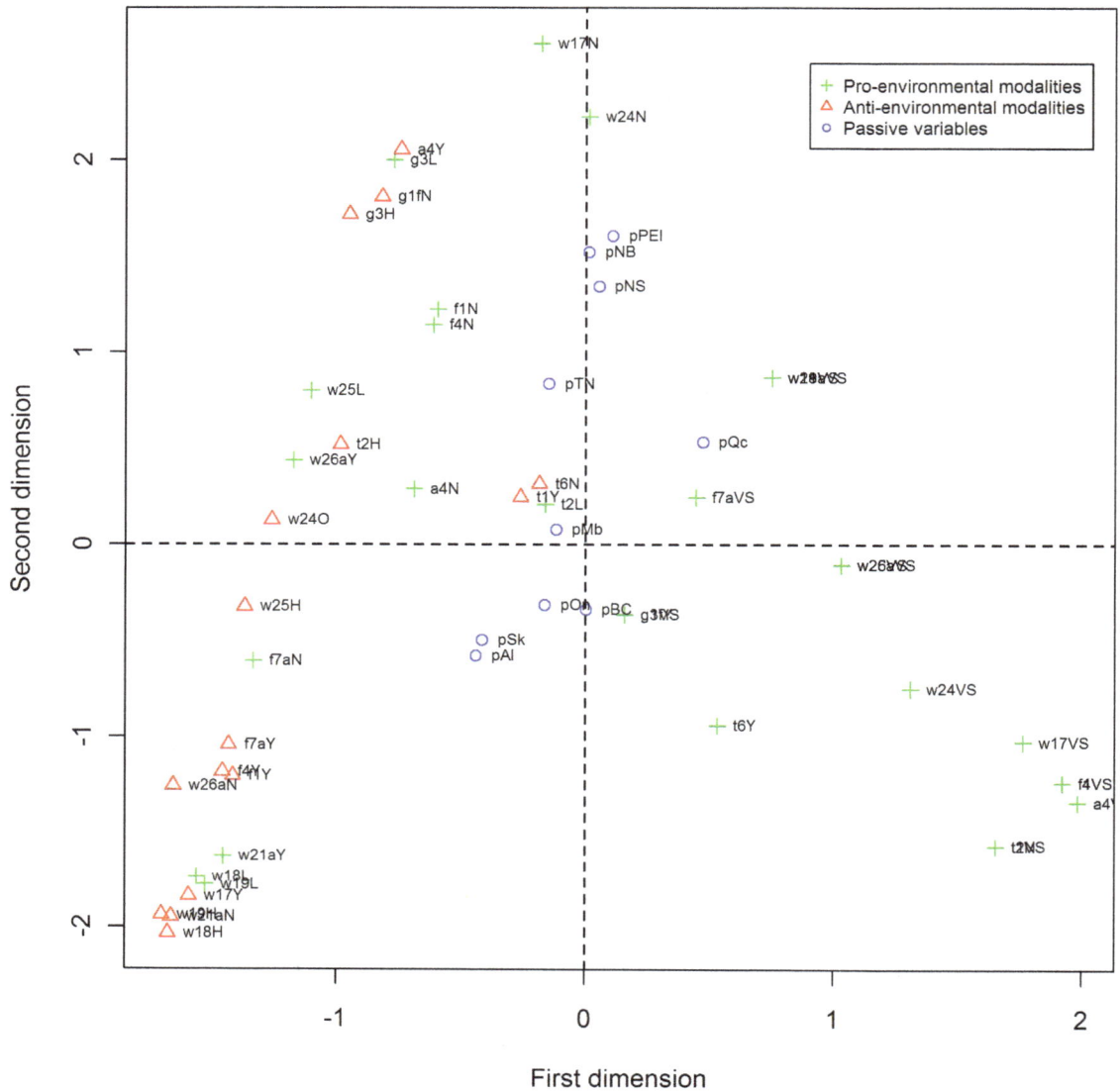

Figure 4. Map representation of the MCA results on the 16 questions of the group B (Outdoor).

Table 2. Explained inertia by each dimension for group B: Outdoor Index, 2007.

Dimension	Inertia	Inertia (%)	Cumulative inertia (%)	scree plot
1	0,2173	62,1	62,1	**************************
2	0,0434	12,4	74,5	*****
3	0,0162	4,6	79,1	**
4	0,0129	3,7	82,8	*
5	0,0111	3,2	86,0	*
6	0,0094	2,7	88,7	*
7	0,0066	1,9	90,6	*
8	0,0061	1,7	92,3	*
9	0,0041	1,2	93,5	
45	0	0,0	100,0	

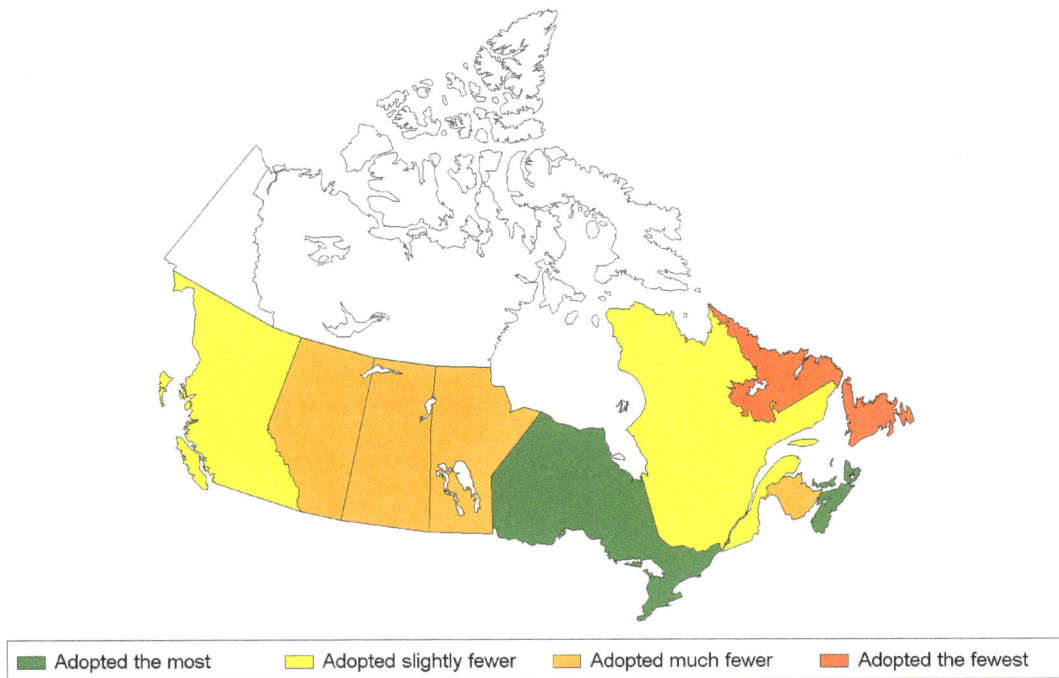

Figure 5. Provinces' classification according to the four categories of pro-environmental behaviours, Indoor Index, 2007. Legend: from left to right – British-Columbia, Alberta, Saskatchewan, Manitoba, Ontario, Quebec, New-Brunswick, Nova-Scotia. Prince-Edward-Island is North of the two latter provinces and Newfoundland-and-Labrador is located North-East of Quebec.

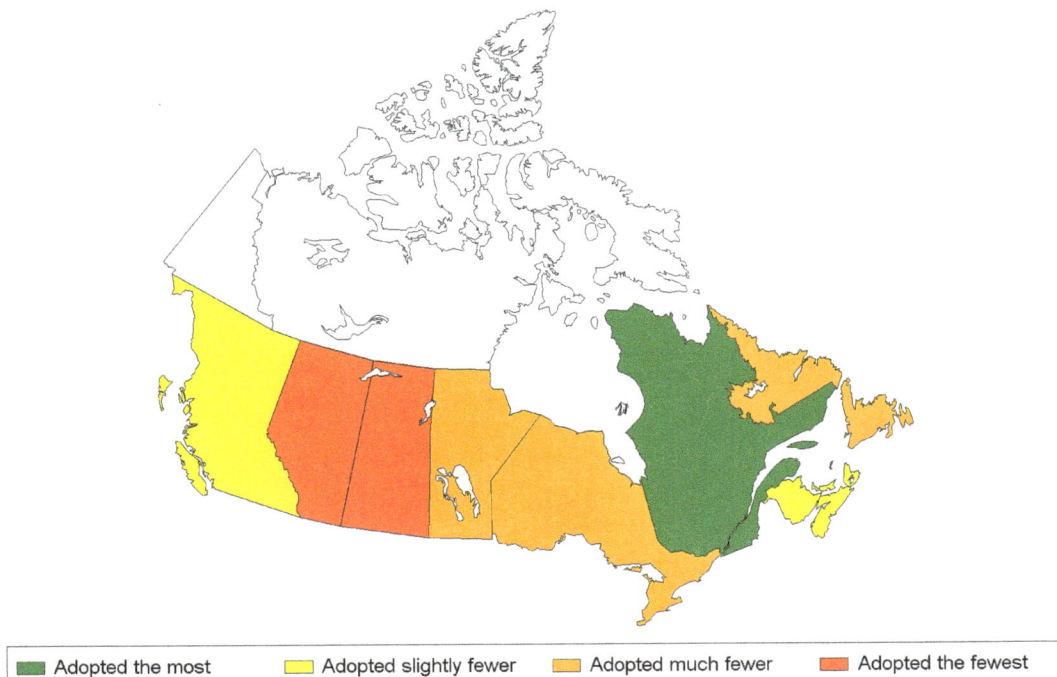

Figure 6. Provinces' classification according to the four categories of pro-environmental behaviours, Outdoor Index, 2007. Legend: from left to right – British-Columbia, Alberta, Saskatchewan, Manitoba, Ontario, Quebec, New-Brunswick, Nova-Scotia. Prince-Edward-Island is North of the two latter provinces and Newfoundland-and-Labrador is located North-East of Quebec.

Table 3. Provinces' coordinates on the Indoor Index, 2007.

Provinces	Coordinates	Categories[a]
Prince-Edward-Island	−0,0262	++
Nova-Scotia	−0,0179	++
Ontario	−0,0130	++
British-Columbia	−0,0055	+
Quebec	−0,0029	+
Alberta	0,0159	−
Manitoba	0,0225	−
Saskatchewan	0,0363	−
New-Brunswick	0,0381	−
Newfoundland-and-Labrador	0,0853	−

[a]Categories are: adopted the most pro-environmental behaviours (++), adopted slightly fewer (+), adopted much fewer (−) and adopted the fewest (−).

same households that adopt pro-environmental behaviours on both indices. Higher income households may be more able to maintain and repair their housing and also invest in environmentally friendly products such as water and energy efficient appliances or fixtures, which can be more expensive than their regular counterparts [34;35]. Access to such products may contribute to the better classification on the Indoor Index for higher income households. On the other hand, lower income households may be less willing to pay water taxes linked to consumption levels, or to buy chemical products for their lawn or garden. Furthermore, those lower income households live more frequently in apartments where they do not have a yard, and they also own fewer recreational vehicles (data not shown). All these factors likely weigh in on the higher classification attributed to lower versus higher income households on the Outdoor Index.

In Canada, income is usually positively associated to educational attainment [36]. Also, the number of persons in a household will influence the household income. In the HES database, there was a significant correlation between households' income and education level as well as one with the households' income and the number of persons in the households (data not shown). This may explain why the indices by educational level and by number of

Table 4. Provinces' coordinates on the Outdoor Index, 2007.

Provinces	Coordinates	Categories[a]
Quebec	0,1038	++
Prince-Edward-Island	0,0261	+
Nova-Scotia	0,0123	+
New-Brunswick	0,0052	+
British-Columbia	0,0012	+
Manitoba	−0,0243	−
Newfoundland-and-Labrador	−0,0273	−
Ontario	−0,0350	−
Saskatchewan	−0,0887	−
Alberta	−0,0962	−

[a]Categories are: adopted the most pro-environmental behaviours (++), adopted slightly fewer (+), adopted much fewer (−) and adopted the fewest (−).

persons in a household are similar to the one by household income. Any one of these three socio-demographic variables could potentially be used as a surrogate for the other two for future data collection for following Index trends over time.

Studies have shown that water meters with appropriate pricing are an incentive to reduce water consumption [2]. The US EPA estimated a 20% reduction in water consumption with universal metering [37] and a Canadian study also estimated a similar reduction according to structured water pricing [38]. While our results showed that households with water meters tended to score higher on the Indoor Index, households without a water meter scored higher on the Outdoor Index, which is in contrast to the other studies. We estimated that only 9% of households living in an apartment have water meters as opposed to 58% for all other types of dwellings in Canada (data not shown). As stated earlier, a household living in an apartment passively adopts more outdoor pro-environmental behaviours for lack of a lawn or garden to maintain with only a few having a water meter, possibly explaining the discrepancy between our results and those of other studies.

Factors that can lead to pro-environmental behaviours

There is a wide variety of measures or instruments than can be introduced by governments to influence households behaviours, from economic instruments to direct regulation, labeling, information campaigns and provision of environment-friendly public goods such as public transportation or bicycle paths [39].

This study has identified factors which seem to influence the uptake of beneficial environmental behaviours at the household level. Investment in infrastructure is one of them. The physical or material possibility to act pro-environmentally must indeed be available [40], such as what might be needed for Newfoundlanders to improve their recycling profile.

Regulation is frequently used to efficiently influence the environmental impacts of household decision-making [39] and in our study it also seems to be an important incentive for the adoption of pro-environmental behaviours. This was seen both in the case of building codes requiring the installation of water efficient shower heads and toilets, and in the case of the ban on pesticides for lawn care. In Ontario, the ban of cosmetic pesticides decreased significantly the concentration of some pesticides, mainly herbicides, in the majority of streams under surveillance near urban areas with limited agriculture activities [41].

To encourage a reduction in water consumption, both price and non-price policies should be used. Volumetric water charges are associated with both water-saving behaviours and adoption of water-efficient devices [39]. However, in a study in several OECD countries, Canada had the highest proportion of households that did not know how they were charged for residential water consumption, thus reducing the price effect on water-saving behaviours [39]. In our study, presence of water meters was an incentive to water-saving behaviours only for indoor behaviours. Climate was also another factor that could be influential. Hence, public information on the environmental impact of water consumption and on measures households can adopt to save water should be combined to economic measures according to the OECD [39] and this study.

Other than governmental measures, household characteristics may play a role in the adoption of environment-related behaviours such as income, household composition and dwelling characteristics [39]. According to the OECD survey, low income households and tenants households make fewer financial investments in water efficiency, as can be expected. Grants targeted at those households to correct the economic imbalance are thus recommended by the agency. Moreover, our study showed

Table 5. Coordinates and categories of pro-environmental behaviours for other socio-demographic variables, Indoor and Outdoor Indices, 2007.

	Indoor Index		Outdoor Index	
	Coordinates	Categories[a]	Coordinates	Categories[a]
Household income				
Less than $40,000	0,0253	−	0,1533	++
$40,000 to less than $80,000	−0,0080	+	−0,0133	−
$80,000 and over	−0,0273	++	−0,1567	−
Highest education level				
Secondary diploma or less	0,0228	−	0,0927	++
Postsecondary certificate or diploma	−0,0030	+	−0,0134	−
University	−0,0159	++	−0,0451	−
Dwelling type				
Apartment	0,0370	−	0,4335	++
Others	−0,0144	++	−0,1501	−
Number of persons in the dwelling				
One	0,0260	−	0,2058	++
Two	−0,0073	+	−0,0255	−
Three	−0,0104	++	−0,0723	−
Four or more	−0,0163	++	−0,1383	−
Water meter				
Yes	−0,0201	++	−0,1682	−
No	0,0123	−	0,1497	++
Year the dwelling was built				
Before 1946	−0,0089	+	0,0034	+
Between 1946 and 1960	−0,0040	+	−0,0130	−
Between 1961 and 1977	−0,0009	+	0,0025	+
Between 1978 and 1983	−0,0054	+	−0,0218	−
Between 1984 and 1995	−0,0084	+	−0,0364	−
Between 1996 and 2000	−0,0088	+	−0,0434	−
Between 2001 and 2005	−0,0099	++	−0,0948	−
2006 or latter	0,0042	+	0,0384	+

[a]Categories are: adopted the most pro-environmental behaviours (++), adopted slightly fewer (+), adopted much fewer (−) and adopted the fewest (−).

households from both income groups (high or low) or dwelling type (owned or rented) have to improve their act in different domains and that programs should target them accordingly. In short, Canadians remain very dependent for many such actions on where they live and what the climate brings to their yards, or not.

Limits of the study

We used data from a survey that has been created to address the needs of Statistics Canada and the federal government. Thus, we were limited to its content. The questionnaire does not cover all behaviours that can impact the environment and public health. Also, the indices developed here measure the behaviours available in the survey and retained after the analysis by an expert group for their potential positive impacts on health, and not all existing pro-environmental behaviours. The classification could have been different if other behaviours had been included.

Three themes of the survey were not covered by the indices, namely dwelling characteristics, motor vehicle and indoor environment. However, we believe they would not have much impact in the indices. First, there were no behaviours measured in the dwelling characteristics theme and some of the characteristics were included as passive variables in the indices (e.g. year the building was built). The same happened for the motor vehicle theme (focused on the characteristic of the car), yet we used another theme to include the number of vehicles owned by the households in the outdoor index. For the indoor environment theme, only 2 of the 5 questions measured behaviours and they both concerned the type of chemical products used to clean windows and the dwelling. Although every small action is important for the environment, those questions were excluded as some other practices, such as agriculture, use similar products in much larger quantities [42].

A good standing in the classification does not mean that there is no place for improvement. Indeed, a high proportion of households that adopt pro-environmental behaviours on one question can compensate for a lower proportion on another question of the same index. Also, the provinces were compared to each other and not classified in relation to a gold standard.

Furthermore, it was the MCA that attributed the weight for each modality. Thus, a modality with a higher weight has more impact in the index. For instance, all four recycling questions had the highest contribution to the indoor index. Further studies should investigate if the inclusion of only one of those recycling question or a composite index of those four questions would be more appropriate. The same reasoning should also be applied to questions related to the watering of the lawn or of the garden. Households without a garden or a lawn are rewarded for every question on that subject which at the end can impact greatly the province classification. For example, they were not only rewarded for not watering their lawn, but they were also rewarded for the question concerning the watering duration and the number of time they water. Because the MCA attributed the weights, those household without a garden or lawn had a higher 'reward' that households with a garden or lawn that did not water them.

One general limit of MCA is that the first dimension usually explains a low proportion of the total inertia in the data set and the other dimensions explain less than the first [18]. In this study, the first dimension explained 33% and 62% of the total inertia for the indoor and outdoor index respectively. By using only the first dimension, these indices might not properly reflect all of the behaviours, especially for the indoor one. However, using more than one dimension to build the index would not considerably increase the total inertia explained but would in return increase its complexity. Composite indices are indeed built to simplify the analysis.

Because of the study design, based on households, it was also not possible to evaluate the impact of personal attributes, like age and gender, on the adoption of environmental behaviours. The association between environmental behaviours and age is not clear. Studies have observed all possible trends, from older people adopting more pro-environmental behaviours to the opposite trends or no trend at all [43–45]. Also, women would be more likely to take pro-environmental actions than men, although some studies have found the opposite depending on behaviour and region [43–46]. In our study we found that socio-demographic characteristics like household income and a higher level of education did not have the same influence on outdoor behaviours compared to indoor behaviours. Hence, some differences could also be expected between indoor and outdoor behaviours for age and gender.

Because the survey was not meant to measure attitudes or values, we cannot associate the classification of the province to any difference in values or perception. However, others studies have showed cultural differences across Canadian provinces [47–51]. In Canada, French speaking people are at majority in the province of Quebec but a minority in the rest of Canada as opposed to English speaking Canadian that are a majority in the rest of Canada [52;53]. Several studies have observed differences of values and attitudes in terms of personality, political perspective, priorities and social issues between English-Canadians and French-Canadians [47–51]. Differences in those values could also explain some differences in the adoption of pro-environmental behaviours but further studies are required to confirm it.

Attitudes and values can also be different in immigrants compared to the native born. The former usually have a smaller ecological footprint [54–57]. For example, several studies, mostly from United States, have observed that immigrants have lifestyles that are less demanding on the environment: they consume less, possess fewer luxury items like SUVs, they carpool or use public transportation more often and live in smaller houses [54–57]. In 2006, around 55% of all Canadian immigrants were in Ontario, followed by 18% in British-Columbia and 14% in Quebec [58]. British-Columbia and Quebec had a good classification on both indices. However, we could not estimate the impact of immigra-

tion on these classifications, as immigration rules and influx have changed significantly over the last decades [58].

Despite those limits, the indices still give a good idea of the global adoption of pro-environmental behaviours with potential positive impacts on health in Canada and remain easy to explain and understand. The main sectors in which households can have an impact are covered by the indices, like air and soil quality as well as water conservation. The weighting methods used (i.e. MCA) are also more appropriate to assign weights as opposed to an equal weights or expert opinion approach that are often criticized for being arbitrary or simplistic [22]. Others similar indices could be created as the survey is performed every two years. The results obtained with the 2007 indices could serve as the baseline for surveillance purposes, as the survey has been more comprehensive since that date.

Conclusion

MCA was successfully applied in creating Indoor and Outdoor composite Indices of environmental health relevance based on a readily available periodic Statistics Canada dataset. The Indices cover a good range of environmental themes at the household level and the analysis, particularly the indices weights obtained in the MCA, could be applied to similar surveys worldwide (as baseline weights) enabling temporal trend comparisons for recurring themes. Results uncovered provincial patterns of pro-environmental behaviours adoption with certain provinces scoring consistently higher and others consistently lower, as well as the associations between socio-demographic factors and the indices. Much more than voluntary measures, this study shows that existing regulations, dwelling type, household composition and income as well as climate are the major factors determining pro-environmental behaviours.

Supporting Information

Figure S1   Map representation of the MCA results on the 55 questions with extreme responses.

Table S1   Results of the MCA on the 55 questions without extreme responses (exploratory analysis). **Legend:** N/A: Results are not available for supplementary variables. Qlt: Quality (i.e. the sum of the squared correlations for the first two dimensions in this case). Inr: Inertias. K: Principal coordinates for the first dimension. Cor: Squared correlation with the first dimension. Ctr: Contributions of the modality to the explained inertia of the first dimension. All cells are multiplied by 1000. Results are the same on rows and on column when a Burt table is used.

Table S2   Results of the MCA for the Indoor Index, 2007. **Legend:** N/A: Results are not available for supplementary variables. Qlt: Quality (i.e. the sum of the squared correlations for the first two dimensions in this case). Inr: Inertias. K: Principal coordinates for the first dimension. Cor: Squared correlation with the first dimension. Ctr: Contributions of the modality to the explained inertia of the first dimension. All cells are multiplied by 1000. Results are the same on rows and on column when a Burt table is used.

Table S3   Results of the MCA for the Outdoor Index, 2007. **Legend:** N/A: Results are not available for supplementary variables. Qlt: Quality (i.e. the sum of the squared correlations for the first two dimensions in this case). Inr: Inertias. K: Principal coordinates for the first dimension. Cor: Squared correlation with the first dimension. Ctr: Contributions of the modality to the explained inertia of the first dimension. All cells are multiplied by 1000. Results are the same on rows and on column when a Burt table is used.

Table S4   Questions and responses selected for the Indoor Index, 2007.

Table S5   Questions and responses selected for the Outdoor Index, 2007.

References

1. Natural Resources Canada (2011) Energy Efficiency trends in Canada, 1990 to 2009Ottawa (On)54 p.
2. Environment Canada (2011) Ottawa (On)Municipal Water Use Report24 p.
3. Mustapha I, Tait M, Trant D (2012) Human Activity and the Environment. Waste management in Canada. Statistics CanadaOttawa (On)Report No: 16-201-x, 46 p.
4. Milito AC, Gagnon G (2008) Greenhouse gas emissions-a focus on Canadian households. EnviroStats 2(4):3–6.
5. Statistics Canada (2008) Human Activity and the Environment: Annual Statistics 2007 and 2008Ottawa (On)159 p.
6. Environment Canada (2013) Wise Water Use. Available: http://www.ec.gc.ca/eau-water/default.asp?lang = En&n = F25C70EC-1. Accessed 6 may 2014.
7. Statistics Canada (2009). Households and the environment, 2007.Ottawa (On)report no: 11-526-x,102 p.
8. Hardie D, Alasia A (2009) Domestic Water Use: The relevance of Rurality in Quantity Used and Perceived Quality. Rural and Small Town Canada Analysis Bulletin 7(5):1–31.
9. Mustapha I (2013) Composting by households in Canada. EnviroStats 7(11):1–6.
10. Lynch MF, Hofmann N (2007) Canadian lawns and gardens: Where are they the greenest? EnviroStats 1(2): 9–14.
11. Birrell C (2008) Energy-efficient holiday lights. EnviroStats 2(4):19–20.
12. Nelligan T (2008) Household's use of water and wastewater services. EnviroStats 2(4):17–8.
13. Babooram A (2008) Canadian participation in an environmentally active lifestyle. EnviroStats 2(4):7–12.
14. Nardo M, Saisana M, Saltelli A, Tarantola S, Hoffman A, et al. (2005) Handbook on Constructing Composite Indicators: Methodology and User GuideParis (Fr)Organisation for Economic Co-operation and Development Publishing162 p.
15. Esty DC, Levy M, Srebotnajk T, de Sherbinin A (2005) 2005 Environmental Sustainability Index: Benchmarking National Environmental Stewardship. New Haven: Yale Center for Environmental Law & Policy, 403 p.
16. Statistics Canada (2010) Microdata User Guide – Households and the environment survey, 2007Ottawa (On)53 p.
17. Stewart DW, Shamdasani PN, Rook DW (2007) Focus Groups: Theory and Practise. 2nd ed.Thousand OaksSAGE Publications200 p.
18. Greenacre M (2007) Correspondence analysis in practice. 2nd EditionNew YorkChapman & Hall/CRC284 p.
19. Dossa LH, Buerkert A, Schlecht E (2011) Cross-Location Analysis of the Impact of Household Socioeconomic Status on Participation in Urban and Peri-Urban Agriculture in West Africa. Hum Ecol Interdiscip J 39(5): 569–581.
20. Charreire H, Casey R, Salze P, Kesse-Guyot E, Simon C, et al. (2010) Leisure-time physical activity and sedentary behaviour clusters and their associations with overweight in middle-aged French adults. Int J Obes (Lond) 34(8):1293–1301.
21. Cortinovis I, Vella V, Ndiku J (1993) Construction of a socio-economic index to facilitate analysis of health data in developing countries. Soc Sci Med 36(8): 1087–1097.
22. Howe LD, Hargreaves JR, Huttly SR (2008) Issues in the construction of wealth for the measurement of socio-economic position in low-income countriesEmerg Themes Epidemiol 5(3): 14 p.
23. Asselin LM (2002) Composite indicator of Multidimensional Poverty - TheoryQuébec (Qc)Institut de Mathématique Gauss33 p.
24. Greenacre M, Nenadic O (2010) Package 'ca' - Simple, Multiple and Joint Correspondence AnalysisR project, 20 p.
25. Munro A (2010) Recycling by Canadian Households, 2007.Statistics Canada, Ottawa (On)34 p.
26. Oaks(2012) Province and Territory Water Efficiency and Conservation Policy Information. Available: http://www.allianceforwaterefficiency.org/2012-Province-Information.aspx. Accessed 13 November 2013.
27. Kinkead J, Boardley A, Kinkead M (2006) An analysis of Canadian and other water conservation practices and initiativesMississauga (On)Canadian Council of Ministers of the Environment274 p.
28. Gibbons WD (2008) Who uses water-saving fixture in the home? EnviroStats 2(3): 8–12.
29. Statistics Canada (2011) CANSIM Table 027-0017: Canada Mortgage and Housing Corporation, mortgage loan approvals, new residential construction and existing residential properties, monthly. Available: http://www5.statcan.gc.ca/cansim/a26?lang = eng&retrLang = eng&id = 0270017&paSer = &pattern = &stByVal = 1&p1 = 1&p2 = 37&tabMode = dataTable&csid = . Accessed 13 November 2013.
30. Bonsal B, Koshida G, O'Brien EG, Wheaton E (2013). Droughts. Available: http://www.ec.gc.ca/inre-nwri/default.asp?lang = En&n = 0CD66675-1&offset = 8&toc = hide . Accessed 13 july 2012.
31. Environment Canada (2010) Water and climate change. Available: http://www.ec.gc.ca/eau-water/default.asp?lang = En&n = 3E75BC40-1. Accessed 13 November 2013.
32. Mekis É, Vincent LA (2011) An overview of the second generation adjusted daily precipitation dataset for trend analysis in Canada. Atmosphere-Ocean 49(2): 163–77.
33. Ministère du Développement durable, de l'Environnement, de la Faune et des Parcs (2011) The pesticides Management Code - Highlights. Available: http://www.mddep.gouv.qc.ca/pesticides/permis-en/code-gestion-en/index.htm. Accessed 13 July 2012.
34. Canada mortgage and Housing Corporation (2013) Reducing energy cost. Available: https://www.cmhc-schl.gc.ca/en/inpr/afhoce/afhoce/afhostcast/afhoid/opma/reenco/index.cfm. Accessed 22 November 2013.
35. BChydro (2013) Buy, build, or rent an efficient home. Available: http://www.bchydro.com/powersmart/residential/guides_tips/green-your-home/whole_home_efficiency/energy_efficient_home.html. Accessed 22 November 2013.
36. Human Resources and Skills Development Canada (2007) What difference does learning make to financial security. Indicators of Well-Being – Special ReportGovernment of Canada14 p.
37. U.S. Environmental Protection Agency (1998) Washington (DC)Water Conservation Plan Guidelines208 p.
38. Reynaud A, Renzetti S, Villeneuve M (2005) Residential water demand with endogenous pricing: The Canadian CaseWater Resour Res 41(w11409)11 p.
39. OECD (2013) Greening Household Behaviour: Overview from the 2011 Survey, OECD Studies on Environmental Policy and Household Behaviours.OECD Publishing306 p.
40. Kollmuss A, Agyeman J (2002) Mind the Gap: Why Do People Act Environmentally and What Are the Barriers to Pro-Environmental Behaviour? Environmental Education Research Aug;8(3):239.
41. Todd A, Struger J (2014) Changes in acid herbicide concentrations in urban streams after a cosmetic pesticides ban. Challenges, 5:138–151.
42. Environment Canada (2013) Ammonia Emissions. Available: https://www.ec.gc.ca/indicateurs-indicators/default.asp?lang = en&n = FE578F55-1. Accessed 30 January 2014.
43. Mainieri T, Barnett EG, Valdero TR, Unipan JB, Oskamp S (1997) Green buying: The influence of environmental concern on consumer behavior. The Journal of Social Psychology Apr;137(2):189–204.
44. Melgar N, Mussio I, Rossi M (2013) Environmental Concern and Behavior: Do Personal Attributes Matter?Facultad de Ciencias Sociales, Universidad de la Republica; 21 p.
45. Xiao C, Hong D (2010) Gender differences in environmental behaviours in China. Population and Environment Sep;32(1):88–104.
46. Lopez A, Torres CC, Boyd B, Silvy NJ, Lopez RR (2007) Texas Latino College Student Attitudes Toward Natural Resources and the Environment. Journal of Wildlife Management Jun;71(4):1275–80.
47. Baer DE, Curtis JE (1984) French Canadian-English Canadian Differences in Values: National Survey Findings. Canadian Journal of Sociology/Cahiers canadiens de sociologie 9(4):405–27.
48. Baillargeon JP (1994) The Cultural Practices of Anglophones in Quebec. Recherches Sociographiques May;35(2):255–71.
49. Gibson KL, McKelvie SJ, Man AF (2008) Personality and Culture: A Comparison of Francophones and Anglophones in Québec. The Journal of Social Psychology Apr;148(2):133–65.

Acknowledgments

The authors thank Mr. Yves Lafortune of Statistics Canada for relevant comments on a preliminary version of this study and Ms Sandra Owens for her contribution to the redaction of this article. Also, thanks to Mr. Gaston Quirion of Laval University Library for facilitating access to the Statistics Canada survey database.

Author Contributions

Conceived and designed the experiments: MC BA DB PG. Analyzed the data: MC BA. Wrote the paper: MC BA DB PG.

50. Wu Z, Baer DE (1996) Attitudes toward family and gender roles: A comparison of English and French Canadian women. Journal of Comparative Family Studies 27(3):437–52.

51. Young N, Dugas E (2012) Comparing climate change coverage in Canadian English and French-language print media: environmental values, media cultures, and the narration of global warming. Canadian journal of sociology 37(1):25–54.

52. Corbeil JP (2012) Ottawa (On)French and the francophonie in Canada. Census in brief no. 1, Statistics Canada12 p.

53. Corbeil JP (2012) Linguistic Characteristics of Canadians. Language, 2011 Census of Population, Statistics CanadaOttawa (On)22 p.

54. Atiles JH, Bohon SA (2003) Camas Calientes: Housing Adjustments and Barriers to Social and Economic Adaptation among Georgia's Rural Latinos. Southern Rural Sociology 19(1):97–122.

55. Blumenberg E, Smart M (2010) Getting by with a little help from my friends and family: immigrants and carpooling. Transportation May;37(3):429–46.

56. Bohon SA, Stamps K, Atiles JH (2008) Transportation and Migrant Adjustment in Georgia. Population Research and Policy Review Jun;27(3):273–91.

57. Price CE, Feldmeyer B (2012) The Environmental Impact of Immigration: An Analysis of the Effects of Immigrant Concentration on Air Pollution Levels. Population Research and Policy Review Feb;31(1):119–40.

58. Statistics Canada (2011) Immigration in Canada: A portrait of the Foreign-born Population, 2006 Census: Data tables, figures and maps. Available: http://www12.statcan.ca/census-recensement/2006/as-sa/97-557/tables-tableaux-notes-eng.cfm. Accessed 30 January 2014.

15

Habitat Fragmentation and Species Extirpation in Freshwater Ecosystems; Causes of Range Decline of the Indus River Dolphin (*Platanista gangetica minor*)

Gill T. Braulik[1,2,3]*, **Masood Arshad**[2], **Uzma Noureen**[2], **Simon P. Northridge**[1]

1 Sea Mammal Research Unit, Scottish Oceans Institute, University of St. Andrews, St. Andrews, Fife, United Kingdom, **2** World Wildlife Fund-Pakistan, Lahore, Pakistan, **3** Wildlife Conservation Society, Zanzibar, United Republic of Tanzania

Abstract

Habitat fragmentation of freshwater ecosystems is increasing rapidly, however the understanding of extinction debt and species decline in riverine habitat fragments lags behind that in other ecosystems. The mighty rivers that drain the Himalaya - the Ganges, Brahmaputra, Indus, Mekong and Yangtze - are amongst the world's most biodiverse freshwater ecosystems. Many hundreds of dams have been constructed, are under construction, or are planned on these rivers and large hydrological changes and losses of biodiversity have occurred and are expected to continue. This study examines the causes of range decline of the Indus dolphin, which inhabits one of the world's most modified rivers, to demonstrate how we may expect other vertebrate populations to respond as planned dams and water developments come into operation. The historical range of the Indus dolphin has been fragmented into 17 river sections by diversion dams; dolphin sighting and interview surveys show that river dolphins have been extirpated from ten river sections, they persist in 6, and are of unknown status in one section. Seven potential factors influencing the temporal and spatial pattern of decline were considered in three regression model sets. Low dry-season river discharge, due to water abstraction at irrigation barrages, was the principal factor that explained the dolphin's range decline, influencing 1) the spatial pattern of persistence, 2) the temporal pattern of subpopulation extirpation, and 3) the speed of extirpation after habitat fragmentation. Dolphins were more likely to persist in the core of the former range because water diversions are concentrated near the range periphery. Habitat fragmentation and degradation of the habitat were inextricably intertwined and in combination caused the catastrophic decline of the Indus dolphin.

Editor: Tom Gilbert, Natural History Museum of Denmark, Denmark

Funding: World Wildlife Fund-Pakistan -(http://www.wwfpak.org/) funded the interview surveys with support through the Pakistan Wetlands Project. The US Marine Mammal Commission (www.mmc.gov) provided funding to GB for reporting and analysis in grant number E4047595. The funders had no role in study design, data collection and analysis, decision to publish, or preparation of the manuscript.

Competing Interests: The authors have declared that no competing interests exist.

* Email: gillbraulik@downstream.vg

Introduction

Fresh waters are experiencing declines in biodiversity far greater than those in the most affected terrestrial ecosystems [1,2]. Dam construction has dramatically increased habitat fragmentation and degradation in freshwaters, which is likely to have incurred a large unredeemed extinction debt [3]. However, this debt is not yet well quantified or understood, as metapopulation ecology in freshwaters has lagged behind similar work in other habitats, such as tropical forests [4,5,6]. A fundamental, yet unanswered question for conservation biology is how rapidly freshwater species disappear from river fragments and which factors influence the extinction of freshwater species in habitat patches. We investigate this issue using the example of the highly fragmented Indus River system and its endemic, endangered freshwater cetacean.

The great rivers that drain the Himalaya are amongst the world's most biodiverse freshwater ecosystems, but they are increasingly under threat as the emerging nations of China, India, and Pakistan and the countries of Southeast Asia scramble to harness hydropower and provide water for expanding agrarian economies, in the midst of increasing water scarcity and climatic uncertainty [7,8,9]. Many hundreds of new dams and other water development projects are planned or under construction in the region, including mainstem and tributary mega-dams, run-of-the-river hydroelectric schemes, irrigation barrages, and inter-basin water transfers [10,11]. The Himalayan region will soon have the highest concentration of dams in the world [12]. The combined effects of these activities are predicted to cause rapid and escalating hydrological change and habitat fragmentation that will negatively impact riverine biodiversity and ecosystem services [9,13].

The freshwater cetaceans that inhabit the largest Himalayan rivers, the Indus, Ganges, Brahmaputra, Yangtze, Mekong, and Ayeyarwady, collectively form one of the world's most endangered groups of mammals, each listed as endangered or critically endangered on the IUCN RedList [14], and, in the case of the Yangtze River dolphin (*Lipotes vexillifer*), the species has probably been extinct since the mid-2000's [15]. River dolphins are iconic species that can serve as charismatic flagships for conservation of freshwater ecosystems; but they are poorly understood and increasingly threatened. This is exemplified by the demise of the Yangtze River dolphin which disappeared so quickly that a comprehensive evaluation of the causes of its decline was not conducted until after it was presumed extinct [16].

We examine the pattern, and causes of range decline of the Indus dolphin (*Platanista gangetica minor*), an obligate freshwater cetacean endemic to the Indus River system, which, after more than 150 years of barrage (gated-dam) construction and removal of water to feed irrigated agriculture, is one of the most fragmented and modified rivers in the world. The distribution of the Indus dolphin was carefully documented in the 1870's, (just prior to the first major barrage being constructed) and at that time the dolphin inhabited the entire lower Indus River system from the delta with the Indian Ocean, to the foothills of the Himalayas in what is now India and Pakistan [17]. By the early 1990's, Indus dolphins had undergone an 80% reduction in range, having been extirpated from the upper and lower reaches of the Indus and four of the largest tributaries [18]. They are now confined to five contiguous 'river sections' on the Indus mainstem in Pakistan, separated by barrages, and in the Beas River, in India (Fig. 1) [19,20]. Details of when dolphins were extirpated from different parts of their former range are vague, and the causes not clearly understood. However, the construction of twenty irrigation barrages between 1886 and 1971 (gated-dams used for water diversion) that fragmented the dolphins historical range into 17 river sections (numbered 1–17 on Fig. 1), and large-scale water abstraction for irrigation rendering many sections of river almost dry for many months, certainly played a role (see Fig. 2) [19,21].

In this paper we document the spatial and temporal dynamics of the Indus River dolphin range decline, and then use a series of regression models to determine the causes of the spatial pattern of decline, the timing of subpopulation extirpation, and the speed of subpopulation disappearance after habitat fragmentation. Greater understanding of how the Indus dolphin has responded to the presence of dams and water diversions within its habitat demonstrates how we may expect other vertebrate populations to respond as planned dams and water developments come into operation elsewhere. The results provide important and relevant insights into factors influencing species extinction in fragmented riverscapes.

Methods

Last Dolphin Sighting Date

The status of each of the six extant dolphin subpopulations is fairly well understood [19,22], but there is little information on when or why dolphins disappeared from the 11 river sections where dolphins are presumed extirpated (Fig. 1). In 2007, we compiled historical dolphin sighting dates and locations from areas in which dolphins are believed to have been extirpated, by conducting community interviews, and a review of historical literature. GB received clearance from the Pakistan Home Department for the survey, and as there was no appropriate ethics or review board in Pakistan to provide approval no other permits were required. Depending on location, dolphins are believed to have disappeared between 20 and 80 years ago, so the short structured interviews targeted elderly riverside inhabitants old enough to have seen dolphins in their lifetimes. The objective of the interviews was explained to potential interviewees and they provided their verbal, rather than written, consent to participate as the majority were illiterate. All interviews were anonymous. A calendar of significant local historical events was compiled to assist informants in recalling dates correctly. A last dolphin sighting date

(LDSD) was allocated to each river section based on the most recent dolphin sighting that we identified. We did not attempt to identify the exact extinction date of each subpopulation [23], but used the LDSD as a general indicator of when dolphins disappeared [24]. Inexact dates were assigned to 5 year intervals, for example, if the date was early 1970s, a date of 1972 was assigned, if it was mid 1970s, 1975 was designated, and if it was late 1970s the date used was 1978.

Identifying the Causes of Range Decline

The following seven explanatory variables that may have contributed to the Indus dolphin range decline were determined for each of the 17 sections of the dolphin's former range:

1. *Fragmentation date* - The year that each river section was created; assigned as the date that the second of the two bounding barrages became operational. For the river section upstream of Harike barrage in India, the isolation date was assigned as the completion date of Hussainiwala barrage which is located only 30 km downstream of Harike.

2. *River length* - The number of river kilometres between two barrages.

3. *Proximity to range edge* - The distance along the river's course from the former dolphin distributional limit recorded by Anderson [17], to the barrage located closest to the range core.

4. *Size of river* - The mean annual discharge in Million Acre Feet reported for each river prior to implementation of the Indus Water Treaty in 1960 [25]. This illustrates original river size prior to large-scale water abstraction.

5. *Confluences* - The number of river confluences within each river section was included as an indicator of favourable habitat, as Indus dolphins occur with higher frequency at confluences [26].

6. *River slope* - The slope within each river section was calculated as the drop in elevation between the up and downstream barrages, or upstream range limit in the case of peripheral segments, divided by the length of river. Slope exerts a direct effect on flow velocity and sediment transport and therefore may influence dolphin habitat.

7. *Dry season river discharge* - River discharge data were obtained for all twelve barrages and two dams north of Guddu on the Indus River system in Pakistan for the period July 2008 to April 2013 (4 years 9 months). The daily discharge below Guddu and Kotri barrages was obtained from October 2010 to April 2013 (~2 ½ years), and discharge below Sukkur barrage was obtained from April 1994 to January 2000, and October 2010 to April 2013 (8 years 3 months). Occasional missing data were interpolated. It is very low flows that are likely to adversely impact dolphins, therefore median daily discharge during the dry season (1st October to 31st March) was determined using the years for which there was an entire dry season's data. Mean monthly discharge was available above and below Harike and Hussainiwala barrages in India from January 2009 to December 2011, and for these two river sections the median of the mean monthly dry season discharge was used. The number of years of data available differed according to barrage but the temporal discharge pattern was predictable and similar across years in each location.

Generalised Linear Models (GLMs) and a survival analysis were used with the seven explanatory variables described above as predictors of the continued presence of river dolphins in each of the 17 river sections. Generalised Additive Models were used in

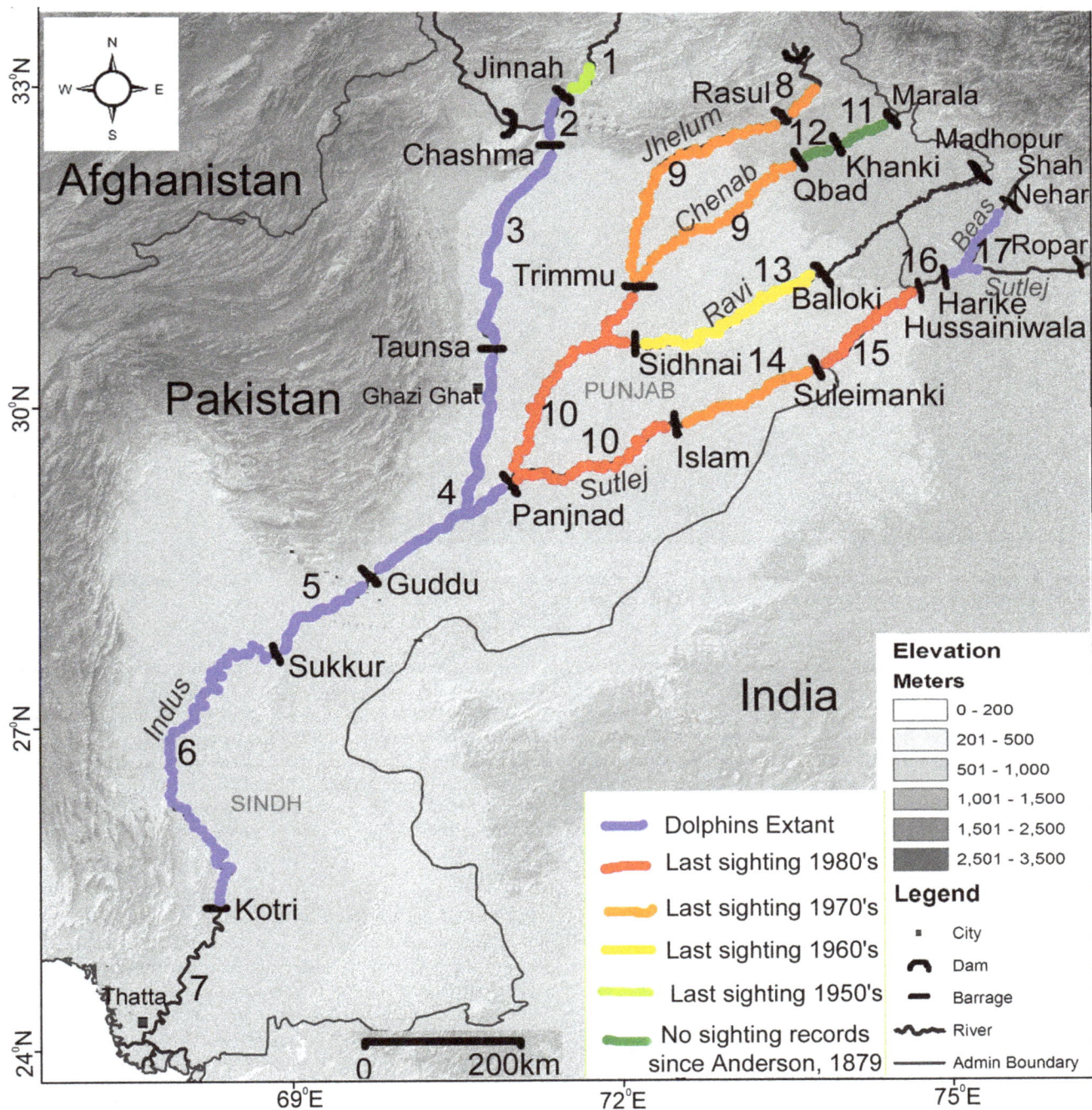

Figure 1. Map of the lower Indus River system. Rivers and barrages are named, and each river section is numbered and coloured according to whether river dolphins are extant, or the approximate date that they were extirpated (see Table A1 for details).

the initial data exploration to visually investigate whether the relationship between the predictor and explanatory variables was linear, and, which type of transformation could be used to best account for non-linearity. Three sets of models were developed, with objectives summarised below:

Spatial pattern of dolphin persistence. The objective of the first set of models was to identify which factors best explained the observed geographic pattern of range decline. The presence or absence of dolphins in each river section was modelled using a GLM and a binomial error distribution, with presently extant

populations coded as 1 and extirpated populations coded as 0. The best fitting models were then used to predict the probability that dolphins are still present in the Harike-Hussainiwala river section on the India-Pakistan border (Fig. 1: no. 16), where dolphin presence is unknown.

Temporal pattern of decline. The second model set included only sections where dolphins have been extirpated and examined which factors influenced when dolphins disappeared. The number of years since dolphins were sighted (as of 2013) was

Figure 2. Aerial photograph of Sukkur Barrage. Image shows the seven canals diverting water out of the river, and demonstrates the dramatically reduced flows downstream (river flow direction right to left).

the response variable modelled using a GLM with a quasi-Poisson error distribution.

Time to extirpation. The third model set used a survival function to investigate which factors influenced the speed with which dolphin populations were extirpated following their isolation between barrages. We used the Kaplan–Meier estimate of survival which allows for the inclusions of censored data, in this instance allowing for the inclusion of river sections where dolphins have not yet been extirpated, as well as those where dolphins have disappeared. Each river section was qualified with a status assignment, where 1 = dolphins extirpated, and 0 = dolphins extant [27]. The time to extirpation was calculated as the number of years between the fragmentation date and either the LDSD if dolphins have been extirpated, or the current year, 2013, where they are still present. Time to extirpation and status together were the predictor variable (the Kaplan-Meier survivorship object) modelled using the 'survreg' function in the survival library of the program R [28]. Both an exponential and a Weibull error distribution were tested, and the Weibull distribution was selected as it provided a significantly better fit to the data (ΔAIC 15.17) [27].

All models were implemented using the program R 2.15.1 [28]. Logit, probit and cloglog link functions were included in global models and the logit function, which resulted in the best fit, applied. Variance Inflation Factors (VIFs) that demonstrate the degree of collinearity between variables were generated from the maximal models and collinear variables removed until VIF scores were less than five [27]. Three two-way interactions that described potentially meaningful relationships between variables (fragmentation date and dry season discharge, fragmentation date and river length, and river length and proximity to range edge) were included, as well as second and third order polynomials of significant variables. The binomial and survival models were simplified using backwards stepwise selection based on Akaike's Information Criteria (AIC). Quasi-Poisson models were selected

on the basis of quasi-AIC (QAIC) scores, and non-significant terms were sequentially dropped based on their levels of significance. Models separated by at least two AIC/QAIC points were assumed to be significantly different [29]. Goodness of fit for the GLMs was measured by determining the proportion of the total deviance explained by the final model. Model plots were examined for non-normality of errors, statistical independence of observations, heteroscedasticity and influential points [30].

Results

Pattern of Range Decline

Historical dolphin sightings were obtained for all river sections formerly occupied by dolphins except for the area downstream of Kotri Barrage to the delta (Fig. 1: no. 7) and the stretch from Harike to Hussainiwala barrage (Fig. 1: no. 16) which is close to the India-Pakistan border. Our focus on retired fishermen and on areas from which dolphins have already been extirpated meant that the communities were forth-coming with information because it was not regarded by them as sensitive. However, there were very few elderly members of each fishing community and our pool of available informants was consequently small (n = 57). 79% of informants were, or had been, full-time commercial fishermen or contractors and the remainder were part-time subsistence fishermen. There was no significant difference in the age of informants interviewed at each barrage (GLM p = 0.7) but older individuals were significantly more likely to have seen dolphins than younger informants (GLM p<0.01). We found no evidence that dolphins persist in any of the Indus tributaries in Pakistan. Of 17 sections of river, dolphins are extant in six, have been extirpated from ten, and in one border area that could not be surveyed (no. 16), dolphin presence or absence is unknown (Fig. 1).

Causes of Range Decline

For each river section, dolphin presence or absence, estimated LDSD, and the time to extirpation were compiled along with the physical characteristics and these data included in each of the models (Table S1). On the upper Chenab River (Fig1: no 11 & 12) the LDSD is that reported by Anderson in 1879. Time to extirpation was not calculated for these two sections reflecting the lack of recent sighting evidence and uncertainty of the extirpation date.

Where dolphins are still extant the median monthly dry season river discharge averaged 30,830 cusecs (ranged 7,224–47,040 cusecs), as compared to an average of 8,022 cusecs (range 0–38,000 cusecs) in locations from which dolphins have been extirpated. In general, sections of river where dolphins are still present were fragmented by barrages later, are further from the range periphery, are of longer length, have a shallower slope and greater dry season discharge than river sections where dolphins are no longer found.

Sixteen river sections, 6 where dolphins are extant and ten where they have been extirpated, were included in the spatial GLM models. The VIFs generated from the full model indicated that river discharge and slope were collinear. Slope was considered to be less important than discharge in explaining dolphin range decline because it has not changed substantially in hundreds of years, and it was therefore removed from further candidate models. The final model that best explained the observed spatial pattern in Indus dolphin range decline retained the explanatory variables dry season river discharge and distance from range edge (Table 1). The probability that an Indus dolphin subpopulation is still extant increases with increasing distance from the range edge, and with increasing dry season river discharge (Fig. 3). The final spatial model predicted only a 2.6% probability that dolphins are still extant in the Harike-Hussainiwala river section in India that was not surveyed for dolphins. This is not unexpected given that this section has very low dry season discharge and is near the periphery of the dolphin's range, both factors that increase the likelihood of subpopulation extirpation. Although the linear arrangement of river segments might suggest a lack of statistical independence, there was no clumping of residuals according to the geographic position of the river segments, and the independence of observations was further shown by the Durbin-Watson test of correlated errors (p = 0.24).

The variables that described the temporal pattern of Indus dolphin subpopulation extirpation were the same as those that influenced the spatial pattern of decline: dry season river discharge and distance from former range limit (Table 2). Within areas from which dolphins have disappeared, they were extirpated earlier from river sections where discharge was lower, and from those sections located near the periphery of the subspecies former range (Fig. 3).

In river sections where dolphins have been extirpated, the mean time from fragmentation to the LDSD was 50 years (SD = 23, range = 9–76). For river sections where dolphins are still extant, the mean time from subpopulation isolation to present (2013) was 57 years (SD = 15, range = 42–86). Thirteen river sections were included in the survival model (number 2, 10, 12 and 16 were excluded because of missing data) and the slope parameter was included. The final survival model retained four variables: median dry season river discharge, isolation date, length of river section and slope. Dolphin subpopulations were extirpated more quickly in sections with low dry season river discharge. Subpopulations persisted longer where the river slope is more gentle (e.g. in the lower reaches) and those that were isolated between barrages a long time ago persisted for longer than those in more recently

Table 1. Summary of spatial range decline model output.

| Model | AIC | Δ AIC | % explained Deviance | n | Deviance | | | | | | |
					Q	Range	Is. Date	L	Conf	Size	USD: Is. Date
1	18.26	1.82	32.6	1	6.91	-	-	-	-	-	-
2	**16.44**	**-**	**50.7**	**2**	**7.88**	**2.84**	-	-	-	-	-
3	17.72	1.28	54.1	3	7.88	2.84	0.72	-	-	-	-
4	19.67	3.23	54.3	4	7.88	2.84	0.68	0.08	-	-	-
5	21.64	5.20	54.4	5	7.88	2.84	0.42	0.31	0.07	-	-
6	23.64	7.20	54.5	6	7.88	2.84	0.42	0.31	0.07	0.0001	0.001
7	25.16	8.72	56.7	7	7.88	2.84	0.79	0.31	0.07	0.0001	0.11

n = number of covariates, Is. Date = Isolation Date, L = Length of river section, Range = Distance from range edge, Size = River size, Conf = confluences, Q = River discharge, USD: Is. Date = Interaction between Isolation Date and River Discharge. Model in bold was the final selected model.

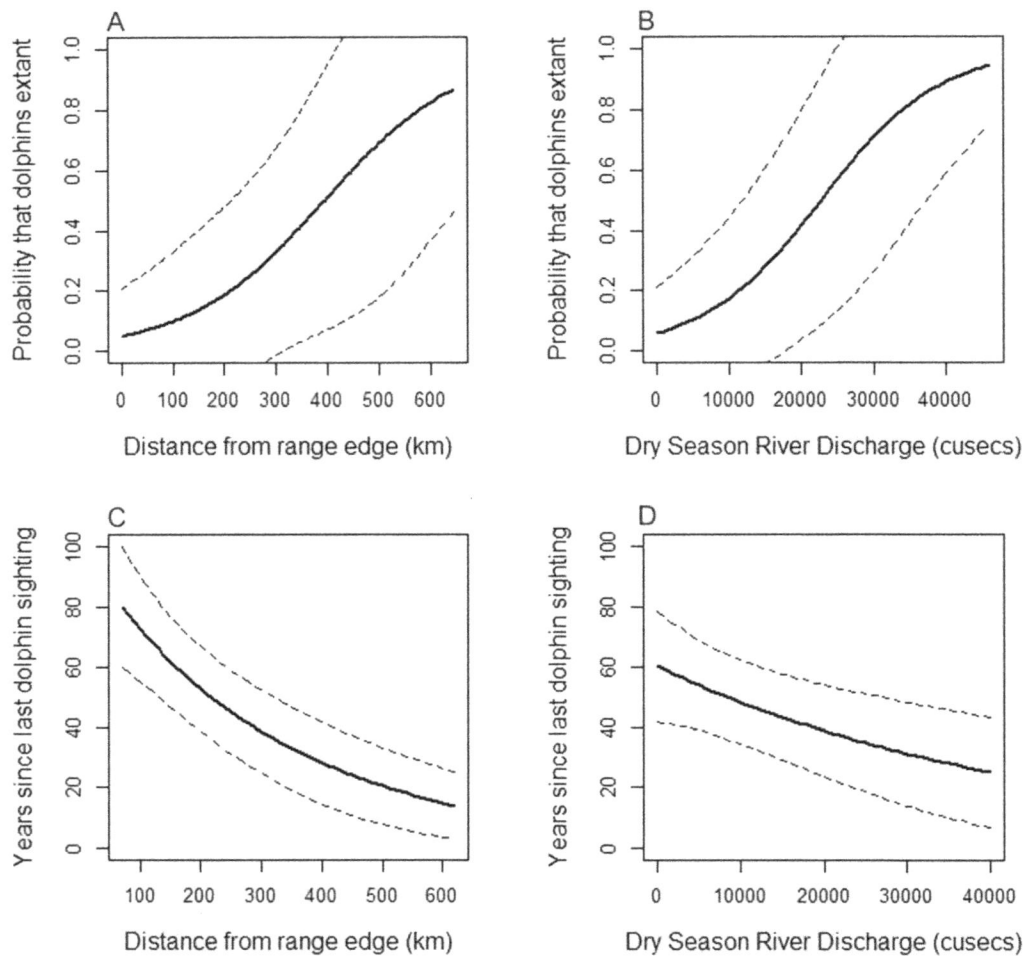

Figure 3. Significant relationships retained in GLM models of the causes of the spatial and temporal pattern of Indus dolphin decline. The figure demonstrates the probability that an Indus dolphin subpopulation is extant according to A) proximity to the edge of the former range and B) median dry season river discharge, and the relationship between the number of years since a dolphin was sighted and C) distance from the historical distributional limit, and D) median dry season river discharge.

subdivided river sections. Fifty years after Indus dolphins were isolated between barrages there is a less than a 50% chance that they will still be extant, and after 100 years this probability drops to 37%.

Discussion

Model evaluation

The river discharge data used in these models were from the last ten years but they explained well the pattern of dolphin decline that occurred decades ago. Although river discharge varies from year to year, and has generally declined, the relative discharge among barrages (e.g. the spatial relationship) has remained constant with the same locations consistently reporting high (e.g. the upper Indus) and low discharge (e.g. Indus tributaries) over time. Therefore, the assumption implicit in this analysis that the present spatial pattern of discharge reflects that present during the period leading up to dolphin extirpation is not unreasonable.

As for terrestrial habitats, such as forest fragments, we would have expected to see a relationship between species extinction and habitat patch size [5]. In fact, length of river section was one of the first variables to be excluded in candidate models. This may be

because only the current configuration of 17 comparatively small habitat fragments were included in the models, and we did not consider the progression of escalating habitat fragmentation and concomitant diminishing fragment size over time. To investigate this, we constructed an additional model considering all 33 river sections that have existed since the onset of barrage construction (Table A2) in a binomial GLM, with dolphins recorded as extirpated (0) or still extent (1) in each river section at the point it was further subdivided. Explanatory variables were 1) Length of river section, 2) Isolation Date, and 3) End Date, taken as the year a new barrage was completed resulting in the sections further subdivision. When considering the entire history of habitat fragmentation, the models showed that dolphins were significantly more likely to be extirpated in smaller fragments (p<0.05), and that this relationship was independent of fragment creation date or duration. The link between species extirpation and habitat fragment size has been clearly established in terrestrial habitats for several species groups [5] but this is one of the first studies to show a similar relationship in riverscapes. It underlines the great importance of maintaining large sections of intact river habitat to sustain tropical aquatic biodiversity.

Habitat Fragmentation and Species Extirpation in Freshwater Ecosystems; Causes of Range Decline of the Indus... 151

Table 2. Summary of temporal range decline model outputs.

| Model | AIC | Δ AIC | % explained Deviance | n | Deviance | | | | | | |
					Q	Range	Is. Date	L	Conf	Size	USD*Is. Date
1	17.57	1.36	56.75	1	-	115.36	-	-	-	-	-
2	**16.21**	-	**76.76**	2	40.67	115.36	-	-	-	-	-
3	17.80	1.59	79.17	3	40.67	115.36	-	-	-	-	4.91
4	17.54	1.33	92.62	4	52.06	115.36	-	-	-	6.66	14.20
5	19.50	3.29	92.85	5	52.06	115.36	-	-	6.60	6.66	8.11
6	21.46	5.25	93.12	6	52.06	115.36	0.75	-	6.60	6.66	7.9
7	23.39	7.18	93.49	7	52.06	115.36	0.60	0.33	9.16	6.66	5.89

n = number of covariates, Is. Date = Isolation Date, L = Length of river section, Range = Distance from range edge, Size = River size, Conf. = confluences, Q = River discharge. Model in bold was the final selected model.

River discharge and distance from range periphery provided a good fit to the range decline data, explaining more than 76% of the deviance in the temporal model and 50% in the spatial model. However, three other aspects that may have also have played a role in the dolphins decline were not included as explanatory variables because of a lack of suitable data. These are a) water quality, b) incidental capture in fishing gear and c) hunting. The possible contributions of these to the Indus dolphin decline are discussed below.

The magnitude of surface water pollution in Pakistan has increased at a dramatic rate over the last ten years and more than 90% of industrial and municipal effluents enter water courses untreated [31]. The Indus tributaries flow through the industrial and agricultural heartland of Pakistan and are more polluted than the Indus River itself which has a greater dilution capacity and passes through remoter areas. There has been no systematic monitoring of river water quality that could provide data for this analysis. However, dolphins had already been extirpated from most areas prior to significant declines in water quality which occurred in the 1980s and 90s, and this asynchronous timing indicates that pollution was not primarily responsible for the dolphins' decline.

Mortality from accidental capture in fishing gear is considered to be the greatest threat to most cetacean populations [32]. In the past, the Indus River main channel was not intensively fished because the water was too swift for easy manoeuvrability of oar-powered boats, and instead fishing focused on side channels and adjacent pools that are rarely used by dolphins [33]. Since 2010 changing fishing practices in Sindh Province have led to an increase in dolphin mortality, however, prior to this there are very few records of incidental capture of dolphins in fishing gear and this is not likely to be a large factor in the decline of the Indus dolphin.

Indus dolphins were killed for food, oil and medicine until the late 1970s when the animal became legally protected [17,34]. Information on dolphin hunting is sparse and un-quantified and records refer only to hunting on the Indus River, where dolphins are still extant. Although it is possible that dolphins were hunted throughout the river system, there is no evidence that this was so, and the fact that they persist in the places that hunting is reported to have been intense, and have disappeared from places where hunting was not reported, suggests that this was unlikely to have been the cause of the subspecies' decline. However, the timing of reported hunting coincides with the period of decline and without more information, it is not possible to completely discount the role of hunting.

For the majority of the year the gates on all barrages are lowered to divert water into canals, and the physical opening is sufficiently small that it would be difficult or impossible for dolphins to pass through the gates and between different sections of river. It has been hypothesized that there may be consistent or frequent movements of dolphins through some barrages and between subpopulations [21,35]. It has also been theorised that due to the high water velocity and turbulence often found within the barrage gates it would be more likely for animals to move down-, rather than up-stream, and that this would lead to the downstream migratory attrition of upstream subpopulations [18,36]. The only evidence of this was obtained in 2009, when, during the annual canal maintenance period, which is one of the brief periods in the year when no water is diverted and barrage gates are fully open for several weeks, one radio-tagged Indus dolphin was recorded to move through the gates on Sukkur barrage in both an up- and down-stream direction (WWF-Pakistan unpublished). Each barrage is quite different in terms of design, location and operation and one dolphin moving through Sukkur barrage does not prove that this occurs at other barrages, or that it is a regular occurrence. Therefore, for the purposes of this study, and in the absence of evidence to the contrary, we assume no significant migration between dolphin subpopulations. It is possible that future research may demonstrate that dolphins do move across some, or all, barrages with regularity and if the movement of individuals is primarily in one direction this would be another important factor to consider in the extinction dynamics. The pattern of subpopulation persistence near the range core, and earlier extirpation of subpopulations near the range periphery could perhaps be partially explained by the consistent downstream migration of animals, but again this pattern is disrupted by the presence of dolphins in the Beas River.

Causes of Indus dolphin range decline

The clear result of this study was the relationship between low dry season river discharge and the decline of the Indus dolphin. Reduced flows directly impact dolphins by reducing the physical space available to them, reducing average water velocity and depth and increasing water temperatures. Flow regulation is also likely to indirectly impact river dolphins due to declines in fish diversity, the dominance of generalist fish species, and increased success of invasive species [37,38]. The dampened flood peaks typically associated with dams and diversions reduce the frequency, extent and duration of floodplain inundation that determines how long fish can gain access to nursery habitat and food. Water abstraction also exacerbates and concentrates existing anthropogenic threats, for example, increasing the concentration of nutrients and pollutants, and concentrating dolphins into deep pools that are also important areas for fishing, thereby increasing the chances of negative human interactions [39]. The altered hydrological regime on the Indus River has likely reduced the complexity of hydrologic and geomorphologic habitat and ultimately also diminished its carrying capacity and ability to support large numbers of aquatic megafauna. To preserve aquatic biodiversity, river management is needed that focuses on restoring both the timing and duration of flood pulses, as well as on maintaining critical minimum flows in the dry season.

The persistence of dolphins in the Beas River, India is likely to be due to the presence of constant water supplies little depleted by diversions. Dolphins in the Beas River occur in an isolated habitat fragment as the river downstream is virtually dry, and only connected with the rest of the river system for a few weeks each year during the monsoon floods. This demonstrates that in the presence of sufficient water, and an absence of other threats, river dolphins can persist for decades even in relatively small fragments of habitat near the periphery of their range. This subpopulation is of conservation importance, as all other Indus dolphins occur in a single river, the Indus, and are therefore at risk from environmentally correlated catastrophic events [40]. However, based on the historical pattern of decline, Indus dolphins are most likely to disappear in the future from locations with low river discharge located closer to the range periphery, meaning that dolphins in the Beas (close to range periphery with moderate discharge) and between Sukkur and Kotri Barrages (with low discharge located a moderate distance from the former range edge) are most at risk.

The date of habitat fragmentation was not selected by any of the models as a strong predictor of whether dolphins are still present. However depleted river discharge and habitat fragmentation by barrages are inextricably intertwined as barrages are responsible for diverting water, and they are a physical barrier that greatly impedes or prevents the dispersal of dolphins out of impacted river reaches.

Population extirpation, core habitat and conservation

Contraction of geographic range is one of the principal characteristics exhibited by declining or threatened species [41,42]. In general, at the periphery of a species geographic range, populations occupy less favourable habitat and occur at lower and more variable densities. Therefore, as a species becomes endangered it is expected that its geographic range will contract inwards, and that populations will persist in the range core until the final stages of decline [43]. However, for many endangered mammals the pattern of range decline is instead dictated by the spread of factors driving the decline, with those populations last impacted, regardless of their location, persisting longer than those that were historically large [41,43]. The range of the Indus dolphin has also contracted inwards, and dolphins persist primarily in what is assumed to be the former range core or higher density area, however this is likely to be because the greatest threat, water extraction, is concentrated in the periphery of the subspecies range. This conclusion is supported by the continued persistence of the Beas River population at the range limit. However, that animals naturally occur at lower density in upstream areas and smaller rivers and are therefore more vulnerable near the range limit is also certainly a factor. The spatial pattern of Indus River dolphin decline is very different from the gradual decline in abundance described for the Yangtze River dolphin [16].

One of the greatest challenges in conservation science involves disentangling the relative contributions of multiple factors in the decline of species, especially when causes interact or vary spatially and temporally with importance [44]. Nevertheless, the primary factor identified in these models (i.e. low dry season discharge due to water diversion at barrages) is well supported, and is also the most salient for informing current ecosystem management. In the Mekong River, numerous stressors such as fisheries bycatch, hunting and habitat destruction have reduced the resident Irrawaddy dolphin (*Orcaella brevirostris*) population to less than 100 individuals [45]. The results of our study suggests that habitat fragmentation and/or flow disruption associated with the many proposed new dams on the Mekong are likely to further drive Irrawaddy dolphin decline potentially leading to local extinction. The Indus dolphin range decline is probably most prescient for the related Ganges dolphin (*Platanista gangetica gangetica*) which occurs in the neighbouring Brahmaputra and Ganges River systems and is subject to the same threats. The Ganges River is fragmented by barrages and flow is severely reduced in many areas, and the Brahmaputra River system is the focus of massive hydropower development [46,47]. The range of the Ganges dolphin has begun to decline especially at the upper limit of the distributional range [48], and if water development continues as planned the range of the dolphin is expected to continue to shrink towards larger habitat fragments with higher discharge that are primarily in downstream locations. The results of this study suggest that other vertebrate populations in other large rivers, such as the Amazon, Orinoco and Ayeyarwady, will also respond with dramatic declines in range when dams and other water developments that reduce discharge, fragment habitat and change the hydrological regime are constructed.

The amount of habitat fragmentation and level of water withdrawals from rivers in Pakistan is extreme, negatively affecting human communities, eroding the delta, destroying fisheries and concentrating pollutants. This study indicates that if water development plans in South Asia and the wider Himalayan region proceed as currently proposed [10,13]and follow the pattern demonstrated by the Indus, the resulting habitat fragmentation and flow disruption will likely cause large declines in resident freshwater cetaceans and other freshwater dependent species. Healthy rivers are of great importance to communities, and it is critical that where developments are planned, environmental flow and impact assessments be conducted that balance human requirements for irrigation water and power with the habitat requirements of the aquatic ecosystem that are vital to humans.

Acknowledgments

We are grateful to the Office of Executive Engineers at Harike and Hussainiwala Canal Divisions and Ropar Headworks, India for allowing us river discharge data. A. Khan and R. Garstang facilitated the fieldwork. Valuable reviews were provided by P. Hammond, M. Lonergan, S. Turvey and G. Ryan.

Author Contributions

Conceived and designed the experiments: GTB. Performed the experiments: GTB MA UN. Analyzed the data: GTB SPN. Contributed to the writing of the manuscript: GTB MA UN SPN.

References

1. Sala OE, Chapin FS, Armesto JJ, Berlow R, Bloomfield J, et al. (2000) Global biodiversity scenarios for the year 2100. Science 287.
2. Dudgeon D, Arthington AH, Gessner MO, Kawabata Z-I, Knowler DJ, et al. (2006) Freshwater biodiversity: importance, threats, status and conservation challenges. Biological Reviews 81: 163–182.
3. Tilman D, May RM, Lehman CL, Nowak MA (1994) Habitat destruction and the extinction debt. Nature 371: 65–66.
4. Strayer DL, Dudgeon D (2010) Freshwater biodiversity conservation: recent progress and future challenges. Journal of the North American Benthological Society 29: 344–358.
5. Gibson L, Lynam AJ, Bradshaw CJA, He F, Bickford DP, et al. (2013) Near-complete extinction of native small mammal fauna 25 years after forest fragmentation. Science 341: 1508–1510.
6. Prugh LR, Hodges KE, Sinclair ARE, Brashares JS (2008) Effect of habitat area and isolation on fragmented animal populations. PNAS 105: 20770–20775.
7. FAO (2012) Coping with water scarcity. An action framework for agriculture and food security. Rome, Italy: FAO Water Reports No.38.
8. World Water Assessment Programme (2012) The United Nations World Water Development Report 4. Managing water under uncertainty and risk. Paris: UNESCO.
9. Dudgeon D (2000) Large-scale hydrological changes in tropical Asia: prospects for riverine biodiversity. Bioscience 50: 793–806.
10. Dutta AP (2010) Reservoir of Dams. Down to Earth May 1–15: 32–39.
11. Verma S, Kampman DA, van der Zaag P, Hoekstra AY (2009) Going against the flow: A critical analysis of inter-state virtual water trade in the context of India's National River Linking Program. Physics and Chemistry of the Earth 34: 261–269.
12. Dharmadhikary S (2008) Mountains of Concrete: Dam building in the Himalayas. Berkeley, CA: International Rivers. 48 p.
13. Ziv G, Baran E, Nam S, Rodríguez-Iturbe I, Levin SA (2012) Trading-off fish biodiversity, food security, and hydropower in the Mekong River basin. PNAS www.pnas.org/cgi/doi/10.1073/pnas.1201423109.
14. IUCN (2013) IUCN Red List of Threatened Species. version 2013.2 <http://www.iucnredlist.org> Downloaded on 6 January 2014.
15. Turvey ST, Pitman RL, Taylor BL, Barlow J, Akamatsu T, et al. (2007) First human-caused extinction of a cetacean species? Biology Letters 3: 537–540.
16. Turvey ST, Barrett LA, Hart T, Collen B, Yujiang H, et al. (2010) Spatial and temporal extinction dynamics in a freshwater cetacean. Proceedings of the Royal Society B 277: 3139–3147.
17. Anderson J (1879) Anatomical and zoological researches: comprising an account of the zoological results of the two expeditions to western Yunnan in 1868 and 1875 and a monograph of the two cetacean genera Platanista and Orcella. Bernard Quaritch, Piccadilly, London.
18. Reeves RR, Chaudhry AA, Khalid U (1991) Competing for water on the Indus Plain: Is there a future for Pakistan's river dolphins? Environmental Conservation 18: 341–349.
19. Status assessment of the Indus River dolphin, Platanista gangetica minor, March-April 2001. Biological Conservation 129: 579–590.
20. Behera SK, Nawab A, Rajkumar B (2008) Preliminary investigations confirming the occurrence of Indus River dolphin Platanista gangetica minor in River Beas, Punjab, India. Journal of the Bombay Natural History Society 105: 90–126.
21. Reeves RR, Leatherwood S (1994) Dams and river dolphins: can they co-exist? Ambio 23: 172–175.
22. Braulik GT, Bhatti ZI, Ehsan T, Hussain B, Khan AR, et al. (2012) Robust abundance estimate for endangered river dolphin subspecies in South Asia. Endangered Species Research 17: 201–215.
23. Collen B, Purvis A, Mace GM (2010) When is a species really extinct? Testing extinction inference from a sighting record to inform conservation assessment. Diversity and Distributions 16: 755–764.
24. Butchart SHM, Stattersfield AJ, Brooks TM (2006) Going or gone: defining 'Possibly Extinct' species to give a truer picture of recent extinctions. Bulletin-British Ornithologists Club 126: 7–24.
25. IUCN (2011) Pakistan Water Gateway: Gateway to water information about Pakistan. www.waterinfo.net.pk.
26. Braulik GT, Reichert AP, Ehsan T, Khan S, Northridge SP, et al. (2012) Habitat use by a freshwater dolphin in the low-water season. Aquatic Conservation: Marine and Freshwater Ecosystems 22: 533–546.
27. Crawley MJ (2007) The R Book. Chichester, England: John Wiley & Sons Ltd. 951 p.
28. R Development Core Team (2012) R: A language and environment for statistical computing. Vienna, Austria: R Foundation for Statistical Computing.
29. Burnham KP, Anderson DR (2002) Model selection and multi- model inference: a practical information-theoretic approach. 2nd edition. New York: Springer-Verlag. 496 p.
30. Fox J (2008) Applied regression analysis and Generalised Linear Models. 2nd edition. Los Angeles: Sage Publications, Inc. 665 p.
31. Directorate of Land Reclamation Punjab (2007) Surface water quality monitoring plan. Lahore: Irrigation and Power Department, Government of Punjab. 27 p.
32. Read AJ (2008) The looming crisis: interactions between marine mammals and fisheries. Journal of Mammalogy 89: 541–548.
33. Khan H (1947) A fishery survey of River Indus. Journal of the Bombay Natural History Society 46: 529–535.
34. Pilleri G (1972) Field observations carried out on the Indus dolphin Platanista indi in the Winter of 1972. Investigations on Cetacea 4: 23–29.
35. Braulik GT (2012) Conservation ecology and phylogenetics of the Indus River dolphin (Platanista gangetica minor). St. Andrews, UK. 267: University of St. Andrews.
36. Reeves RR (1991) Conservation of the bhulan (blind river dolphin) in the Punjab. Natura. 3–22.
37. Nilsson C, Reidy CA, Dynesius M, Revenga C (2005) Fragmentation and flow regulation of the world's large river systems. Science 308: 405–408.
38. Xenopoulos MA, Lodge DM (2006) Going with the flow: using species-discharge relationships to forecast losses in fish biodiversity. Ecology 87: 1907–1914.
39. Kelkar N, Krishnaswamy J, Choudhary S, Sutaria D (2010) Coexistence of fisheries with river dolphin conservation. Conservation Biology 24: 1130–1140.
40. Gilpin ME (1990) Extinction in finite metapopulations living in correlated environments. In: Shorrocks B, Swingland IR, editors. Living in a patchy environment. New York: Oxford University Press. 177–187.
41. Channell R, Lomolino MV (2000) Trajectories to extinction: spatial dynamics of the contraction of geographical ranges. Journal of Biogeography 27: 169–179.
42. Simberloff D (1986) The proximate causes of extinction. In: Raup DM, Jablonski D, editors. Patterns and processes in the history of life. Berlin: Springer-Verlag. 259–276.
43. Lomolino MV, Channell R (1995) Splendid isolation: patterns of geographic range collapse in endangered mammals. Journal of Mammalogy 76: 335–347.
44. Johnson PTJ, McKenzie VJ, Peterson AC, Kerby JL, Brown J, et al. (2010) Regional decline of an iconic amphibian associated with elevation, land-use change and invasive species. Conservation Biology 25: 556–566.
45. Ryan GE, Dove V, Trujillo F, Doherty Jr PF (2011) Irrawaddy dolphin demography in the Mekong River: an application of mark-resight models. Ecosphere 2: 1–15.
46. Bashir T (2010) Ganges river dolphin (Platanista gangetica) seeks help. Current Science 98: 287–288.
47. Baruah D, Hazarika LP, Bakalial B, Borah S, Dutta R, et al. (2012) A grave danger for the Ganges dolphin (Platanista gangetica, Roxburgh) in the Subansiri River due to a large hydroelectric project. Environmentalist 32: 85–90.
48. Sinha RK, Smith BD, Sharma G, Prasad K, Choudhury BC, et al. (2000) Status and distribution of the Ganges Susu (Platanista gangetica) in the Ganges River system of India and Nepal. In: Reeves RR, Smith BD, Kasuya T, editors. Biology and Conservation of Freshwater Cetaceans in Asia. Gland, Switzerland & Cambridge, U.K.: IUCN. 54–61.

16

GFP Transgenic Medaka (*Oryzias latipes*) under the Inducible *cyp1a* Promoter Provide a Sensitive and Convenient Biological Indicator for the Presence of TCDD and Other Persistent Organic Chemicals

Grace Hwee Boon Ng, Zhiyuan Gong*

Department of Biological Sciences, NUS Graduate School for Integrative Sciences and Engineering, National University of Singapore, Singapore and Computation and Systems Biology, Singapore-MIT Alliance, Singapore, Singapore

Abstract

Persistent organic pollutants (POPs) are resistant to environmental degradation and can cause multitude of health problems. Cytochrome P450 1A (Cyp1a) is often up-regulated by POPs through the activation of aryl hydrocarbon receptor (AhR) pathway and is thus usually used as a biomarker for xenobiotics exposure. To develop a convenient *in vivo* tool to monitor xenobiotic contamination in the water, we have established GFP transgenic medaka using the inducible *cyp1a* promoter, *Tg(cyp1a:gfp)*. Here we tested *Tg(cyp1a:gfp)* medaka at three different stages, prehatching embryos, newly hatched fry and adult with 2,3,7,8-tetrachlorodiebnzo-*p*-dioxin (TCDD), a dioxin. While GFP induction was observed in all three stages, newly hatched fry were the most sensitive with the lowest observed effective concentration of 0.005 nM or 16.1 ng/L. The highly sensitive organs included the kidney, liver and intestine. With high concentrations of TCDD, several other organs such as the olfactory pit, tail fin, gills, lateral line neuromast cells and blood vessels also showed GFP expression. In addition, *Tg(cyp1a:gfp)* medaka fry also responded to two other AhR agonists, 3-methylcholanthrene and benzo[a]pyrene, for GFP induction, but no significant GFP induction was observed towards several other chemicals tested, indicating the specificity of this transgenic line. The GFP inducibility of *Tg(cyp1a:gfp)* medaka at both fry and adult stages may be useful for development of high-throughput assays as well as online water monitoring system to detect xenobiotic toxicity.

Editor: Bernhard Ryffel, French National Centre for Scientific Research, France

Funding: This work was supported by the Singapore National Research Foundation under its Environmental and Water Technologies Strategic Research Programme and administered by the Environment and Water Industry Programme Office (EWI) of the PUB, grant number R-154-000-328-272. GHBN acknowledges the support from an NGS graduate scholarship. The funders had no role in study design, data collection and analysis, decision to publish, or preparation of the manuscript.

* E-mail: dbsgzy@nus.edu.sg

Introduction

Polycyclic aromatic hydrocarbons (PAHs) and dioxins are persistent organic pollutant (POPs) which are resistant to environmental degradation and can persist in soils and sediments for decades or even centuries [1]. They can enter into aquatic ecosystem through effluent, atmospheric deposition, petroleum spill, run-off and ground water. With their persistent presence, POPs can be bioaccumulated and biomagnified in animal and human tissues through direct contact and food chains, thus causing health hazards such as carcinogenicity, organs failures and system dysfunction [2–4]. Since POPs include a variety of chemical compounds, the toxicities and affinities of POPs can be highly complicated. Currently it is widely accepted that most of the effects of POPs are mediated by aryl hydrocarbon receptor (AhR) pathway [5], which up-regulates Cyp1a enzyme that catalyzes the oxygenation of PAHs and heterocyclic aromatic amines/amids, and many other agents [6]. Cyp1a is often used as a specific and reliable biomarker for PAH contamination. As such, measurement of ethoxyresorufin-O-deethylase (EROD) activity is commonly performed to indicate the enzymatic activity of Cyp1a induced during chemical exposure [7]. Organisms that inhabit the water bodies, such as fish, are usually sacrificed to obtain their liver tissue for EROD assay.

In this study, we intended to develop a convenient biological tool to monitor PAH exposure using medaka fish (*Oryzias latipes*), which is a hardy teleost fish and is able to survive in a broad range of temperature (4°C to 40°C) as well as in a wide salinity range (0–50% of sea water) [8]. Transgenic fish have long been proposed for the purpose of biomonitoring aquatic contaminants as water sentinels [9,10]. In these biomonitoring fish, they have been genetically modified to respond to contaminants with easily detectable reporter, based on the principle that certain genes are inducible by certain chemical contaminants. Currently, there are already a few stable transgenic fish lines established for detecting certain classes of pollutants [11–15]. For example, *Tg(mvtg1:gfp)* medaka, with its transgene consists of medaka vitellogenin promoter and GFP reporter gene, was established and has shown to induce GFP in the liver of male fish when exposed to estradiol [13]. Here we have developed a transgenic medaka line, *Tg(cyp1a:gfp)* in which GFP is expressed under the inducible *cyp1a* promoter, thus providing a convenient assay for easy detection *in situ* and in real time for the presence of *cyp1a*-inducing chemicals. Here we exposed the transgenic medaka to TCDD, a well studied congener of dioxin commonly used as reference point for Toxic Equivalency Factor (TEF) [2], to determine the sensitivity of the transgenic line as well as the pattern of induced GFP expression. In addition to TCDD, GFP expression in *Tg(cyp1a:gfp)* were also induced by other PAH chemicals such as 3-methylcholanthrene (3-MC) and benzo[a]pyrene (BaP), but not by other categories of chemicals such as mercury chloride, 4-nitrophenol, lindane and bisphenol A, suggesting that GFP induction in *Tg(cyp1a:gfp)* medaka is quite specific to TCDD and members of PAH family. Thus, *Tg(cyp1a:gfp)* medaka will be another convenient and useful tool in environmental monitoring.

Materials and Methods

Fish Husbandry

Hd-rR medaka strain was obtained from National BioResource Project, Japan [16]. Husbandry of medaka fish was according to Kinoshita et al., 2009 [8]. Staging of medaka embryos and fry was mainly based on Iwamatsu, 2004 [17]. Generation of *Tg(cyp1a:gfp)* [previously named *Tg(cyp1a1:gfp)*] transgenic medaka has been described previously [18]. *cyp1a* was used here because there is only one copy of *cyp1a* gene in the medaka genome [19] All experimental protocols were approved by Institutional Animal Care and Use Committee (IACUC) of National University of Singapore (Protocol 079/07).

Toxic Chemical exposure

TCDD, 3-MC and BaP were purchased from Sigma-Aldrich. All stock solutions were prepared in dimethyl sulfoxide (DMSO) (Sigma-Aldrich) and stored at $4°C$. Hemizygous transgenic embryos, fry and fish were used in all chemical exposures for consistency. Five or more fry (1–3 dph [days posthatching]) or embryos (6–9 hpf [hours postfertilization]) were placed in each well of a 6–well dish, containing 5 ml of various concentrations of a testing chemical and incubated at $28°C$. Each experiment was repeated at least three times. For adult treatment, 6 month-old transgenic fish, 5 male and 5 female per tested concentration, were used and chemical solutions were changed every other day. The fish were not fed during the 72-hour chemical exposures. At the end of the experiment, all fish were examined for GFP expression. In all the acute exposure experiments, there was no mortality and no observable abnormality even with the highest concentration of TCDD (5 nM) after 24 hours of exposure.

Epifluorescence microscopy

GFP expression was observed under an inverted fluorescence microscope (Axiovert 200 M, Zeiss) equipped with a digital camera (Axiocam HRc, ZEISS) for capturing GFP fluorescence. Fry or fish were first anaesthetized by 0.1% 2-phenoxyethanol (Sigma-Aldrich) and positioned in 3% methylcellulose (Sigma-Aldrich) for observation. For adult fish, GFP expression was observed and imaged with a stereomicroscope, Olympus MVX10 with a digital camera.

Whole mount *in situ* hybridization

Medaka *cyp1a* cDNA was amplified from nucleotides +1321 to +1821 with reference to the transcription start site (Ensembl ID: ENSORLG00000014421) and subcloned into pGEM-T EASY

(Promega, Madison, WI). Digoxigenin riboprobes of *cyp1a* was synthesized using DIG RNA labeling kit (Roche). Whole mount *in situ* hybridization was performed essentially according to the protocol described in Kinoshita et al., 2009 [15]. The period of proteinase K digestion is either 25 min to maintain surface epithelial integrity or 80 min for better penetration of riboprobes. After *in situ* hybridization, embryos were incubated with alkaline phosphatase-conjugated anti-DIG Fab Fragments (1:7000; Roche) and were visualized with tetrazolium chloride and 5-bromo-4-chloro-3-indolyl-phosphate (Roche).

Results

Selection of *Tg(cyp1a:gfp)* lines for chemical tests

Four *Tg(cyp1a:gfp)* transgenic founders were initially identified and all F1 transgenic progenies have been demonstrated to respond to TCDD treatment by inducing GFP expression in the liver, intestine and kidney [18]. To select a transgenic strain for monitoring xenobiotics, transgenic F1 adults that produced transgenic offspring with low or no GFP background and had a single transgenic insertion were selected, and finally a transgenic family, *Tg(cyp1a:gfp)* 4.2, with the most sensitive response to TCDD and other xenobiotic chemicals was selected. This transgenic strain, referred as *Tg(cyp1a:gfp)*, was used for all chemical exposures experiments in this study and has been maintained for five generations.

Sensitivity and dosage-dependent GFP expression of *Tg(cyp1a:gfp)* medaka in response to TCDD

1–3 dph *Tg(cyp1a:gfp)* medaka fry were exposed to a range of concentrations of TCDD for 24 hours to determine its sensitivity to TCDD (Figure 1). No GFP expression was observed in the vehicle solvent, 0.1% DMSO-treated transgenic fry (Figure 1). About 7% of the transgenic fry responded to 0.005 nM TCDD with GFP expression in the liver (Figure 1I), indicating the lowest observed effective concentration (LOEC) was around 0.005 nM TCDD. With the increasing concentrations of TCDD, induced GFP fluorescence became increasingly intensive and appeared in more tissues/organs. For examples, at 0.05 nM, GFP expression was observed in the kidney, liver and also gut (Figure 1D) and the GFP signal became intensified in these three organs at 0.1 nM (Figure 1E). At 0.5 nM or above, GFP expression could be observed in several other organs such as olfactory pits, caudal fin, gills, neuromast cells along the trunk and around the eyes in majority of the treated transgenic fry (Figure 1F–H, also see Figure 2F,I,L). Some fry also exhibited weak GFP expression in blood vessels along the trunk (Figure 2F), myocardium and skin epithelial cells (Figure 2F,I). Quantitative GFP expression data in these major expressing organs are summarized in Figure 1I. The most sensitive organs were the kidney and liver, followed by the gut (Figure 1I).

Induced GFP expression is consistent with induced endogenous *cyp1a* expression

To determine the endogenous *cyp1a* mRNA expression after TCDD exposure, we performed whole mount *in situ* hybridization with a *cyp1a* antisense probe in TCDD-treated wild type fry. The hybridization signal was observed in major internal organs including the kidney, liver and gut (Figure 2B), similar to the typical GFP expression pattern induced by TCDD in *Tg(cyp1a:gfp)* fry (Figure 2C); in comparison, there was no hybridization signal detected in these organs in the vehicle control (Figure 2A). Similarly, whole mount *in situ* hybridization with sense riboprobes did not have any staining (Figure S1), thus demonstrating the

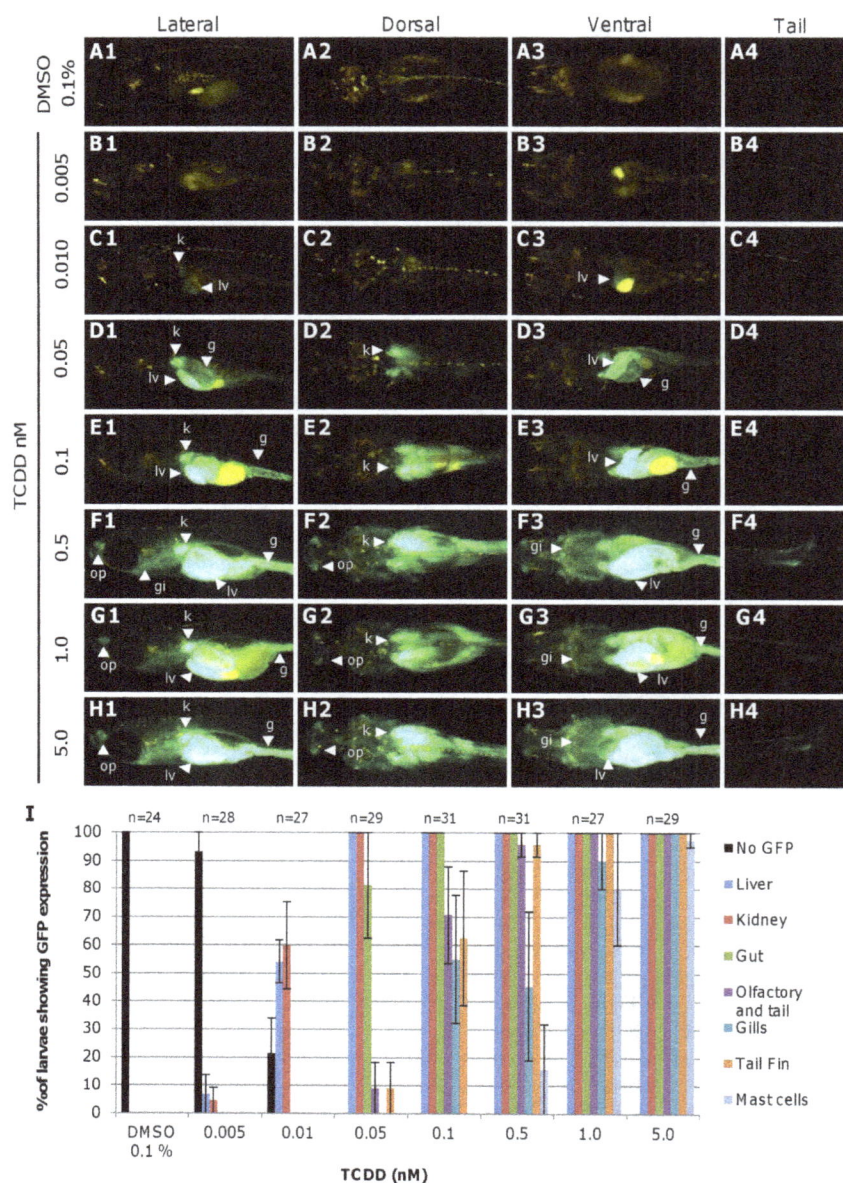

Figure 1. TCDD-induced GFP expression in posthatched *Tg(cyp1a:gfp)* fry. (A–H) GFP expression in *Tg(cyp1a:gfp)* fry at different concentrations of TCDD. 1–3 dph *Tg(cyp1a:gfp)* fry were exposed to TCDD for 24 hours and photographed for GFP expression under a fluorescence microscope. Lateral, dorsal, ventral and tail images were taken from the same representative fry from each concentration group. GFP expressing tissues are pointed by arrowheads. Abbreviations: g, gut; gi, gills; k, kidney; lv, liver; op, olfactory pit. (I) Percentage of fry expressing GFP in specific organs at different concentrations of TCDD. The scoring organs were based on direct observation GFP fluorescence under a fluorescence microscope. Error bars represent standard error. n represents the total number of fry used.

specificity of the *cyp1a* antisense probes. Under a short proteinase K digestion treatment where the integrity of the epithelial surface of *in situ* hybridization sample was maintained, hybridization signals were observed in the neuromast cells along the lateral line, skin, gills and olfactory pits in TCDD-treated fry (Figure 2E,H,K), but not in the vehicle control (Figure 2D,G,J). These corresponded to the sites of GFP expression observed in 5.0 nM TCDD treated transgenic fry (Figure 2F, I, L). Overall, these observations demonstrated that induced GFP expression in *Tg(cyp1a:gfp)* embryos/fry faithfully represent the induced endogenous *cyp1a* expression by TCDD treatment at least in these major expressing organs.

Specificity of *Tg(cyp1a:gfp)* in response to other types of chemicals

To investigate the specificity of *Tg(cyp1a:gfp)* in response to other types of chemicals, transgenic fry were also exposed to other PAHs including 3-MC and BaP. Induced GFP fluorescence was observed in the kidney, liver and gut by BaP treatment (Figure 3B), in comparison with the lack of GFP expression in the vehicle control (Figure 3A). The percentages of fry expressing GFP in these three organs were quite similar in each concentration of BAP (Figure 3D) [0.25 µM (62.5 µg/l) to 3.96 µM (1000 µg/l)]. In 3-MC treatment, although GFP expression was also observed in the kidney, liver and gut (Figure 3C), the liver seemed to be the most sensitive

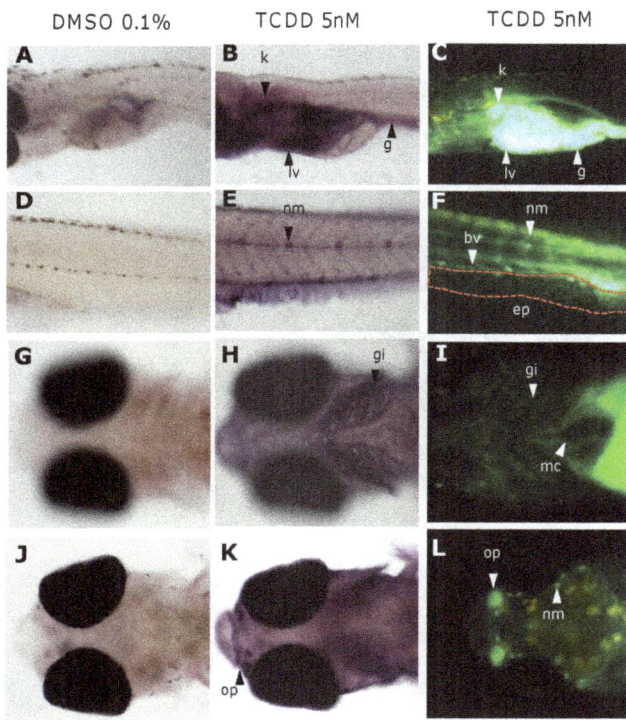

Figure 2. GFP induction is similar to induced endogenous cyp1a transcript in TCDD treatment. (A, B, D, E, G, H, J, K) Whole mount in situ hybridization detection of cyp1a mRNAs in 0.1% DMSO-treated wild type fry (A, D, G, J) and 5 nM TCDD-treated wild type fry (B, E, H, K). (C, F, I, L) GFP expression in 5 nM TCDD treated transgenic fry. Interesting organs are indicated by arrowheads. Abbreviations: bv, blood vessel; ep, epithelia; g, gut; gi, gills; k, kidney; lv, liver; mc, myocardium; nm, neuromast cells; op, olfactory pit.

organ, followed closely by kidney. At the lowest dosage, 23.3 nM (6.25 µg/l), ~50% of the fry had induced GFP expression in the liver while ~30% of the fry showed GFP expression in the gut and kidney (Figure 3E). The percentages of fry expressing GFP in the kidney and liver were higher than those in the gut for higher concentrations of 3-MC (Figure 3D) [46.6 nM (12.5 µg/l) to 372.7 nM (100 µg/l)]. Unlike TCDD exposure, GFP expression was not observed in other organs beyond the liver, kidney and gut for both 3-MC and BAP treatments. We also tested several other types of chemicals, including mercury chloride, bisphenol A (BPA), 4-nitrophenol and lindane, but there was no induced GFP expression observed in the transgenic fry, except that a small number of fry (~20%) showed weak spotty GFP expression in the liver at the high concentration, 53.9 µM and 71.89 µM of 4-nitrophenol (Figure S2). Table 1 summarizes the information of responses of Tg(cyp1a:gfp) to all these chemicals, including TCDD.

Induction of GFP expression in Tg(cyp1a:gfp) embryos

To fully characterize the induction of GFP expression in early Tg(cyp1a:gfp) embryos, hemizygous transgenic embryos were exposed to TCDD continuously from ~6 hpf (stage 10 to stage 11). It was apparent that Tg(cyp1a:gfp) embryos were less sensitive than post-hatching fry in TCDD induction of GFP expression. Up to 0.05 nM, there was no robust GFP fluorescence induced up to 72 hours of exposure. At 1.0 nM of TCDD, GFP fluorescence was observed weakly in the epithelial layer of yolk and embryonic body at 24 hours of exposure (Figure 4A,B). In a few embryos (6/21),

moderate GFP signal was observed in the retina after 24 hours of exposure (Figure 4B,D) but diminished from 2 dpf (Figure 4F,H). At 2 dpf (~48 hours of exposure), weak GFP expression persisted in the epithelial layer of embryonic body and the yolk (Figure 4E,F). By 3 dpf (~72 hours of exposure), GFP was clearly expressed in the liver, kidney tubule, gut and olfactory pits (Figure 4I–L) in all TCDD- treated transgenic embryos. Longer exposure period gave similar observations to that of 3-dpf embryos (Figure 4M–P). In comparison, no GFP expression, except for a transient, faint GFP signal in the mid-brain at 1 dpf (Figure S3), was observed in the vehicle solvent group at any time of TCDD treatment (Figure 4Q–T). In summary, weak induction of GFP expression can be performed in epithelia and yolk within 24 hours of fertilization, but robust induction of GFP expression occurs only after the liver is developed at 3 dpf. These observations may indicate the possible presence of AhR activity in early stages of embryonic development. However, the pre-hatching embryos are relatively insensitive to TCDD induction compared to post-hatching fry, hinting a lower AhR activity in embryos than that in fry.

Induction of GFP expression in Tg(cyp1a:gfp) adults

To test whether GFP induction could be directly observed from live adult Tg(cyp1a:gfp) medaka, 6-month-old adult transgenic fish were treated with three different concentrations (0.1, 0.5 and 2.5 nM) of TCDD for 72 hours. After 24 hours of treatment, 2 male and 2 female Tg(cyp1a:gfp) fish from each of the three TCDD concentration groups were randomly picked and observed. In all the fish observed, GFP was expressed in olfactory pits as well as in the abdomen region corresponding to the kidney, liver and urinary pore positions (Figure 5A) while no GFP signal was observed in non-treated control (Figure 5B). After 72 hours of exposure, increasing GFP fluorescence was observed in the head, body trunk, skin and brightly in kidney, urinary pores, gills region, lens, lips and olfactory pit as well as rib cage in all fish from the three TCDD-treated groups (Figure 5C,E,F) and again there was no GFP signal observed in untreated control fish (Figure 5D). To confirm the GFP expressing organs, we dissected all the fish at the end of the 72-hour exposure in order to view its internal organs. In both TCDD-treated males (Figure 5G,H) and females (Figure 5I,J), the liver was found to have the most intensive GFP signal, next by the gut (Figure 5G,I). The GFP signal was also strong in the head kidney and kidney tubules (Figure 5H,J). Interestingly, the abdomen wall that was dissected out did not fluorescence, indicating that the earlier mentioned GFP signal in the rib cage was due to the strong signal from the internal organs which illuminated the rib cage. This observation indicate the feasibility to use live adult Tg(cyp1a:gfp) fish to develop an in vivo biomonitoring system for TCDD and related chemicals.

Discussion

Sensitivity of Tg(cyp1a:gfp) medaka

Previously we have generated several Tg(cyp1a:gfp) medaka for monitoring TCDD and other POPs [18] and the selection of cyp1a promoter is because cyp1a is the most reliable and robust responsive genes to TCDD [20,21]. From multiple Tg(cyp1a:gfp) founders, here we isolated and characterized a highly sensitive transgenic line, where GFP expression was detected in the liver, kidney, gut and several other tissues/organs when treated with TCDD. In addition, this line has the highest sensitivity towards TCDD among the four Tg(cyp1a:gfp) families that we have maintained and compared. The LOEC of TCDD to evoke GFP expression is around 0.005 nM (Figure 1I) in the hemizygous

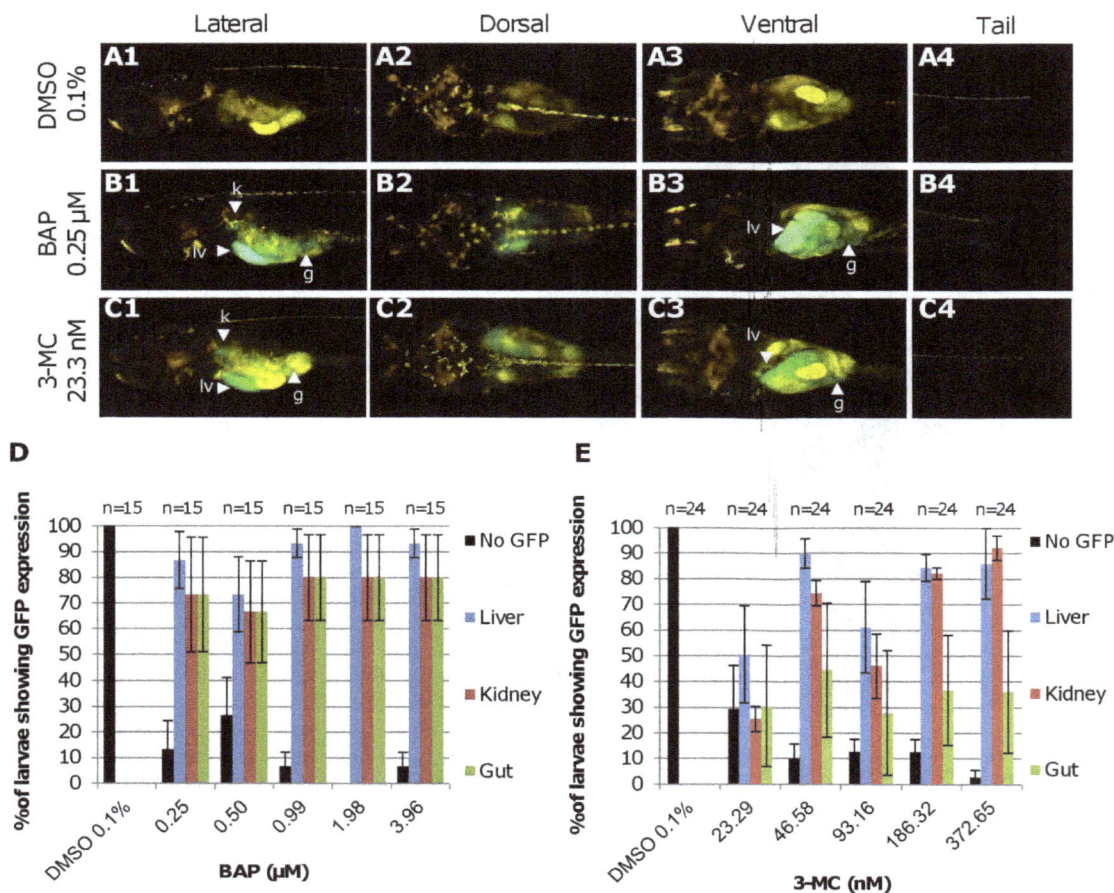

Figure 3. GFP induction of *Tg(cyp1a:gfp)* fry by BAP and 3-MC. (A–C) GFP expression in *Tg(cyp1a:gfp)* fry in vehicle control 0.1% DMSO (a), 0.25 μM BAP (B) and 23.3 nM 3-MC (C). GFP expressing tissues are pointed by arrowheads. Abbreviations: g, gut; k, kidney; lv, liver. (D, E) Percentages of fry expressing GFP in specific organs induced by BAP (D) or by 3-MC (E). Error bars represent standard error. n represents the total number of fry used.

Tg(cyp1a:gfp) 1–3 dph fry. The TCDD detection limit of the traditional EROD assay from cell culture is ranged from 2.4 pM to 1 nM [22,23]. Thus, the LOEC we estimated based on our transgenic fish assay is quite comparable to the best EROD sensitivity reported. In comparison, some limitations exist in the EROD assay. Firstly, unknown chemicals in the sample could act as Cyp1a enzyme antagonists, potentiators or competitors, thus affecting the accuracy of the EROD assay [7,24,25]. Secondly,

EROD assay from cell culture studies has shown to decrease at high concentrations of inducer, presumably due to inhibition of Cyp1a enzyme activity by the high concentration of inducer [23,26,27]. Thirdly, external parameters such as temperature, pH, etc. can also influence the outcome of EROD assay [7,25,28]. Transgenic fish assay can avoid these problems as it monitors TCDD more directly from the upstream transcriptional activity, which is controlled by TCDD-AhR interaction. Thus this

Table 1. Summary of GFP expression induced in Tg(*cyp1a:gfp*) fry with different chemical exposures.

Chemicals	GFP expression	Lowest effective concentration
TCDD	liver, kidney, gut, olfactory pits, gills, blood vessels, neuromast cells, skin, caudal fin, myocardium	0.005–0.01 nM
3-MC	liver, kidney and gut	≤23.3 nM
BaP	liver, kidney and gut	≤247.7 nM
4-nitrophenol	Spotty GFP expression in liver	35.9–53.9 μM
Bisphenol A	No GFP induction	Nil
Mercury chloride	No GFP induction	Nil
Lindane	No GFP induction	Nil

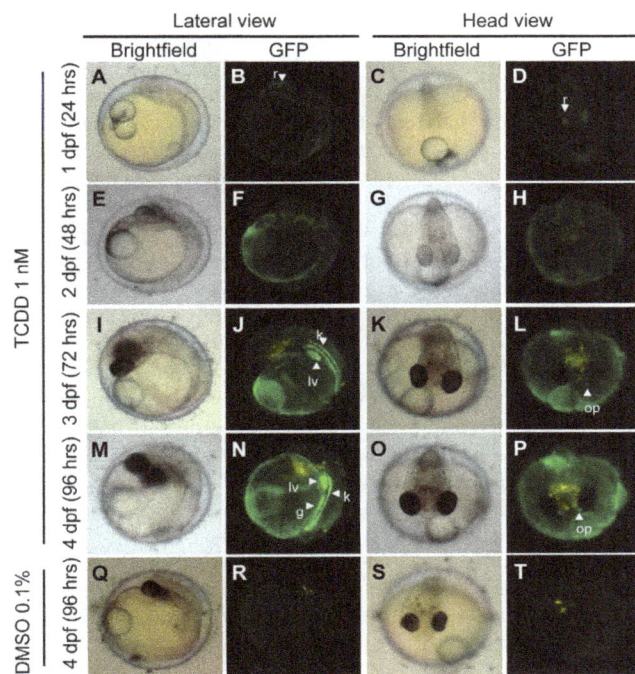

Figure 4. GFP induction in *Tg(cyp1a:gfp)* **embryos by TCDD.** 6–9 hpf hemizygous embryos were treated with 1 nM of TCDD or vehicle solvent control (0.1% DMSO) continuously. Bright field and fluorescence images of lateral and head views were taken from a representative embryo at different time points posttreatment: 24 hrs (A–D), 48 hrs (E–H), 72 hrs (I–L) and 96 hrs (M–P). No induced GFP expression was observed in the control group during the course of 96 hrs of treatment (Q–T). Interesting organs are indicated by arrowheads. Abbreviations: g, gut; k, kidney; lv, liver; op, olfactory pit.

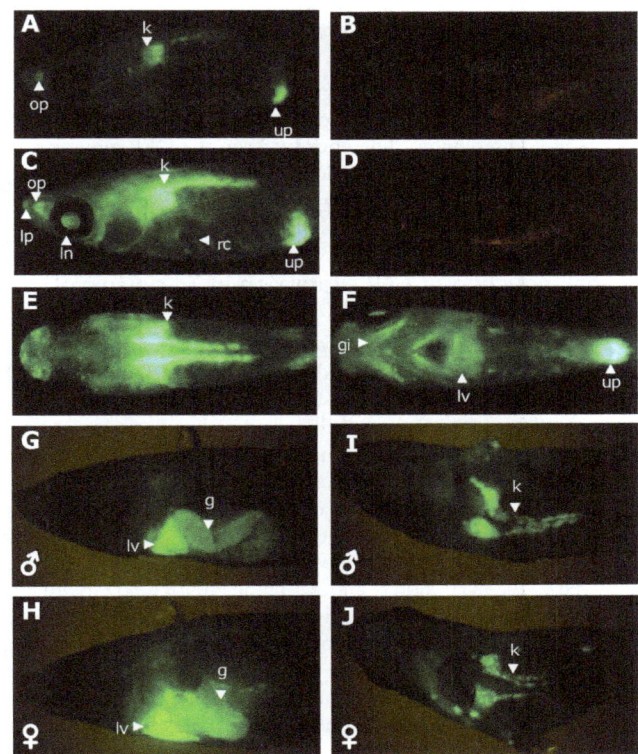

Figure 5. GFP induction of adult *Tg(cyp1a:gfp)* **medaka by TCDD.** (A, B) Lateral views of GFP induction of a representative male *Tg(cyp1a:gfp)* fish at 24 hrs after 2.5 nM TCDD exposure (A) in comparison with an untreated control fish (B). (C, D) Lateral views of GFP induction of a representative male *Tg(cyp1a:gfp)* fish at 72 hrs after 2.5 nM TCDD exposure.(C) in comparison with an untreated control fish (D) (E, F) Dorsal view (E) and ventral view (F) of the same fish from (C). *Tg(cyp1a:gfp)* fish in other TCDD concentrations (0.1 nM and 0.5 nM) displayed similar expression pattern and intensity to those shown here in (A, C, E, F). (G, H) GFP expression in the internal organs of a TCDD-treated male transgenic fish. (I, J) GFP expression in the internal organs of a TCDD-treated female transgenic fish. Lateral view for the liver and gut is shown in (G, I) and ventral view for the kidney after removal of internal digestive organs is shown in (H, J). Abbreviations: g, gut; gi, gills; k, kidney; ln, lens; lp, lips; lv, liver; op, olfactory pit; rc, ribcage; up, urinary pore.

transgenic model provide a better in vivo model for more direct monitoring of these environmental chemicals than EROD assays.

Compared to several cell-based transgenic assays using similar dioxin responsive elements linked luciferase reporter gene [22,24,29,30], our transgenic fish assay is less sensitive as it has been reported in several cell-based assays have the detection limit from 0.8 pM to 1.0 pM of TCDD. The higher sensitivity of the transgenic cell cultures could be due to the direct exposure of cells by the toxicants while transgenic fish have additional physiological barrier against xenobiotics intake. However, our transgenic fish represent an authentic *in vivo* biological model and it is easy to perform by direct observation of easily scorable GFP fluorescence in live fry or adult fish. Moreover, the result is instant and the assay is economical compared to the EROD and transgenic cell-based assays. As fish naturally inhabit in the water environment, the transgenic medaka is an ideal model for developing into an online monitoring systems to report aquatic contamination by TCDD and similar pollutants.

Specificity of *Tg(cyp1a:gfp)* medaka

We have further tested *Tg(cyp1a:gfp)* medaka fry with other known Cyp1a inducer such as 3-MC and BaP. Different from TCDD, BaP and 3-MC evoked GFP expression in the kidney, liver and gut only. TCDD apparently induced the highest and strongest reaction of *Tg(cyp1a:gfp)* among the three chemicals and at a concentration much lower than BaP and 3-MC, consistent with many previous reports that TCDD has the greatest affinity for AhR [20,27,31]. Perhaps, the concentrations of BAP and 3-

MC used in this study may not be high enough to warrant GFP expression in other organs. Other chemicals such as BPA, lindane and mercury chloride, all of which are unknown to be AhR ligands, were tested on *Tg(cyp1a:gfp)* fry and basically there was no induced GFP expression observed even with their lethal dosage and longer exposure periods. Only weak GFP induction was observed in small percentage of fry in high concentrations of 4-nitrophenol. Thus, these observations demonstrated that *cyp1a* promoter based transgenic line is capable of responding specifically to AhR ligands such as TCDD, BAP and 3-MC. However, this transgenic line is unable to predict the exact chemical from its responses.

Although *cyp1a* transcription is primarily regulated by AhR pathway, *cyp1a* expression is also influenced by hormonal factors such as glucocorticoids, insulin and sex hormones [32]. These factors may directly and/or indirectly affect the *cyp1a* expression by crosstalking with other transcription factors. Furthermore, AhR pathway was suggested to be involved in other functions such as immunology and development in addition to the mediation of biological responses to chemicals and pollutants [33]. It has been reported that administration of AhR ligands, such as beta-naphthoflavone, modulates the induction of estrogen-receptor mediated downstream gene, suggesting that AhR pathway does crosstalk with estrogen receptor pathway [32,33,34,35]. However, we observed no obvious difference between the GFP induction of transgenic adult male and female fish towards TCDD treatment, suggesting that the sex hormones did not significantly affect the transgenic responses towards TCDD.

Spatial induction of GFP expression in Tg(cyp1a:gfp) medaka

A major advantage of Tg(cyp1a:gfp) medaka over cell-based transgenic models is the direct detection of tissue-distribution of induced GFP expression. Interestingly, induction of GFP expression was first observed in the liver, kidney and gut by TCDD (Figure 1I) as well by BAP and 3-MC (Figure 3D,E). In situ hybridization (Figure 2B) of TCDD-treated wild type fry demonstrated similar induction of endogenous cyp1a mRNAs in the liver, kidney and gut (Figure 2B). Consistent with our observation, strong immunohistostaining of Cyp1a has also been observed in the liver, kidney and gut lining after TCDD treatment in zebrafish as well as in other fish species [25,36]. It is generally shown that the liver and kidney are major sites of xenobiotic metabolism and excretion, thus resulting in high level of Cyp1a activity. Due to dietary exposure, accumulation of xenobiotics could occur in the gut and induce Cyp1a [25]. However, in our exposure treatments, the fry were not fed yet but the gut was still a major organ for induced GFP expression. Thus, it is possible that gastrointestinal tract also acts as a major organ for detoxification. Interestingly, in 3-MC treatment, the percentage of fry with GFP induction in the gut was not as high as those in BAP and TCDD treatments, signifying differential organ affinity to different chemicals or differential distribution of different chemicals in different organs.

In addition to GFP induction in the kidney, liver and gut in Tg(cyp1a:gfp) fish, GFP was also induced in epithelial surface such as neuromast cells, olfactory pits, gills and skin. These have been confirmed by in situ hybridization (Figure 2E,H,K). Noticeably, neuromast cells, olfactory pits, gills and skins were organs that were in direct contact with toxicants in water; thus, non-enriched TCDD may directly interact with the AhR pathway while in the metabolic organs including the liver, kidney and gut, TCDD could be enriched and thus achieve higher response. It has also been reported from several fish studies that Cyp1a is present in olfactory pits, gills and skins in AhR-ligands treated fishes by immunostaining [25,36]. To our knowledge, our observation that cyp1a gene is inducible by TCDD in neuromast cells is a novel finding.

Fewer organs showed cyp1a mRNA induction by in situ hybridization analysis than those displayed induced GFP expression in Tg(cyp1a:gfp) fish under the same TCDD chemical treatment. Organs that expressed GFP but not detected by in situ hybridization include blood vessels along the trunk of the body (Figure 2F), caudal fin (Figure 1F4,G4,H4) and heart muscles. The difference between the two assays may reflect a detection sensitivity rather than ectopic GFP induction in the transgenic medaka. Consistent with our argument, it has been previously reported that Cyp1a induction is detected in vascular endothelia, myocardium and caudal fin of treated fish in whole mount and section immunostaining [25,36,37]. Nonetheless, GFP expression in the blood vessels, heart muscles and caudal fin was rather weak in comparison to the intense GFP signal in the major organs such as liver, kidney and gut (Figure 2C); thus, whole mount in situ hybridization is unable to detect the low level of expression.

We also noticed a trend that an increasing number of organs express GFP with increasing TCDD concentration. The order of these organs expressing GFP from low to high concentrations of TCDD is as follows: Kidney and liver>Gut>olfactory pits and caudal fin>gills>neuromast cells (Figure 1). Although the phenotypic expression was difficult to quantify with concentration, the order of organs that expressed GFP can be a rough indicator of relative dosage of TCDD.

Tg(cyp1a:gfp) medaka for biomonitoring assays

We have tested Tg(cyp1a:gfp) medaka with TCDD at three different development stages: pre-hatching embryos, post-hatching fry and adult. In all the three stages, GFP was induced within 24 hours of TCDD exposure. However, the induced GFP expression in early embryos was relatively weak compared to the post-hatching fry and adults. This is likely due to both of the presence of thick chorion in pre-hatching embryos, which may affect the permeability of chemicals, and the weak AhR activity before the metabolic organs become mature. Consistent with this, we did observe stronger GFP expression induced by TCDD in the kidney, liver and gut at 3 dpf when these organs begin to develop (stage 19–26) [17]. Although strong GFP expression was observed in the kidney and olfactory pits in the adult transgenic fish after 24 hours of exposure, due to the presence of the thick body wall, GFP expression within abdomen area could not be sensitively observed unless dissected. This problem could be overcome by inserting the cyp1a:gfp construct in the see-through medaka strain [38], which remains transparent in both embryonic and adult stages, thus allowing fluorescence in internal organs to be observed directly without a need to sacrifice the animal. Currently, larvae stage remains the best stage for biomonitoring due to its complete and functional organs development and also the ease of detection in its transparent body at this stage.

Based on the data presented here, it is feasible to develop two biomonitoring systems using the Tg(cyp1a:gfp) medaka. One is to use the newly hatched fry as they offer maximal sensitivity to TCDD induction and the assay can be performed for at least 72 hours without feeding since the newly hatched fry can rely on their yolk reserve in the first few days after hatching. The assay can be used for large scale of chemical screening for xenobiotic activity because of the easy availability of a large number of fry. Another assay is based on the adult and it can be developed into both laboratory-based and on-site monitoring systems for both acute and chronic exposure. For on-site monitoring, medaka offer additional advantages as medaka fish are very hardy and can adapt in a broad salinity (up to 50% sea water) and a wide range of temperature from near 4°C to 40°C [15,39]. Thus, Tg(cyp1a:gfp) transgenic medaka can be used in broad environmental conditions compared to most other fish species including the popular zebrafish model.

Supporting Information

Figure S1   Whole mount _in situ_ hybridization with _cyp1a_ sense probe in newly hatched fry in 0.1% DMSO vehicle solvent control (A) or 5 nM TCDD (B).

Figure S2   Weak GFP induction by _Tg(cyp1a:gfp)_ fry by 4-nitrophenol. (A) Lack of detectable GFP expression in a fry in the control (egg water) group. (B) Spotty GFP expression in the liver from a fry in the 4-nitrophenol treatment group. (C) Percentages of fry expressing GFP in the liver induced by 4-nitrophenol. Abbreviation: lv, liver.

Figure S3   Transient and consitutive GFP expression in the mid-brain of *Tg(cyp1a:gfp)* embryos at 1 dpf. (A) Brightfield view. (B) Fluorescent view. (C) Merged view. The GFP signal is diminished by 2 dpf. White arrowheads point to the positions of GFP expression and astrisks denote eyes.

Author Contributions

Conceived and designed the experiments: GHBN ZG. Performed the experiments: GHBN. Analyzed the data: GHBN ZG. Contributed reagents/materials/analysis tools: GHBN ZG. Wrote the paper: GHBN ZG.

References

1. Weber R, Gaus C, Tysklind M, Johnston P, Forter M, et al. (2008) Dioxin- and POP-contaminated sites–contemporary and future relevance and challenges: overview on background, aims and scope of the series. Environ Sci Pollut Res Int 15: 363–393.
2. Marinković N, Pašalić D, Ferenčak G, Gršković B, Stavljenić Rukavina A, et al. (2010) Dioxins and human toxicity. Arh Hig Rada Toksikol 61: 445–453.
3. King-Heiden TC, Mehta V, Xiong KM, Lanham KA, Antkiewicz DS, et al. (2012) Reproductive and developmental toxicity of dioxin in fish. Mol Cell Endocrinol 354: 121–138.
4. Srogi K (2007) Monitoring of environmental exposure to polycyclic aromatic hydrocarbons: a review. Environ Chem Lett 5: 169–195.
5. Zhou H, Wu H, Liao C, Diao X, Zhen J, et al. (2010) Toxicology mechanism of the persistent organic pollutants (POPs) in fish through AhR pathway. Toxicology mechanisms and methods 20: 279–286.
6. Ma Q, Lu AYH (2007) CYP1A induction and human risk assessment: an evolving tale of in vitro and in vivo studies. Drug Metab Dispos 35: 1009–1016.
7. Whyte J, Jung R (2000) Ethoxyresorufin-O-deethylase (EROD) activity in fish as a biomarker of chemical exposure. Crit Rev Toxicol 30: 347–570.
8. Kinoshita M, Murata K, Naruse K, Tanaka M (2009) Medaka: Biology, Management and Experimental protocols. Iowa: Wiley-Blackwell. 444 p.
9. Lele Z, Krone PH (1996) The zebrafish as a model system in developmental, toxicological and transgenic research. Biotechnol Adv 14: 57–72.
10. Carvan MJ, Dalton TP, Stuart GW, Nebert DW (2000) Transgenic zebrafish as sentinels for aquatic pollution. Ann N Y Acad Sci 919: 133–147.
11. Legler J, Broekhof J, Brouwer A, Lanser P, Murk A, et al. (2000) A Novel in Vivo Bioassay for (Xeno-)estrogens Using Transgenic Zebrafish. Environ Sci Technol 34: 4439–4444.
12. Blechinger SR, Warren JT, Kuwada JY, Krone PH (2002) Developmental toxicology of cadmium in living embryos of a stable transgenic zebrafish line. Environ Health Perspect 110: 1041–1046.
13. Kurauchi K, Nakaguchi Y, Tsutsumi M, Hori H, Kurihara R, et al. (2005) In vivo visual reporter system for detection of estrogen-like substances by transgenic medaka. Environ Sci Technol 39: 2762–2768.
14. Zeng Z, Shan T, Tong Y, Lam SH, Gong Z (2005) Development of estrogen-responsive transgenic medaka for environmental monitoring of endocrine disrupters. Environ Sci Technol 39: 9001–9008.
15. Wu YL, Pan X, Mudumana SP, Wang H, Kee PW, et al. (2008) Development of a heat shock inducible gfp transgenic zebrafish line by using the zebrafish hsp27 promoter. Gene 408: 85–94.
16. Sasado T, Tanaka M, Kobayashi K, Sato T, Sakaizumi M, et al. (2010) The National BioResource Project Medaka (NBRP Medaka): an integrated bioresource for biological and biomedical sciences. Exp Anim 59: 13–23.
17. Iwamatsu T (2004) Stages of normal development in the medaka *Oryzias latipes*. Mech Dev 121: 605–618.
18. Ng GHB, Gong Z (2011) Maize Ac/Ds transposon system leads to highly efficient germline transmission of transgenes in medaka (*Oryzias latipes*). Biochimie 93: 1858–1864.
19. Kasahara M, Naruse K, Sasaki S, Nakatani Y, Qu W, et al. (2007) The medaka draft genome and insights into vertebrate genome evolution. Nature 447: 714–719.
20. Mandal PK (2005) Dioxin: a review of its environmental effects and its aryl hydrocarbon receptor biology. J Comp Physiol [B] 175: 221–230.
21. Handley-Goldstone HM, Grow MW, Stegeman JJ (2005) Cardiovascular gene expression profiles of dioxin exposure in zebrafish embryos. Toxicol Sci 85: 683–693.
22. Sanderson JT, Aarts JM, Brouwer A, Froese KL, Denison MS, et al. (1996) Comparison of Ah receptor-mediated luciferase and ethoxyresorufin-O-deethylase induction in H4IIE cells: implications for their use as bioanalytical tools for the detection of polyhalogenated aromatic hydrocarbons. Toxicol Appl Pharmacol 137: 316–325.
23. Zhou B, Liu C, Wang J, Lam PK, Wu RS, et al. (2006) Primary cultured cells as sensitive in vitro model for assessment of toxicants–comparison to hepatocytes and gill epithelia. Aquat Toxicol 80: 109–118.
24. Elskens M, Baston DS, Stumpf C, Haedrich J, Keupers I, et al. (2011) CALUX measurements: statistical inferences for the dose-response curve. Talanta 85: 1966–1973.
25. Sarasquete C, Segner H (2000) Cytochrome P4501A (CYP1A) in teleostean fishes. A review of immunohistochemical studies. Sci Total Environ 247: 313–332.
26. Hahn ME, Woodward BL, Stegeman JJ, Kennedy SW (1996) Rapid assessment of induced cytochrome P4501 a protein and catalytic activity in fish hepatoma cells grown in multiwell plates: Response to TCDD, TCDF, and two planar PCBS. 15: 582–591.
27. Hahn ME (2002) Biomarkers and bioassays for detecting dioxin-like compounds in the marine environment. Sci Total Environ 289: 49–69.
28. Noury P, Geffard O, Tutundjian R, Garric J (2006) Non destructive in vivo measurement of ethoxyresorufin biotransformation by zebrafish prolarva: development and application. Environ Toxicol 21: 324–331.
29. Aarts JM, Denison MS, Cox MA, Schalk MA, Garrison PM, et al. (1995) Species-specific antagonism of Ah receptor action by 2,2′,5,5′-tetrachloro- and 2,2′,3,3′4,4′-hexachlorobiphenyl. Eur J Pharmacol 293: 463–474.
30. Nagy SR, Sanborn JR, Hammock BD, Denison MS (2002) Development of a green fluorescent protein-based cell bioassay for the rapid and inexpensive detection and characterization of ah receptor agonists. Toxicol Sci 65: 200–210.
31. Bugiak B, Weber LP (2009) Hepatic and vascular mRNA expression in adult zebrafish (Danio rerio) following exposure to benzo-a-pyrene and 2,3,7,8-tetrachlorodibenzo-p-dioxin. Aquat Toxicol 95: 299–306.
32. Monostory K, Pascussi JM, Kóbori L, Dvorak Z (2009) Hormonal regulation of CYP1A expression. Drug Metab Rev 41: 547–572.
33. Swedenborg E, Pongratz I (2010) AhR and ARNT modulate ER signalling. Toxicol 268: 132–138.
34. Bussmann UA, Sáez JMP, Bussmann LE, Baranao JL (2013) Aryl hydrocarbon receptor activation leads to impairment of estrogen-driven chicken vitellogenin promoter activity in LMH cells. Comp Biochem Phys C 157: 111–158.
35. Yan ZH, Lu GH, He JJ (2012) Reciprocal inhibiting interactive mechanism between the estrogen receptor and aryl hydrocarbon receptor signaling pathways in goldfish (Carassius auratus) exposed to 17 beta-estradiol and benzo[a]pyrene. Comp Biochem Phys C 156: 17–23.
36. Zodrow JM, Stegeman JJ, Tanguay RL (2004) Histological analysis of acute toxicity of 2,3,7,8-tetrachlorodibenzo-p-dioxin (TCDD) in zebrafish. Aquat Toxicol 66: 25–38.
37. Andreasen EA, Spitsbergen JM, Tanguay RL, Stegeman JJ, Heideman W, et al. (2002) Tissue-specific expression of AHR2, ARNT2, and CYP1A in zebrafish embryos and larvae: effects of developmental stage and 2,3,7,8-tetrachlorodibenzo-p-dioxin exposure. Toxicol In Vitro 68: 403–419.
38. Wakamatsu Y, Pristyazhnyuk S, Kinoshita M, Tanaka M, Ozato K, et al. (2001) The see-through medaka: a fish model that is transparent throughout life. Proc Natl Acad Sci U S A 98: 10046–10050.
39. Inoue K, Takei Y (2003) Asian medaka fishes offer new models for studying mechanisms of seawater adaptation. Comparative Biochemistry and Physiology Part B: Biochemistry and Molecular Biology 136: 635–645.

17

Synergistic Removal of Pb(II), Cd(II) and Humic Acid by Fe₃O₄@Mesoporous Silica-Graphene Oxide Composites

Yilong Wang[1], Song Liang[2], Bingdi Chen[1], Fangfang Guo[1], Shuili Yu[2], Yulin Tang[2]*

1 The Institute for Biomedical Engineering and Nano Science, Tongji University School of Medicine, Shanghai, P. R. China, **2** State Key Laboratory of Pollution Control and Resource Reuse, College of Environmental Science and Engineering, Tongji University, Shanghai, P. R. China

Abstract

The synergistic adsorption of heavy metal ions and humic acid can be very challenging. This is largely because of their competitive adsorption onto most adsorbent materials. Hierarchically structured composites containing polyethylenimine-modified magnetic mesoporous silica and graphene oxide (MMSP-GO) were here prepared to address this. Magnetic mesoporous silica microspheres were synthesized and functionalized with PEI molecules, providing many amine groups for chemical conjugation with the carboxyl groups on GO sheets and enhanced the affinity between the pollutants and the mesoporous silica. The features of the composites were characterized using TEM, SEM, TGA, DLS, and VSM measurements. Series adsorption results proved that this system was suitable for simultaneous and efficient removal of heavy metal ions and humic acid using MMSP-GO composites as adsorbents. The maximum adsorption capacities of MMSP-GO for Pb(II) and Cd (II) were 333 and 167 mg g^{-1} caculated by Langmuir model, respectively. HA enhances adsorption of heavy metals by MMSP-GO composites due to their interactions in aqueous solutions. The underlying mechanism of synergistic adsorption of heavy metal ions and humic acid were discussed. MMSP-GO composites have shown promise for use as adsorbents in the simultaneous removal of heavy metals and humic acid in wastewater treatment processes.

Editor: Stephen J. Johnson, University of Kansas, United States of America

Funding: This work was supported by National Natural Science Foundation of China (numbers 51003077, 51173135, 21007048), Nano-tech Foundation of Shanghai (11 nm0506100), and the Fundamental Research Funds for the Central Universities. The funders had no role in study design, data collection and analysis, decision to publish, or preparation of the manuscript.

Competing Interests: The authors have declared that no competing interests exist.

* E-mail: tangyulin@tongji.edu.cn

Introduction

Heavy metal ions and humic acid (HA) in underground water pose a severe threat to public health and ecological systems.[1–3] A great deal of effort has been made to develop effective adsorbents with different chemical compositions, microstructures, and surface functionalities for the removal of heavy metal ions and organic pollutants from water.[4–9] In this way, the development of adsorbents with high adsorption capacity, low toxicity, and efficient separation has attracted considerable interest. [10,11].

In the past decades, silica-based mesoporous materials, which are robust inorganic solids, have shown good potential in water treatment applications due to their large specific surface area (200–1500 m^2 g^{-1}) and accessible surface functionalization. [12,13] Mesoporous silica groups containing amino and thiol groups have been extensively used to accumulate the metal ions from aqueous solutions.[14–20] Recently, magnetic nanoparticles have been introduced in mesoporous silica adsorbents, contributing to direct enhancement of the adsorption of metal ions and possible recovery by magnetic separation [21–23].

Recently, the development of bifunctionalized adsorbents capable of enabling simultaneous removal of two kinds of environmental pollutants has become an emerging frontier in the field of water treatment. [24] However, synergistic adsorption of heavy metal ions and humic acid molecules is still a challenge because of their competitive adsorption onto most adsorbents.[25–27] To the best of our knowledge, there have been few studies that have evaluated the use of functionalized magnetic mesoporous silica and graphene oxide in the synergistic adsorption of two kind of pollutants. The high surface-to-volume ratio and surface functionalities of the graphene oxide (GO) and PEI-grafted magnetic mesoporous silica (MMSP) microspheres can be both used to this effect. The addition of GO can further improve the colloidal stability of the MMSP.

Herein, we report a facile method for preparation of magnetic mesoporous silica-graphene oxide (MMSP-GO) composites and describe our investigation of the synergistic adsorption of HA and heavy metal ions, specifically Pb(II) and Cd(II). First, the magnetic mesoporous silica microspheres were synthesized and functionalized with PEI molecules, followed by chemical conjugation with the carboxyl groups on GO sheets. The functionalization of the mesoporous silica microspheres containing PEI molecules was found to both facilitate the adsorption of HA and provide sites for chemical reaction with graphene oxide sheets. The adsorption isotherm of Pb(II) and Cd(II) by MMSP-GO composites were studied. And the underlying mechanism of synergistic adsorption of heavy metal ions and humic acid were discussed. The magnetic cores of MMSP microspheres were also embedded to facilitate easy separation using magnetic fields.

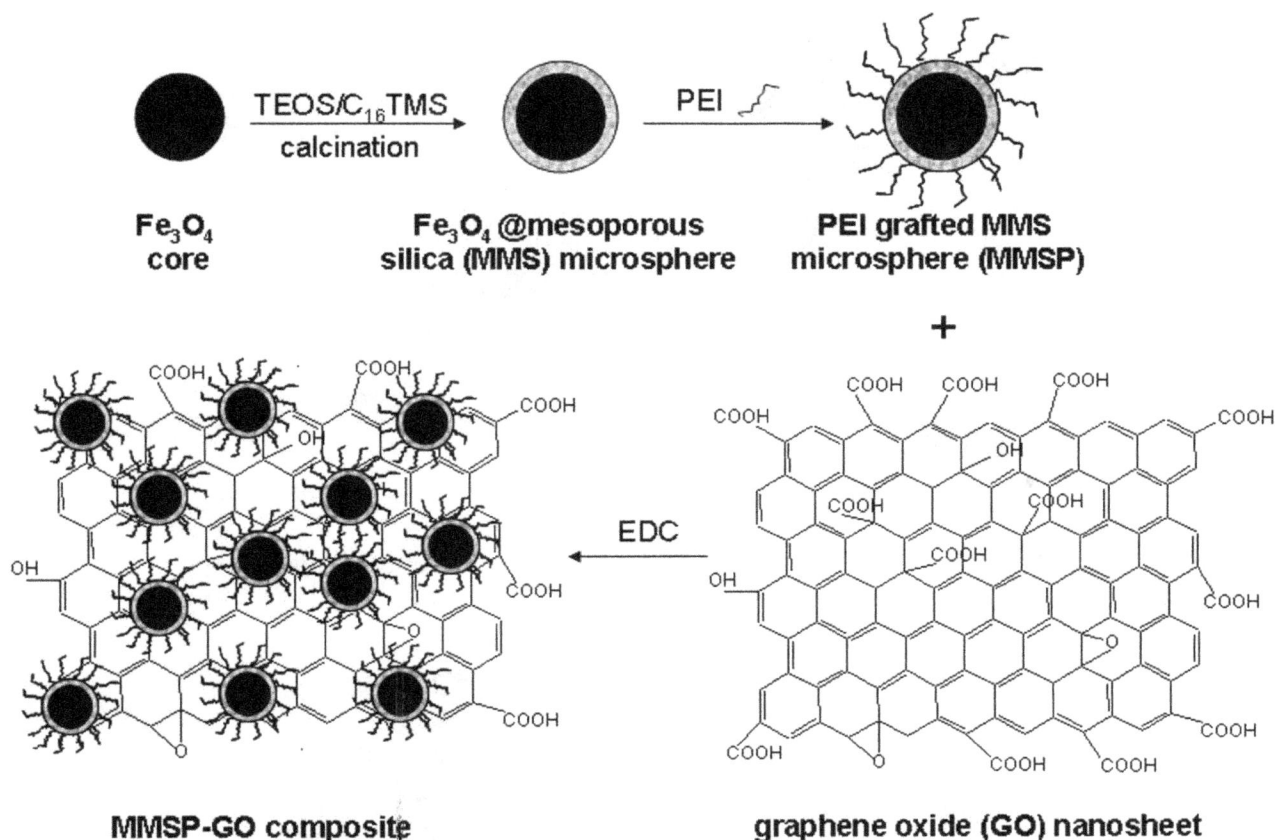

Figure 1. Scheme of the preparation pathway. Schematic illustration of pathway for preparation of polyethylenimine-modified magnetic mesoporous silica and graphene oxide (MMSP-GO) composites.

Experimental procedures

2.1. Materials

Pb(NO$_3$)$_2$, CdCl$_2 \cdot 2.5$H$_2$O, iron (III) chloride hydrate (FeCl$_3 \cdot 6$H$_2$O), sodium acetate, ethylene glycol, ammonium hydroxide (NH$_4$OH, 28 wt%), and hydrochloric acid (37 wt% aqueous solution) were purchased from Shanghai Reagent Company (China). Tetraethoxysilane was purchased from Sigma-Aldrich (U.S.). Hexadecyltrimethoxysilane (C$_{16}$TMS) and polyethyleneimide (PEI, 99%, Mw = 1800) were purchased from Alfa Aesar (U.S.). Humic acid was purchased from Aldrich Chemical Company (China) and treated using the method described in our previous experiment. [28] Graphene oxide was purchased from Nanjing XF Nano Company (China). All these reagents were used without further purification. The deionized (DI) water used for the preparation of reagents was purified by Millipore reverse osmosis (RO).

2.2. Instruments

The TEM samples were dispersed in DI water and dried onto carbon-coated copper grids before examination. Transmission electron microscope (TEM) images were obtained with a Philips Tecnai 20 transmission electron microscope. Scanning electron microscope (SEM) images were taken using a JEOL SM4800 scanning electron microscope. FTIR spectra were obtained with a Bruker Tensor 27. The TGA data was obtained with a thermogravimetric analyzer (NETZSCH TG209) conducted under nitrogen atmosphere from ambient to 1173 K with the

rate of heating at 283 K min^{-1}. The magnetic properties were measured at room temperature using a Vibrating Sample Magnetometer (VSM 7407, Lake Shore, USA). The surface zeta potentials of the microspheres were measured using a DLS Particle Size analyzer (Zetasizer Nano-ZS, Malvern, UK.).

The concentrations of total Cd (II) and Pb (II) were determined by an ICP-AES (ICP-optima 2001DV, Perkin-Elmer, USA). Humic acid was determined using a UV-vis spectrophotometer at 254 nm wavelength. The pH of the solutions was measured using a pH meter (pHS-3C model, Leici, China).

2.3. Preparation of MMSP-GO Composites

2.3.1. Synthesis of PEI-modified magnetic mesoporous silica microspheres. *Synthesis of Fe$_3$O$_4$ microspheres.* The magnetic microspheres were prepared using a solvothermal reaction. [29] Then 2.16 g of sodium acetate and 0.86 g of FeCl$_3 \cdot 6$H$_2$O were dissolved in 30 mL of ethylene glycol under vigorous magnetic stirring. The homogeneous yellow solution was transferred to the Teflon-lined stainless-steel autoclave and heated to 200°C for nearly 8 h. After that, the autoclave was cooled to room temperature. The products were washed three times with ethanol and DI water and redispersed into DI water for use.

Synthesis of core-shell structure magnetic mesoporous silica composite microspheres. First, 4 mL of Fe$_3$O$_4$ aqueous dispersion (about 100 mg mL^{-1}) was treated with 0.1 M HCl aqueous solution under sonication for 20 min. The treated Fe$_3$O$_4$ microspheres were dispersed in the mixture of ethanol and water (v/v = 70/30) for a while. Ammonium hydroxide was added to adjust pH value

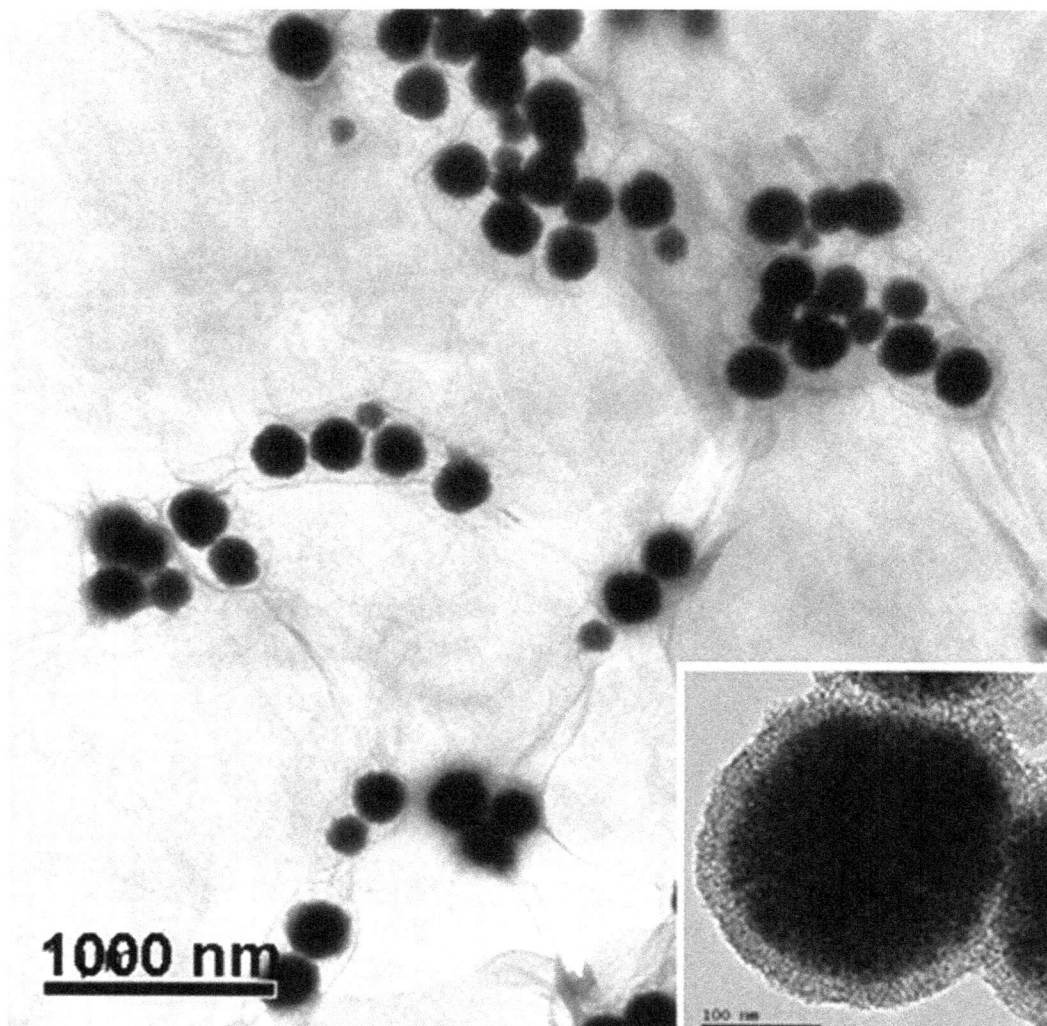

Figure 2. TEM images of the composites. TEM image of polyethylenimine-modified magnetic mesoporous silica and graphene oxide (MMSP-GO) composites (inset is the enlarged TEM image of an individual Fe_3O_4@mesoporous silica microsphere on the GO sheet).

of the dispersion to about 9.5. Then 0.12 mL of TEOS was poured into the reactor with vigorous stirring at room temperature. The reaction lasted for about 24 h. The extra reactants were washed four times with ethanol and removed from Fe_3O_4@$SiO2$ microspheres. Then 150 mL of the mixture of ethanol and DI water (v/v = 88/12) was treated with 7.2 mL of ammonium hydroxide. Then 5 mL of ethanol solution of TEOS and $C_{16}TMS$ (molar ratio of 4.7:1) was dropped into the reaction system at the rate of one drop every 10 seconds. The mixture was stirred at constant temperature for 6 h and then the reaction was stopped. The product was washed twice with ethanol and DI water and then calcinated at 550°C for 6 h, producing magnetic mesoporous silica composite microspheres.

Surface modification of magnetic mesoporous silica composite microspheres with PEI molecules. In 35 mL of DI water, 20 mg of as-synthesized magnetic mesoporous silica microspheres was dispersed under sonication. Then 30 mg of PEI (Fw = 1800) was added to the aqueous dispersion with vigorous mechanical stirring and sonication. After 30 min, the resulting MMSP were washed three times with DI water.

2.3.2. Synthesis of MMSP-GO composites. In a 50 mL centrifuge tube, 20 mg of GO sheets were dispersed in 30 mL of DI water with 1 h sonication. Into this yellow-brown homogeneous dispersion, 0.080 mL of newly-produced EDC aqueous solution at a concentration of 4.0 mg mL^{-1} was added. This mixture was vortexed for 8 min. Then 2 mg of PEI functionalized magnetic mesoporous silica microspheres were introduced. The reaction lasted for 2.5 h in an end-over-end reactor and the final products were purified by washing three times with DI water, producing MMSP-GO composites.

2.4. Adsorption of Pb(II)/Cd(II) and HA

Batch experiments were employed to evaluate Pb(II) and Cd(II) or HA adsorption characteristics under various conditions. All the adsorption experiments were conducted at room temperature (25±1°C). The average values of triplicate measurements were reported and all standard errors were smaller than 5%.

The batch adsorption procedure consisted of (a) distributing 1.0 mg of MMSP-GO or MMSP through 10 mL water solution containing selected concentrations of Pb(II) and Cd(II) in a series of 20 mL glass tubes; (b) adjusting the pH to 2.0–9.0 using 0.1 M

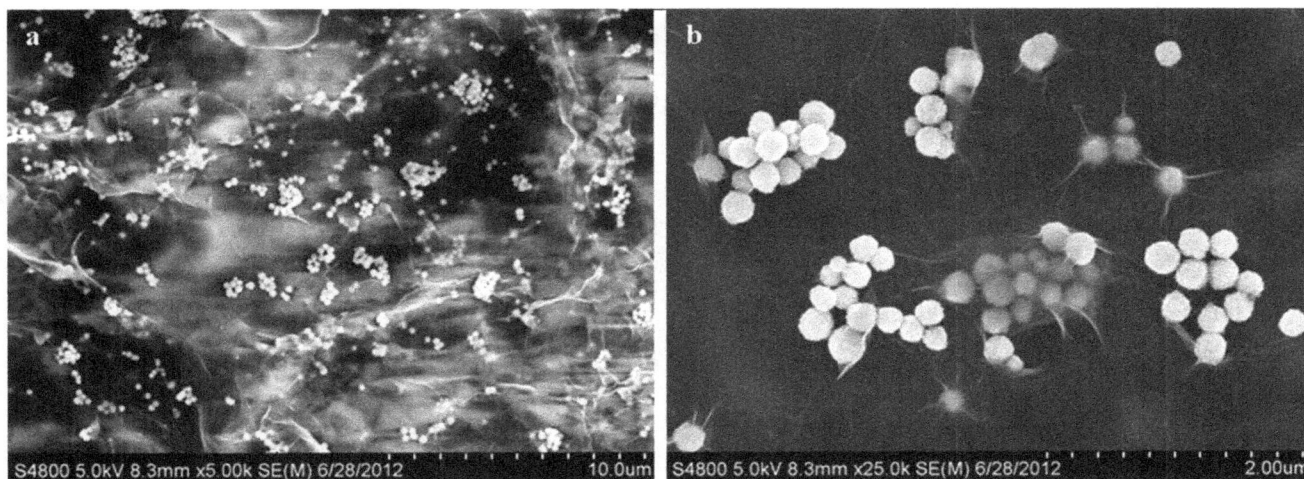

Figure 3. SEM images of the composites. SEM images of polyethylenimine-modified magnetic mesoporous silica and graphene oxide (MMSP-GO) composites.

stock of HNO_3 and NaOH solution; (c) adding HA stock solution to each of the tubes to specific preselected concentrations; (d) sealing and shaking all tubes in an incubator at 150 rpm; (e) separating the suspension using an external magnetic field; (f) detection of the residual Pb(II) and Cd(II) concentrations with ICP-AES and detection of the residual concentration of HA with an UV-vis spectrophotometer; (g) calculation of the amounts of heavy metal and HA adsorbed using Eq. (1):

$$q_t = \frac{(C_0 - C_t)V}{M} \qquad (1)$$

Here q_t (mg/g) is the amount of heavy metal or HA adsorbed at time t, C_0 (mg L^{-1}) is the initial heavy metal or HA concentration. C_t (mg L^{-1}) is the heavy metal or HA concentration at time t. V (mL) is the volume of heavy metal or HA solution, and M (g) is the adsorbent mass.

Results and Discussion

3.1. Material Characterizations

The typical synthesis of the MMSP-GO composites is shown schematically in Figure 1. Fe_3O_4@mesoporous silica (MMS) core-shell microspheres were first prepared using standard methods of mesoporous silica shell formation via a sol-gel process based on the magnetic microspheres obtained through a solvothermal reaction. [29] $C_{16}TMS$, a coupling agent, was used as the porogen during the silica coating on the magnetic cores, which could be easily removed by calcination. Then monodispersed MMS core-shell microspheres with regular mesoporous silica shells were produced. In order to connect the magnetic mesoporous composites to the graphene oxide sheets, PEI branched molecules were used to modify the surface of the mesoporous microspheres by electrostatic adsorption. [30] The dramatically change in the surface potential of the mesoporous microspheres from -15 mV to $+37$ mV under the neutral conditions proved that the modification was successful. Next, the formation of MMSP-GO composites was carried out by

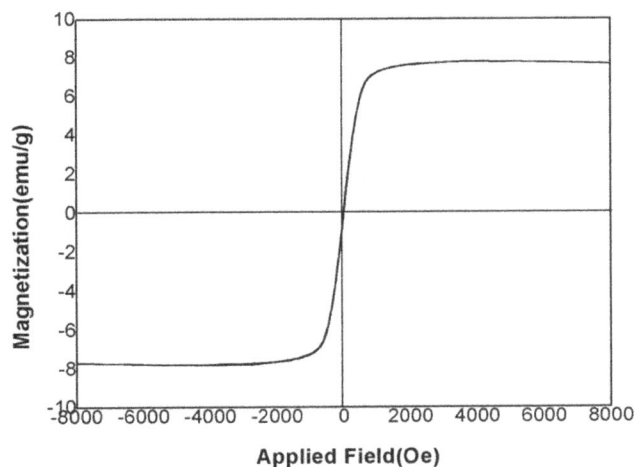

Figure 4. Magnetic property of the composites. Magnetic property of the polyethylenimine-modified magnetic mesoporous silica and graphene oxide (MMSP-GO) composites.

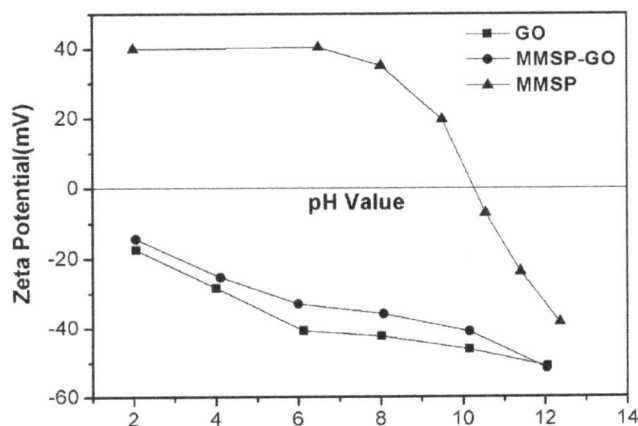

Figure 5. pH-zeta potential relation curves. pH-zeta potential relation curves of polyethylenimine-modified magnetic mesoporous silica and graphene oxide (MMSP-GO) composites, MMSP microspheres, and GO nanosheets, respectively.

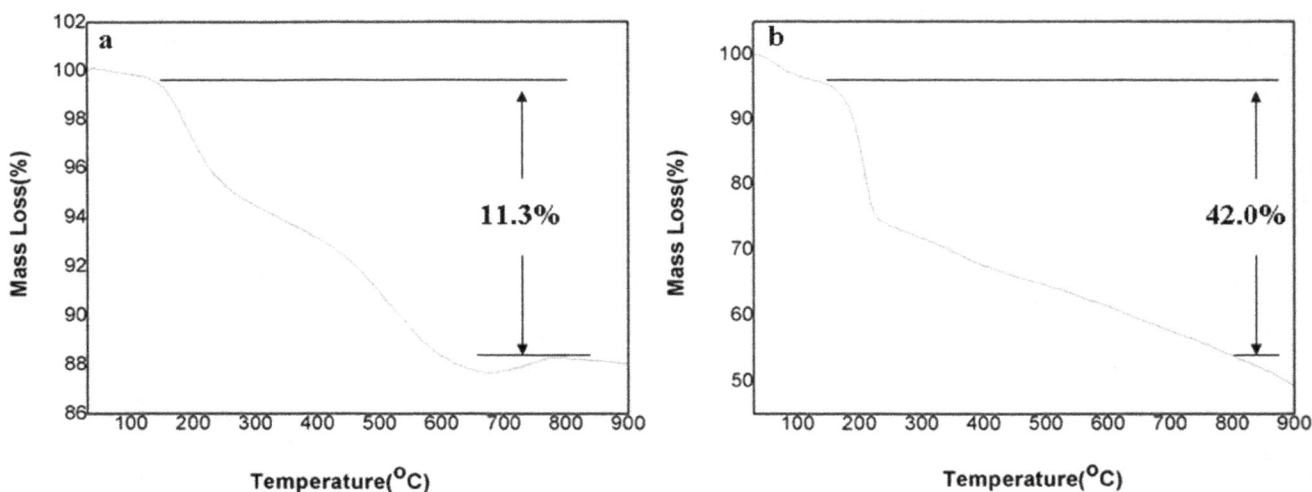

Figure 6. TG analysis curves. TG analysis of (a) polyethylenimine-modified magnetic mesoporous silica and graphene oxide (MMSP-GO) composites and (b) polyethylenimine-modified magnetic mesoporous silica microspheres.

chemical reaction of MMSP and original GO nanosheets in the presence of 1-ethyl-3-(3-dimethyaminopropyl) carbodiimide (EDC) (Figure 1). In this case, PEI molecules played bilateral roles one of them serving as the binding bridge for conjugation and the other to improve the adsorption affinity of the MMSP microspheres to the pollutant matter. [28].

Figure 2 shows a TEM image of MMSP-GO composites. The monodispersed MMSP microspheres were distributed on the GO surface. Some microspheres adhered to each other, leaving most of the GO surface exposed. There was little aggregation or multilayer accumulation of the MMSP microspheres, implying good connections between the microspheres and the GO sheets. The average diameter of the MMSP microspheres was 260 nm. The microstructures of the MMSP microspheres were shown in the inset of Figure 2. There was an obvious mesoporous silica shell on the Fe_3O_4 core. The loading density of MMSP microspheres on

the GO surface could be well controlled maintaining the balance between strong adsorption properties and magnetic resonance.

Figure 3 shows SEM images of MMSP-GO composites at different magnifications. In Figure 3a, the low-magnification SEM image indicates that the MMSP microspheres were distributed homogeneously across the whole surface of the GO. As shown, the MMSP microspheres were firmly anchored on both sides of the wrinkled GO nanosheets (Figure 3b). These GO layers may play a critical role in preventing MMSP microspheres from aggregating in solution.

Figure 4 gives magnetic hysteresis loops for MMSP-GO composites measured at 300 K. The profile of the magnetization curve of the MMSP-GO composites showed their typical superparamagnetic properties. The saturation magnetization of the MMSP-GO was 7.5 emu g^{-1}, which ensured effective magnetic separation after adsorption equilibrium.

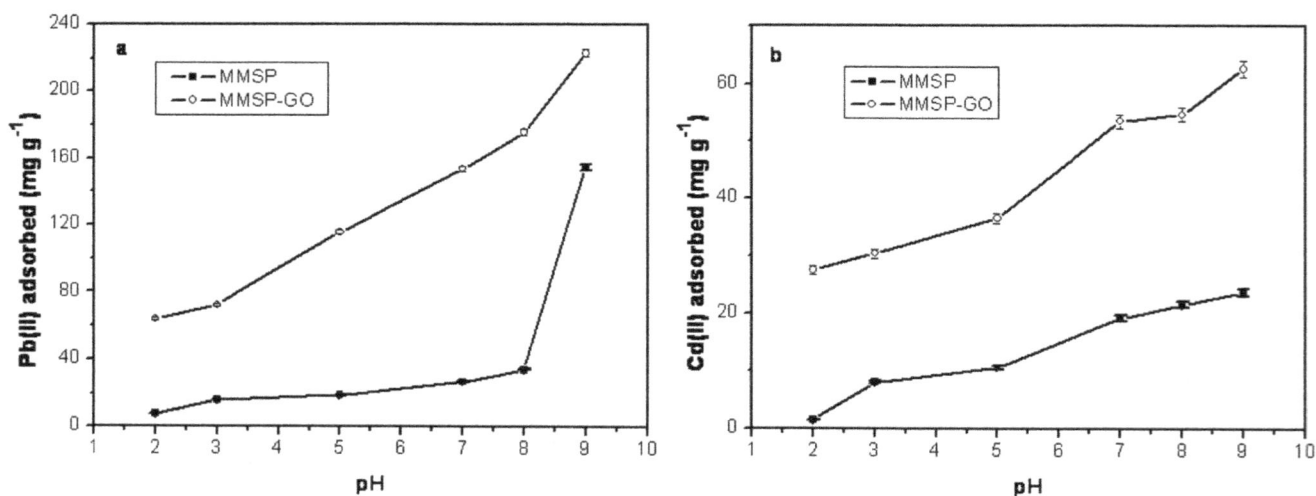

Figure 7. Comparison of adsorptive capacities of heavy metal ions by different materials. a. Effects of pH on adsorption of Pb(II) by polyethylenimine-modified magnetic mesoporous silica (MMSP) and polyethylenimine-modified magnetic mesoporous silica and graphene oxide (MMSP-GO) (Initial concentrations: 20 mg L^{-1}; adsorbent loading: 100 mg L^{-1}); b. Effects of pH on adsorption of Cd(II) by MMSP and MMSP-GO (Initial concentrations: 20 mg L^{-1}; adsorbent loading: 100 mg L^{-1}).

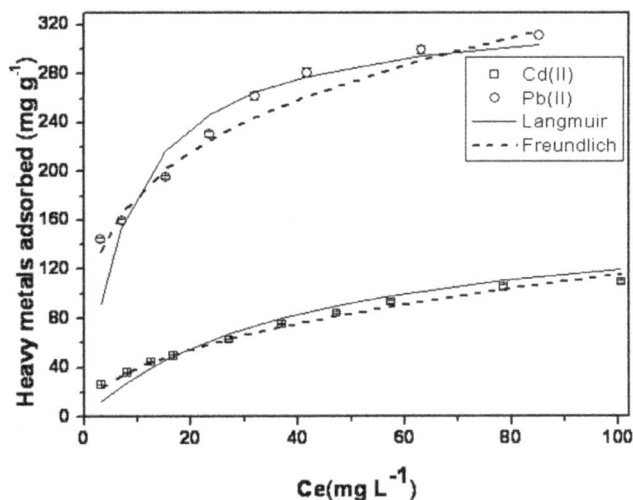

Figure 8. Pb(II) and Cd(II) adsorption isotherms on MMSP-GO. Pb(II) and Cd(II) adsorption isotherms on MMSP-GO (adsorbent loading: 100 mg L^{-1}; pH: 7.1; contact time: 24 h). The solid lines are Langmuir model simulation, and the dotted lines are Freundlich model simulation.

MMSP microspheres within a given temperature range was about 11.3% due to the removal of PEI molecules. The weight loss of GO sheets in this range due to the groups of COOH, OH, C = O, and epoxide was about 44.5%. [27] The combined analysis of these TGA results of the composites with each component indicated that the mass partition of the MMSP microspheres was about 7.5%, which was consistent with the reaction stoichiometry.

3.2. Investigation of Adsorption of Pb(II)/Cd(II) by MMSP-GO or MMSP

We performed batch adsorption experiments with Pb(II) and Cd(II) using MMSP-GO composites and MMSP microspheres. The adsorption capacities of Pb(II) and Cd(II) both increased as pH value increased, peaking at pH 9.0. The same trend was observed for each heavy metal ion. The adsorptive property of MMSP-GO composites was better than that of MMSP microspheres. This was because the GO nanosheets had a stronger absorptive ability than MMSP microspheres (Figure S1). For Pb(II) removal, the maximum adsorption at an equilibrium between MMSP-GO composites and MMSP microspheres was 220 and 152 mg g^{-1}, respectively, at an initial metal ion concentration of 20 mg L^{-1}. For Cd(II) removal, the maximum adsorption at equilibrium of MMSP-GO composites and MMSP microspheres was 65 and 23 mg g^{-1}, respectively (Figure 7). All results indicated that the composites containing GO were able to adsorb more heavy metal ions than those without GO. By comparison, the adsorptive capacity of Pb(II) was higher than that of Cd(II), which resulted from the increased utilization of amino groups on MMSP. [31] At the same time, GO has a greater saturation adsorptive capacity for Pb(II) than Cd(II). [32] It was reported that electrostatic interaction maybe the predominant driven force for adsorption of heavy metal ions by GO materials. [27] In this case, majority of Pb(II) and Cd(II) were adsorbed by GO sheets due to the electrostatic interaction between the GO with negative surface charges and cations in the broad pH range.

3.3. Adsorption Isotherms

Adsorption isotherm is important for determining the adsorption behavior of an adsorbent. Pb(II) and Cd(II) adsorption isotherm of MMSP-GO were investigated at pH 7.1. The

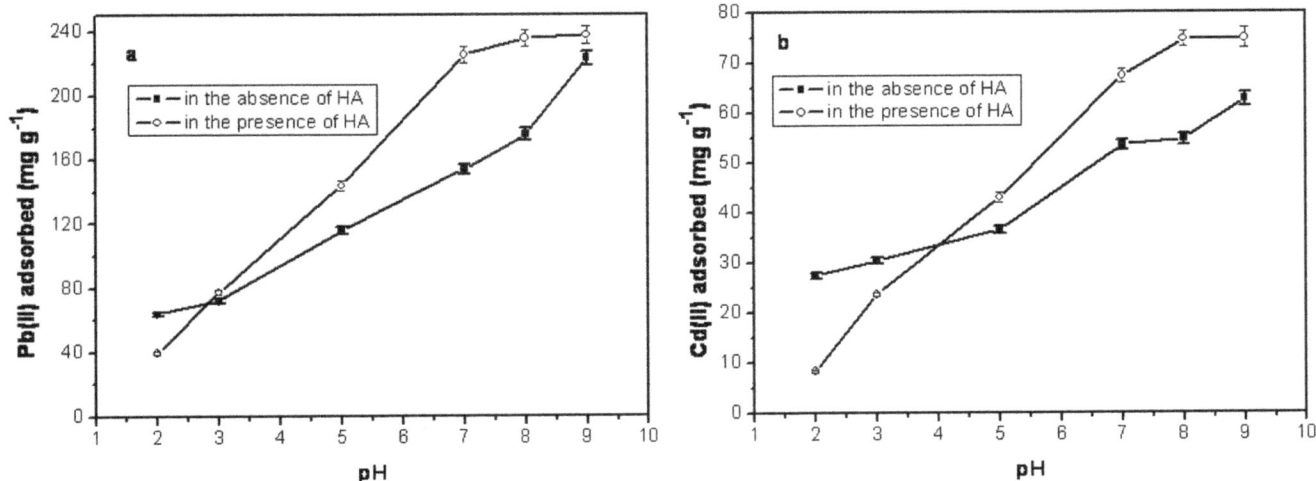

Zeta potentials of MMSP-GO composites and corresponding two components of GO and MMSP as a function of solution pH are shown in Figure 5. The point of zero charge (pH$_{pzc}$) of MMSP was 10.4 because of the existence of large numbers of PEI molecules. Points of zero charge (pH$_{pzc}$) of MMSP-GO composites and GO nanosheets approached 1.0. This was because both of them had many oxygen-containing groups on their surfaces. This showed that a majority of the functional groups of GO were still present after the formation of the composites.

To calculate the composition partition of microspheres to GO nanosheets, we performed TGA analysis of the MMSP-GO composites and MMSP microspheres (Figure 6). The big mass loss of the composites (42.0%) at temperature between 150 to 800°C was due to removal of oxygen-containing functional groups of GO and PEI molecules on MSP microspheres. The loss of mass of

Figure 9. Effects of HA on the adsorption of heavy metal ions by the composites. Effects of HA on the adsorption of (a) Pb(II) and (b) Cd(II) on polyethylenimine-modified magnetic mesoporous silica and graphene oxide (MMSP-GO) composites.

maximum Pb(II) and Cd(II) adsorption amounts on MMSP-GO within the tested concentration range in Figure 8. The equilibrium data were fitted by Langmuir and Freundlich model.

$$\text{Langmuir}: \frac{C_e}{q_e} = \frac{C_e}{q_{max}} + \frac{1}{q_{max}b} \qquad (2)$$

$$\text{Freundlich}: \log q_e = \frac{\log C_e}{n} + \log k_f \qquad (3)$$

where, q_{max} (mg g^{-1}) is the theoretical maximum heavy metals adsorption amounts, q_e (mg g^{-1}) is the equilibrium adsorption amount at heavy metals equilibrium concentration C_e (mg L^{-1}), where K_f is Freundlich coefficient characteristic of the adsorption affinity of the adsorbent, and n is the linearity index. The validity of isotherm models used in this study is assessed by correlation coefficient (R^2). The fitted results of all isotherm models were investigated in this study are presented in the Table S1. The results showed that Langmuir model with R^2 higher than 0.99 fitted better than Freundlich model, indicating that Pb(II) and Cd(II) adsorption on MMSP-GO can be considered to be a monolayer adsorption process. The abundant oxygen-containing functional groups on the surface of graphene oxide made the adjacent oxygen atoms available to bind metal ions [27]. At the same time, the amino groups of MMSP had a strong affinity towards metal ions, and the possible adsorption mechanism could be reasoned out by the coordinate interactions. [33].

3.4. Adsorption of Pb(II)/Cd(II) by MMSP-GO Influenced by HA

We also investigated the influence of HA at an initial concentration of 10 mg L^{-1} on the adsorption of heavy metal ions onto MMSP-GO composites within a pH range of 2.0 to 9.0. For both types of metal ions of Pb(II) and Cd(II), the adsorption capacity of the ion in the presence of HA was greater than in the absence of HA for MMSP-GO composites after a certain pH value (Figure 9). The two curves crossed at about pH 3.0 for Pb(II) and about pH 4.0 for Cd(II). Under pH neutral conditions, about 72 mg g^{-1} more Pb(II) was adsorbed in the presence of HA than in its absence, and about 14 mg g^{-1} more Cd(II) was adsorbed. The amounts HA adsorbed by MMSP-GO composites in the presence of Pb(II) or Cd(II) were about 83 and 87 mg g^{-1}, respectively, under neutral conditions. The adsorption capacity of HA molecules remained roughly the same no matter which heavy metal ions were present. This implied that there was a specific site for HA adsorption on the MMSP-GO and that this site was not influenced by metal ion adsorption. We reported previously that PEI-modified MMSP microspheres could adsorb HA molecules efficiently across a wide pH range. [28] Nanoparticles coated with HA showed enhanced adsorption of heavy metal ions, such as Pb(II).[34–37] As with GO nanosheets, at pH>7.0, adsorption of Cd(II)/Pb(II) was hindered by the presence of 10 mg L^{-1} HA in the solution because the HA would bind to the adsorbent, competing with the metal ions. [27] On the contrary, the enhanced adsorption of the metal ions onto MMSP-GO composites in the presence of HA molecules may be attributed to the joint effects of the PEI-modified MMSP microspheres and the GO sheet. When HA is not present during adsorption, the metal ions may have been mainly adsorbed onto GO nanosheets, accompanied by only some adsorption onto MMSP. In contrast, in the presence of HA, most of the HA was trapped by the MMSP microspheres, and so metal ions were able to interact with the HA

on the composite surface. Moreover, adsorption of HA molecules could neutralize the positive surface charge of the MMSP and enhance adsorption of heavy metal ions. In this way, in addition to the adsorption of metal ions onto the GO surface, extra metal ions could be adsorbed onto MMSP microspheres mediated by HA on their surfaces. We examined the sorption of HA by the composites through FTIR measurements. Description of the major transmittance bands in FTIR spectra of HA, MMSP-GO composites, MMSP-GO/HA/Pb^{2+} and MMSP-GO/HA/Cd^{2+} complex are presented in Figure S2. By comparison with bulk HA and pure MMSP-GO composites, the strengthened peak intensity and occurrence of new peak of bands at 1580, 2700 and 3900 cm^{-1} diminished markedly for both MMSP-GO/HA/Pb^{2+} and MMSP-GO/HA/Cd^{2+}, which could result from the strong interactions of COOH, Phenolic OH, and OH of aliphatic alcohol with MMSP-GO composites. This result proved successful adsorption of HA by the composites.

Conclusions

A facile and reproducible method was developed and used to prepare magnetic mesoporous silica-graphene oxide composites (MMSP-GO) with hierarchical structures and unique properties. These composites were suitable for synergistic adsorption of heavy metal ions and humic acid molecules. The magnetic mesoporous silica microspheres were synthesized and functionalized with PEI molecules, which provided abundant amine groups for chemical conjugation with the carboxyl group on GO sheets and enhanced affinity between pollutants and the mesoporous silica. Adsorption results recorded under different conditions indicated that this experiment was sufficient to prove that this system was capable of simultaneous removal of heavy metal ion and humic acid using MMSP-GO composites as the adsorbent. The underlying mechanism of synergistic adsorption of heavy metal ions and humic acid were discussed. This kind of composite adsorbent was able ot make use of both the large specific surface area and surface functionalities of GO and mesoporous structures. This allowed them to efficiently adsorb the heavy metal ions and HA, and the adsorption of HA onto the composites enhanced the adsorption efficiency of Pb(II) and Cd(II).

Supporting Information

Figure S1 a. Effects of pH on adsorption of Pb(II) by MMSP and GO (Initial concentrations: 20 mg L^{-1}; adsorbent loading: 100 mg L^{-1}); b. Effects of pH on adsorption of Cd(II) by MMSP and GO (Initial concentrations: 20 mg L^{-1}; adsorbent loading: 100 mg L^{-1}).

Figure S2 FTIR spectra of bulk HA (a), MMSP-GO composites (b), MMSP-GO/HA/Pb(II) complex (c), and MMSP-GO/HA/Cd(II) complex (d).

Table S1 Constants and correlation coefficients of Pb(II) and Cd(II) adsorption by langmuir and Freundlich model.

Author Contributions

Conceived and designed the experiments: YLW YLT. Performed the experiments: YLW SL FFG YLT. Analyzed the data: YLW BDC SLY YLT. Contributed reagents/materials/analysis tools: YLW YLT. Wrote the paper: YLW YLT.

References

1. Prelot B, Einhorn V, Marchandeau F, Douillard J-M, Zajac J (2012) Bulk hydrolysis and solid-liquid sorption of heavy metals in multi-component aqueous suspensions containing porous inorganic solids: Are these mechanisms competitive or cooperative? Journal of Colloid and Interface Science 386: 300–306.

2. Zhang XG, Minear RA (2002) Characterization of high molecular weight disinfection byproducts resulting from chlorination of aquatic humic substances. Environmental Science & Technology 36: 4033–4038.

3. Campbell L, Dixon DG, Hecky RE (2003) A Review Of Mercury in Lake Victoria, East Africa: Implications for Human and Ecosystem Health. Journal of Toxicology and Environmental Health, Part B 6: 325–356.

4. Larraza I, López-Gónzalez M, Corrales T, Marcelo G (2012) Hybrid materials: Magnetite-Polyethylenimine-Montmorillonite, as magnetic adsorbents for Cr(VI) water treatment. Journal of Colloid and Interface Science 385: 24–33.

5. Guo J, Chen S, Liu L, Li B, Yang P, et al. (2012) Adsorption of dye from wastewater using chitosan-CTAB modified bentonites. Journal of Colloid and Interface Science 382: 61–66.

6. Hua M, Zhang S, Pan B, Zhang W, Lv L, et al. (2012) Heavy metal removal from water/wastewater by nanosized metal oxides: A review. Journal of Hazardous Materials 211–212: 317–331.

7. Jung JH, Lee JH, Shinkai S (2011) Functionalized magnetic nanoparticles as chemosensors and adsorbents for toxic metal ions in environmental and biological fields. Chemical Society Reviews 40: 4464–4474.

8. Zhang Y, Xu S, Luo Y, Pan S, Ding H, et al. (2011) Synthesis of mesoporous carbon capsules encapsulated with magnetite nanoparticles and their application in wastewater treatment. Journal of Materials Chemistry 21: 3664–3671.

9. Sun DD, Lee PF (2012) TiO_2 microsphere for the removal of humic acid from water: Complex surface adsorption mechanisms. Separation and Purification Technology 91: 30–37.

10. Yokoi T, Kubota Y, Tatsumi T (2012) Amino-functionalized mesoporous silica as base catalyst and adsorbent. Applied Catalysis a-General 421: 14–37.

11. Wang F, Tang Y, Zhang B, Chen B, Wang Y (2012) Preparation of novel magnetic hollow mesoporous silica microspheres and their efficient adsorption. Journal of Colloid and Interface Science 386: 129–134.

12. Walcarius A, Mercier L (2010) Mesoporous organosilica adsorbents: nanoengineered materials for removal of organic and inorganic pollutants. Journal of Materials Chemistry 20: 4478–4511.

13. Sierra I, Perez-Quintanilla D (2013) Heavy metal complexation on hybrid mesoporous silicas: an approach to analytical applications. Chemical Society Reviews, Advance Article. doi:10.1039/C2CS35221D.

14. Feng X, Fryxell GE, Wang LQ, Kim AY, Liu J, et al. (1997) Functionalized monolayers on ordered mesoporous supports. Science 276: 923–926.

15. Lam KF, Yeung KL, McKay G (2006) A rational approach in the design of selective mesoporous adsorbents. Langmuir 22: 9632–9641.

16. Lam KF, Chen XQ, McKay G, Yeung KL (2008) Anion Effect on Cu^{2+} Adsorption on NH_2-MCM-41. Industrial & Engineering Chemistry Research 47: 9376–9383.

17. Burleigh MC, Markowitz MA, Spector MS, Gaber BP (2001) Amine-functionalized periodic mesoporous organosilicas. Chemistry of Materials 13: 4760–4766.

18. Yoshitake H, Yokoi T, Tatsumi T (2002) Adsorption of chromate and arsenate by amino-functionalized MCM-41 and SBA-1. Chemistry of Materials 14: 4603–4610.

19. Liu AM, Hidajat K, Kawi S, Zhao DY (2000) A new class of hybrid mesoporous materials with functionalized organic monolayers for selective adsorption of heavy metal ions. Chemical Communications 0: 1145–1146.

20. Li GL, Zhao ZS, Liu JY, Jiang GB (2011) Effective heavy metal removal from aqueous systems by thiol functionalized magnetic mesoporous silica. Journal of Hazardous Materials 192: 277–283.

21. Yokoi T, Tatsumi T, Yoshitake H (2004) Fe^{3+} coordinated to amino-functionalized MCM-41: an adsorbent for the toxic oxyanions with high capacity, resistibility to inhibiting anions, and reusability after a simple treatment. Journal of Colloid and Interface Science 274: 451–457.

22. Chen XQ, Lam KF, Zhang QJ, Pan BC, Arruebo M, et al. (2009) Synthesis of Highly Selective Magnetic Mesoporous Adsorbent. Journal of Physical Chemistry C 113: 9804–9813.

23. Wang P, Lo IMC (2009) Synthesis of mesoporous magnetic gamma-Fe_2O_3 and its application to Cr(VI) removal from contaminated water. Water Research 43: 3727–3734.

24. Tokuyama H, Hisaeda J, Nii S, Sakohara S (2010) Removal of heavy metal ions and humic acid from aqueous solutions by co-adsorption onto thermosensitive polymers. Separation and Purification Technology 71: 83–88.

25. Yang SB, Hu J, Chen CL, Shao DD, Wang XK (2011) Mutual Effects of Pb(II) and Humic Acid Adsorption on Multiwalled Carbon Nanotubes/Polyacrylamide Composites from Aqueous Solutions. Environmental Science & Technology 45: 3621–3627.

26. Sheng GD, Li JX, Shao DD, Hu J, Chen CL, et al. (2010) Adsorption of copper(II) on multiwalled carbon nanotubes in the absence and presence of humic or fulvic acids. Journal of Hazardous Materials 178: 333–340.

27. Zhao GX, Li JX, Ren XM, Chen CL, Wang XK (2011) Few-Layered Graphene Oxide Nanosheets As Superior Sorbents for Heavy Metal Ion Pollution Management. Environmental Science & Technology 45: 10454–10462.

28. Tang YL, Liang S, Yu SL, Gao NY, Zhang J, et al. (2012) Enhanced adsorption of humic acid on amine functionalized magnetic mesoporous composite microspheres. Colloids and Surfaces a-Physicochemical and Engineering Aspects 406: 61–67.

29. Xu XQ, Deng CH, Gao MX, Yu WJ, Yang PY, et al. (2006) Synthesis of magnetic microspheres with immobilized metal ions for enrichment and direct determination of phosphopeptides by matrix-assisted laser desorption ionization mass spectrometry. Advanced Materials 18: 3289–3293.

30. Xu X, Song C, Andrésen JM, Miller BG, Scaroni AW (2003) Preparation and characterization of novel CO_2 "molecular basket" adsorbents based on polymer-modified mesoporous molecular sieve MCM-41. Microporous and Mesoporous Materials 62: 29–45.

31. Hao S, Zhong Y, Pepe F, Zhu W (2012) Adsorption of Pb^{2+} and Cu^{2+} on anionic surfactant-templated amino-functionalized mesoporous silicas. Chemical Engineering Journal 189–190: 160–167.

32. Deng XJ, Lu LL, Li HW, Luo F (2010) The adsorption properties of Pb(II) and Cd(II) on functionalized graphene prepared by electrolysis method. Journal of Hazardous Materials 183: 923–930.

33. Pang Y, Zeng GM, Tang L, Zhang Y, Liu YY, et al. (2011) PEI-grafted magnetic porous powder for highly effective adsorption of heavy metal ions. Desalination 281: 278–284.

34. Liu JF, Zhao ZS, Jiang GB (2008) Coating Fe_3O_4 Magnetic Nanoparticles with Humic Acid for High Efficient Removal of Heavy Metals in Water. Environmental Science & Technology 42: 6949–6954.

35. Chen QQ, Yin DQ, Zhu SJ, Hu XL (2012) Adsorption of cadmium(II) on humic acid coated titanium dioxide. Journal of Colloid and Interface Science 367: 241–248.

36. Wu PX, Zhang Q, Dai YP, Zhu NW, Dang Z, et al. (2011) Adsorption of Cu(II), Cd(II) and Cr(III) ions from aqueous solutions on humic acid modified Ca-montmorillonite. Geoderma 164: 215–219.

37. Lin JW, Zhan YH, Zhu ZL (2011) Adsorption characteristics of copper (II) ions from aqueous solution onto humic acid-immobilized surfactant-modified zeolite. Colloids and Surfaces a-Physicochemical and Engineering Aspects 384: 9–16.

18

Nitrophenol Chemi-Sensor and Active Solar Photocatalyst Based on Spinel Hetaerolite Nanoparticles

Sher Bahadar Khan[1,2]*, **Mohammed M. Rahman**[1,2], **Kalsoom Akhtar**[3], **Abdullah M. Asiri**[1,2], **Malik Abdul Rub**[1,2]

1 Center of Excellence for Advanced Materials Research (CEAMR), King Abdulaziz University, Jeddah, Saudi Arabia, **2** Chemistry Department, Faculty of Science, King Abdulaziz University, Jeddah, Saudi Arabia, **3** Division of Nano Sciences and Department of Chemistry, Ewha Womans University, Seoul, Korea

Abstract

In this contribution, a significant catalyst based on spinel $ZnMn_2O_4$ composite nanoparticles has been developed for electro-catalysis of nitrophenol and photo-catalysis of brilliant cresyl blue. $ZnMn_2O_4$ composite (hetaerolite) nanoparticles were prepared by easy low temperature hydrothermal procedure and structurally characterized by X-ray powder diffraction (XRD), field emission scanning electron microscopy (FESEM), X-ray photoelectron spectroscopy (XPS), Fourier transform infrared (FTIR) and UV-visible spectroscopy which illustrate that the prepared material is optical active and composed of well crystalline body-centered tetragonal nanoparticles with average size of $\sim$38$\pm$10 nm. Hetaerolite nanoparticles were applied for the advancement of a nitrophenol sensor which exhibited high sensitivity (1.500 μAcm^{-2} mM^{-1}), stability, repeatability and lower limit of detection (20.0 μM) in short response time (10 sec). Moreover, hetaerolite nanoparticles executed high solar photo-catalytic degradation when applied to brilliant cresyl blue under visible light.

Editor: Andrew C. Marr, Queen's University Belfast, United Kingdom

Funding: This work was funded by the Deanship of Scientific Research (DSR), King Abdulaziz University, Jeddah, under grant No. (130-031-D1433). The authors, therefore, acknowledge with thanks DSR technical and financial support. The funders had no role in study design, data collection and analysis, decision to publish, or preparation of the manuscript.

Competing Interests: The authors have declared that no competing interests exist.

* E-mail: sbkhan@kau.edu.sa

Introduction

Environmental pollution has received extensive interest because of the uncertain consequences on human health and living organisms [1,2]. Industrial activity causes many environmental problems by discharging vast quantities of toxic compounds into the environment. Many hazardous waste sites have been created by gathering pollutants in soil and water for many years [3,4]. Therefore inspection of health risky pollutants in water and environment is an urgent demand for pollution controlling option. Thus easy, efficient and low cost processes for the recognition and detoxification of organic pollutants in aqueous solutions are needed to check and protect water resources and food supplies.

Several instrumental techniques exist for the recognition of organic pollutants, but are of less significance on the basis of efficiency and expense. However, sensor technology plays an important role in environmental safety and water treatment [5,6]. Thus for environmental and health safety, it is essential to produce simple, reliable, and inexpensive sensors for the detection of p-nitrophenol (p-NP) in water, because nitrophenols are hazardous and toxic pollutants with inhibitory and biorefractory nature and have adverse effect on living organisms. Nitrophenols have been widely used in the manufacture of pesticides, dyes and pharmaceuticals. p-NP is a toxic derivative of the parathion insecticide and is carcinogenic, hazardous, mutagenic and toxic (cytotoxic and embryo-toxic) to mammals [7]. Due to high solubility and stability of p-NP in water, it has been found in freshwater, marine environments and has been detected in industrial wastewaters.

Thus detection and monitoring of nitrophenols are crucial for environmental pollution control and industrial applications. Various chromatographic and spectroscopic techniques have been utilized for the detection and finding of hazardous solvents but their use is limited due to complication and sluggishness. Electrochemical sensors have achieved immense interest in the recognition and quantification of unsafe compounds since it is uncomplicated and rapid operation, response and recognition [8]. The sensitivity and selectivity of electrochemical sensor strongly dependent on the size, structure and properties of electrode materials and thus semiconductor nanostructured materials have received much importance and has extensively been utilized as a redox mediator in various sensors [9].

Photo-catalysis is also one of the low cost processes for detoxification of lethal organic compounds but the lack of active catalysts excluded this process from a wide range of applications. Among different photo-catalysts, TiO_2 and ZnO have proven to act as dynamic photocatalysts and play pivotal roles in the detoxification of lethal organic compounds [10,11]. However these photo-catalysts only encourage photo-catalysis upon irradiation by UV light because they absorb in the UV region. For solar photocatalysis, a photo-catalyst must promote photo-catalysis by irradiation with visible light because solar spectrum consists 46% of visible light while the UV light is only 5–7% of the solar spectrum. Thus it is an urgent demand to develop an active photocatalyst which can promote photo-catalysis in the visible region. Manganese oxides with various crystalline structures have been explored for various applications such as catalysis and electrode materials for lithium cathodes [12,13]. However, to enhance the various properties of manganese oxide to meet the increasing needs for different applications, the features of manganese oxide must be modified. One of the main noteworthy methods to amend the characteristics of these nanomaterials is the introduction of doped materials in the parent system because recently doped metal oxides have shown excellent properties in various applications [14].

Therefore in this study, spinel (AB_2O_4) like zinc doped manganese oxide (spinel hetaerolite) has been synthesized by a simple low-temperature hydrothermal process and characterized by XRD, FESEM, XPS, FTIR, XPS and UV-vis. spectroscopy. From an application point of view, spinel hetaerolite was investigated as a sensor substance for the discovery of p-NP and solar photo-catalyst for the degradation of brilliant cresyl blue under visible light. To the best of our knowledge, this is the first detail of a spinel heterolite as a nitrophenol sensor and solar photo-catalyst for brilliant cresyl blue.

Methods

1. Synthesis of hetaerolite nanoparticles

An equi-molar aqueous solution was prepared by dissolving 1.36 g of $ZnCl_2$ and $MnCl_2.4H_2O$ in 100 ml distilled water and further titrated by NH_4OH solution to increase the pH above 10.0. This high basic solution was then shifted to a Teflon autoclave and kept at 150.0°C for 16 hours in an oven. Finally a black product was obtained by washing the precipitate. The precipitate was dried and calcined at 400.0°C for 5 hours.

2. Possible growth of hetaerolite nano-particles

In the case of manganese oxides formation, it has been reported that manganese gets reduced to Mn^{3+} or Mn^{2+} states and generates lamellae structure which then appear in the form of rod like structure to reduce the surface energy [15–17]. After hydrothermal process it crystallizes to form MnO-OH nanorods. The calcination of MnO-OH gives Mn_3O_4 due to dehydration and oxygen out diffusion at high temperature [18]. However the incorporation of a Zn precursor has great influence on the reduction of manganese. It enhances the reduction process of Mn by first reducing to Mn^{+3} state and further reduce Mn^{+3} to Mn^{+2} state which result in the formation of $ZnMn_2O_4$ nanoparticles. Schematically probable growth process of $ZnMn_2O_4$ nanoparticles is given in Figure 1.

3. Characterization

JEOL Scanning Electron Microscope (JSM-7600F, Japan) was used to analyze the morphology of the prepared material while the crystallography was studied by X'Pert Explorer, PANalytical X-ray diffractometer. The XPS spectrum was recorded in the range of 0 to 1350 eV by using a Thermo Scientific K-Alpha KA1066

spectrometer (Germany). For compositional, optical and degradation studies of the nanomaterial, FT-IR and UV spectrum were record by a PerkinElmer (spectrum 100) FT-IR spectrometer and PerkinElmer (Lambda 950) UV-visible spectrometer. For sensing study, I–V measurements were carried out using an electrometer (Kethley, USA).

4. Fabrication of chemical sensor

Hetaerolite nanoparticles were coated as a thin film on the surface of a glassy carbon electrode (GCE, surface area 0.0314 cm^2) and dried at 60.0°C for 12 hours. Time delaying and response time were 5.0 sec and 10.0 sec, respectively. The sensing ability of the hetaerolite nanoparticle modified GCE was evaluated using nitrophenol in a similar way as we published earlier [19,20].

5. Solar photo-catalytic degradation

Solar photo-catalytic activity of hetaerolite nanoparticles was evaluated through degradation of brilliant cresyl blue under sun light. The dye is stable under visible light irradiation in the absence of a photo-catalyst. In photo-catalytic degradation, two different 100.0 mL, 1×10^{-4} M of brilliant cresyl blue solutions were taken in different beakers and adjusted the pH to 5, 8 and 10, by drop wise addition of 0.2 M NaOH solution under vigorous stirring. 0.1193 g and 0.1132 g catalyst were then added into the dye reaction solutions. The solution was then irradiated under sunlight at constant stirring and 4–5 mL of dye solutions were pipetted out at regular interval and measured the absorbance at $\lambda_{max} = 595.0$ nm using UV visible spectrophotometer. Dye degradation without catalyst was also studied under visible light to check any degradation of dye.

Results and Discussion

1. Physiochemical characterization of hetaerolite nanoparticles

FESEM was utilized to explore the morphology and size of hetaerolite. The morphology of hetaerolite was illustrated by FESEM which are shown in Figure 2 (a–c). It is evident from the FESEM images that the synthesized material is composed of particles having average diameter of $\sim$38±10 nm. These nanoparticles were prepared in large quantity possessing a spherical shape.

The crystallographic information of the hetaerolite nanoparticles were corroborated by X-ray diffraction (Figure 2 (d)). All the characteristic diffraction peaks coincided with those for well-crystalline distinct body-centered tetragonally structured of $ZnMn_2O_4$. X-ray diffraction peaks of the samples are in outstanding agreement and excellent accordance with the JCPDS card no. 077-0470 [21]. According to the JCPDS card, the synthesized product is a body-centered tetragonal phase $ZnMn_2O_4$ with cell parameters of a = 5.72 Å and c = 9.24 Å and space group of 141/amd(141). All diffraction peaks were only related to $ZnMn_2O_4$ without any impurity peaks and thus the synthesized product therefore consist of pure $ZnMn_2O_4$ crystals. The size of hetaerolite nano-particles (38 nm) suggested by FESEM was also verified and supported by Scherrer formula.

$$D = 0.9\lambda/\beta\cos\theta$$

Where λ is the wavelength of X-ray radiation, β is the full width at half maximum (FWHM) of the peaks at the diffracting angle θ.

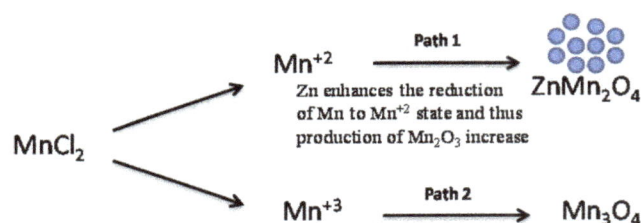

Figure 1. Possible growth mchanism of hetaerolite nanoparticles.

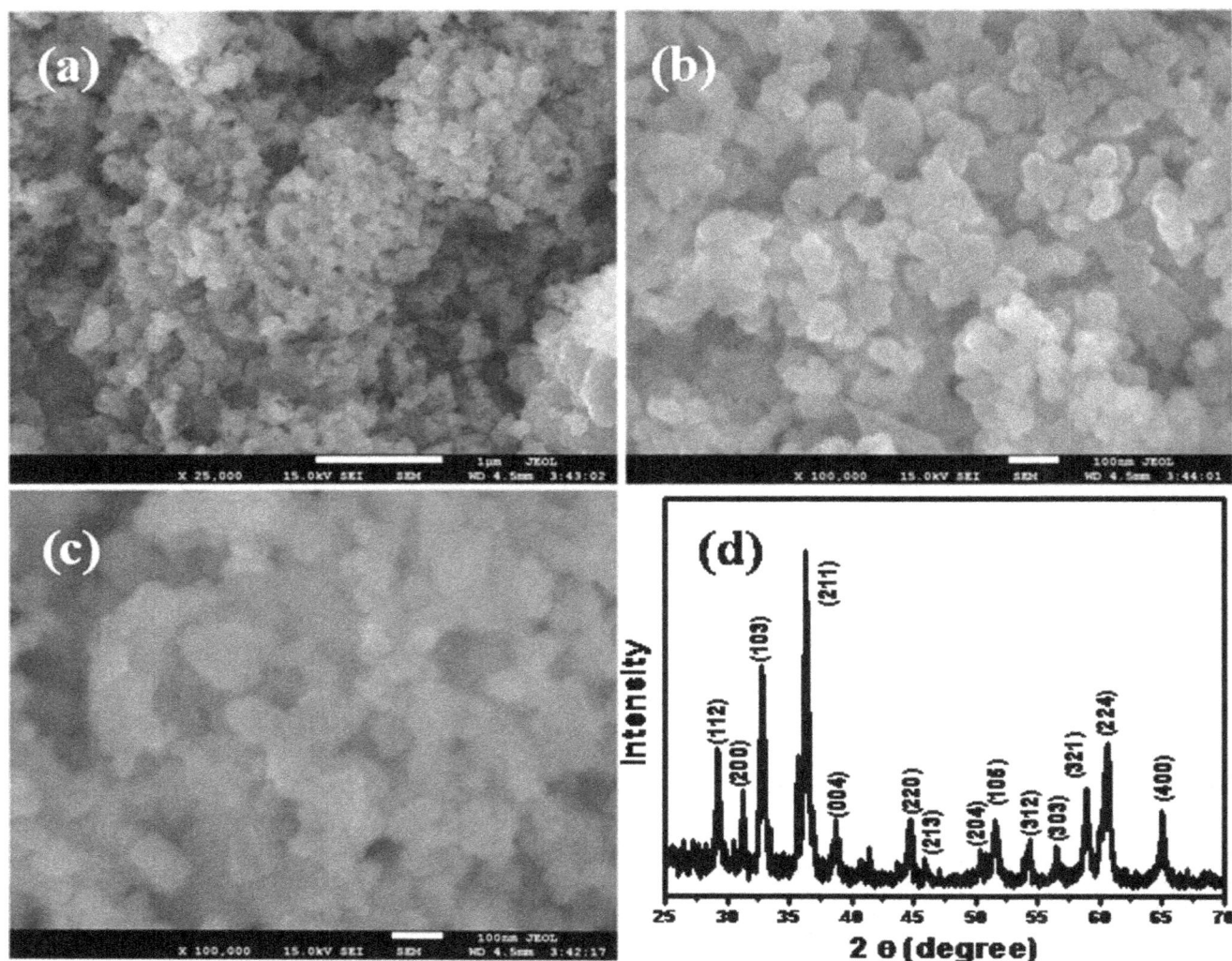

Figure 2. Typical low and high-resolution FESEM images (a–c) and XRD pattern (d) of the synthesized hetaerolite nano-particles.

XPS was analyzed to study the composition of hetaerolite nano-particles and the graphs of hetaerolite nano-particles are depicted in Figure 3. Hetaerolite nano-particles XPS spectrum exhibited peaks at 530.6, 642.6, 654.2, 1022.1 and 1045.1 eV which are responsible for O 1 s, Mn $2p_{3/2}$, Mn $2p_{1/2}$, Zn $2p_{3/2}$ and Zn $2p_{1/2}$ peaks, respectively. The spin–orbit splitting between Mn $2p_{3/2}$ and Mn $2p_{1/2}$ is 11.6 eV while 23.0 eV is observed among Zn $2p_{3/2}$ and Zn $2p_{1/2}$. These results are comparable to the values report earliar [18,22]. The above results are consisted with XRD which confirms the synthesis of $ZnMn_2O_4$.

The chemical structure of the hetaerolite nano-particles was also examined by FTIR analysis which was recorded in the range of $400{\sim}4000$ cm^{-1} and shown in Figure 4 (a). The very intense bands observed at 510 and 621 cm^{-1} were attributed to M–O (M = Zn, Mn) and M-O-M bonds, respectively. Supplementary peaks centered at 3417 and 1625 cm^{-1} were assigned to H_2O absorbed from the environment [23].

The photocatalytic performance of a photocatalyst can be evaluated by its optical properties, one of the essential requirements for a photocatalyst. Thus optical properties of hetaerolite nano-particles were scrutinized by using a UV-Vis. spectropho-tometer and the spectrum is shown in Figure 4 (b). In UV/visible absorption method, energy band gap of nanomaterials can be acquired by analyzing their optical absorption. UV–vis absorption spectra exhibited a broad absorption peak and showed band gap energy equal to 2.91 eV which is calculated by Tauc's formula. The relation between absorption coefficient (α) and the incident photon energy ($h\upsilon$) is given by the equation [24].

$$\alpha h\upsilon = A(h\upsilon - E_g)^n$$

where A is a constant, E_g is the band gap of the material and the exponent n depends on the type of transition, n = 1/2, 2, 3/2 and 3 corresponding to allowed direct, allowed indirect, forbidden direct and forbidden indirect, respectively. Taking n = 1/2, we have calculated the direct energy band gap from the $(\alpha h\upsilon)^{1/n}$ vs. $h\upsilon$ plots (Figure 4 (c).

2. Chemical sensing properties

The hetaerolite nano-particles were utilized for the recognition of p-nitrophenol in aqueous media in order to study their chemical sensing properties toward p-nitrophenol [25,26]. I–V technique was used to measure the electrical response hetaerolite nano-

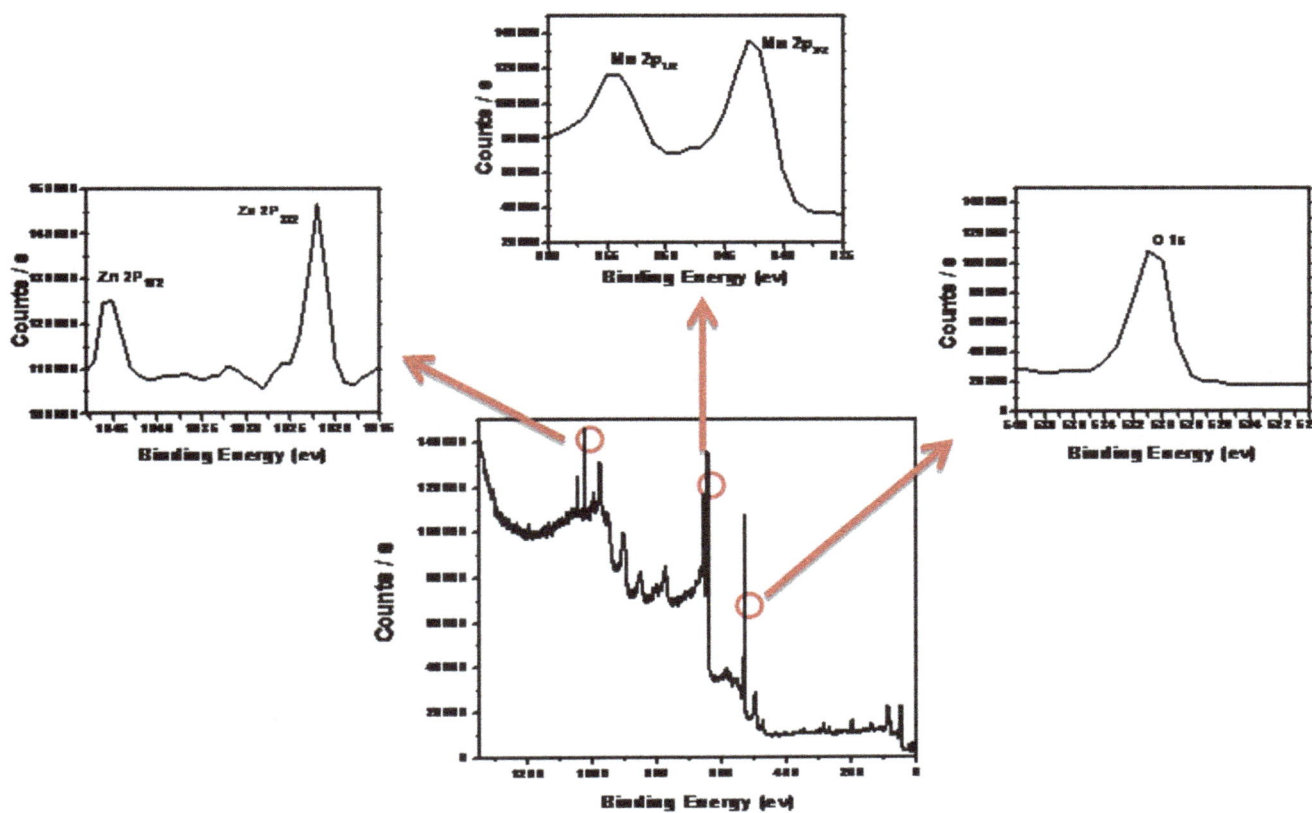

Figure 3. Typical XPS spectrum of the synthesized hetaerolite nano-particles.

particles sensor for p-nitrophenol which is shown in Figure 5 and Figure 6.

I–V curves for the glassy carbon electrode without coating (without hetaerolite nano-particles) and after coating (with hetaerolite nano-particles) were measured and shown in Figure 6 (a). The hetaerolite coated glassy carbon electrode (black square) illustrates not as much current response in contrast to the naked glassy carbon electrode (gray square), which might be attributed to resistance originated by hetaerolite nano-particles along with binders coated on the electrode surface [27].

Figure 4 (b) illustrates electrical reactions of the hetaerolite nano-particles without p-nitrophenol (gray-dotted line) and with 100.0 μL p-nitrophenol (dark-dotted line) in 0.1 M phosphate buffer solution (pH = 7.0). It is observed from the Figure 6 (b) that by adding the target chemical, hetaerolite nano-particles demonstrate a noteworthy enhancement in electrical current that reveals the sensitivity of hetaerolite nano-particles to p-nitrophenol. Consequently by injection of analyte, the augmentation in electrical response shows that hetaerolite nano-particles exhibit a fast and sensitive reply to the target chemical which might be due to speedy redox reaction (electron exchange) and fine electrocatalytic oxidation properties of the hetaerolite sensor. p-Nitrosophenol generally undergo reduction and produces p-hydroxylaminophenol. In second step, oxidation of p-hydroxylaminophenol takes place which give rise to 4-nitrosophenol and the subsequent reversible reduction.

The influence of p-nitrophenol concentration on the electrical reaction of hetaerolite nano-particles was examined by consecutive addition of p-nitrophenol in the range of 50.0 μM to 1.0 M into 0.1 M PBS solution (pH = 7.0) and the graph is portrayed in

Figure 6 (c). Enhancement of the electrical current with rising p-nitrophenol concentration is observed which designates that the hetaerolite nano-particles conductivity was enhanced by the increase in the concentration of the target chemical [28]. The mechanism of p-nitrophenol sensing is graphically shown in Figure 5.

Calibration curve (Figure 6 (d)) was plotted from the difference in target concentration. The calibration curve portrays two sensitivity areas; the region at inferior concentrations (physisorption process) is linear up to 0.005 M with correlation coefficient (R) of 0.7599. The sensitivity is determined from the slope of the lesser concentration section of calibration curve, which is 1.500 μA.cm^{-2}.mM^{-1}. The linear dynamic range reveals from 50.0 μM to 0.05 M and the detection limit was estimated, based on signal to noise ratio (S/N), to be 20.0 μM. Above 0.005 M concentration the sensor become saturated due to chemisorption method which could be due to the lack of free hetaerolite nano-particles sites for p-nitrophenol adsorption [28].

3. Photocatalysis

3.1. Solar photocatalytic performance of hetaerolite nanoparticles.
The solar photo-catalytic performance of hetaerolite nanoparticles was evaluated by degrading brilliant cresyl blue under solar light [29]. In this study two sets of photo-catalytic reaction were carried out utilizing hetaerolite nanoparticles. Irradiation of brilliant cresyl blue under visible light degraded a little amount of dye at pH 7 without the catalyst (hetaerolite nanoparticles) indicating a photolysis reaction. Photo-catalytical degradation of brilliant cresyl blue solution was carried out in the presence of hetaerolite nano-particles under visible light irradia-

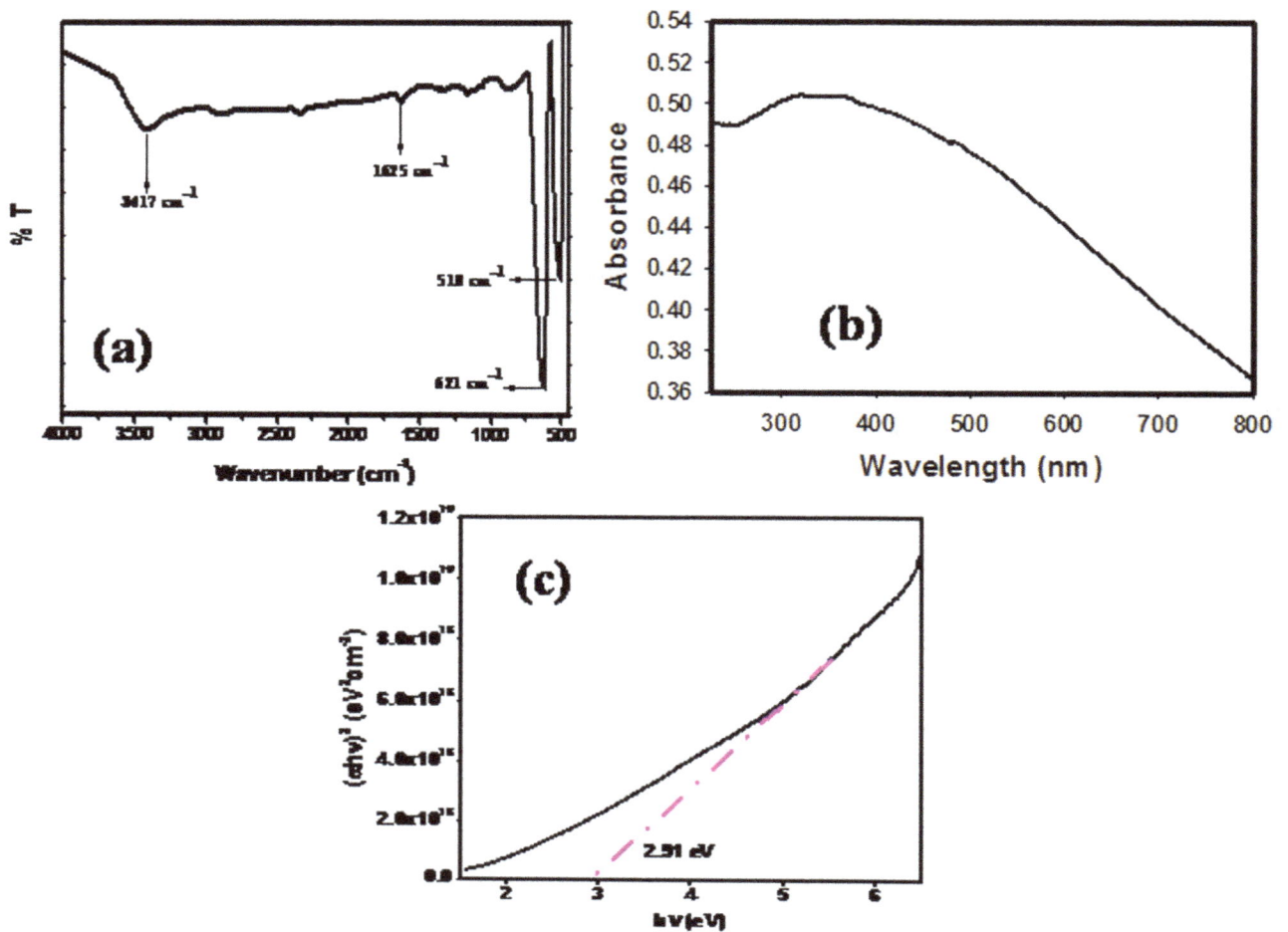

Figure 4. Typical FT-IR (a), UV-vis (b) spectrum and (ahυ)² vs hυ graph (c) of the synthesized hetaerolite nano-particles.

tion at different pHs and the effects of pH on the photo-catalytic degradation of brilliant cresyl blue studied under solar light irradiation [6]. Hetaerolite nanoparticles showed efficient activity for degradation of brilliant cresyl blue at different pH under solar light irradiation.

An aqueous suspension of brilliant cresyl blue was irradiated with solar light in the presence of hetaerolite nanoparticles and this led to the alteration of absorbance with irradiation time [8]. Figure 7 (a) display transformation in absorption for the photo-catalytic degradation of brilliant cresyl blue at different time intervals which showed decline in absorption strength. It is clear that the absorbance at 595 nm steadily reduces with increase in irradiation time. Figure 7 (b) illustrates the transformation in absorbance with change in irradiation time for the brilliant cresyl blue in the presence and absence of hetaerolite nano-particles. Irradiation of brilliant cresyl blue solution in the presence of hetaerolite nano-particles leads to decline in absorption intensity. Figure 7 (c) demonstrates % degradation of brilliant cresyl blue in the presence and absence of hetaerolite nano-particles with respect to irradiation time (min). Degradation (%) graph shows that 27%, 52% and 83% of brilliant cresyl blue is degraded at pH 5, 8 and 10, correspondingly in the presence of hetaerolite nano-particles after 120 minutes of irradiation time while in the absence of hetaerolite nano-particles, no apparent loss of brilliant cresyl blue might be seen.

Hetaerolite nanoparticles were further compared with TiO_2 which is a dynamic photo-catalyst and displayed an imperative role in the detoxification of polutants [9]. However TiO_2 only promotes photo-catalysis upon irradiation by UV light because it absorbs only in the UV region. Hetaerolite nanoparticles promote

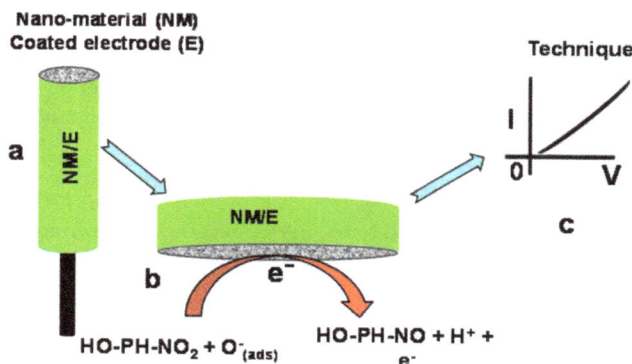

Figure 5. Schematic views of (a) fabricated AgE with hetaerolite nanoparticles and conducting binders, (b) reaction occurred at fabricated AgE (c) I–V detection technique.

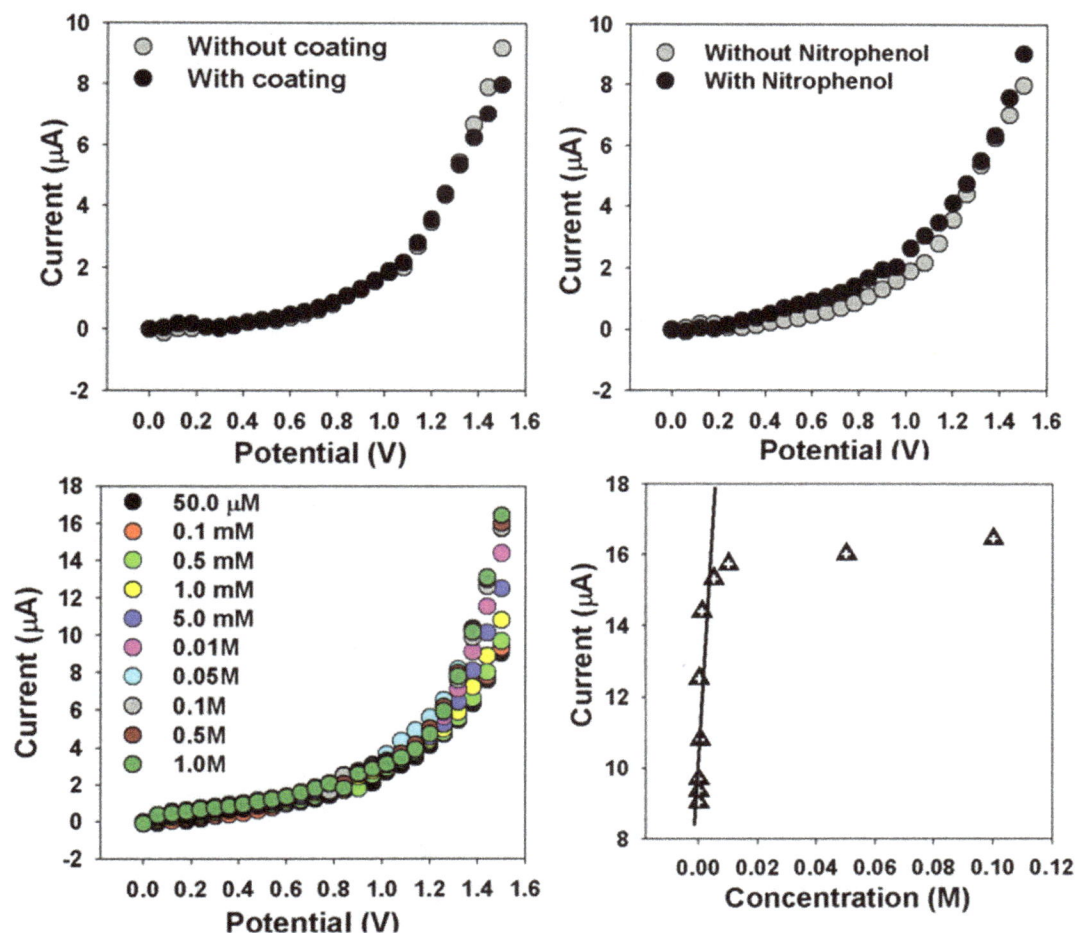

Figure 6. I–V characterization of hetaerolite nano-particles (a) Comparison of with and without hetaerolite nano-particles coating on GCE, (b) Comparison of with and without p-nitrophenol injection, (c) Concentration variation of p-nitrophenol, and (d) calibration plot.

photo-catalysis by irradiation with visible light because the solar spectrum consists of 46% visible light while the UV light is only 5–7% in the solar spectrum. Thus it is vital to grow an active photocatalyst which can promote photo-catalysis in the visible region.

3.2. Effect of pH. The influence of pH on the visible light photocatalytic degradation of brilliant cresyl blue was examined in pH range 5–10 and compared with the brilliant cresyl blue degradation at various pH in the presence of hetaerolite nanoparticles (Figure 7). It exhibited the same trend of degradation at all pH but exhibited high photo-catalytic degradation at pH 10 as compared to pH 5 and 8. The results showed that the rate of decomposition of brilliant cresyl blue increases with increase in pH and at pH 10, brilliant cresyl blue was degraded 83% in the presence of hetaerolite nano-particles. The photocatalytic performance of hetaerolite nano-particles was attributed to the surface electrical properties, which facilitate the brilliant cresyl blue adsorption. It is beneficial for the promotion of a visible light generated charge carrier i.e. electron to the surface which leads to formation of hydroxide radical. Thus high pH makes the surface of hetaerolite nano-particles negatively charged and also supports the formation of OH^- radicals [30]. Both these factors favor the attraction and degradation of brilliant cresyl blue. It is clear from the results that pH has a substantial influence on the photocatalytic degradation of brilliant cresyl blue and hetaerolite nano-

particles synthesized by a very straightforward synthesis process exhibit considerable solar photo-catalytic activity towards brilliant cresyl blue degradation. This demonstrates the applicability of the solar photo-catalyst towards organic pollutants and the potential to extend the applications to environmental pollutants.

3.3. Reaction kinetics of photo-degradation. In order to realize the degradation behavior, we studied the degradation pattern of brilliant cresyl blue by Langmuir–Hinshelwood (L–H) model. Langmuir–Hinshelwood (L–H) model well defines the relationship among the rate of degradation and the preliminary concentration of brilliant cresyl blue in photo-catalytic reaction [31]. The rate of photo-degradation was calculated by using Equation 1:

$$r = -dC/dt = K_r KC = K_{app}C \qquad (1)$$

Where r represents the degradation rate of brilliant cresyl blue, K_r is the reaction rate constant, K is the equilibrium constant, C is the reactant concentration. When C is very diminutive, then KC is insignificant and Equation 1 turns into first order kinetic. Consider preliminary conditions (t = 0, C = C_0), Equation 1 turn into Equation 2.

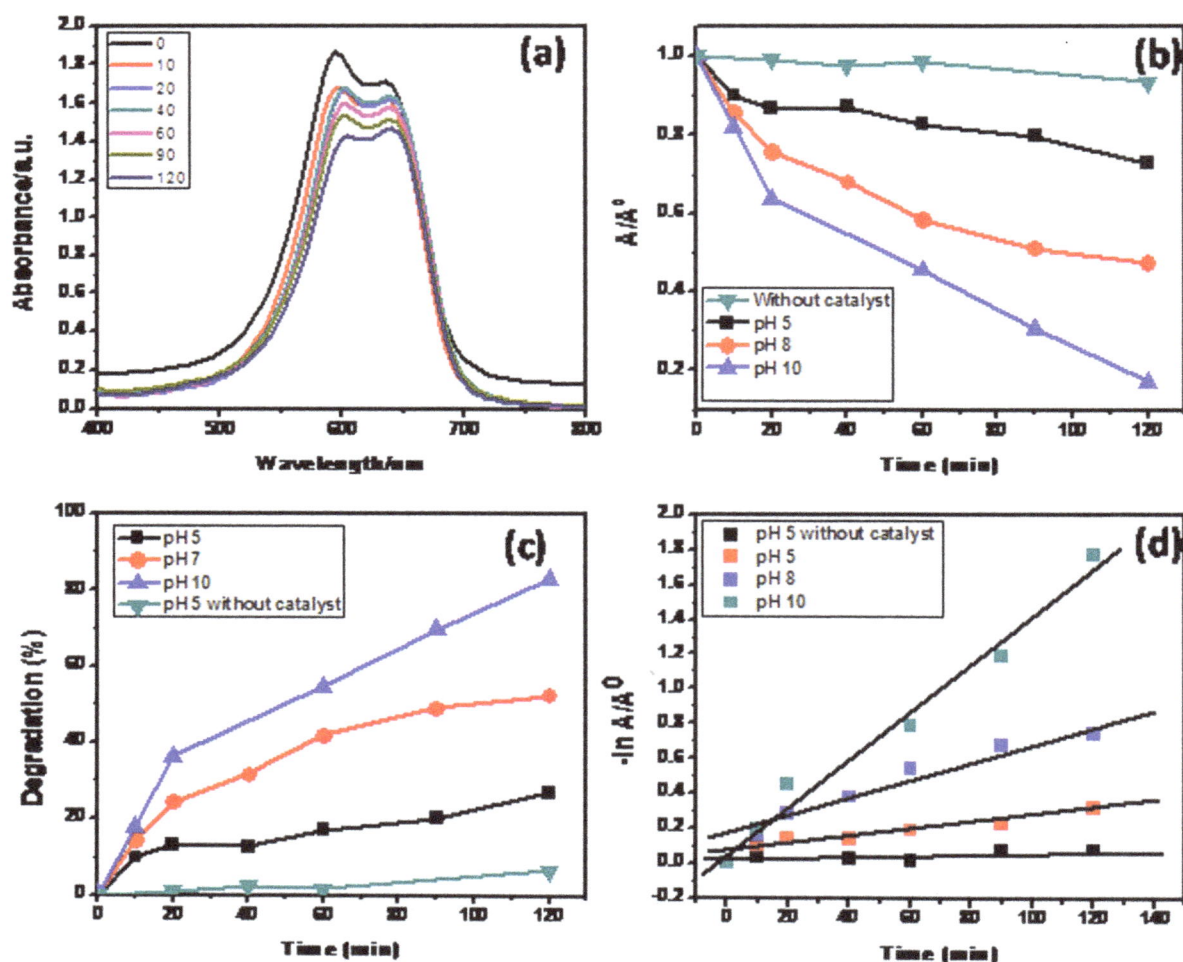

Figure 7. Typical plot for **(a)** Change in the absorption spectrum of brilliant cresyl blue **(b)** Change in absorbance vs irradiation time for brilliant cresyl blue at different pH, **(c)** % degradation vs irradiation time and **(d)** reaction kinetic for brilliant cresyl blue at different pH in the presence of hetaerolite nanoparticles.

$$-\ln C/C_0 = kt \qquad (2)$$

Half-life, $t_{1/2}$ (in min) is

$$t_{1/2} = 0.693/k \qquad (3)$$

Figure 7 (d) demonstrated degradation of brilliant cresyl blue which pursued first-order kinetics (plots of $\ln(C/C_0)$ vs time showed linear relationship). First-order rate constants, calculated from the slopes of the $\ln(C/C_0)$ vs time plots and the half-life of the degraded brilliant cresyl blue can simply determined by Equation 3 [30,31]. Rate constant for hetaerolite nano-particles were found to be 0.000526 min^{-1} ($t_{1/2} = 1317.5$ min), 0.00212 min^{-1} ($t_{1/2} = 326.9$ min), 0.00596 min^{-1} ($t_{1/2} = 116.3$ min) and 0.0136 min^{-1} ($t_{1/2} = 51.0$ min). Thus the kinetic study revealed that hetaerolite nano-particles is a proficient photo-catalyst for degradation of organic noxious wastes.

3.4. Mechanism of photo-degradation. In the current study, a heterogeneous photo-catalysis method was employed for the degradation of brilliant cresyl blue. Briefly, when hetaerolite nano-particles were exposed to light having energy the same or superior to the band gap of nanoparticles, the development of an electron and hole pair occurs on the surface of hetaerolite nano-particles. If charge partition is sustained, subsequently the electron hole pair reacts with brilliant cresyl blue in the presence of oxygen. Hydroxyl radicals (OH$^{\bullet}$) and superoxide radical anions (O$_2^{\bullet-}$) are assumed to be the major degrading mediators (oxidizing species)

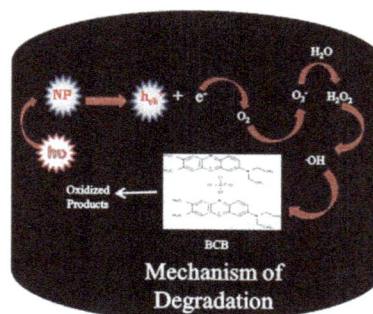

Figure 8. Mechanism of photodegradation of brilliant cresyl blue in the presence of hetaerolite nano-particles.

and oxidative reactions result in the degrading (oxidation) of the brilliant cresyl blue. The whole mechanism of photo-activity of hetaerolite nano-particles is depicted in Figure 8 [30,31].

Conclusion

Well-crystalline body-centered tetragonal hetaerolite nano-particles based p-nitrophenol electrochemical sensor has been fabricated. Hetaerolite nano-particles were produced by a hydrothermal process. The featured structural characterizations proved that the manufactured hetaerolite is optically active, well-crystalline, body-centered tetragonal nanoparticles. Sensing toward p-nitrophenol was executed which showed excellent sensitivity and limit of detection with quick response time. Moreover the solar photo-catalytic property of hetaerolite nanoparticles was utilized in the degradation of brilliant cresyl blue. Thus it is concluded that the hetaerolite nanoparticles are an attractive

sensor material and active photo-catalyst for accomplishing a proficient chemical sensor and photo-catalyst for application within water resources and health monitoring.

Acknowledgments

This work was funded by the Deanship of Scientific Research (DSR), King Abdulaziz University, Jeddah, under grant No. (130-031-D1433).

Author Contributions

Conceived and designed the experiments: SBK MMR KA MAA. Performed the experiments: SBK MMR KA MAA MAR. Analyzed the data: SBK MMR KA MAA. Contributed reagents/materials/analysis tools: SBK MMR KA MAA. Wrote the paper: SBK MMR KA MAA MAR.

References

1. Jamal A, Rahman MM, Khan SB, Faisal M, Akhtar K, et al. (2012) Cobalt doped antimony oxide nano-particles based chemical sensor and photo-catalyst for environmental pollutants. App Surf Sci 261: 52–58.
2. Stanca SE, Popescu IC, Oniciu L (2003) Biosensors for phenol derivatives using biochemical signal amplification. Talanta 61: 501–507.
3. Khan SB, Akhtar K, Rahman MM, Asiri AM, Seo J, et al. (2012) Thermally and mechanically stable green environmental composite for chemical sensor applications. New J Chem 36: 2368–2375.
4. Jain RK, Kapur M, Labana S, Lal B, Sharma PM, et al. (2005) Microbial diversity: Application of microorganisms for the biodegradation of xenobiotics. Curr Sci 89: 101–112.
5. Rahman MM, Jamal A, Khan SB, Faisal M (2011) Characterization and applications of as-grown b-Fe$_2$O$_3$ nanoparticles prepared by hydrothermal method. J Nanoparticle Res 13: 3789–3799.
6. Faisal M, Khan SB, Rahman MM, Jamal A (2011) Synthesis, characterizations, photocatalytic and sensing studies of ZnO nanocapsules. Appl Surf Sci 258: 672–677.
7. Banik RM, Prakash MR, Upadhyay SN (2008) Microbial biosensor based on whole cell of Pseudomonas sp. for online measurement of p-Nitrophenol. Sens Actuat B 131: 295–300.
8. Khan SB, Rahman MM, Akhtar K, Asiri AM, Seo J, et al. (2012) Novel and sensitive ethanol chemi-sensor based on nanohybrid materials. Int J Electrochem Sci 7: 4030–4038.
9. Rahman MM, Khan SB, Jamal A, Faisal M, Asiri AM (2012) Fabrication of a methanol chemical sensor based on hydrothermally prepared α-Fe$_2$O$_3$ codoped SnO$_2$ nanocubes. Talanta 95: 18–24.
10. Khan SB, Faisal M, Rahman MM, Jamal A (2011) Low-temperature growth of ZnO nanoparticles: Photocatalyst and acetone sensor. Talanta 85: 943–949.
11. Faisal M, Khan SB, Rahman MM, Jamal A (2011) Smart chemical sensor and active photo-catalyst for environmental pollutants. Chem Engineer J 173: 178–184.
12. Qamar MM, Lofland SE, Ramanujachary KV, Ganguli AK (2009) Magnetic and photocatalytic properties of nanocrystalline ZnMn$_2$O$_4$. Bull Mater Sci 32: 231–237.
13. Jamal A, Rahman MM, Khan SB, Faisal M, Asiri AM, et al. (2013) Hydrothermally preparation and characterization of un-doped manganese oxide nanostructures: Efficient photocatalysis and chemical sensing applications). Micro and Nanosystems 5: 22–28.
14. Rahman MM, Jamal A, Khan SB, Faisal M (2011) Fabrication of highly sensitive ethanol chemical sensor based on Sm-doped Co$_3$O$_4$ nanokernels by a hydrothermal method. J Phys Chem C 115: 9503–9510.
15. Jha A, Thapa R, Chattopadhyay KK (2012) Structural transformation from Mn3O4 nanorods to nanoparticles and band gap tuning via Zn doping. Mater Res Bull 47: 813–819.

16. Du J, Gao Y, Chai L, Zou G, Li Y, et al. (2006) Hausmannite Mn~3O~4 nanorods: synthesis, characterization and magnetic properties. Nanotech 17: 4923–4928.
17. Wang X, Li Y (2003) Synthesis and formation mechanism of manganese dioxide Nanowires/Nanorods. Chem Eur J 9: 300–306.
18. Yang J, Zeng JH, Yu SH, Yang L, Zhou GE, et al. (2000) Formation process of CdS nanorods via solvothermal route. Chem Mater 12: 3259–3263.
19. Khan SB, Rahman MM, Akhtar K, Asiri AM, Alamry KA, et al. (2012) Copper oxide based polymer nanohybrid for chemical sensor applications. Int J Electrochem Sci 7: 10965–10975.
20. Rahman MM, Khan SB, Faisal M, Asiri AM, Tariq MA (2012) Detection of aprepitant drug based on low-dimensional un-doped iron oxide nanoparticles prepared by a solution method. Electrochimica Acta 75: 164–170.
21. Zhang P, Li X, Zhao Q, Liu S (2011) Synthesis and optical property of one-dimensional spinel ZnMn$_2$O$_4$ nanorods. Nanoscal Res Lett 6: 323–331.
22. Apte SK, Naik SD, Sonawane RS, Kale BB, Pavaskar N, et al. (2006) Nanosize Mn3O4 (Hausmannite) by microwave irradiation method. Mater Res Bull 41: 647–654.
23. Faisal M, Khan SB, Rahman MM, Jamal A, Umar A (2011) Ethanol chemi-sensor: Evaluation of structural, optical and sensing properties of CuO nanosheets. Mater Lett 65: 1400–1403.
24. Rahman MM, Jamal A, Khan SB, Faisal M (2011) Highly sensitive ethanol chemical sensor based on Ni-doped SnO$_2$ nanostructure materials. Biosens Bioelectron 28: 127–134.
25. Ansari SG, Ansari ZA, Seo HK, Kim GS, Kim YS, et al. (2008) Urea sensor based on tin oxide thin films prepared by modified plasma enhanced CVD. Sensors Actuators B 132: 265–271.
26. Ansari SG, Ansari ZA, Wahab R, Kim YS, Khang G, et al. (2008) Glucose sensor based on nano-baskets of tin oxide templated in porous alumina by plasma enhanced CVD. Biosens Bioelectron 23: 1838–1842.
27. Ansari SG, Wahab R, Ansari ZA, Kim YS, Khang G, et al. (2009) Effect of nanostructure on the urea sensing properties of sol–gel synthesized ZnO. Sensors Actuators B 137: 566–73.
28. Khan SB, Rahman MM, Jang ES, Akhtar K, Han H (2011) Special susceptive aqueous ammonia chemi-sensor: Extended applications of novel UV-curable polyurethane-clay nanohybrid. Talanta 84: 1005–1010.
29. Khan SB, Faisal M, Rahman MM, Jamal A (2011) Exploration of CeO$_2$ nanoparticles as a chemi-sensor and photo-catalyst for environmental applications. Sci Tot Environ 409: 2987–2992.
30. Mohapatra L, Parida KM (2012) Zn–Cr layered double hydroxide: Visible light responsive photocatalyst for photocatalytic degradation of organic pollutants Separat Purif Technol 91: 73–80.
31. Parida KM, Mohapatra L (2012) Characterization of inverted pyramidal hollow cathode microplasma devices operating in reactive gases for maskless scanning plasma etching. Chem Engineer J 179: 131–139.

19

Cloud-Enabled Microscopy and Droplet Microfluidic Platform for Specific Detection of *Escherichia coli* in Water

Alexander Golberg[1]⑨, Gregory Linshiz[2,3,4]⑨, Ilia Kravets[5], Nina Stawski[2,3], Nathan J. Hillson[2,3,4], Martin L. Yarmush[1,6], Robert S. Marks[7,8,9], Tania Konry[10]*

1 Centre for Engineering in Medicine, Massachusetts General Hospital, Harvard Medical School, Shriners Burns Institute, Boston, Massachusetts, United States of America, 2 Fuels Synthesis Division, Joint BioEnergy Institute, Emeryville, California, United States of America, 3 Physical BioSciences Division, Lawrence Berkeley National Labs, Berkeley, California, United States of America, 4 DOE Joint Genome Institute, Walnut Creek, California, United States of America, 5 Department of Computer Science, Technion Institute of Technology, Haifa, Israel, 6 Department of Biomedical Engineering, Rutgers University, New Jersey, United States of America, 7 Department of Biotechnology Engineering, The National Institute of Biotechnology in Negev, Ben Gurion University, Beer-Sheva, Israel, 8 School of Materials Science and Engineering, Nanyang Technological University, Singapore, 9 NRF CREATE program for Nanomaterials in Energy and Water Management, Singapore, 10 Department of Pharmaceutical Sciences, School of Pharmacy Bouvé College of Health Sciences, Northeastern University, Boston, Massachusetts, United States of America

Abstract

We report an all-in-one platform – ScanDrop – for the rapid and specific capture, detection, and identification of bacteria in drinking water. The ScanDrop platform integrates droplet microfluidics, a portable imaging system, and cloud-based control software and data storage. The cloud-based control software and data storage enables robotic image acquisition, remote image processing, and rapid data sharing. These features form a "cloud" network for water quality monitoring. We have demonstrated the capability of ScanDrop to perform water quality monitoring via the detection of an indicator coliform bacterium, *Escherichia coli*, in drinking water contaminated with feces. Magnetic beads conjugated with antibodies to *E. coli* antigen were used to selectively capture and isolate specific bacteria from water samples. The bead-captured bacteria were co-encapsulated in pico-liter droplets with fluorescently-labeled anti-*E. coli* antibodies, and imaged with an automated custom designed fluorescence microscope. The entire water quality diagnostic process required 8 hours from sample collection to online-accessible results compared with 2–4 days for other currently available standard detection methods.

Editor: James P. Brody, University of California, Irvine, United States of America

Funding: AG and MY acknowledge Shriners Foundation Grant #85120-BOS for the support of this work. The portion of this work conducted by the Joint BioEnergy Institute, and the U.S. Department of Energy Joint Genome Institute, was supported by the Office of Science, Office of Biological and Environmental Research, of the U.S. Department of Energy (Contract No. DE-AC02–05CH11231). This research is funded in part by the Singapore National Research Foundation and the publication is supported under the Campus for Research Excellence And Technological Enterprise (CREATE) program for the project 'Nanomaterials for Energy and Water Management'. The funders had no role in study design, data collection and analysis, decision to publish, or preparation of the manuscript.

Competing Interests: The authors have declared that no competing interests exist.

* E-mail: t.konry @neu.edu

⑨ These authors contributed equally to this work.

Introduction

Worldwide water-associated infectious diseases are a major cause of morbidity and mortality [1]. It is estimated that 4.0% of global deaths and 5.7% of the global disease burden are caused by waterborne diseases [1–4]. Common waterborne diseases include diarrhea (bacterial, viral and parasitic), schistosomiasis, trachoma, ascariasis, and trichuriasis [1–4]. Low income countries are particularly vulnerable to waterborne diseases because of their under-developed infrastructure and poor water management [5–14]. Water and sewage distribution systems in high income societies also require pollutant and microorganism monitoring [15].

Escherichia coli, found in mammalian feces [16], has been a biological indicator for water quality since the 19th century [16]. Testing for the presence of *E. coli* is obligatory for current water management systems [17–19]. Herein, we report a comprehensive system – ScanDrop – for the rapid and specific identification of *E. coli* in drinking water.

The identification of bacteria in a water sample includes two major steps: 1) the capture of target bacteria from the water sample, and 2) the identification of the captured bacteria. Traditional methods for *E. coli* detection include culture, fermentation, enzyme-linked immunosorbent (ELISA), and polymerase chain reaction (PCR) assays [20,21]. These traditional methods have disadvantages including long identification times (2–4 days), and/or high labor and reagent costs [20,21]. Despite high costs, rapid tests are necessary to enable quick responses to putative contamination threats. Recently, novel sensors and assays for rapid pathogen detection have been developed, including the capture of whole pathogen cells or molecular fragments for further amplification and identification [22–27], with detection methods utilizing a variety of transducing technologies (optical, electro-chemical, surface plasmon resonance and piezoelectric) [27–40]. Many of these newer methods remain expensive and/or require sophisticated instrumentation, and most have yet to reach the market place. Therefore, there remains a need for alternative platforms for the detection of bacteria in water samples.

It remains challenging to inexpensively perform water quality control testing at multiple locations along a distribution system, and to rapidly process and share the test results. To address these challenges, we have developed the ScanDrop platform. ScanDrop is a self-contained detection platform that enables the online control of water testing at multiple locations along the distribution system. ScanDrop integrates live-bacteria capturing and detection, droplet microfluidics, automated fluorescence microscopy, and cloud-based data management and sharing. Droplet microfluidics, applied in ScanDrop, is an emerging application of microelectromechanical systems (MEMS) technology, where assay reagents and biological sample are confined to the pico-liter reactors, composed of water in oil emulsion [41–43]. Small volumes, rapid reagent mixing and non-complex droplet control make droplet microfluidics an attractive choice for the next-generation of high-throughput assays [41–43] and herein detection of bacteria in water samples.

In this work, we demonstrate ScanDrop's capability to detect live *E. coli* in water samples. Magnetic beads, conjugated with specific antibodies, were used to quickly and effectively capture *E. coli* from contaminated water. The captured bacteria were then encapsulated into pico-liter droplets containing fluorescently labeled antibodies, for subsequent detection using a proprietary automated optical fluorescence signal registration system. Imaging system control was facilitated by leveraging a cloud-based laboratory automation system, coined Programing a Robot, PR-PR [44]. We envision that multiple ScanDrop systems could be dispatched at multiple locations to form a cloud-enabled water quality assessment network. Each system could be managed in real-time from a remote control center. Such a network could potentially reduce the infrastructure, management, and labor costs required to perform multiple sample analysis and rapidly share results.

Results and Discussion

Bead-based *E. coli* capture and detection assay

Herein the isolation of bacteria and detection are conducted utilizing simple magnetic bead based immunoassay thus no bacteria agar plate cultivation step is necessary to identify a presumptive positive sample. This approach saves considerable time and resources. In our approach, magnetic beads conjugated with anti-*E. coli* antibodies are added to a water sample (**Fig. 1**). Within 10 min, the magnetic beads have captured the bacteria (if any) from the water sample. The beads are then concentrated with a simple magnet (**Fig. 1**), and a single immunoassay step labels the captured bacteria with a fluorescent antibody for subsequent detection (**Fig. 1**). Detection protocols are integrated into a droplet microfluidic device to reduce reagent volume and enhance reaction rates.

ScanDrop Sensor

The ScanDrop sensor consists of two major components: 1) a droplet microfluidic device for bacteria labeling, and 2) a portable fluorescent optical system for signal detection and sharing.

Droplet microfluidic device. To reduce reagent volumes and detection times, we designed a pico-liter droplet microfluidic chip. The design of poly(dimethylsiloxane) (PDMS) microfluidic device is shown in **Fig. 2A**, the generation of monodisperse droplets in a micro-channel through shearing flow at a flow-focusing zone in **Fig. 2B**, and the resulting droplet array in **Fig. 2C**. Three perpendicular inlet channels form a nozzle, (**Fig. 2A rectangle**), independent syringe pumps controlling flow rates for the oil, beads, and fluorescently labeled secondary antibodies streams. Each droplet in the array co-encapsulates fluorescently labeled anti-*E. coli* antibodies with captured bacteria (if any), to generate a localized fluorescent signal for subsequent detection. The chip enables the generation and incubation of 10^3 droplets with ~100 micron diameter (~520 pL). The advantages of this droplet-based array technique include the physical and chemical isolation of beads in droplets, and the rapid and efficient mixing of the reagents that occurs inside droplets providing fast reaction rates [45–48]. Importantly, this nano-liter microenvironment also enables gas exchange for bacterial viability if further studies are required [45–48]. Previous works in the field of droplet microfluidics showed that the chance to find a cell or a bead inside droplet follows Poisson distribution [49,50]. This puts certain theoretical limitations of the limit of detection of bacteria of droplet microfluidic system. Clausell-Tormos et al. showed that decreasing the number of cells in aqueous solution that is converted to droplets to less than 10^6 cell/mL reduced the probability to find droplets with encapsulated cells and increased the number of empty droplets [50]. Therefore, at the end of incubation time we need to get at least 10^6 CFU/mL of bacteria. The relation between the bacteria in the analyzed sample and the bacteria subjected to encapsulation after enrichment is as follows:

$$2^n N_0 = N_d$$

Where N_o (CFU/mL) is the initial load of bacteria, n is the number of generations in enrichment phase, N_d (CFU/mL) is the concentration of bacteria subjected to encapsulation after enrichment.

Given the generation time of 20 min for *E. coli* in the optimum cultivation condition, to get 10^6 CFU/mL at the end of 6 hours of incubation (18 generations), proposed in our assay, the initial concentration should be at least 3.5 CFU/mL.

To demonstrate the feasibility of our ScanDrop system for multiplex analysis, we co-encapsulated red florescent protein (RFP) and green florescent protein (GFP) expressing *E. coli* in the same droplet (**Fig. 2D**). Capturing on a bead and latter encapsulation for detection of two different bacteria in the same microenvironment will enable multiplex future studies using several types of beads conjugated with different antibodies that bind different target bacteria and different fluorescent tags. The probability of capturing two different objects in a single droplets were analyzed in [49].

Optical system

The schematic for the portable optical system for fluorescent signal detection in the droplet microfluidic device is presented in **Fig. 3A**. The system enables remote microscope control as well as simultaneous top and inverted image registration (**Fig. 3B**). The top camera allows for whole chip bright field imaging, while the bottom camera allows for fluorescence imaging with $10\times$ magnification. This combination allows for high-throughput droplet imaging. A robotic stage is used to scan the array of multiple droplets, with an XY microscope scanning range of 45 mm×45 mm and a resolution of 5 μm/step (10 mm/sec). Z-axis focus capabilities include 15 mm travel with 1 μm/step, at 2 mm/sec. The ScanDrop optical system is controlled by Python scripts which can be automatically generated by PR-PR.

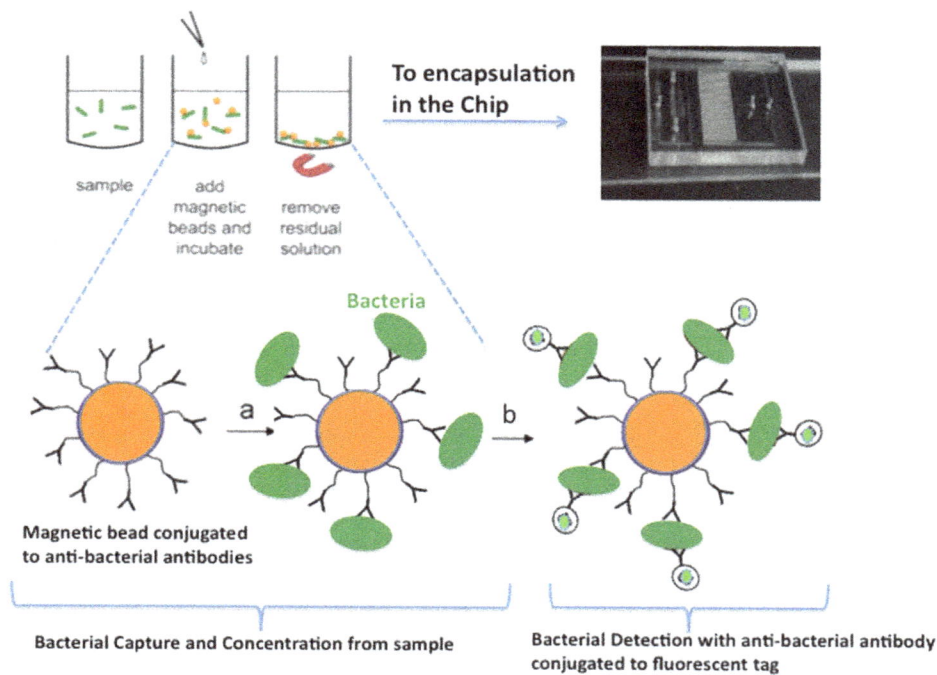

Figure 1. Bacteria capturing and detection assay. Magnetic bead capture of *E. coli* from enriched water samples, and downstream chip encapsulation for fluorescent labeling and detection. 1L of water is passed through a 0.22 μm filter, which is then incubated for 6 hr in LB media. Dynabeads® MAX anti-*E. coli* O157 are added to the resulting cell culture ("sample"), incubated for 20 min, and concentrated via magnet. The beads (potentially conjugated with bacteria) are then co-encapsulated with secondary fluorescently labeled anti-*E. coli* antibodies in the chip and incubated up to 1 hour before imaging.

PR-PR cloud-based Laboratory Automation System, and Data Management and Sharing

In this work, we have further developed PR-PR, a biology-friendly high-level language for laboratory automation [44], to control ScanDrop's automated microscopy system and enable ScanDrop to be promptly and easily adjusted to changes in experimental protocol. In PR-PR, transfer of a material (*e.g.*, a liquid) or system component (*e.g.*, a robotic arm) is described by a

Source, Destination, Quantity, and Method. For ScanDrop, the Source is the initial coordinates of the microscope stage (XY) and lens (Z), the Destination is the final target coordinates of interest, the Quantity is the number of pictures that should be taken, and the Method specifies imaging parameters: light, filters, and delay between image capture. PR-PR inputs a script for ScanDrop automated microscope control (such as that presented in **Fig. S2**) and outputs a Python script that can directly operate the

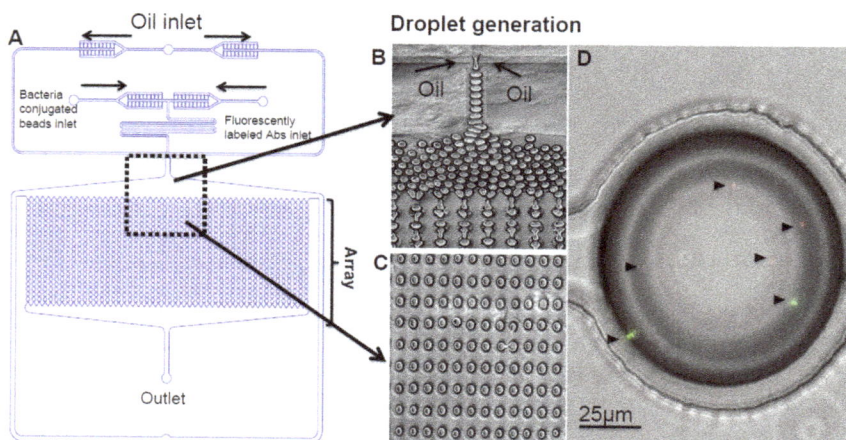

Figure 2. Droplet microfluidic device for bacteria monitoring: A) Schematic representation of ScanDrop droplet microfluidic chip and fluid control system. **B)** Droplet generator. **C)** Droplet incubation array (up to 10³ droplets can be incubated simultaneously). **D)** Co-encapsulation of GFP- and RFP-expressing *E. coli* inside a single droplet (20× magnification, Zeiss microscopic imaging). Arrows indicate single bacteria cells.

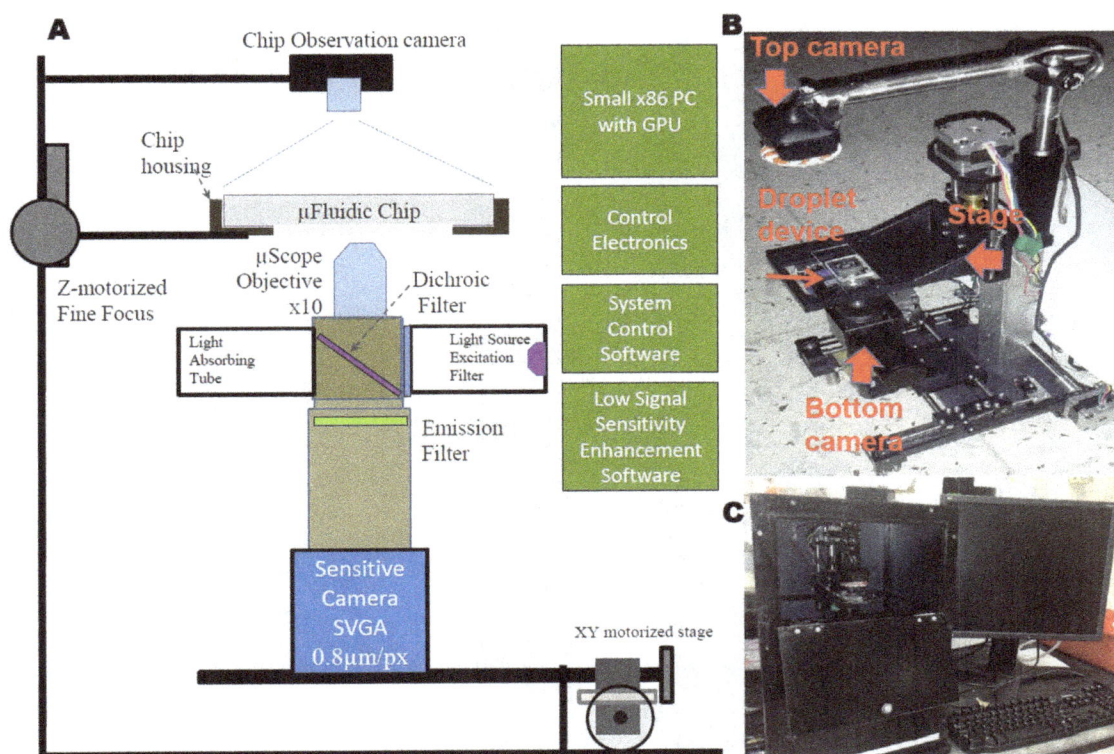

Figure 3. Portable fluorescence microscope system: A) Microscope design scheme. **B)** Robotic stage and two cameras (arrows) for chip observation and data acquisition. **C)** ScanDrop system set up, including 2 pressure pumps to create droplets, a microscope imaging system, internet access, and a monitor for local data viewing.

ScanDrop automated microscope system. The PR-PR script protocol and the resulting data for each experiment are stored in a local folder within the ScanDrop sensor and can shared between users via Dropbox.

Detection of RFP-expressing *E. coli* in drinking water

As a positive control, we tested ScanDrop for the detection of 150 CFU/mL of RFP-expressing *E. coli* in drinking water (**Fig. 4**). The overall assay for *E. coli* detection is divided into three steps: enrichment, capture, and detection. For enrichment, 1L of contaminated water sample was filtered and the filter with captured bacteria was incubated for 6 hours in the LB medium. For capturing, the enriched solution was mixed with Dynabeads® MAX anti-*E. coli* O157 for 10 min and separated by magnet.

Fig. 4B shows the droplet-based microfluidic chip used to perform the immunoassay described in **Fig. 1**. For the detection step, beads conjugated to bacteria captured from contaminated water sample were co-encapsulated with secondary FITC fluorescently labeled anti-*E. coli* antibodies in the droplet array and incubated up to 1 hr in the chip array at room temperature. **Fig. 4C** shows a representative droplet, and **Fig. 4D** shows the green florescence signal detected in a single droplet containing *E. coli* capture on the bead and tagged by secondary FITC labeled antibodies. The presence of RFP expressing *E. coli* in water was confirmed by PCR (**Fig. 5 C, D**).

Detection of fecal *E. coli* in drinking water

Next, we tested the ScanDrop system for the detection of fecal *E. coli* in drinking water. We contaminated the water with rat feces and applied the droplet detection assay described above. **Fig. 5A**

presents the procedure flow for the detection of fecal *E. coli* in water. **Fig. 5B** shows a representative resulting image, with fluorescence signal indicating *E. coli* contamination. We confirmed the ScanDrop detection with PCR (**Fig. 5 C,D**), which clearly showed that the water samples were contaminated with *E. coli*. The results from the ScanDrop tests were uploaded to Dropbox cloud data storage.

The work presented here demonstrates the potential of automated microscale systems for water quality analysis. To detect *E. coli* in water samples, we developed and demonstrated a bead-based immuno-assay performed with a droplet microfluidic device to reduce reagent volume and enhance reaction rates. We integrated the microfluidic assay with a portable imaging system and remote control automation software. We demonstrated ScanDrop system capabilities through the detection of model coliform bacteria, *E. coli*, in feces–contaminated drinking water. Our successful multiplex detection assay results suggest that simultaneous multiple bacteria detection, using several types of beads conjugated with different antibodies that bind different target bacteria, will be possible with further development. The ScanDrop platform decreased reagent volumes, (the full chip uses 520 nL of reagents, while conventional assay require at least 10 µL of reagents) and allows for results within 8 hours from the time of water sampling. Our results demonstrate that a combination of droplet microfluidics with low cost optics and cloud network can provide a flexible and efficient alternative for pathogen detection in drinking water. The ScanDrop platform has the potential to significantly improve water diagnostics, particularly in low income countries where the infrastructure does not yet exist [51,52].

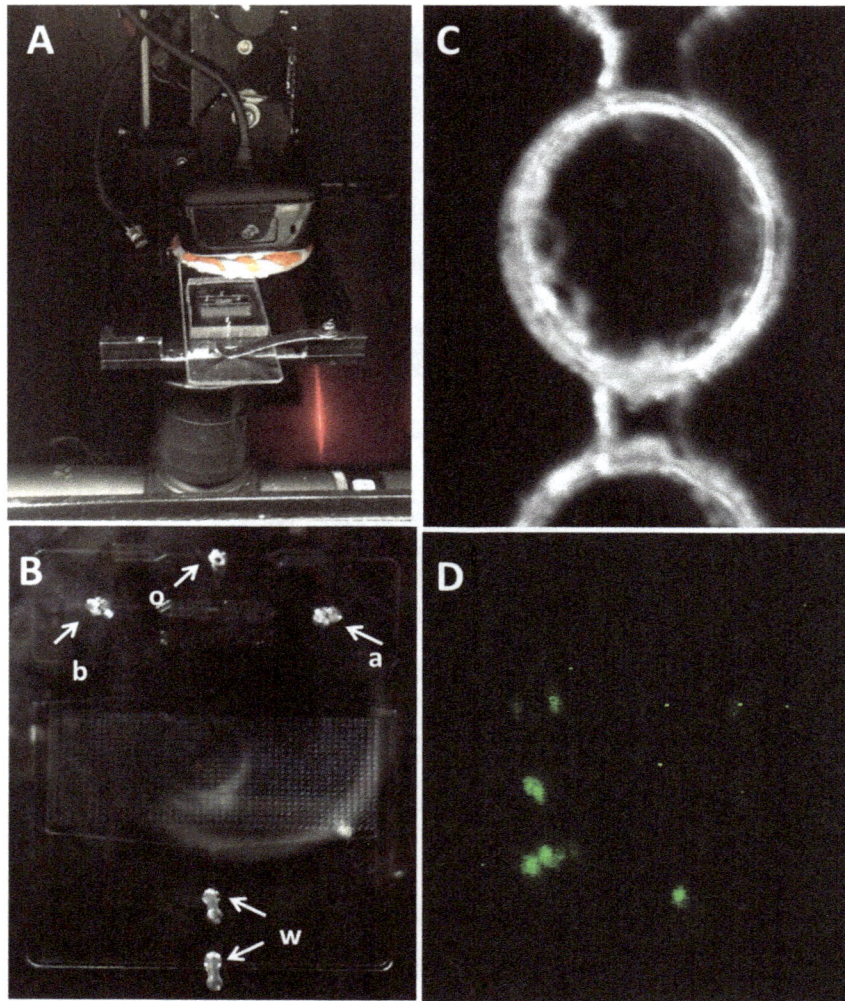

Figure 4. ScanDrop: *E. coli* detection: **A**) Microfluidic chip operating inside the imaging system. **B**) Droplet array as viewed from the top camera. Arrows indicate tubing/chip connection locations as follows: o- oil (inlet); b- beads conjugated with *E. coli* (inlet); a- fluorescently labeled antibody (inlet); w- waste (outlets). **C**) Droplet image as seen from the top camera with white LED illumination. **D**) Antibody green fluorescence indicates the presence of RFP-expressing *E. coli* (positive control).

Conclusions

We developed the ScanDrop platform for *E. coli* detection in water. The platform uses magnetic beads to capture bacteria, droplet microfluidics to encapsulate the captured bacteria with fluorescent antibodies, low cost portable optics for signal detection, PR-PR to facilitate microscopy control and data acquisition, and cloud-based storage for results sharing. The use of droplet microfluidics increases reaction kinetics and reduces reagent volumes (lowering the cost per test), the developed florescence microscopy system allows for data generation in multiple locations, and PR-PR facilitates ScanDrop control. A schematic illustration of an envisioned ScanDrop network for water quality analysis is shown in **Fig. 6**. The ScanDrop network would consist of 1) the PR-PR laboratory automation system and cloud-based data storage for remote control, image capturing, and result sharing; and 2) ScanDrop sensor stations deployed at multiple water distribution locations. The control station would perform image analysis for multiple sensors and shares the test results in the real time with multiple end users. This ScanDrop network could contribute to more rapid, cost-effective, and continuous water quality monitoring systems, with centralized facilities simultaneously monitoring multiple water sampling sites without complex imaging or data processing infrastructure.

Materials and Methods

ScanDrop optics system

A custom made, motorized, dual view, computerized portable microscopy system was designed for droplet microfluidic imaging (R&D Engineering Solutions, Netania, Israel). The dual view system was used for the simultaneous imaging of the whole chip (top view camera) and specific droplets (bottom view camera). The top view camera includes: 1280×768 resolution, color sensor, auto/computer-controlled focus, manually configurable [83×50 mm - 30×18 mm] field of view, 640×480 region of interest (ROI), and zoom functionality. The bottom view (microscope) camera includes: 752×582 resolution, monochrome 8.6 µm×8.3 µm pixels sensor, and a 10× objective. A single 3W 468 nm light emitting diode (LED) was used for florescence

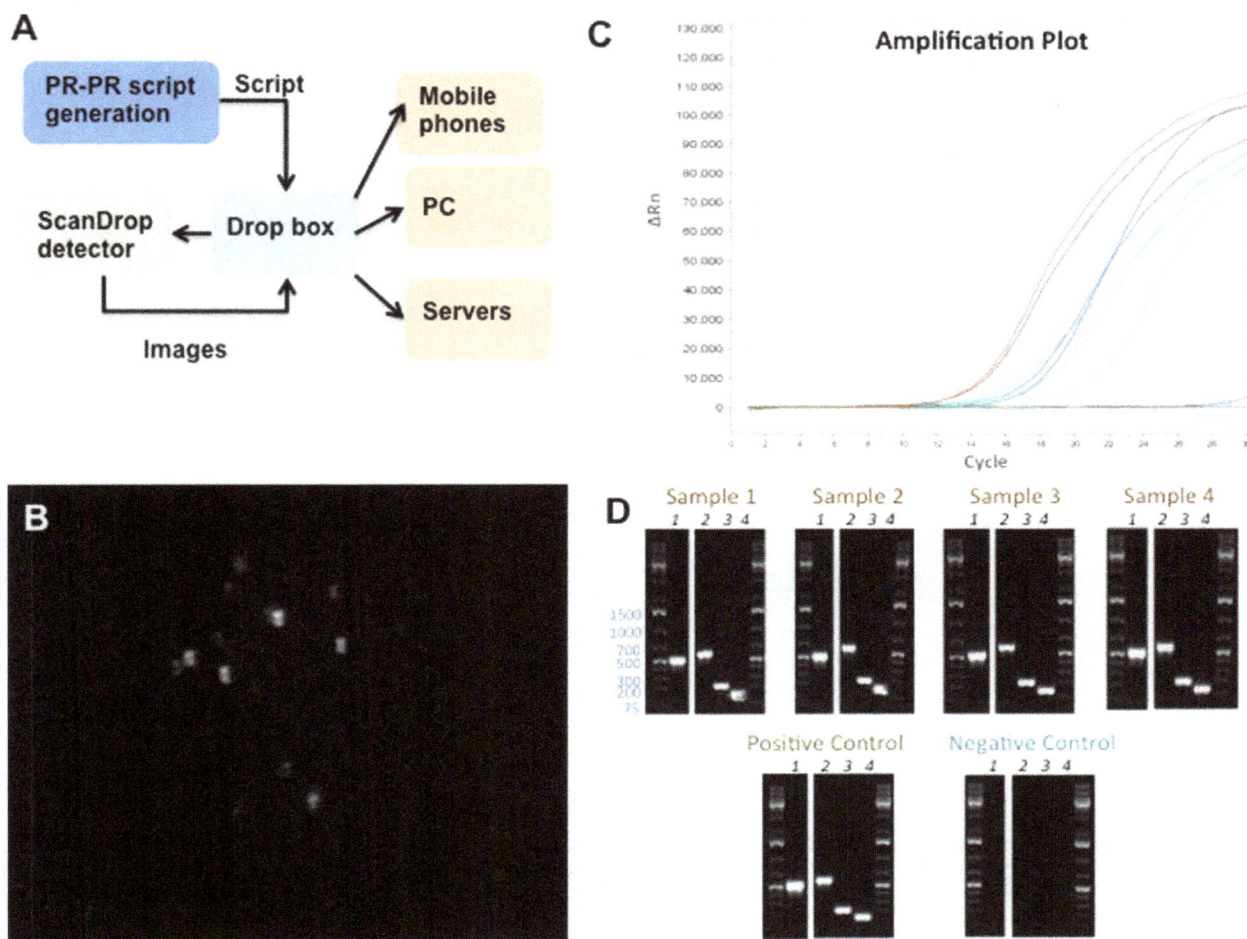

Figure 5. Detection of fecal *E. coli* in drinking water: A) ScanDrop detection network. PR-PR generates a Python script to control the ScanDrop detector. The Python script is uploaded to Dropbox, and then run on the ScanDrop detector. The captured images are uploaded to Dropbox, and then distributed to various devices. **B)** Representative ScanDrop image demonstrating fecal *E. coli* detection in drinking water. **C)** Real-time PCR amplification plot for contaminated sample 3, and positive and negative controls. Red curves indicate amplification of primary *16S rRNA* locus, cyan amplification of secondary *16S rRNA* locus, green *tuf* locus, and blue *uidA* locus. Sample 3 and the positive control amplified similarly. The negative control did not amplify. **D)** Gel electrophoresis analysis of PCR reaction products for the four contaminated samples, along with positive and negative controls. For all samples, gel lanes correspond to the amplification of loci as follows: lane 1 - *16S rRNA* primary locus, lane 2 - *16S rRNA* secondary locus, lane 3 - *tuf*, lane 4 – *uidA*.

excitation. A 41017 - Endow GFP/EGFP bandpass fluorescence filter set (Chroma Inc., VT) was used for florescence detection. Top illumination was made by a single 30 mW white LED for chip observation and microscope camera positioning. An embedded ×86 dual core computer with HDMI display port outputs (CompuLab, Israel) was used for the local control of the system. An embedded computer runs custom software, which allows full control of the microscope, including XY position, focus, illumination, image acquisition and enhancement. A system can be controlled manually by the human operator via a standard PC console (keyboard, mouse and monitor). Alternatively, a system can be controlled programmatically via a program written in Python. We further improved the programmatic control aspect of our system by leveraging PR-PR [44], whereby a PR-PR microscope control script (**Fig. S2**) is translated into a Python script that can control the microscope system, as described above. Python script deployment and image retrieval across distributed microscope systems was performed with the Dropbox cloud-based storage service.

Bacterial strains and plasmids

Plasmids pFAB_SchPMK36GFP and pFAB_SchPMK36RFP (unpublished results, Vivek Mutalik, Drew Endy, and Adam Arkin; see **Fig. S1**), both carrying a kanamycin resistance marker, were transformed into *E. coli* BW25113. These bacterial strains and plasmids, along with their associated information (*e.g.*, annotated Genbank-format DNA sequence files), have been deposited in the public instance of the JBEI Registry [53] (https://public-registry.jbei.org; corresponding Part IDs JPUB_001327-001329). For transformation, 1 µL pFAB_SchPMK36GFP or pFAB_SchPMK36RFP was mixed on ice with 40 µL chemically competent *E. coli* BW25113. The mixture was incubated on ice for 20 min, then placed at 42°C for 45 s (heat shock), and then returned to ice. 200 µL SOC media was then added to each tube of transformed cells and incubated with agitation at 37°C for 30 min. 100 µL of each transformation mixture was plated on solid LB media (Sigma-Aldrich, MO) supplemented with 30 µg/mL kanamycin (Sigma-Aldrich, MO) and then cultured at 37°C.

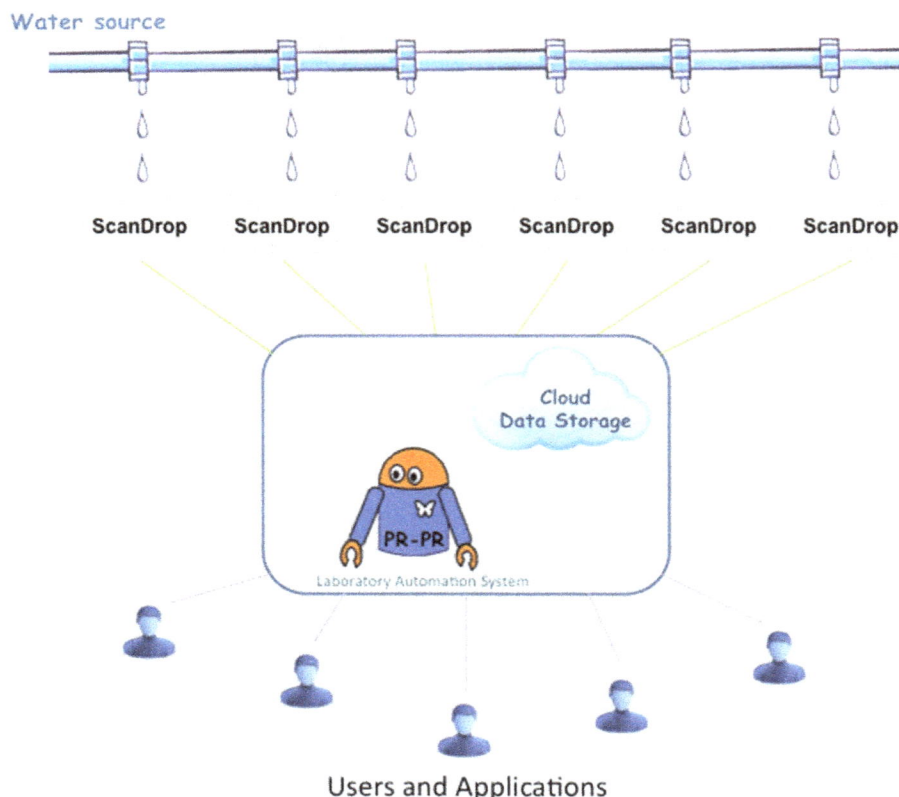

Figure 6. Schematic overview of a ScanDrop cloud-based water quality assessment system. The ScanDrop detector network is enabled by PR-PR and cloud-based data storage. Users send requests for water quality assessment at different locations in the distribution system. ScanDrop detectors perform the tests, the results are stored in the cloud, and the collected data is shared between users and applications.

Microfluidic device for droplet generation

The droplet microfluidic flow focusing device mask was fabricated by soft lithography. Negative photo resist SU-8 2100 (MicroChem, Newton, MA) was deposited onto clean silicon wafers to a thickness of 150 µm, and patterned by exposure to UV light through a transparency photomask (CAD/Art Services, Bandon, OR). To manufacture consumable devices, Sylgard 184 poly(dimethylsiloxane) (PDMS) (Dow Corning, Midland, MI) was mixed with cross-linker (ratio 10:1) and poured onto the photoresist pattern, degassed thoroughly and cured for 12 hours at 75°C. After curing, the PDMS devices were peeled off the wafer and bonded to glass slides after oxygen-plasma activation of both surfaces. The microfluidic device was composed of two parts: 1) a droplet forming nozzle (channel cross section $6.25 \cdot 10^{-8}$ m^2) and 2) a 10^3 droplets storage array (channel cross section $3.13 \cdot 10^{-7}$ m^2). The bonded microfluidic channels were treated with Pico-Sur^{TM} 2 (Dolomite Microfluidics, UK) by filling the channels with 10 µL of the solution as received and then flushing with air. This treatment was done to improve the wetting of the channels with mineral oil in the presence (1% w/w) of the surfactant (span80). 1 mL syringes were used to load the fluids into the devices through Tygon Micro Bore PVC Tubing 100f, 0.010" ID, 0.030" OD, 0.010" Wall (Small Parts Inc, FL). Individual syringe pumps (Harvard Apparatus, USA) were used to control the flow rates of oil and other reagents. To form droplets, the flow-rate-ratio of water-to-oil was adjusted to $Q_w/Q_o = 1$.

Droplet microfluidics multiplex detection assay

E. coli expressing GFP or RFP were incubated for 12 hours at 37°C in LB media (Sigma-Aldrich, MO) to 10^6 CFU/mL and encapsulated into droplets. Fluorescence images were captured on a Zeiss 200 Axiovert microscope using an AxioCAM MRm digital camera and AxioVision 4.8 software at 20× magnification. Each experiment consisted of 4 repeats.

ScanDrop detection of *E. coli* in water

1 L drinking water was spiked with RFP-expressing *E. coli* to 150 CFU/mL. The spiked water was filtered through a 0.22 µm filter (Corning Inc., NY), and the filter was then inoculated in 10 mL LB media (Sigma-Aldrich, MO) and incubated for 6 hr at 37°C. 20 µL of Dynabeads® MAX anti-*E. coli* O157 beads (Life Technologies, CA) were added to 1.5 mL of the incubation media and further incubated for 20 min on a rotating stage at room temperature (RT). Beads with captured bacteria were separated by magnet and resuspended in 400 µL of Phosphate Buffered Saline (PBS). The resuspended solution was co-encapsulated 500:1 with green fluorescently labeled anti-*E. coli* antibody (FITC, ab30522, Abcam, MA) in droplet reactors inside the chip positioned on the ScanDrop robotic stage. After a further 1 hr of incubation at RT, images were taken from different locations on the chip. The objective position movements were controlled via PR-PR, and the generated images were automatically uploaded to Dropbox. Each experiment consisted of 4 repeats.

Table 1. Primers and loci used for PCR detection of *E. coli*.

Locus	Primers	T_m	Amplicon Length
16S rRNA (primary)	ECA75F - GGAAGAAGCTTGCTTCTTTGCTGAC	60°C	544 bp
	ECR619R- AGCCCGGGGATTTCACATCTGACTTA		
16S rRNA (secondary)	16E1 - GGGAGTAAAGTTAATACCTTTGCTC	60°C	583 bp
	16E2 - TTCCCGAAGGCACATTCT		
tuf gene	TEcol553 - TGGGAAGCGAAAATCCTG	58°C	258 bp
	TEcol754 - CAGTACAGGTAGACTTCTG		
UidA gene	UAL - TGGTAATTACCGACGAAAACGGC	62°C	147 bp
	UAR - ACGCGTGGTTACAGTCTTGCG		

Drinking water contaminated with rat feces

Fresh feces were collected from rat cages in the animal facility of Massachusetts General Hospital. 1.5 g feces was mechanicaly homogenized in 1 L of drinking water. The contaminated water was filtered twice through a 40 μm filter (BD Falcon™, BD Biosciences, CA). The detection of E.coli in permeate was done by ScanDrop assay (as described in the previous section) and by Real-Time PCR (as described in the following section). Each experiment consisted of 4 repeats.Detection of RFP expressing and fecal origin E. coli by PCR. We chose four primer sets (**Table 1**) to detect E. coli in the prepared drinking water contaminated with rat feces. The primers targeted specific sequences from different loci in the E. coli genome: two primer sets for 16S rRNA, one for *tuf*, and one for *uidA* [54–57]. For each primer set, we tested four contaminated samples and a positive and a negative control. Negative controls contained water only, and positive controls contained water supplemented with *E. coli* BW25113. After enrichment of the microbial population (described above), 5 μL of enriched culture was added to 45 μL H_2O. All samples were incubated 15 min at 98°C and then diluted in additional 100 μL H_2O. Each 30 μL PCR reaction contained 10 μL of the diluted cell lysate (as template), 10 μL of 3× qPCR master mix (H_2O 3.3 μL, 5× Phusion HF 6 μL, dNTP 100 mM 0.25 μL, Phusion DNA Polymerase (NEB) 0.3 μL, SYBR® Green II 200× (Molecular Probes) 0.15 μL), and a pair of primers at 5 pmol each. PCR reactions were subjected to thermal cycling (3 min at 95°C, and then 30 cycles of 30 s at 95°C, 30 s at 58°C, and 30 s at 72°C, with a final hold step at 10°C) in a StepOnePlus™ Real-Time PCR System (Life Technologies, CA). We tracked the amplification curves and stopped the PCR amplifications after most reactions plateaued (**Fig. 5c**). We analyzed PCR fragments using electrophoresis by running the PCR products in 1% agarose gels (**Fig. 5d**). Each experiment consisted of 4 repeats.

PR-PR software availability

PR-PR is open-source software under the BSD license and is freely available from GitHub (https://github.com/jbei/prpr), and is also available through its web interface on the public PR-PR webserver (http://prpr.jbei.org).

Supporting Information

Figure S1   Plasmid maps for pFAB_SchPMK36GFP and pFAB_SchPMK36RFP.

Figure S2   Representative PR-PR script for ScanDrop. LOCATION declarations define microscope stage (XY) and lens (Z) locations. TRANSFER commands specify the starting and destination locations, the number of pictures to capture, and the capture parameters. In a single TRANSFER statement, multiple sequential destinations can be defined by location offset and number of repetitions.

Acknowledgments

The authors thank Vivek Mutalik for the gift of plasmids pFAB_SchPMK36GFP and pFAB_SchPMK36RFP.

Author Contributions

Conceived and designed the experiments: AG GL TK. Performed the experiments: AG GL IK NS. Analyzed the data: AG GL TK. Contributed reagents/materials/analysis tools: AG GL IK NS NH TK MY. Wrote the paper: AG GL NH TK. Read the manuscript: MY RM Supervised the work: MY RM.

References

1. Yang K (2012) Global distribution of outbreaks of water-associated infectious diseases. PLoS Negl Trop Dis, 2012. 6(2): p. e1483.

2. Lewin S (2007), Estimating the burden of disease attributable to unsafe water and lack of sanitation and hygiene in South Africa in 2000. S Afr Med J, 2007. 97(8 Pt 2): p. 755–62.

3. Kosek M, Bern C, Guerrant RL(2000) The global burden of diarrhoeal disease, as estimated from studies published between 1992 and 2000. Bull World Health Organ, 2003. 81(3): p. 197–204.

4. Pruss A (2002) Estimating the burden of disease from water, sanitation, and hygiene at a global level. Environ Health Perspect, 2002. 110(5): p. 537–42.

5. Cann KF (2013) Extreme water-related weather events and waterborne disease. Epidemiol Infect, 141(4): p. 671–86.

6. Teschke K(2010) Water and sewage systems, socio-demographics, and duration of residence associated with endemic intestinal infectious diseases: a cohort study. BMC Public Health, 10: p. 767.

7. Sarkar R(2013) Burden of childhood diseases and malnutrition in a semi-urban slum in southern India. BMC Public Health, 2013. 13: p. 87.

8. Jafari N (2011) Prevention of communicable diseases after disaster: A review. J Res Med Sci, 2011. 16(7): p. 956–62.

9. Subbaraman R(2013) The social ecology of water in a Mumbai slum: failures in water quality, quantity, and reliability. BMC Public Health, 2013. 13: p. 173.

10. Levy K(2012) Rethinking indicators of microbial drinking water quality for health studies in tropical developing countries: case study in northern coastal Ecuador. Am J Trop Med Hyg, 2012. 86(3): p. 499–507.

11. Stoler J (2012) When urban taps run dry: sachet water consumption and health effects in low income neighborhoods of Accra, Ghana. Health Place, 2012. 18(2): p. 250–62.

12. Gwimbi P (2011) The microbial quality of drinking water in Manonyane community: Maseru District (Lesotho). Afr Health Sci, 2011. 11(3): p. 474–80.

13. Mandeville KL(2009) Gastroenterology in developing countries: issues and advances. World J Gastroenterol, 2009. 15(23): p. 2839–54.

14. Al-Bayatti KK, Al-Arajy KH, Al-Nuaemy SH(2012), Bacteriological and physicochemical studies on Tigris River near the water purification stations within Baghdad Province. J Environ Public Health, 2012. 2012: p. 695253.

15. Ashbolt N, Grabow W, Snozzi M(2001) Indicators of Microbial Water Quality, in Water Quality: Guidelines, Standards and Health, L. Fewtrell and J. Bartman, Editors. IWA Publishing: London,UK. p. 289–316.

16. Edberg SC(2000) Escherichia coli: the best biological drinking water indicator for public health protection. Symp Ser Soc Appl Microbiol, (29): p. 106S–116S.

17. (2006) EPA, Total Coliform Rule Requirements. Available: http://water.epa. gov/lawsregs/rulesregs/sdwa/tcr/regulation.cfm. Accessed Nov 25 2013.

18. WHO (2006), Guidlines for drinking-water Quality. Third Edition.

19. Leclerc, H., Relationships between common water bacteria and pathogens in drinking-water, in Heterotrophic Plate Counts and Drinking-water Safety, J. Bartman, et al., Editors. 2003, IWA Publishing: London, UK. p. 80–118.

20. (2005) StandardMethods, For the examination of water and wastewater. 9020 Quality Assurance/Quality control. Available: http://www.standardmethods. org/store/ProductView.cfm?ProductID = 538. Accessed Nov 25 2013

21. Mairhofer J, Roppert K, Ertl P (2009) Microfluidic systems for pathogen sensing: a review. Sensors (Basel), 2009. 9(6): p. 4804–23.

22. Kao MC, Durst RA (2010) Detection of Escherichia coli Using Nucleic Acid Sequence-Based Amplification and Oligonucleotide Probes for 16S Ribosomal RNA. Analytical Letters. 43(10–11): p. 1756–1769.

23. Xiao XL, (2013) Detection of viable but nonculturable Escherichia coli O157:H7 using propidium monoazide treatments and qPCR. Can J Microbiol. 59(3): p. 157–63.

24. Langer V, Niessner R, Seidel M (2013) Stopped-flow microarray immunoassay for detection of viable E. coli by use of chemiluminescence flow-through microarrays. Anal Bioanal Chem, 2013. 399(3): p. 1041–50.

25. Rompre A (2002) Detection and enumeration of coliforms in drinking water: current methods and emerging approaches. J Microbiol Methods, 2002. 49(1): p. 31–54.

26. Liu Y (2008) Detection of viable but nonculturable Escherichia coli O157:H7 bacteria in drinking water and river water. Appl Environ Microbiol. 74(5): p. 1502–7.

27. Noble RT, Weisberg SB (2005) A review of technologies for rapid detection of bacteria in recreational waters. J Water Health. 3(4): p. 381–92.

28. Setterington EB, Alocilja EC (2011) Electrochemical Biosensor for Rapid and Sensitive Detection of Magnetically Extracted Bacterial Pathogens. Biosensors. 2(1): p. 15–31.

29. Bogomolny E, Swift S, Vanholsbeeck F, (2013) Total viable bacterial count using a real time all-fibre spectroscopic system. Analyst. 138(14): p. 4112–9.

30. Allevi RP (2013) Quantitative analysis of microbial contamination in private drinking water supply systems. J Water Health. 11(2): p. 244–55.

31. Johnson PE (2006) Fountain Flow cytometry, a new technique for the rapid detection and enumeration of microorganisms in aqueous samples. Cytometry A. 69(12): p. 1212–21.

32. Sun H(2009) Nano-silver-modified PQC/DNA biosensor for detecting E. coli in environmental water. Biosens Bioelectron. 24(5): p. 1405–10.

33. Abu-Rabeah K (2009) Highly sensitive amperometric immunosensor for the detection of Escherichia coli. Biosensors and Bioelectronics. 24: p. 3461–3466.

34. Liebes Y (2009), Immobilization strategies of Brucella particles on optical fibers for use in chemiluminescence immunosensors. Talanta. 80: p. 338–345.

35. Liebes Y, Marks R, Banai M (2009) Chemiluminescent optical fiber immunosensor detection of Brucella cells presenting smooth A antigen. Sensors and Actuators B. 140: p. 568–576.

36. Abdulhalim I (2009)Surface enhanced fluorescence from metal sculptured thin films with application to biosensing in water. App Phys Lett 94: p. 063206.

37. Abdulhalim, I., Zourob M., and Lakhtakia A., Surface plasmon resonance sensors-a mini review. J Electromagnetism 2008. 28(3): p. 213–242.

38. Karabchevsky A (2012) Microspot biosensing based on surface enhanced fluorescence from Nano-STFs. J NanoPhotonics, 2012. 6(6): p. 061508–1.

39. Shalabney A, Abdulhalim I (2011) Sensitivity enhancement methods for surface plasmon sensors. Lasers and Photonics Reviews. 5: p. 571–606.

40. Zourob M, Elwary S, Turner A (2008) Principles of Bacterial Detection - Biosensors, Recognition Receptors and Microsystems. NY: Springer.

41. Niu X, deMello AJ (2012) Building droplet-based microfluidic systems for biological analysis. Biochem Soc Trans. 40(4): p. 615–23.

42. Choi K (2012) Digital microfluidics. Annu Rev Anal Chem (Palo Alto Calif). 5: p. 413–40.

43. Guo MT (2012) Droplet microfluidics for high-throughput biological assays. Lab Chip. 12(12): p. 2146–55.

44. Linshiz G (2012) PaR-PaR Laboratory Automation platform. ACS Synthetic Biology,. 2(5): p. 216–222.

45. Baroud CN, Gallaire F, Dangla R (2010) Dynamics of microfluidic droplets. Lab on Chip,. 10: p. 2032–2045.

46. Bringer MR (2004) *Microfluidic systems for chemical kinetics that rely on chaotic mixing in droplets*. Philosophical Transactions of the Royal Society of London. Series A: Mathematical, Physical and Engineering Sciences. 362(1818): p. 1087–1104.

47. Handique K, Burns MA (2001) Mathematical modeling of drop mixing in a slit-type microchannel. Journal of Micromechanics and Microengineering. 11: p. 548–554.

48. Sarrazin F(2007) Mixing characterization inside microdroplets engineered on a microcoalescer. Chemical Engineering Science. 62(4): p. 1042–1048.

49. Mazutis L(2013) Single-cell analysis and sorting using droplet-based micro-fluidics. Nature Protocols. 8: p. 870–891.

50. Clausell-Tormos J(2008) Droplet-based microfluidic platforms for the encapsu-lation and screening of Mammalian cells and multicellular organisms. Chem Biol, 2008. 15(5): p. 427–37.

51. Reyes DR (2002) Micro total analysis systems. 1. Introduction, theory, and technology. Anal Chem. 74(12): p. 2623–36.

52. Buffi N(2011) Development of a microfluidics biosensor for agarose-bead immobilized Escherichia coli bioreporter cells for arsenite detection in aqueous samples. Lab Chip, 2011. 11(14): p. 2369–77.

53. Ham TS(2012) Design, implementation and practice of JBEI-ICE: an open source biological part registry platform and tools. Nucleic Acids Res, 2012. 40(18): p. e141.

54. Maheux AF(2009) Analytical comparison of nine PCR primer sets designed to detect the presence of Escherichia coli/Shigella in water samples. Water Res, 2009. 43(12): p. 3019–28.

55. Tsen HY, Lin CK, Chi WR (1998) Development and use of 16S rRNA gene targeted PCR primers for the identification of Escherichia coli cells in water. J Appl Microbiol. 85(3): p. 554–60.

56. Tantawiwat S (2005) Development of multiplex PCR for the detection of total coliform bacteria for Escherichia coli and Clostridium perfringens in drinking water. Southeast Asian J Trop Med Public Health. 36(1): p. 162–9.

57. Sabat G (2000) Selective and sensitive method for PCR amplification of Escherichia coli 16S rRNA genes in soil. Appl Environ Microbiol, 2000. 66(2): p. 844–9.

20

Using Magnetically Responsive Tea Waste to Remove Lead in Waters under Environmentally Relevant Conditions

Siang Yee Yeo[1], Siwon Choi[2], Vivian Dien[3], Yoke Keow Sow-Peh[4], Genggeng Qi[5], T. Alan Hatton[2], Patrick S. Doyle[2], Beng Joo Reginald Thio[1]*

1 Engineering Product Development, Singapore University of Technology and Design, Singapore, Singapore, **2** Department of Chemical Engineering, Massachusetts Institute of Technology, Cambridge, Massachusetts, United States of America, **3** Department of Materials Science and Engineering, Massachusetts Institute of Technology, Cambridge, Massachusetts, United States of America, **4** Science Department, Hwa Chong Institution (High School), Singapore, Singapore, **5** Kaust-Cornell Center for Energy and Sustainability, Cornell University, Ithaca, New York, United States of America

Abstract

We report the use of a simple yet highly effective magnetite-waste tea composite to remove lead(II) (Pb^{2+}) ions from water. Magnetite-waste tea composites were dispersed in four different types of water–deionized (DI), artificial rainwater, artificial groundwater and artificial freshwater–that mimic actual environmental conditions. The water samples had varying initial concentrations (0.16–5.55 ppm) of Pb^{2+} ions and were mixed with the magnetite-waste tea composite for at least 24 hours to allow adsorption of the Pb^{2+} ions to reach equilibrium. The magnetite-waste tea composites were stable in all the water samples for at least 3 months and could be easily removed from the aqueous media via the use of permanent magnets. We detected no significant leaching of iron (Fe) ions into the water from the magnetite-waste tea composites. The percentage of Pb adsorbed onto the magnetite-waste tea composite ranged from ~70% to 100%; the composites were as effective as activated carbon (AC) in removing the Pb^{2+} ions from water, depending on the initial Pb concentration. Our prepared magnetite-waste tea composites show promise as a green, inexpensive and highly effective sorbent for removal of Pb in water under environmentally realistic conditions.

Editor: Zhi Zhou, National University of Singapore, Singapore

Funding: This work was supported in part by the Singapore University of Technology and Design (SUTD) Start-Up Research Grant SRG EPD 2012 022 and the SUTD-MIT International Design Centre Research Grant IDG11200105/IDD11200109, together with financial support from the Singapore-MIT Alliance (SMA). The funders had no role in study design, data collection and analysis, decision to publish, or preparation of the manuscript.

Competing Interests: The authors have declared that no competing interests exist.

* E-mail: reginaldthio@sutd.edu.sg

Introduction

As industrialization continues worldwide, heavy metal contamination of the environment is becoming an increasing concern for its adverse health effects on human society. Heavy metals are released into the environment primarily via one of four main anthropogenic processes: the mining and smelting industries; manufacturing; agriculture; and fossil fuel burning power plants, vehicles, ships and aircrafts. Most heavy metals are toxic to humans. Specifically, lead (Pb) is a major cause of neurological diseases that impair basic mobility functions, and lead to growth defects in children [1]. Leakages of Pb ions into the water bodies generally occur as a result of corrosion of Pb-containing plumbing systems such as pipes and fittings, and surface run-offs of Pb-based materials (e.g. paints). Pb concentration levels in water systems around some industrial areas have been found to reach as high as 500 ppm [2], while the World Health Organization (WHO) recommends Pb concentrations below 0.01 ppm for safe drinking water [3].

The most common material used for heavy metal removal in contaminated waters is powdered activated carbon (AC) [4]. The high adsorption capacity of AC for the heavy metal ions can be attributed to its microporosity and high surface area to mass ratio

[5]. However, the use of AC for water treatment can result in very high costs economically and environmentally. High temperatures of up to 800°C [6] are required to produce or "activate" the carbonaceous material, leading to high energy and capital costs [7]. This has provided motivation to search for a less expensive, more efficient natural adsorbent for removing heavy metals from wastewater.

Tea is very readily available in much of the world and has been proven to exhibit a high removal efficiency and sorption capacity for Pb^{2+} [8,9]. The surface of tea leaves contains many polar, aliphatic and aromatic function groups [10] that allow it to successfully adsorb contaminants such as Pb^{2+}, Ag^+, Cu^{2+} and Al^{3+} ions [8]. Waste tea from local restaurants and cafes could be utilized to create a secondary use for a product that is already widely consumed in our societies. Magnetization of the tea waste (via the synthesis of a magnetite or Fe_3O_4 coating) can also serve as a more efficient, easier and faster way of removing the adsorbent-contaminant complex in place of filtration or flocculation. However, iron (Fe) can serve as a micronutrient that promotes algal bloom in inland and coastal waters [11]. Hence it is critical that the introduction of magnetite-waste tea composites as sorbents for Pb removal in contaminated waters do not accidentally create

Table 1. Concentrations of main electrolytes and dissolved organic content in the artificial waters used in the experiments.

	Units	Artificial Freshwater	Artificial Groundwater	Artificial Rainwater	DI Water
pH		8.4	7.1	4.2	6.3
TOC	µM C	5200	1564	–	–
Na^+	mg/L	50	45	19.4	–
Ca^{2+}	mg/L	26.5	87	22.4	–
Cl^-	mg/L	124.6	155	60.7	–
HCO_3^-	mg/L	–	120	–	–

algal blooms with the release of large amounts of Fe ions into the waters.

The objective of this study is to quantify the Pb^{2+} sorption capacity of magnetically responsive waste tea (magnetite-waste tea composite) under environmental conditions similar to those found in natural waters which are typically used for drinking needs. These include but are not limited to rainwater, groundwater and freshwater. Numerous studies [8,12,13,14] have been published that demonstrated the effectiveness of tea leaves as sorbents for Pb^{2+}. However, their studies were performed in acidic media with varying initial Pb concentrations of up to 400 ppm and in the absence of natural organic matter (NOM); such conditions are not representative of actual real-world conditions. Further, $[Pb^{2+}]$ at concentrations above 100 ppm have been reported to precipitate at pH >5.3 [8], making pH effect sorption studies difficult if not impossible. For these reasons in this work, we varied the initial $[Pb^{2+}]$ from 0.1–5 ppm in artificial waters to allow us to understand the Pb sorption capacity and performance of the magnetite-waste tea composite in the simulated environmentally-relevant conditions.

Materials and Methods

Materials

Tea waste (Lipton Red Tea and Dilmah Ceylon Black Tea) were obtained from the Hwa Chong Institution (HCI) and Singapore University of Technology and Design (SUTD) cafeterias, washed with hot boiling distilled water 3 times and then rinsed with deionized (DI) water 7 times to remove any traces of impurities. The tea dusts were then dried overnight in an oven at

70°C, and subsequently stored in plastic vials in dessicators for use as sorbents in the sorption experiments.

Iron(II) sulfate (Chemicon), iron(III) chloride (Scharlau), hydrochloric acid (Honeywell), ethanol (Merck, Germany), Darco G-60 100 mesh activated carbon, AC (Sigma-Aldrich) and aqueous ammonia (Honeywell) were all analytical grade reagents and used as-is without further purification. The stock standard solution of 1000 ppm $Pb(NO_3)_2$ was of atomic absorption spectroscopy (AAS grade) and purchased from Perkin Elmer (Waltham, MA).

Synthesis and Characterization of Magnetite-waste Tea Composite

The synthesis of the magnetite-waste tea composite was performed using the method adapted from a previous study involving the preparation of magnetic ragweed pollen grains [15]. First, 13.31 g of $FeCl_3 \cdot 6 H_2O$, 27.77 g of $FeSO_4 \cdot 7 H_2O$, 45 ml of DI water, 5 ml of 5 M HCl and 5 ml of ethanol were mixed in a 100 ml beaker, followed by stirring and heating to 90°C, to make sure all Fe salts were fully dissolved. Second, 1 g of tea waste powder was dispersed in 30 ml of the Fe solution and stirred for 48 hours at room temperature on a magnetic stirrer to allow for the complexation of tea with the Fe salts. After the 2 days, the remaining iron solution was filtered off using a vacuum pump and the tea was rinsed briefly with DI water. Third, the filtered tea was then transferred into a clean beaker and 20 ml of 25% (w/v) aqueous ammonia was added. After 2 hours, the magnetite-waste tea was again filtered to remove the ammonia and rinsed with DI water. Finally, the magnetite-waste tea composite was placed on glass petri dishes and left to dry overnight in the oven at 60°C, and then subsequently stored in plastic vials. The synthesis steps for

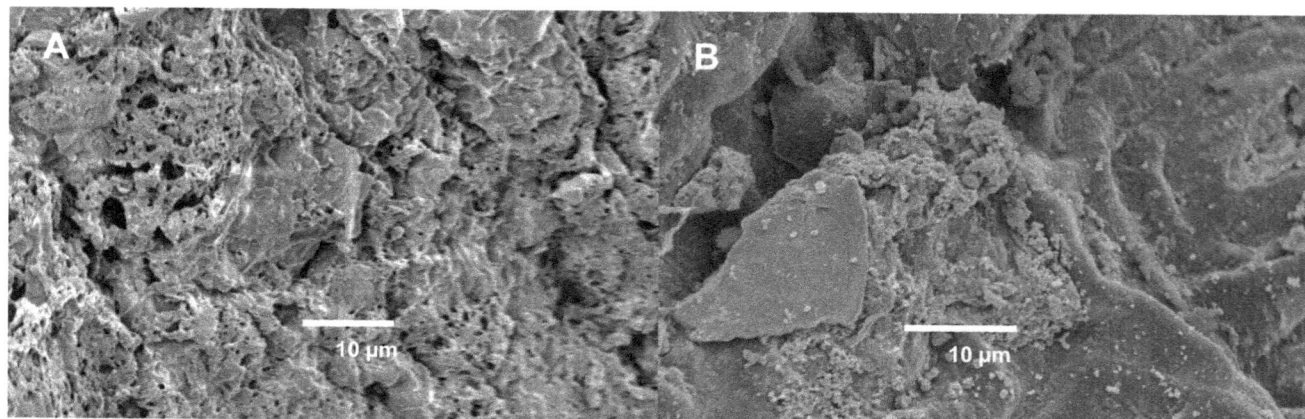

Figure 1. SEM images of waste tea powder (A) without and (B) with magnetite coating.

Table 2. Key parameters of the tea powders (unmagnetized tea and magnetite-waste tea composite) measured using the nitrogen physisorption isotherms.

	Units	Unmagnetized tea	Magnetite-waste tea composite	AC
BET surface area	m²/g	1.04	6.00	3350
Pore size	Å	160	125	32

magnetite are similar to the magnetite-waste tea composites except that the tea powders were not added.

The morphologies of both uncoated tea waste and magnetite-waste tea were obtained using a scanning electron microscope (SEM JEOL-6060SEM, JEOL Ltd, Japan). All samples for SEM were sputter-coated with platinum prior to imaging.

Vibrating Sample Magnetometry (VSM Model 1660, ADE) was used to obtain the magnetization (*M-H*) curve of magnetite-waste tea composite powders at room temperature. Each sample was prepared by fixing 20–80 mg of the powder onto a glass coverslip and fixing the coverslip to the VSM sample holder. All parts were held together by double-sided adhesive tape.

Thermogravimetric Analyses (TGA) of both uncoated and magnetite-waste tea composite powders were performed on a TGA Q50 (TA Instruments). 5 to 15 mg of each sample was taken for measurement under a constant nitrogen flow rate of 100 mL/min. The temperature was raised from 25°C to 980°C at a rate of 20°C/min.

Nitrogen physisorption isotherms were measured at 77 K using a Micromeritics ASAP 2020 analyzer. The samples (uncoated tea and magnetite-waste tea composite powders) were degassed at 323 K under vacuum for 24 hours prior to the measurements. The specific surface areas of the samples were calculated by the Brunauer-Emmett-Teller (BET) method [16]. The pore sizes of the samples were calculated using the Barrett–Joyner–Halenda (BJH) model [17].

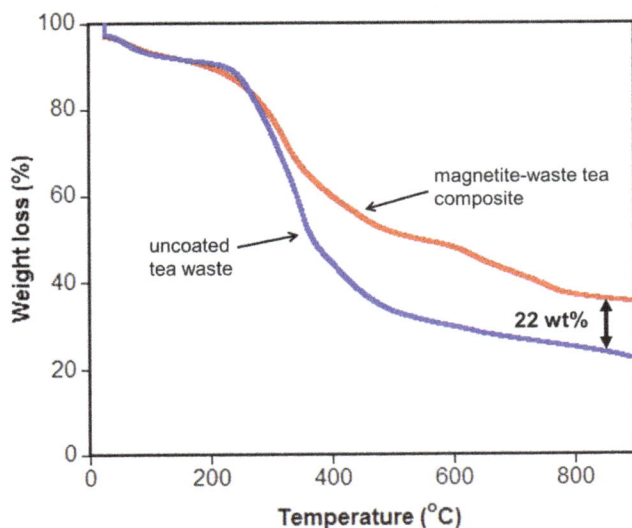

Figure 3. Magnetization curve of the magnetite-waste tea composite at room temperature. The points were obtained from VSM measurement, while the solid line was the best fit from the Langevin function assuming the composite particles were monodisperse.

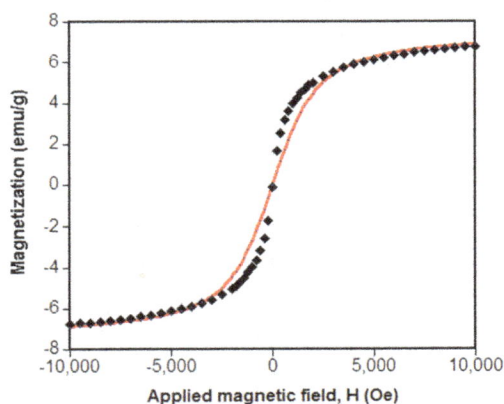

Figure 2. TGA analyses of the uncoated and magnetite-waste tea composite showing that the magnetite coating accounts for about 22% of the total mass of the composite.

Preparation of Artificial Waters

DI water (Mili-Q, Millipore) with a resistivity of 18.2 MΩ.cm was used to prepare the artificial waters, with the appropriate amounts of key electrolytes and dissolved organic contents [18,19,20]. All three of the artificial waters (rainwater, groundwater and freshwater) were prepared by dissolving similar amounts of the salts (NaCl, $CaCl_2$, $MgCl_2$ and $NaHCO_3$) commonly found in real natural waters while organic content was represented using humic acid (Sigma-Aldrich). About half the given mass of humic acid is made up of carbon [21]. The concentrations of the key electrolytes and organic carbon in the various artificial waters are given in Table 1.

Fe Leaching and Magnetite-waste Tea Composite Stability Tests

Iron (Fe) has been shown in studies to be a critical nutrient in promoting harmful algal blooms [11,22]. In order to determine the stability of the magnetite coating on the tea, leaching tests were performed to observe the amount of Fe that would leach out of the magnetite-waste tea composite in all 4 water types. 50 mg of magnetite-waste tea composite was dispersed in 10 ml of water sample in a 15 ml centrifuge tube (BD Falcon), and left on a shaker for 3 consecutive days. 3 samples (triplicates) were removed after 24 hours (Day 1), 48 hours (Day 2), and 72 hours (Day 3) respectively, for each water type. The samples were then centrifuged at 8000 rpm for 5 minutes. A calibration curve of Fe standards (1, 2, 5 ppm) was run on the Flame-Atomic Absorption

Figure 4. Testing the magnetic stability of magnetite-waste tea composite after 8 days in water.

Spectrometer (AAS, AA-6300 Shimadzu), and the water samples were analyzed on the AAS for iron content.

Batch Adsorption Tests

Batch sorption experiments to evaluate the efficiency of removal of Pb^{2+} ions by the tea wastes (both uncoated and magnetite coated), magnetite and AC were carried out in 15 mL centrifuge

tubes (BD Falcon) at room temperature. 30 mg of each sorbent was added to 10 mL of water with varying initial Pb^{2+} concentrations. Starting Pb^{2+} concentrations ranged from 0.16–5.55 ppm; these values are representative of the typical Pb^{2+} concentrations in environmentally realistic Pb–contaminated waters [23,24]. Samples were placed on an orbital shaker (Stuart) for 24 hours to allow the adsorption to achieve equilibrium and to fully disperse the adsorbents in the artificial waters. The samples were then centrifuged (Beckman Coulter) at 8000 rpm for 5 minutes prior to sampling by an atomic absorption spectrophotometer (AAS, Shimadzu AA-6300) for lead content. A calibration curve of Pb standards (0.1, 1, 2 and 5 ppm) was run on the AAS with each batch of sorption experiments. All batch adsorption tests were carried out at least thrice. One-way analysis of variance (ANOVA) at 95% confidence level (Microsoft Excel 2010) was used to check for statistically significant differences in the batch Pb^{2+} sorption experiments between the four different sorbents under the same solution conditions. Differences in Pb^{2+} sorption results with p-value <0.05 are considered to be statistically significant.

Zeta Potential Measurements

The electrophoretic mobility (EPM) of the various sorbents (tea wastes, magnetite and AC) was characterized using Laser Doppler velocimetry (Malvern Zetasizer Nano ZS-90), with the EPMs converted to ζ-potentials using the Smoluchowski equation [25]. Before each measurement, the sample was subjected to ultrasonic treatment for several minutes to disperse the sorbents in suspension thoroughly. 5 measurements were conducted for each sample.

Results and Discussion

Coating and Characterization of Magnetite onto Tea Waste

The magnetite-waste tea composite was successfully synthesized by using the co-precipitation of Fe(II) and Fe(III) salts in aqueous ammonia and is shown in Figure 1. The BET surface areas for the uncoated tea, magnetite-waste tea composite and activated carbon (AC) were measured to be 1.04, 6.00 and 3350 m^2/g respectively, while the pore sizes were calculated to be 160, 125 and 328 Å. Table 2 summarizes the key parameters of the sorbents measured using the nitrogen physisorption isotherms.

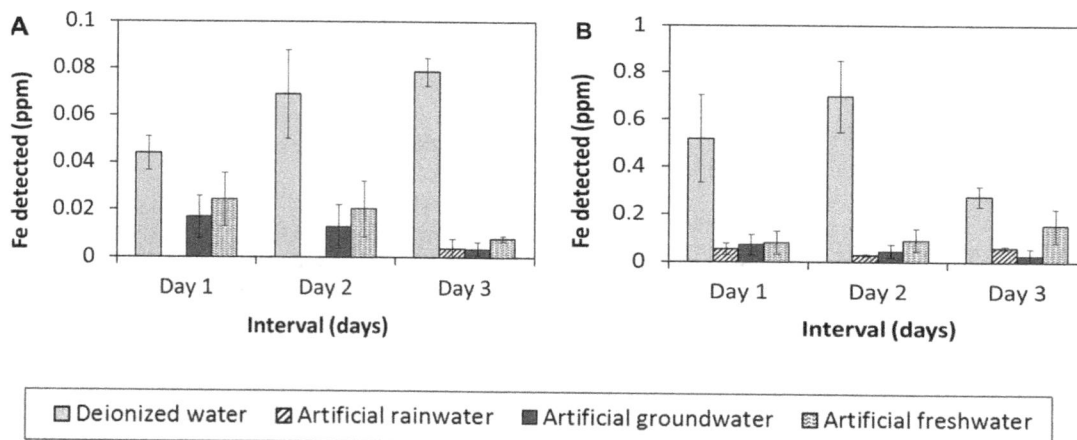

Figure 5. Amount of Fe leached out into the four waters from uncoated tea (A) and magnetite-waste tea composite (B).

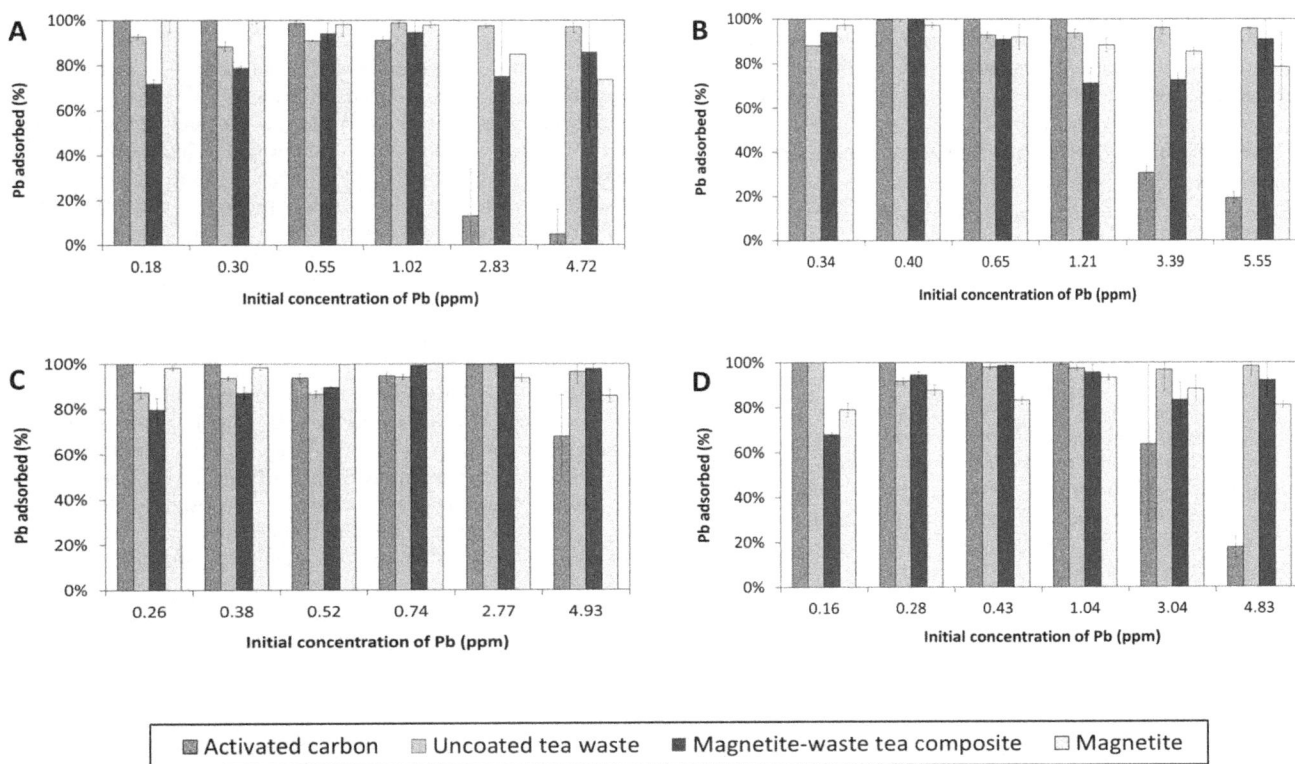

Figure 6. Percentage of Pb adsorbed onto the four sorbents for each of the four different waters: (A) deionized water, (B) rainwater, (C) groundwater and (D) freshwater.

Thermogravimetric analysis (TGA) of the uncoated and magnetite-waste tea composite (Figure 2) showed that the magnetite coating contributed about 22 wt% to the total mass of the magnetite-waste tea composite.

Magnetic Property of the Magnetite-waste Tea Composite

The magnetic behavior of the magnetite-waste tea composite was analyzed by VSM at 298 K and the externally applied magnetic field was cycled between -10 and 10 kOe. Figure 3 shows that the magnetite-waste tea composite is superparamagnetic,

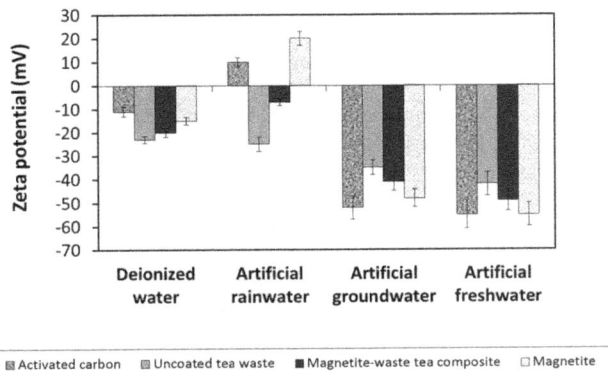

Figure 7. ζ-potentials of the various sorbents dispersed in the artificial waters.

given the absence of a hysteresis loop. The saturation magnetization (M_s) of the magnetite-waste tea composite is 7 emu/g. Since magnetite accounts for about 22 wt% of the total mass of the composite, the M_s value of the magnetite without the tea waste can be calculated as 31.8 emu/g. Bulk Fe_3O_4 has a M_s value of 92 emu/g [26]. The smaller M_s value of the magnetite in the waste tea complex compared with the bulk Fe_3O_4 value can be attributed to the reduced primary grain size of the magnetite coating on the tea [26] and the interaction of organics from the tea with the iron oxide, both of which decrease M_s [27].

The magnetization M of the magnetite-waste tea composite assuming a monodisperse particle diameter d in the direction of an applied magnetic field H can be modeled using the Langevin function $L(\alpha)$.

$$\frac{M}{\phi M_S} = \coth(\alpha) - \frac{1}{\alpha} \equiv L(\alpha) \qquad (1)$$

where ϕ is the solid volume fraction, M_s is the saturation magnetization of the bulk magnetite, $\alpha = \pi\mu_0 M_s H d^3 / 6kT$, μ_0 is the vacuum permeability, k is the Boltzmann constant and T is the temperature [28].

From this equation, the solid volume fraction of magnetite in the magnetite-waste tea composite can be obtained by fitting the Langevin function to the experimental magnetization curve. The best fit solid line in Figure 3 was obtained when $\phi = 0.08$. This indicates that while magnetite makes up 22% by mass of the composite it occupies only 8% of the solid's volume. Tea waste is less dense than magnetite.

Figure 4 illustrates the easy separation of the magnetite-waste tea composite from water via the use of a rare earth permanent magnet. The composite is stable in the artificial waters and remains attracted to a magnet for up to at least 3 months.

Fe Leaching from the Magnetite-waste Tea Composite into the Artificial Waters

Figure 5 shows that uncoated tea leached low concentrations of Fe in all 4 types of waters (≤ 0.08 ppm) while the magnetite-waste tea composite leached Fe concentrations of ≤ 0.7 ppm. The magnetite-waste tea composite has more Fe leaching due to the magnetite layer coated onto the waste tea dusts during the magnetization procedure. According to WHO guidelines [29], approximately 0.5–50 ppm of Fe can be found naturally in fresh waters. Our data show that the magnetite coating of tea does not cause significant leaching of Fe into the waters in concentrations sufficiently high that it will pose a health hazard. The concentrations of Fe leaching from both uncoated tea and magnetite-waste tea were similar to Fe concentrations found in the natural waters.

Choice of Solution pH

In this study, the pH values of the artificial waters were adjusted to be 4.2 for rainwater, 7.1 for groundwater and 8.4 for freshwater while the $[Pb^{2+}]$ was varied from 0.15–5.5 ppm. The pH value of DI water was measured to be 6.3. From previous studies of Pb sorption onto tea [8,14], the optimum pH fell in the range of 4–6. Both Liu et al. and Amarasinghe et al. observed in their studies that for $[Pb^{2+}]>100$ ppm and at pH >5.3, Pb^{2+} ions begin to hydrolyze and precipitate out, making it impossible to conduct sorption studies when pH exceeds 6. In addition, the adsorption of metal ions at low pH is known to be poor as there would be competition with the H^+ ions for binding to the active sites on the tea surface. Their studies were conducted in solution pH 4–5.5 with $[Pb^{2+}]$ varying from 0–400 ppm. However, the Pb sorption onto tea under such conditions is not environmentally realistic. Ambient waters have pH values between 7–8, with the exception of rainwater which is typically acidic and has a pH ~ 4 [20]. Typical $[Pb^{2+}]$ in highly contaminated waters range from 0.1 to 0.7 ppm [23,24]. The WHO health based guideline limit for maximum Pb in drinking water is 0.01 ppm [3]. This work thus investigates the potential of using magnetized tea waste under environmentally relevant operating conditions to remove Pb^{2+} ions from aqueous solutions.

Effect of Sorbent Type

AC is the most commonly used sorbent for removal of both inorganic and organic contaminants in polluted waters. Experiments were conducted to compare the Pb^{2+} sorption performance of AC with those of the uncoated tea and magnetite-waste tea composites. Magnetite's capability as a sorbent was also tested. The results are presented in Figure 6. AC shows a reduced Pb^{2+} percentage removal in all 4 waters compared to magnetite and both uncoated and magnetite-waste tea when the initial Pb concentrations were above 2 ppm. AC has a larger surface area compared to both the uncoated and magnetite-waste tea. The observed reduction in Pb sorption by AC could be partly due to the agglomeration of Pb^{2+} ions covering the pores on the surface of AC, thus blocking free Pb^{2+} ions in the solution from binding to the AC surface.

The removal/recovery of simple Pb–tea sorbents from the treated waters poses a challenge. Magnetization of the sorbents (via the coating of magnetite) allows a quick and easy removal/

recovery of used sorbents via the simple application of an external magnetic field.

Among the 4 sorbents tested, uncoated tea has the highest sorption for Pb^{2+} by adsorbing at least 86% of a range of varying initial $[Pb^{2+}]$ for all waters. While the uncoated tea powder does not have a very porous structure to provide a high surface area for adsorption of Pb^{2+} (1.04 m^2/g versus 3350 m^2/g for AC), tea dusts/leaves are known to contain numerous functional groups [10], of which the amine group ($-NH_2$), aromatic and aliphatic carbons, and carboxylic carbons ($-COOH$) play an active role in sorption of Pb [8]. Our results (Figure 6) indicate that both the uncoated tea and magnetite-waste tea composite have significant differences ($p<0.05$) in Pb sorption performance compared to AC in twenty out of twenty-four of the solution conditions. These differences can be partially attributed to the smaller BET surface areas on the teas compared to AC, limiting the number of available sorption sites for Pb^{2+} to bind, in addition to the variations in binding affinity of the sorbents for Pb^{2+} between AC, uncoated tea and magnetite-waste tea composite. In all four water types, both the uncoated tea and magnetite-waste tea composite adsorbed more Pb^{2+} than AC when the initial $[Pb^{2+}]$ exceeds 2 ppm while AC is a more effective sorbent when initial $[Pb^{2+}]$ is less than 2 ppm. The uncoated waste tea's Pb^{2+} sorption capacity is similar to that of other studies [8,10,14] which show that tea wastes can be a potential low cost alternative to other adsorbents such as AC. From Figure 6, the results for Pb^{2+} sorption by the magnetite-waste tea composite samples are comparable to that of uncoated tea for most cases, with the exception of a decrease of $\sim 30\%$ in Pb^{2+} sorption for freshwater with an initial $[Pb^{2+}]$ of 0.16 ppm.

Effect of Water Type

The percentage of Pb removal by the magnetite-waste tea composite in all the four waters was at least 70%. The sorption of Pb by the composite was higher in both artificial groundwater and artificial freshwater ($>80\%$) compared to DI and artificial rainwater ($>70\%$). The humic acid content found in freshwater and groundwater (Table 1) could be a main contributing factor to this increase in Pb sorption and removal. Numerous studies have reported that humic acid acts as a chelating agent and forms complexes with metal ions [30,31,32]. Humic acid contains phenolic hydroxyl ($-OH$) and carboxylic groups ($-COOH$) which are critical for the formation of the Pb-humate complex [33,34]. The formation of these Pb-humate complexes could have aided the increase of Pb sorption onto the surface of magnetite-waste tea complexes. Furthermore the presence of Ca^{2+} ions in groundwater and freshwater can act as a bridge [21,35] between the negatively charged Pb-humate complex and the similarly charged magnetite-waste tea composite (Figure 7), enhancing the adsorption of Pb.

Pb sorption for uncoated tea waste ranged from 86–100% while Pb sorption for magnetite ranged from 73–100%. Pb sorption by AC displayed the same trend, with 100% sorption for initial $[Pb^{2+}]$ <1.5 ppm, and a decrease in Pb adsorption at initial concentrations of 3–5 ppm for all waters. The presence or absence of humic acids did not affect the Pb sorption by the uncoated tea waste, AC or magnetite. Figure 7 shows that all four types of sorbents were negatively charged in water with some exceptions in artificial rainwater. This suggests that electrostatic charge and humic acid are insufficient in explaining the increase of Pb adsorption by the magnetite-waste tea composite, and that enhanced Pb sorption requires the presence of humic substances and both the iron oxide and organics on the tea surface. In many systems involving natural organic matter (NOM), ζ-potential does not adequately predict adsorption. Other mechanisms such as steric interactions can enhance or impede adsorption even under electrostatically repulsive or attractive conditions [36].

Conclusions

This study shows the effective use of magnetite-waste tea as an inexpensive and green sorbent to remove Pb in contaminated waters under environmentally relevant conditions. Our data from the Fe leaching tests indicate that both uncoated tea and magnetite-waste tea composite release very low concentrations of Fe into the waters (<0.08 and 0.7 ppm respectively), and at levels that do not exceed the Fe concentrations found naturally in the environment. The magnetic attraction tests demonstrated that the magnetite-waste tea composite remains stable and retains its magnetic properties while being dispersed in the simulated natural waters for at least 3 months. This allows for easy separation of the Pb-laden sorbents from treated waters. The magnetite-waste tea composite is a better material than AC in removing Pb from waters with initial Pb concentrations >2 ppm. The presence of humic acid in artificial freshwater and groundwater enhances the sorption of Pb by the magnetite-waste tea composite. Given the

availability of tea wastes worldwide, magnetically responsive tea wastes have considerable potential to be used as a low cost sorbent in treating Pb bearing waters. The mass of iron oxide coating on the magnetite-waste tea composite can be optimized to reduce cost and wastage of the Fe salts without sacrificing its magnetic performance. Finally, both the Fe and Pb can be recovered by incineration of the magnetite-waste tea composite.

Acknowledgments

We thank Ying Xia for her assistance with the operation of the AAS, and Hwa Chong Institution students Bram Lim, Gregory Tan and Khang Zhie Phoong for their help with preparing some of the artificial waters. We are also grateful to Prof. Caroline A. Ross and Dr. Dong Hun Kim for allowing us the use and guidance in the operation of the VSM in the Ross lab at MIT.

Author Contributions

Conceived and designed the experiments: TAH PSD BJRT. Performed the experiments: SYY SC VD YKSP GQ. Analyzed the data: TAH PSD BJRT. Contributed reagents/materials/analysis tools: YKSP TAH PSD BJRT. Wrote the paper: SYY SC VD GQ TAH PSD BJRT.

References

1. Rosen JF (1992) Effects of low levels of lead exposure. Science 256: 294–295.
2. Li W, Zhang LB, Peng JH, Li N, Zhang SM, et al. (2008) Tobacco stems as a low cost adsorbent for the removal of Pb(II) from wastewater: Equilibrium and kinetic studies. Ind Crop Prod 28: 294–302.
3. WHO (2003) Lead in drinking-water. Background document for preparation of WHO Guidelines for drinking-water quality. Geneva.
4. Corapcioglu MO, Huang CP (1987) The adsorption of heavy metals onto hydrous activated carbon. Water Research 21: 1031–1044.
5. Pyrzynska K, Bystrzejewski M (2010) Comparative study of heavy metal ions sorption onto activated carbon, carbon nanotubes, and carbon-encapsulated magnetic nanoparticles. Colloids and Surfaces A: Physicochem Eng Aspects 362: 102–109.
6. Sabio E, Gonzalez E, Gonzalez JF, Gonzalez-Garcia CM, Ramiro A, et al. (2004) Thermal regeneration of activated carbon saturated with p-nitrophenol. Carbon 42: 2285–2293.
7. Moreno-Castilla C, Rivera-Utrilla J, Joly JP, Lopez-Ramon MV, Ferro-Garcia MA, et al. (1995) Thermal regeneration of an activated carbon exhausted with different substituted phenols. Carbon 33: 1417–1423.
8. Liu N, Lin D, Lu H, Xu Y, Wu M, et al. (2009) Sorption of Lead from Aqueous Solutions by Tea Wastes. J Environ Qual 38: 2260–2266.
9. Jin CW, Zheng SJ, He YF, Zhou GD, Zhou ZX (2005) Lead contamination in tea garden soils and factors affecting its bioavailability. Chemosphere 59: 1151–1159.
10. Lin DH, Pan B, Zhu LZ, Xing BS (2007) Characterization and phenanthrene sorption of tea leaf powders. J Agric Food Chem 55: 5718–5724.
11. Gobler CJ, Sanudo-Wilhelmy SA (2001) Effects of organic carbon, organic nitrogen, inorganic nutrients, and iron additions on the growth of phytoplankton and bacteria during a brown tide bloom. Mar Ecol Prog Ser 209: 19–34.
12. Tee TW, Khan ARM (1988) Removal of lead, cadmium and zinc by waste tea leaves. Environ Technol Lett 9: 1223–1232.
13. Ahluwalia SS, Goyal D (2005) Removal of Heavy Metals by Waste Tea Leaves from Aqueous Solution. Engineering in the Life Sciences 5: 158–162.
14. Amarasinghe BMWPK, Williams RA (2007) Teawaste as a low cost adsorbent for the removal of Cu and Pb from wastewater. Chemical Engineering Journal 132: 299–309.
15. Thio BJR, Clark KK, Keller AA (2011) Magnetic pollen grains as sorbents for facile removal of organic pollutants in aqueous media. Journal of Hazardous Materials 194: 53–61.
16. Brunauer S, Emmett PH, Teller E (1938) Adsorption of Gases in Multimolecular Layers. J Am Chem Soc 60: 309–319.
17. Barrett EP, Joyner LG, Halenda PP (1951) The Determination of Pore Volume and Area Distributions in Porous Substances. I. Computations from Nitrogen Isotherms. J Am Chem Soc 73: 373–380.
18. Keller AA, Wang H, Zhou D, Lenihan HS, Cherr G, et al. (2010) Stability and Aggregation of Metal Oxide Nanoparticles in Natural Aqueous Matrices. Environ Sci Technol 44: 1962–1967.

19. Thio BJR, Montes M, Mahmoud MA, Lee D-W, Zhou D, et al. (2012) Mobility of Capped Silver Nanoparticles under Environmentally Relevant Conditions. Environ Sci Technol 46: 6985–6991.
20. Hu GP, Balasubramanian R, Wu CD (2003) Chemical characterization of rainwater at Singapore. Chemosphere 51: 747–756.
21. Chen KL, Elimelech M (2007) Influence of humic acid on the aggregation kinetics of fullerene (C60) nanoparticles in monovalent and divalent electrolyte solutions. J Colloid and Interface Sci 309: 126–134.
22. Cosper EM, Garry RT, Milligan AJ, Doall MH (1993) Iron, selenium, and citric acid are critical to the growth of the 'brown tide' microalga, Aureococcus anophagefferens. In: Smayda TJ, Shimizu, Y., editor. Toxic phytoplankton blooms in the sea. Amsterdam: Elsevier Science BV. 667–673.
23. Abdel-Halim SH, Shehata AMA, El-Shahat MF (2003) Removal of lead ions from industrial waste water by different types of natural materials. Water Research 37: 1678–1683.
24. Makokha AO, Mgheweno LR, Magoha H, Nakajugo A, Wekesa JM (2008) Environmental lead pollution and contamination in food around Lake Victoria, Kisumu, Kenya. Afr J Environ Sci Technol 2: 349–353.
25. von Smoluchowski M (1903) Contribution to the theory of electro-osmosis and related phenomena. Bull Int Acad Sci Cracovie 3: 184–199.
26. Xuan S, Wang Y-X, Yu JC, Leung KC-F (2009) Tuning the Grain Size and Particle Size of Superparamagnetic Fe3O4 Microparticles. Chem Mater 21: 5079–5087.
27. Shen L, Laibinis PE, Hatton TA (1999) Bilayer Surfactant Stabilized Magnetic Fluids: Synthesis and Interactions at Interfaces. Langmuir 15: 447–453.
28. Suh SK, Yuet K, Hwang DK, Bong KW, Doyle PS, et al. (2012) Synthesis of Nonspherical Superparamagnetic Particles: In Situ Coprecipitation of Magnetic Nanoparticles in Microgels Prepared by Stop-Flow Lithography. J Am Chem Soc 134: 7337–7343.
29. WHO (2011) Guidelines for drinking-water quality, fourth edition. In: WHO, editor.
30. Coles CA, Yong RN (2006) Humic acid preparation, properties and interactions with metals lead and cadmium. Engineering Geology 85: 26–32.
31. Barancikova G, Makovnikova J (2003) The influence of humic acid quality on the sorption and mobility of heavy metals. Plant Soil Environment 49: 565–571.
32. Stevenson FJ (1994) Humus Chemistry: John Wiley & Sons.
33. Adekunle IM, Arowolo TA, Ndahi NP, Bello B, Owolabi DA (2007) Chemical Characteristics of Humic Acids in Relation to Lead, Copper and Cadmium Levels in Contaminated Soils from South West Nigeria. Annals of Environmental Science 1: 23–24.
34. Stevenson FJ (1976) Stability Constants of Cu2+, Pb2+ and Cd2+ Complexes with Humic Acids. Soil Science Society of America Journal 40: 665–672.
35. Yoon S-H, Lee C-H, Kim K-J, Fane AG (1998) Effect of calcium ion on the fouling of nanofilter by humic acid in drinking water production. Water Research 32: 2180–2186.
36. Chen KL, Elimelech M (2008) Interaction of Fullerene (C60) Nanoparticles with Humic Acid and Alginate Coated Silica Surfaces: Measurements, Mechanisms, and Environmental Implications. Environ Sci Technol 42: 7607–7614.

21

Dryland Soil Hydrological Processes and Their Impacts on the Nitrogen Balance in a Soil-Maize System of a Freeze-Thawing Agricultural Area

Wei Ouyang[1]*, Siyang Chen[2], Guanqing Cai[1], Fanghua Hao[1]

1 School of Environment, State Key Laboratory of Water Environment Simulation, Beijing Normal University, Beijing, China, **2** Marine Monitoring and Forecasting Center of Zhejiang, Hangzhou, China

Abstract

Understanding the fates of soil hydrological processes and nitrogen (N) is essential for optimizing the water and N in a dryland crop system with the goal of obtaining a maximum yield. Few investigations have addressed the dynamics of dryland N and its association with the soil hydrological process in a freeze-thawing agricultural area. With the daily monitoring of soil water content and acquisition rates at 15, 30, 60 and 90 cm depths, the soil hydrological process with the influence of rainfall was identified. The temporal-vertical soil water storage analysis indicated the local *albic* soil texture provided a stable soil water condition for maize growth with the rainfall as the only water source. Soil storage water averages at 0–20, 20–40 and 40–60 cm were observed to be 490.2, 593.8, and 358 m^3 ha^{-1}, respectively, during the growing season. The evapo-transpiration (ET), rainfall, and water loss analysis demonstrated that these factors increased in same temporal pattern and provided necessary water conditions for maize growth in a short period. The dry weight and N concentration of maize organs (root, leaf, stem, tassel, and grain) demonstrated the N accumulation increased to a peak in the maturity period and that grain had the most N. The maximum N accumulative rate reached about 500 mg m^{-2}d^{-1} in leaves and grain. Over the entire growing season, the soil nitrate N decreased by amounts ranging from 48.9 kg N ha^{-1} to 65.3 kg N ha^{-1} over the 90 cm profile and the loss of ammonia-N ranged from 9.79 to 12.69 kg N ha^{-1}. With soil water loss and N balance calculation, the N usage efficiency (*NUE*) over the 0–90 cm soil profile was 43%. The soil hydrological process due to special *soil* texture and the temporal features of rainfall determined the maize growth in the freeze-thawing agricultural area.

Editor: Guoping Zhang, Zhejiang University, China

Funding: We are grateful for assistance with the data requirements of the Bawujiu Farm in Heilongjiang Province. The research discussed in this paper benefited from financial support provided by the National Natural Science Foundation of China (Grant No. 41371018, 51121003), the Supporting Program of the "Twelfth Five-year" Plan for Sci & Tech Research of China (2012BAD15B05), and The Fundamental Research Funds for the Central Universities. The funders had no role in study design, data collection and analysis, decision to publish, or preparation of the manuscript.

Competing Interests: The authors have declared that no competing interests exist.

* Email: wei@itc.nl

Introduction

During the agricultural tillage management, the maximum crop production and minimum diffuse nitrogen (N) loading are the priority issues that need to be considered at the same time [1][2]. It is essential to understand the fates of soil water (SW) and N in agricultural systems in order to attain higher crop yields and N usage efficiency [3]. In developing countries, there is much political and commercial pressure for controlling the N application with the sacrifice of the crop harvest [4]. The dryland agriculture in the Sanjian Plain, Northeast China, is a key food base and the most water limited agricultural zone in China [5]. In this freeze-thawing area, the soil N and water efficiency are the keys for dryland tillage sustainability due to the short growing season. However, there are few reports about the dryland soil hydrological process and N use efficiency in freeze-thawing agricultural areas [6].

The soil nutrient N is a major limiting factor for crop growth and production, which is directly related to the N fertilisation and soil N background level [7]. The soil microbial community, which

affects the soil N cycle [8], is less active in this freeze-thawing area. The application of chemical N is a basic agricultural practice in worldwide, which causes excessive discharge of N to the aquatic environment [9]. The European Union agricultural landscape contributes about 55% of the diffuse pollution for water eutrophication. This is a major emerging environmental issue in the developing counties [10]. The soil eco-hydrolegical process is an essential channel for N transport in the soil-crop system.

Plant organs have different allocation rates with N during growth, and the canopy green area has higher N absorption than the other organs during the jointing period [11]. The N accumulations in the upper leaves of the canopy also follow the eco-hydrological pattern in farmland [12]. Some models have been developed to simulate the N absorption in different organs during crop growth with the impacts of the solar radiation, rainfall, temperature, soil hydrology and N concentration [13]. With the temporal pattern of N accumulation in crop organs, the N cycle and efficiency can also be analyzed. The diffuse N discharge from agricultural systems depends on diverse factors, including the climatic, soil properties and agronomic features [14][15]. The tillage practice, soil hydrological process and N usage efficiency are the main potential factors to consider when the goal is to reduce the agricultural diffuse N pollution [16]. The dryland maize farmland in a freeze-thawing agricultural area presents different N discharge patterns due to the special hydrological process [17].

The soil hydrological process is the bridge for diffuse N loss from soil to water bodies [18]. The nitrate-N is the dominant type of diffuse N in the soil profile and it can move swiftly downstream flow with the rainfall through the soil pathways [19]. The soil hydrological process is also closely related to the soil texture, which is the medium for water movement and crop growth. In the freeze-thawing dryland, the soil has short-term, transitional phase and rainfall is the only water source for N movement with crop growth [20]. The soil pattern in the study area is meadow *albic* soil (*Albic Luvisols*) and has a 20–30 cm depth impermeable stratum. The *albic* soil provides a stable soil water (SW) condition in the crop root zone and interfaces with the soil hydrological process. With the consideration of high N efficiency, it is necessary to analyze the special pattern of soil hydrology and its impact on the N efficiency [21]. Evaluating the response of N loss to the combination of rainfall and the soil hydrological process can help to identify the interactions of water with N and reduce diffuse N pollution.

The general objective is to maximize the N efficiency in tillage practices. Typically, of about 20%–70% of N is lost from soil-crop systems and this loss can cause other environmental issues [22]. The quality of soil properties will decline under the combination of intensive application of N fertilizer and low N efficiency; this in turn, will also deliver some risk to water [23]. There are several methods to improve the N efficiency, including accurate fertilizer application and irrigation [24][25]. The dryland soil hydrological process in a freeze-thawing agricultural area presents a special pattern and also affects the N efficiency [26]. This study aims (1) to identify the temporal-vertical dynamics of dryland SW within an *albic* soil layer; (2) to express the dry weight and N accumulation of maize organs in dryland of a freeze-thawing agricultural area; (3) to explore the variation of soil loss, potential N loss, and N use efficiency (*NUE*) in dryland during the growing season.

Materials and Methods

2.1 Ethics statement

The field experiment was conducted in the farmland of Bawujiu Farm, which has been rented by Wang Dongli for 50 years. He is also the contact person for future permission. No specific permissions were required for these locations. The field studies did not involve endangered or protected species. This study did not involve vertebrate. The specific location of our study is $133°\,50'$–$134°\,33'$ E, $47°\,18'47°50'$ N.

2.2 Study area description

The case study area was conducted on a farm in the northeastern part of China. The farm has a long history and on the east is adjoined to Russia (Fig. 1). The observation site is located in the southern part of the Farm and the elevation is 48 m. The only water source for dryland tillage is precipitation and maize is the dominant crop. This area the temperate continental monsoon climate has an average annual precipitation of 588 mm and an average yearly temperature of 2.94 °C [27]. From October to the following April, the temperature is below zero and, as a result, the soil freeze-thawing process affects the tillage process. The crop growing season is from May to October.

2.3 Field monitoring and soil sample collection

In order to identify soil physical-chemical properties, soil samples at four depths (0–15, 15–30, 30–60 and 60–90 cm) were collected in 2010. The samples from three points chosen at random in each layer were mixed into a composite soil sample and placed in plastic bags for lab measurements. The soil samples were collected before sowing and after harvest. Soil particle size distributions were measured by laser diffraction after the removal of organic residue (MasterSizer S, Malvern Instruments, Malvern, UK). The soil's total nitrogen (TN) and organic carbon (OC) concentrations were measured with a CHN Elemental Analyser (Euro VectorS.P.A EA3000, 136 Milan, Italy) [28]. The ammonia N in rainfall was analyzed with the aid of Nessler's reagent and the nitrate-N concentration was determined by the Ultraviolet Spectro Photometric Method. The soil's pH was determined with a pH meter (METTLER TOLEDO, Switzerland) after mixing the fresh soil with deionised water (1:2.5, w/v) [24]. The soil N stock at four depths was calculated with the following equation,

$$Stock_i = BD_i * H_i * C_i * 10^5 \qquad (1)$$

where $Stock_i$ is the soil N storage at layer i (kg/m^2), BD_i is the soil bulk density at layer i (g/cm^3), H_i is the soil depth at layer i (cm), and C_i is the soil N concentration at layer i (mg/kg).

For the soil hydrological process monitoring, the SW collector head (suction cup: Teflon and quartz; OD 21 mm×L 95 mm; porosity: 2 μm; conductivity: 3.31×10^{-6} mm sec^{-1}, PRENART, Danmark) was set at 15, 30, 60 and 90 cm depths. It was covered with a mixed solution of quartz powder and stirred, which can cause a 0.05 MPa vacuum pressure and in direct contact with the soil. The collector tail was connected with PVC pipe and the water samples were collected in bottles by a portable manual vacuum pump. The soil water storage was calculated with soil volumetric water content at different depths. The mean soil water storage at each layer was defined as the average of SW in the upper and lower soil layers. The evapo-transpiration in dryland is calculated with the Bowen ratio and energy balance (BREB). The SW balance was calculated with the following equation [29],

$$SL = R - I_r - ET - \Delta S \qquad (2)$$

where SL is the soil water loss (mm), R is the precipitation (mm), ET is the evapo-transpiration (mm), I_r is the canopy interception (mm), and ΔS is the soil water storage change (mm).

The local meteorological characteristics were monitored with a ZENO station (Coastal, Seattle, WA, USA). Automated soil temperature sensors (Thermistor, Coastal, Seattle, WA, USA) and soil volumetric water content sensors (TDR type, Coastal, Seattle, WA, USA) were placed in each of the four different soil layers. The monitoring sensors were repeated at double at horizontal 0.5 m space.

2.4 Crop sample collection and measurement

The local dryland is intensively cultivated with the aid of mechanized equipment and the maize (*Zea mays L.*) planting density is 75,000 plants ha^{-1}. The detailed maize planting date, harvest date, relevant tillage and fertilization practices are listed in Table 1. The maize samples in 2011 were collected monthly and the sample in 2010 were only collected in October due to the limited experimental conditions. The detailed process of sample collection and monitoring of crop features (height (CH), leaf area index (LAI) and root depth (RD)) are listed in a previously published paper [24]. The maize samples were taken to the lab

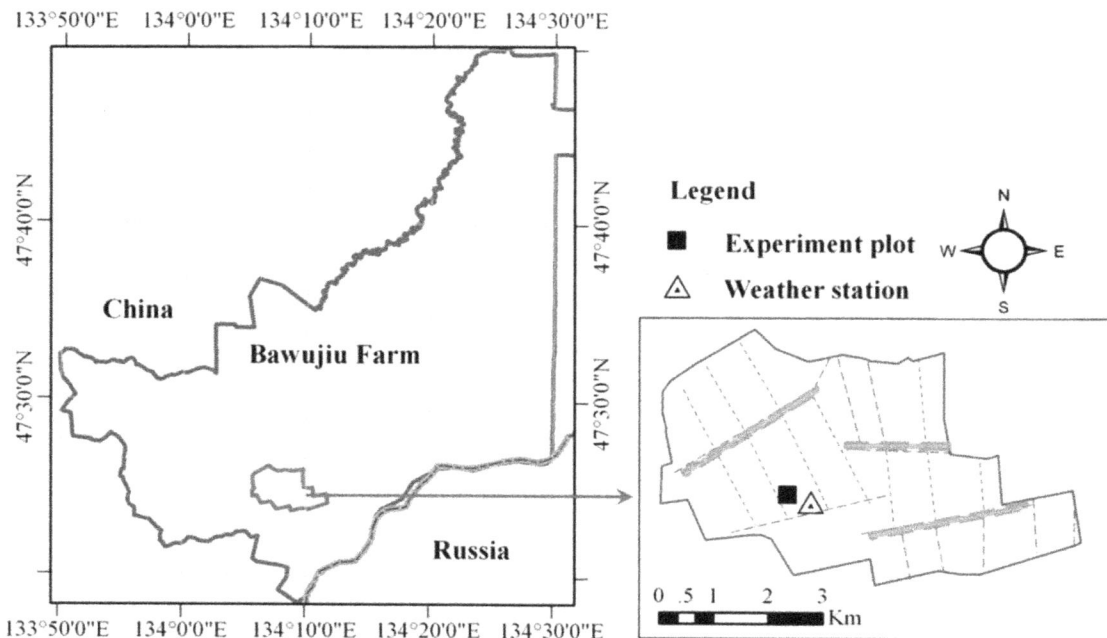

Figure 1. Location of the experimental dryland site in Northeast of China.

immediately and the root, leaves, stems, tassels, and grains were separated. The samples were washed clean and then dried for 48 h at 65°C. The dry weight of organ samples was determined after fixing the temperature for 30 min at 105°C. The dried maize samples were crushed and sieved for crop organ N content (%) analysis (Vario EL, Elementar Co. Ltd., Germany). The N uptake amount by crop was calculated based on crop dry weight and plant N content (%).

2.5 Nitrogen supply and loss calculation

The N use efficiency (NUE) in a crop-soil system is defined as the ratio of N uptake (N_c) by a crop to N supply (N_s) in a system [30]. The N_s was defined as the total available N to the crop, which included fertilizer N (N_f), mineralized N (N_m), initial mineralized N in soil ($N_{min\ initial}$), N fixed by soil (N_x) and N deposition (N_d) from rainfall. The N_s is calculated as:

$$N_s = N_f + N_{\min\ initial} + N_m + N_x + N_d \tag{3}$$

The field N_m is calculated with the crop-soil N equation [31]:

$$N_m + (N_{\min initial} - N_{\min final}) = (N_l + N_c + N_{gl}) - (N_f + N_d + N_x) \tag{4}$$

where N_m is the mineralized N, $N_{min\ initial}$ is the initial mineralized N concentration in soil, $N_{min\ final}$ is the final mineralized N concentration in soil, N_l is the soil loss N, N_{gl} is the N loss in gas. N_c is the N crop absorption, N_x is the N fixed by soil, and N_d is the N deposition.

Table 1. Dryland tillage practices and fertilization information.

Crop year	Planting date	Harvest date	Tillage prior to planting	Main fertilisation (kg ha^{-1})
2010	8/Jun.	9/Oct.	Tilling	525
2011	30/May.	5/Oct.		525
Detailed fertilisation data and amounts * (kg ha^{-1})				
2010	2011	N	P	K
8/Jun.	30/May.	71.25	58.65	27.23
27/Jun.	20/Jun.	34.50	13.80	10.02
5/Jul.	1/Jul.	29.45		

*Based on computation of the N, P, and potassium (K) elements.

The mineralized N is calculated with the following equation,

$$N_m = (N_l + N_c) - (N_f + N_d) - (N_{\min\,initial} - N_{\min\,final}) \quad (5)$$

The potential soil water N loss (N_l) is defined as the total loss of nitrate and ammonia, which was estimated with the N concentration in SW and the water volume in dryland [32]. Based on the SW loss and the N concentration in SW at different depths, the soil N loss by water was calculated.

2.6 Data analyses

Data analysis was performed with Sigma plot 10.0 and SPSS 16.0 software. The differences in dry weight and their percentages on crop organs (including root, leaf, stem, tassel, grain) in different growing stages were tested with ANOVA. The difference in the crop N uptake allocation proportion, the N accumulation in different organs, and the mineralized N concentration in different soil layers were statistically analyzed by ANOVA. The multiple comparisons were analyzed with the Duncan method.

Results

3.1 Temporal-spatial distribution of soil water in dryland

With the soil-maize water monitoring system, the daily precipitation, SW acquisition rate, and SW content in the entire growing period was determined (Fig. 2). The monthly precipitation in the years 2010 and 2011 displayed a similar pattern, except for the lower value in September 2010. The precipitation occurred mainly between July and September, which was also the prime period for maize growth. With the light precipitation in April and May, the maize seed was sowed in moist soil. The precipitation in August and September of 2010 was 64 mm and 120 mm less than in the corresponding months of 2011. The cumulative precipitation from April to September in 2011 was 575 mm, which was 141 mm more than that in 2010. The precipitation was the only water source for the dryland in this freeze-thawing agricultural area, which directly impacted the SW acquisition rate. The SW could be collected only after the rainfall and the vertical difference in the SW acquisition rate was slight (around 0.002 cm³ s⁻¹). In the entire monitoring period, the biggest acquisition rate occurred at the 90 cm depth after the rainfall in early September. The direct reason for it was that only smaller amount of the water can move deeply in the vertical direction after the absorption by roots.

The SW content increased in a stable manner during the growth period and also coincided with the accumulative precipitation. In all of June, the SW content at the 30 cm depth was the highest, the reason for this was that uptake of water in growing roots resulted in water accumulation in the rhizosphere. Due to the 73.6 mm rainfall on July 3, the SW content at the 30 cm depth increased by 20% to 0.30 cm³ cm⁻³. In the next two days, the SW content was transported in the vertical direction and it increased to 0.30 cm³ cm⁻³ in the deep layer. This vertical distribution also confirmed the retention role of roots in the early growing season. The surface SW content increased significantly after August due to the precipitation and then was transported to the deep soil layers. Later, the impact of the crop roots decreased and the soil texture was the main factor for SW transport in September.

In order to further analyse SW content fluctuation in the temporal-vertical dimensions, a statistical analysis was performed on the data from the soil layers during the growing season (30 May–30 September) (Table 2). The SW content averages at the 15 cm and 60 cm depths were very close with values of 0.245 and 0.230 cm³ cm⁻³, respectively. The mean SW content at the 30 cm

and 90 cm depths were higher than at the other two layers with values of 0.272 cm³ cm⁻³ and 0.280 cm³ cm⁻³, respectively. The close relationship between the SW acquisition rate and the SW content that was shown in Fig. 2, was also confirmed by the statistical results. The SW acquisition rate had the same vertical pattern as the SW content and the higher values appeared at the 30 and 90 cm depths. Before maize seeding (May 30), the SW could be collected when the SW content reached about 0.25 cm³ cm⁻³. After maize seeding, the SW could be collected when it reached a content of over 0.30 cm³ cm⁻³. The SW acquisition rate was greatly and primarily influenced by rainfall.

3.2 Temporal variation of soil water storage

Based on the monitoring of the SW characteristics, the soil volumetric water content (SVW) and soil water storage (SWS) in the top 30 and 90 cm were calculated, respectively (Fig. 3). The SVW and SWS of the 0–90 cm and 0–30 cm depth showed similar temporal trends. In the vertical dimension, the SVW of the same two depths (0–30 cm and 0–90 cm) had similar trends in most periods, except in the first month which revealed a difference due to thawing. The thawing process from the surface to the deep layer and the tillage in preparation for sowing in the top soil decreased the SVW at 0–30 cm depth. There was almost no difference in the SVW between the two depths from mid-May to August. Before the crop harvest period in September, the deeper soil had a higher SVW content than the surface layer due to less root activity and strong precipitation. The SWS of the 0–30 cm depth had a seasonal variation and had a more direct relationship with precipitation events. Before the rainfall on 1 July, the SWS remained at a stable level from mid-May to the end of June. The SWS had a periodic decrease when no rainfall occurred in late July. The SWS reached its summit with the large rainfall in the later of August.

The SWS was calculated with the vertical variation in the SVW. The statistical characteristics of SWS and their proportions at each depth are listed in Table 3. The soil water storage averages at 0–20, 20–40 and 40–60 cm were 490.2, 593.8, and 358 m³ ha⁻¹. The vertical differences indicated the middle level has a higher soil water content as a consequence of soil texture and crop root zone influences. Because precipitation was the only water source for dryland, the SWS at the three layers in the dry period was smaller than the average for the entire growing season, which also revealed the impact of maize ET on SWS. The SWS in the vertical dimension had the same trend in the entire period and the middle layer had the largest value (548.9 m³ ha⁻¹). The differences in SWS between the two periods become greater when going from the surface to the deeper layer, an observation derived from the rainfall contribution to the SWS in deeper layer.

Based on daily monitoring data, the accumulative patterns of evapo-transpiration (ET), rainfall, and water loss in the growing season were compared (Fig. 4). The ET was at a low level in the first month of the growing period and then assumed a stable increasing rate beyond the 50th day after sowing. The accumulated SW loss was consistent with the fluctuation in the rainfall. The ET and rainfall stayed with the SW loss over the first 22 days. The accumulated rainfall amount increased by 79.8 mm; a large increase occurred on the 35th day when the ET and water loss increased slightly. The rainfall increased intensively after 80 days, which also caused an increase in the SW loss. The accumulated amounts of rainfall increased from 269.3 mm after 80 days to 440.9 mm after 120 days, and the SW loss increased from 47.9 mm after 75 days to 126.3 mm after 120 days. The temporal pattern demonstrated that the SW loss and ET of dryland were strongly correlated with the rainfall. The ET was also a maize

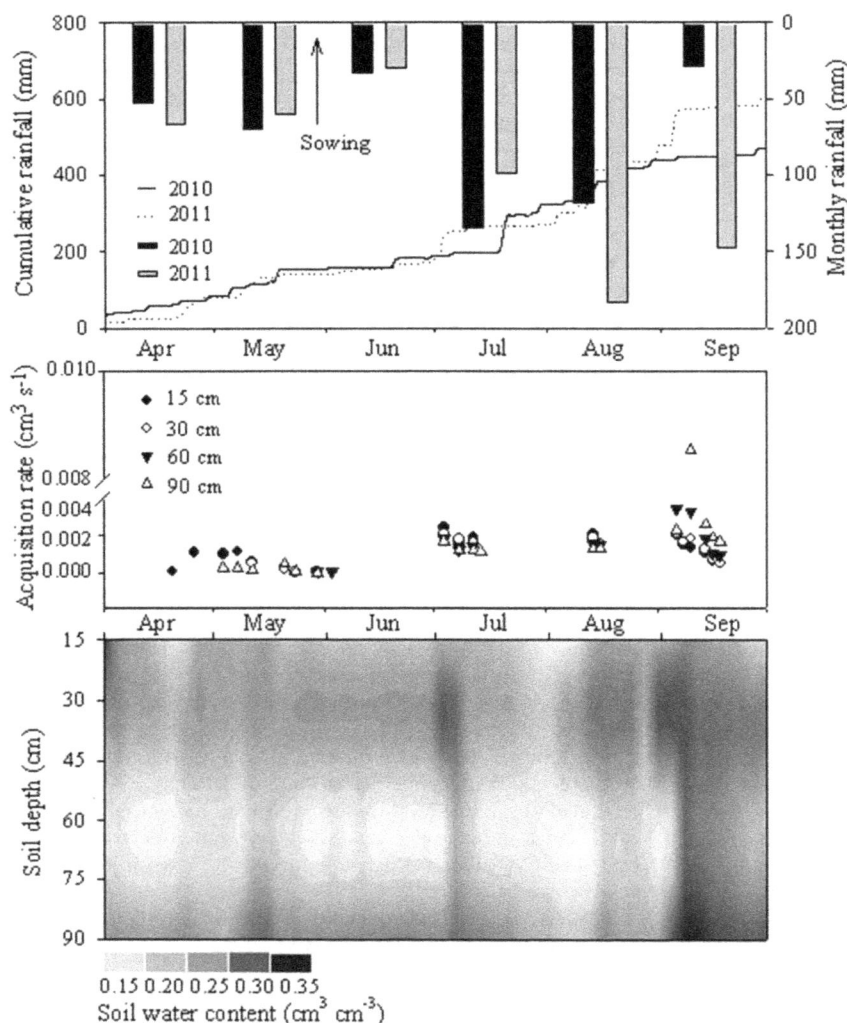

Figure 2. Daily precipitation in 2010 and 2011, and acquisition rate, and soil hydrological pattern of dryland in 2011.

growth feature and its similar pattern with rainfall also proved it was the dominant factor for crop growth.

3.3 Variation of crop biomass in the growing period

The dry weight of the different maize organs (root, leaf, stem, tassel, and grain) in growing stages were analyzed (Fig. 5). Also shown in the Fig. 5 are the percentages of the total dry weight comprised by each organ in each of four growing seasons. The dry organ weight varied greatly from the seedling stage to maturity, a development period which is relatively short in this freeze-thawing area. The temporal pattern of dry weight in 2011 revealed the changes that occurred over four growing sections. The dry weight of grains increased significantly in the grain filling and maturity sections. The dry weight percentage did not change as intensively as the weight, and it was also found that the grains comprised nearly half of the maize weight in the maturity period. With the N concentration in dried organs, the N absorption amount was quantified along with the organ dry weight. Due to logistical limitations, the crop organ sample in 2010 was only collected in the maturity period. The grain dry weights were larger in 2010 than in 2011. Comparing the dry weights of the organs in the

maturity stage of the two years, the dry weights of the roots and tassel were significantly but their percentages differed little.

3.4 Nitrogen uptake by crop and accumulation

The temporal N dynamics of crop uptake was then analyzed with the samples obtained from the maize organs (Fig. 6). As the maize grew, the N uptake in leaf first increased and then decreased with the maximum occurring in the jointing period. From grain filling to maturity, N uptake by leaf diminished insignificantly with a relative flat trend compared with that from the jointing to grain filling stages. As maize grew N uptake in stem decreased significantly decrease from 1 to 2 with a probable transfer of much N to tassel. The difference of N uptake by stem in the grain filling and maturity stages was insignificant, but the N uptake by grain increased significantly ($P < 0.001$). The N uptake by leaf decreased insignificantly, which indicated that a large amount of N was transferred to tassel in the early time period and then transferred from tassel to grain (N uptakes by tassel in grain filling and maturity were 8.12 g kg^{-1} and 3.57 g kg^{-1}, respectively). There were insignificant differences in N uptakes of stem and leaf during the maturity stage in 2010 and 2011. The N uptake by grain in 2010 was 9.1% more than that in 2011.

Table 2. Statistical analysis of soil water content and acquisition rate at four depths during maize growing period.

Depth (cm) Index	Soil water content ($cm^3\ cm^{-3}$)				Soil water acquisition rate ($cm^3\ s^{-1}$)			
	15	30	60	90	15	30	60	90
Minimum	0.210	0.235	0.205	0.240	0.00008	0.00005	0.00009	0.00003
Maximum	0.308	0.333	0.306	0.342	0.00241	0.00208	0.00324	0.00856
Mean	0.245	0.272	0.230	0.280	0.00125	0.00114	0.00140	0.00167
Std. Error	0.018	0.020	0.024	0.024	0.00077	0.00078	0.00091	0.00220
Coefficient of variance	0.074	0.072	0.106	0.088	0.615	0.685	0.648	1.320

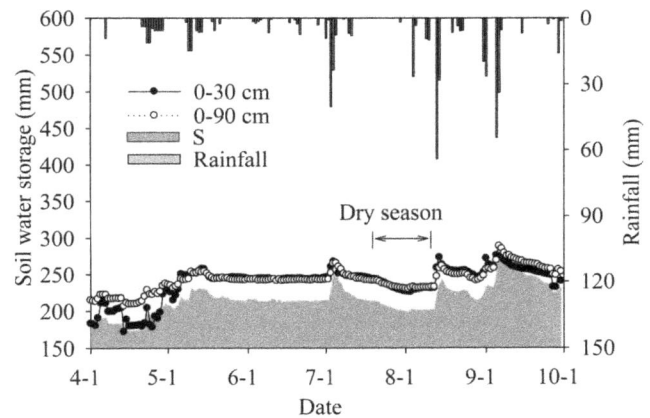

Figure 3. Temporal patterns of daily rainfall, soil water storage (SWS) and volumetric water content at 30 and 90 cm depths.

The accumulated N in different organs gradually increased in the early stages and reached a summit in the maturity period. From the jointing to grain filling period, the N in stem increased about 4.23 times and the amount in leaf increased less than five times. The N amounts in crop organs were not only related to the N uptake efficiency, but they also depended on the organ dry weight. According to the analysis in the maturity period, grain accumulated most of N, which occurred in a short time section. The total N amount of maize organs in 2010 was 309.9 kg ha^{-1}, which was larger than in 2011; the difference in grain was the dominant factor. The differences of N accumulation in leaf and stem during the maturity stages in the years 2010 and 2011 were insignificant.

Seasonal dynamics of N accumulative rates in organs during the growth period were analyzed (Fig. 7). From June to August, the N accumulative rates of stem and leaf increased rapidly and then declined. When the stem and leaf accumulated less N, the absorption efficiency in grain increased to 383 mg m^{-2}d^{-1}. The leaf was the most effective organ in uptaking N and the accumulation rate was about 500 mg m^{-2}d^{-1}.

3.5 Variations of soil N stocks in the growing season

The N concentrations at four depths before sowing and after harvest were analyzed, which demonstrated the temporal-vertical soil N dynamics. The N (nitrate and ammonia) concentrations at different depths were first analyzed after the growing season (Table 5). Before the sowing, the ammonia-N stock at four depths ranged from 18.8 to 21.3 kg N ha^{-1} with no occurrence of a significant difference. After the harvest, the deeper layer experienced more loss than the surface layer. The middle two layers had close values and the 12.5 kg N ha^{-1} in top layer was the largest value. The nitrate-N concentration in soil was much larger than the ammonia-N. At the beginning of the growing season, the nitrate-N concentration was more than 121 kg ha^{-1} at all depths and had slight differences at 60–90 cm depth. The nitrate-N concentration decreased to 90 kg ha^{-1} after the harvest. The upper two layers had similar patterns and the concentrations ranged between 88–90 kg ha^{-1}. There was no significant difference in the concentration in the deeper two layers and their concentrations ranged from 63.2 to 68.0 kg ha^{-1}. The nitrate-N was more variable than ammonia-N and had a larger stock gap.

Table 3. Dryland soil water storage and its proportion at three depths.

Depth/cm	Item	Entire growth season (30 May.–30 Sep.)	Dry season (20 Jul.–12 Aug.)
0–20	SWS/m^3 ha^{-1}	490.2±34.6	441.1±19.5
	Proportion/%	23.4±1.04	23.3±0.4
0–40	SWS/m^3 ha^{-1}	1084±62.9	990.0±55.9
	Proportion/%	51.8±2.6	52.0±0.76
0–60	SWS/m^3 ha^{-1}	1442±111	1303.0±70.6
	Proportion/%	68.8±1.6	68.5±0.55

3.6 Nutrient loss and supply in soil-crop system

Based on field water monitoring and SW content in the 0–90 cm profile, the soil water loss (SL) and the leached N were calculated (Table 6). The rainfall and evapo-transpiration (ET) were calculated in the four stages of the maize growing season. The SL had a closer correlation with rainfall than with ET. The rainfall was nearly 100 mm in the jointing stage, but the SL was 11.5 mm due to the intensive ET. Moving from the grain filling stage to the maturity stage, the ET increased and more SL occurred under the higher rainfall event. The leached N was also calculated with the N concentration in soil water. The nitrate-N was the dominant N lost from the 90 cm soil profile, especially in the grain filling and maturity stages. The leached ammonia-N had similar temporal patterns and the largest amount was 0.18 kg ha^{-1}.

Based on the N monitoring of rainfall and the survey of N fertilization, the N in rainfall and fertilizer was calculated. With the SWS and N concentration in SW in 0–90 cm profile, the *NUE* in the maize-soil system was calculated (Table 7). The N input from rainfall and fertilizer in the entire maize growing season was 16.5 and 120 kg ha^{-1}, respectively. The soil N supply in the entire growth period was about 641 kg ha^{-1}. With the nitrate and ammonia concentrations in SW and water volume, the N loss was estimated at 15.6 kg ha^{-1}. Based on the N balance, it was calculated that the *NUE* was 43%. On the other hand, the excessive N from dryland was an important potential source for N discharge to the environment.

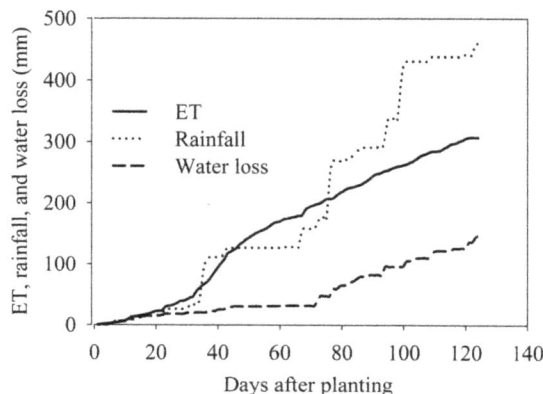

Figure 4. Cumulative evapo-transpiration (ET), rainfall, and water loss throughout maize growth period.

Discussion

4.1 The dryland soil hydrological process in a freeze-thawing area

Rainfall is the only water source in dryland and it has a direct impact on SW content and storage [33]. The field monitoring done in four stages as part of our study indicated that the temporal patterns of SW content and acquisition rate closely followed the rainfall amount (Fig. 2). The soil roughness in a cultivated area is increased by tillage, which enhances the vertical transport of SW [34]. The SL was 146.7 mm (32% of rainfall in the growing season). However, the vertical distribution of SW content and soil water storage demonstrate that the water has a short vertical movement range. The SW can be sampled only after the rainfall events, but the acquisition rates showed slight vertical differences. These two findings both demonstrated that the vertical movement of water is slow even with large volumes and the speed can affect the range.

Compared with the long growing periods in other dryland agricultural areas, the local maize growing season is short and has a higher efficiency in the use of water and N [35]. A proper SW condition in the root zone is essential for crop growth in the maize seeding and grain filling stages. The SWS can remain at a stable level (about 1000 m^3 ha^{-1} of 40 cm profile) even in the dry season without precipitation (Fig. 3). All of these factors indicate that the *albic* soil texture can provide a positive water condition for crop growth [24]. The SW also controls surficial N movements in the dryland and provides a strong signal for crop absorption [36]. Crop transpiration will increase when the maize is near its maturity period, and the precipitation normally increases and meets this need during the same period (Fig. 4). The combined conditions of special soil texture and precipitation patterns directly determine the dryland soil hydrological pattern, which is the critical factor for maize growth in this freeze-thawing area.

4.2 Crop growth and N accumulation differences in plant organs

The quantification and assessment of the organ biomass in crops can help in understanding the impacts of SW and N on crop growth. This information is also necessary for assessing N leaching [37]. The crop growth is measured by the organ dry weight (Fig. 5), which indicates the crop growth rate is higher than in other drylands under warming temperatures. The yearly precipitation in 2011 was more than in 2000 (Fig. 2), but did not lead to the higher grain dry weight in 2011 (Fig. 5). The difference proves that a higher maize yield is based on a combination of climate, soil and tillage. The grain filling period in this freeze-thawing area occurs in August, at a time when temperature and precipitation are at

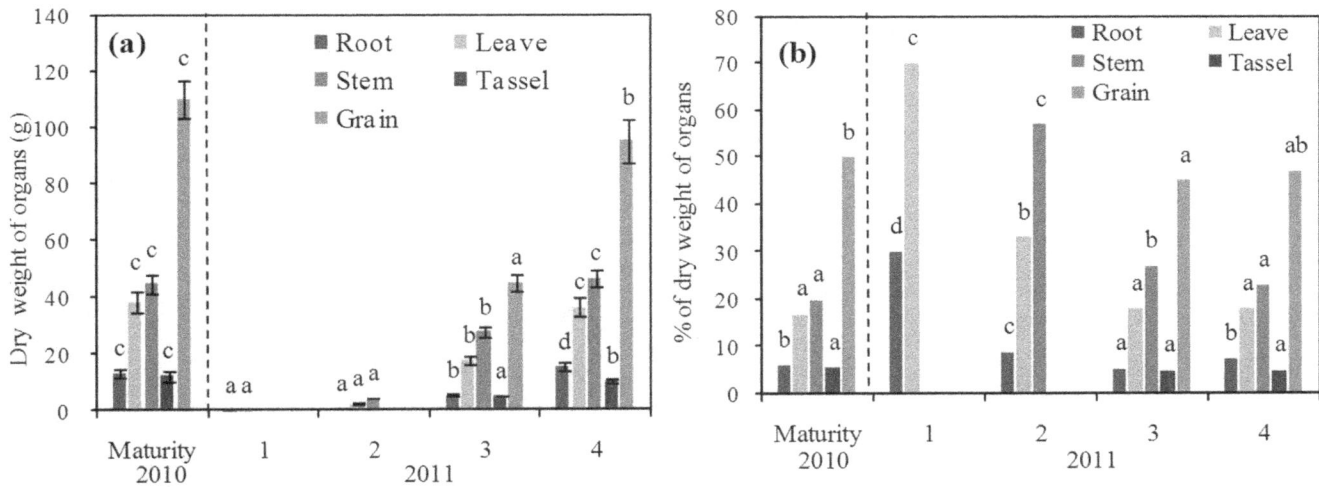

Figure 5. Dry weights (a) and percentage of maize organ weights (b) at different growth stages (Same letters in same organ between different growth stages indicate no significant difference. 1-Seedling, 2-Jointing, 3-Grain filling, 4-Maturity).

their highest levels. The active ET facilitates the transport of water and N from soil to crop, which finally causes a higher grain filling performance [38].

The absorbed N by maize organs in the entire growth period had different allocation rates (Fig. 7), which is determined by the photosynthesis of the organs and the eco-hydrological process. The crop accumulative rate in its canopy, stem and grain varied significantly, which means the N absorption capability for tissue production of dedicated organs shifts due to the N flux from soil to above ground biomass [39]. The soil N concentration in the vertical direction proved that the subsurface has a significant variance due to the root uptake. The accumulation rates of different organs over time also follow the typical sigmoid curve [40]. The temporal patterns of N in maize organs also indicate the N absorption is not only related to the soil N concentration, but that there is a complicated systemic feedback signal within the soil-crop system [41]. The SW availability, ET and N concentration are the key signaling factors. As the soil N is transported in crops with water, there are plenty of field observations and models to identify the growth stress reaction to the SW and N conditions

[42][43]. This information provides detailed guidelines for N and water adjustment in the soil-crop system.

4.3 Implication for N leach control and soil N optimization

The N uptake efficiency of a crop is the main component for *NUE* and the temporal patterns in Table 4 indicated that the highest amount of grain uptake of N occurs in the graining filling and maturity periods. The maize uptake of N will increase until the optimum levels of water and N in soils are reached [44]. Overlaying this information with the leached N loading, it was found that the intensive N uptake and loss occurred at same time. The strong ET and rainfall in that time period were the reason for the N transport in the crop-soil system [45]. The soil N gap analysis revealed that the observed N gap was similar in the vertical direction, which also demonstrates that the soil N can move among the soil depths with hydrological processes and maintain the N concentration balances [3]. The temporal N analysis in crop organs, soil and leached water explain the maize growth in this freeze-thawing area, which also provides a guideline for N fertilization while, also considering SW conditions.

The N pollution to groundwater due to excessive N fertilizer application is an important environmental concern [46]. The

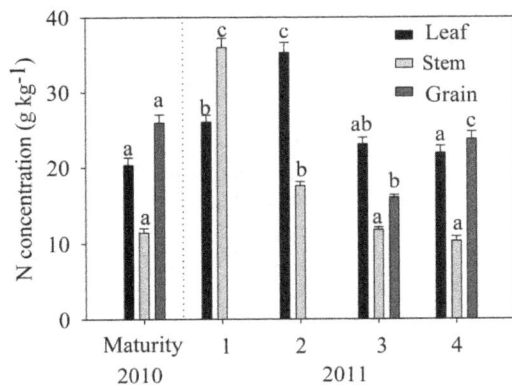

Figure 6. Seasonal dynamics of N concentration in maize organs (Same letters in same crop organs indicates no significant change between different growth stages. 1-Seedling, 2-Jointing, 3-Grain filling, 4-Maturity).

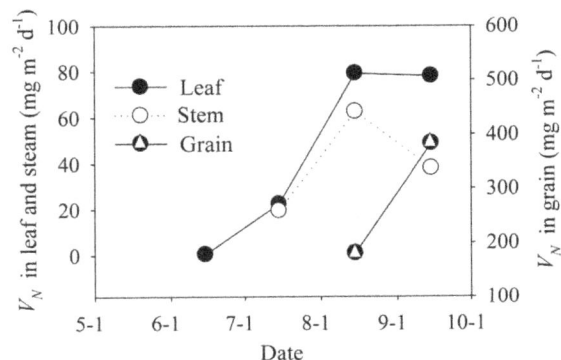

Figure 7. Seasonal dynamics of N accumulative rates (VN) in maize organs.

Table 4. Temporal pattern of N amount (kg ha^{-1}) of maize organs during growing season*.

Year	Growth stage	Leaf	Stem	Grain
2010	Mature	58.6f	38.3ef	213de
2011	Seedling	0.14a		
	Jointing	6.85c	5.81a	
	Grain filling	30.6d	24.6d	54.1b
	Maturity	60.0f	36.0e	169c

*Same lowercase letter in one column indicats no significant difference (P<0.001); Same uppercase letter in one row indicates no significant difference (P<0.001).

Table 5. Mineral N stocks (kg ha^{-1}) and variations at each depth of growing season.

Depth (cm)	Before sowing		After harvest		Stock gap	
	NO$_3^-$-N	NH$_4^+$-N	NO$_3^-$-N	NH$_4^+$-N	NO$_3^-$-N	NH$_4^+$-N
0–15	147d	21.3b	90.1d	12.5c	−48.9c	−8.64bc
15–30	144d	21.2b	88.1d	8.12b	−53.8cd	−11.2c
30–60	130cd	19.3b	68.0c	9.02b	−65.3e	−9.79c
60–90	121c	18.8b	63.2c	5.74a	−56.8d	−12.69c

Same letters in one column indicates no significant difference (P<0.001).

Table 6. Amounts of soil water loss (SL) and N leached below maize root zone (90 cm).

Growth stage	Period	R	ET	SL	leached N (kg ha^{-1})	
		(mm)			NO$_3^-$-N	NH$_4^+$-N
Seedling	30 May to 29 Jun.	29.8	44.6	19.6	0.03	0.01
Jointing	30 Jun. to 25 Jul.	96.5	119.0	11.5	3.36	0.12
Grain filling	26 Jul. to 23 Aug.	165	67.3	48.6	5.53	0.18
Maturity	24 Aug. to 30 Sep.	168	76.6	67.5	6.26	0.10

yearly *NUE* is about 43% in this area, which is a relatively high value in comparison with similar dryland agriculture areas [47]. The N leached from farmland mainly depends on the difference between the N obtained by the soil-water-crop system and the N uptake by the crop [48]. As Table 6 shows, the leached N is mainly related to the loss of SW, which is the gap between rainfall and ET. The rainfall in August 2011 was above the long term monthly value, which caused the highest SW and N loss during the growing

season. Under the normal climatic conditions, the *NUE* in this area can reach 50%. The nitrate-N is the dominant form of leached N to the groundwater, which is a direct consequence of fertilization and soil nitrification [49]. Models have been developed to optimize the fertilizer application to farmland based on a consideration of the soil hydrological process, which can help to maximize the crop yield with minimal discharge to the environment [50]. The model with local characteristic parameters,

Table 7. The N balance and usage efficiency (*NUE*) in crop-soil system during the maize growth period.

Rainfall	Fertilizer	Initial soil N		Final soil N		N loss		Soil supply N	*NUE*
(kg N ha^{-1})		(kg N ha^{-1})		(kg N ha^{-1})		(kg N ha^{-1})		(kg N ha^{-1})	(kg kg^{-1})
		NO$_3^-$-N	NH$_4^+$-N	NO$_3^-$-N	NH$_4^+$-N	NO$_3^-$-N	NH$_4^+$-N		
16.5	120	542.9±29.76	80.6±4.29	309.4±18.4	35.4±4.37	15.2	0.41	641±14.4	0.43±0.03

should be applied in the tillage practice to obtain an improved *NUE*.

Conclusion

The results of this study have shown that combined conditions of soil texture, soil hydrology and precipitation were basic factors for maize growth in the freeze-thawing area in Northeast China. The rainfall significantly affected the pattern of SW content and the allocation of N in the maize organs. Precipitation was the only water source for the dryland, but the variation in SW acquisition rates and SW content remained at a stable level due to the special *albic soil* texture. The dry weight and N accumulation of maize organs in the entire growing period proved that the soil hydrological condition was also enhanced.

The N loss was dependent on both the hydrological process and the N in the water, which was then decided by the rainfall water depth and the applied N. The NO_3-N was the dominant leachate, which was closely correlated with the amount of applied N fertilizer. The *NUE* was about 43%, which was also related to the initial soil N and maize uptake. In order to improve the *NUE* and decrease the amounts of leached nitrogen, it would be better to improve the fertilizer application amounts while giving consideration to rainfall patterns and N accumulation rates.

Author Contributions

Conceived and designed the experiments: WOY. Performed the experiments: SYC GQC. Analyzed the data: WOY SYC. Contributed reagents/materials/analysis tools: FHH. Wrote the paper: WOY.

References

1. Buckley C, Carney P (2013) The potential to reduce the risk of diffuse pollution from agriculture while improving economic performance at farm level. Environ Sci Policy 25(1): 118–126.
2. Quemada M, Baranski M, Nobel-de Lange MNJ, Vallejo A, Cooper JM (2013) Meta-analysis of strategies to control nitrate leaching in irrigated agricultural systems and their effects on crop yield. Agric Ecosyst Environ, 174(15): 1–10.
3. Ferrant S, Oehler F, Durand P, Ruiz L, Salmon-Monviola J, et al. (2011) Understanding nitrogen transfer dynamics in a small agricultural catchment: Comparison of a distributed (TNT2) and a semi distributed (SWAT) modeling approaches. J Hydrol 406(1–2): 1–15.
4. Panda RK, Behera S (2003) Non-point source pollution of water resources: Problems and perspectives. J Food Agric Environ 1(3–4): 308–311.
5. Wang XB, Dai KA, Wang Y, Zhang XM, Zhao QS, et al. (2010) Nutrient management adaptation for dryland maize yields and water use efficiency to long-term rainfall variability in China. Agric Water Manage 97(9): 1344–1350.
6. Yang YG, Xiao HL, Wei YP, Zhao LJ, Zou SB, et al. (2011) Hydrologic processes in the different landscape zones of Mafengou River basin in the alpine cold region during the melting period. J Hydrol 409(1–2): 149–156.
7. Grzebisz W (2013) Crop response to magnesium fertilization as affected by nitrogen supply. Plant Soil 368(1–2): 23–39.
8. Matejek B, Huber C, Dannenmann M, Kohlpaintner M, Gasche R, et al. (2010) Microbial N turnover processes in three forest soil layers following clear cutting of an N saturated mature spruce stand. Plant Soil 337(1–2): 93–110.
9. Moreau P, Ruiz L, Vertes F, Baratte C, Delaby L, et al. (2013) CASIMOD'N: An agro-hydrological distributed model of catchment-scale nitrogen dynamics integrating farming system decisions. Agric Syst 118: 41–51.
10. Kersebaum KC, Steidl J, Bauer O, Piorr HP (2003) Modelling scenarios to assess the effects of different agricultural management and land use options to reduce diffuse nitrogen pollution into the river Elbe. Phys Chem Earth 28: 537–545.
11. Hirel B, Le Gouis J, Ney B, Gallais A (2007) The challenge of improving nitrogen use efficiency in crop plants: towards a more central role for genetic variability and quantitative genetics within integrated approaches. J Exp Bot 58: 2369–2387.
12. Pask AJD, Sylvester-Bradley R, Jamieson PD, Foulkes MJ (2012) Quantifying how winter wheat crops accumulate and use nitrogen reserves during growth. Field Crop Res 126(14): 104–118.
13. Bonato O, Schulthess F, Baumgärtner J (1999) Simulation model for maize crop growth based on acquisition and allocation processes for carbohydrate and nitrogen. Ecol Model 124(1): 11–28.
14. Binder DL, Dobermann A, Sander DH, Cassman KG (2002) Biosolids as nitrogen source for irrigated maize and rainfed sorghum. Soil Sci Soc Am J 66(2): 531–543.
15. Favaretto N, Norton LD, Johnston CT, Bigham J, Sperrin M (2012) Nitrogen and Phosphorus Leaching as Affected by Gypsum Amendment and Exchangeable Calcium and Magnesium. Soil Sci Soc Am J 76(2): 575–585.
16. Lam QD, Schmalz B, Fohrer N (2012) Assessing the spatial and temporal variations of water quality in lowland areas, Northern Germany. J Hydrol 438–439(17): 137–147.
17. Friend AD, Stevens AK, Knox RG, Cannell MGR (1997) A process-based, terrestrial biosphere model of ecosystem dynamics (Hybrid v3.0). Ecol Modell 95(2–3): 249–287.
18. Dorioz JM, Ferhi A (1994) Non-point pollution and management of agricultural areas: Phosphorus and nitrogen transfer in an agricultural watershed. Water Res 28(2): 395–410.
19. Keeney DR, Follett RF (1991) Managing Nitrogen for Groundwater Quality and Farm Profitability, SSSA, Madison, WI. 1–7.
20. Kim DG, Vargas R, Bond-Lamberty B, Turetsky MR (2012) Effects of soil rewetting and thawing on soil gas fluxes: a review of current literature and suggestions for future research. Biogeoscience 9(7): 2459–2483.
21. Lewis DR, McGechan MB, McTaggart IP (2003) Simulating field-scale nitrogen management scenarios involving fertiliser and slurry applications. Agr Syst 76(1): 159–180.
22. Dawson JC, Huggins DR, Jones SS (2008) Characterizing nitrogen use efficiency in natural and agricultural ecosystems to improve the performance of cereal crops in low-input and organic agricultural systems. Field Crop Res 107: 89–101.
23. Schroder JL, Zhang HL, Girma K, Raun WR, Penn CJ, et al. (2011) Soil Acidification from Long-Term Use of Nitrogen Fertilizers on Winter Wheat. Soil Sci Soc Am J 75(3): 957–964.
24. Caliskan S, Ozkaya I., Caliskan ME, Arslan M (2008) The effects of nitrogen and iron fertilization on growth, yield and fertilizer use efficiency of soybean in a Mediterranean-type soil. Field Crop Res 108(2): 126–132.
25. Razzaghi F, Plauborg F, Jacobsen SE, Jensen CR, Andersen MN (2012) Effect of nitrogen and water availability of three soil types on yield, radiation use efficiency and evapotranspiration in field-grown quinoa. Agric Water Manage 109: 20–29.
26. Hao FH, Chen SY, Ouyang W (2013) Temporal rainfall patterns with water partitioning impacts on maize yield in a freeze-thaw zone. J Hydrol 486(12): 412–419.
27. Ouyang W, Huang HB, Hao FH, Shan YS, Guo BB (2012) Evaluating spatial interaction of soil property with non point source pollution at watershed scale: the phosphorus indicator in Northeast China. Sci Total Environ 432, 412–421.
28. Jackson JE (1991) A user's guide to principal components. Wiley-Interscience, New York.
29. Moroizumi T, Hamada H, Sukchan S, Ikemoto M (2009) Soil water content and water balance in rainfed fields in Northeast Thailand. Agric Water Manage 96(1): 160–166.
30. Huggins DR, Pan L (1993) Nitrogen efficiency component analysis: an evaluation of cropping system differences in productivity. Agronomy J 85: 898–905.
31. Meisinger JJ, Randall GW (1991) Estimating nitrogen budgets for soil-crop systems. In: Follet, R.F., et al. (Eds.), Managing Nitrogen for Groundwater Quality and Farm Profitability. SSSA, Madison,WI, USA, 85–124.
32. Vázquez N, Pardo A, Suso ML, Quemada M (2006) Drainage and nitrate leaching under processing tomato growth with drip irrigation and plastic mulching. Agric Ecosyst Environ 112(4): 313–323.
33. Tanveer SK, Wen XX, Lu XL, Zhang JL, Liao YC (2013) Tillage, Mulch and N Fertilizer Affect Emissions of CO2 under the Rain Fed Condition. PLOS ONE: 8(9): e72140.
34. Dörner J, Horn R (2009) Direction-dependent behaviour of hydraulic and mechanical properties in structured soils under conventional and conservation tillage. Soil Till Res 102(2): 225–232.
35. Murungu FS, Chiduza C, Muchaonyerwa P, Mnkeni PNS (2011) Mulch effects on soil moisture and nitrogen, weed growth and irrigated maize productivity in a warm-temperate climate of South Africa. Soil Till Res 112(1): 58–65.
36. Gheysari M, Mirlatifi SM, Homaee M, Asadi ME, Hoogenboom G (2009) Nitrate leaching in a silage maize field under different irrigation and nitrogen fertilizer rates. Agric Water Manage 96(6): 946–954.
37. Wegehenkel M, Mirschel W (2006) Crop growth, soil water and nitrogen balance simulation on three experimental field plots using the Opus model—A case study. Ecol Modell 190(1-2): 116–132.
38. Katerji N, Campi P, Mastrorilli M (2013) Productivity, evapotranspiration, and water use efficiency of corn and tomato crops simulated by AquaCrop under contrasting water stress conditions in the Mediterranean region. Agric Water Manage 130: 14–26.
39. Strullu L, Cadoux S, Preudhomme M, Jeuffroy MH, Beaudoin N (2011) Biomass production and nitrogen accumulation and remobilisation by Miscanthus × giganteus as influenced by nitrogen stocks in belowground organs. Field Crop Res 121(3): 381–391.

40. Shoji S, Gandeza AT, Kimura K (1991) Simulation of crop response to polyolefin-coated urea: II. Nitrogen uptake by corn. Soil Sci Soc Am J 55(5): 1468–1473.

41. Nacry P, Bouguyon E, Gojon A (2013) Nitrogen acquisition by roots: physiological and developmental mechanisms ensuring plant adaptation to a fluctuating resource. Plant Soil, 370(1–2): 1–29

42. Kovács GJ (2005) Modelling of adaptation processes of crops to water and nitrogen stress. Phys Chem Earth, Parts A/B/C 30(1–3): 209–216.

43. Overman AR, Scholtz RV (2013) Accumulation of Biomass and Mineral Elements with Calendar Time by Cotton: Application of the Expanded Growth Model. PLOS ONE 8(9): e72810.

44. Kerbiriou PJ, Stomph TJ, Van Der Putten PEL, Van Bueren ETL, Struik PC (2013) Shoot growth, root growth and resource capture under limiting water and N supply for two cultivars of lettuce (Lactuca sativa L.). Plant Soil 371(1–2): 281–297.

45. Asseng S, Turner NC, Keating BA (2001) Analysis of water- and nitrogen-use efficiency of wheat in a Mediterranean climate. Plant Soil 233(1): 127–143.

46. Fang QX, Yu Q, Wang EL, Chen YH, Zhang GL, et al. (2006) Soil nitrate accumulation, leaching and crop nitrogen use as influenced by fertilization and irrigation in an intensive wheat-maize double cropping system in the North China Plain. Plant Soil 284(1–2): 335–350.

47. Ouyang W, Wei XF, Hao FH (2013) Long-term soil nutrient dynamics comparison under smallholding land and farmland policy in northeast of China. Sci Total Environ 450–451(15): 129–139.

48. Hall JA, Bobe G, Hunter JK, Vorachek WR, Stewart WC, et al. (2013) Effect of Feeding Selenium-Fertilized Alfalfa Hay on Performance of Weaned Beef Calves. PLOS ONE 8(3): e58188.

49. Lteif A, Whalen JK, Bradley RL, Camire C (2010) Nitrogen transformations revealed by isotope dilution in an organically fertilized hybrid poplar plantation. Plant Soil 333(1–2): 105–116.

50. Sierra J, Brisson N, Ripoche D, Noel C (2003) Application of the STICS crop model to predict nitrogen availability and nitrate transport in a tropical acid soil cropped with maize. Plant Soil 256(2): 333–345.

22

A Multicompartment Approach - Diatoms, Macrophytes, Benthic Macroinvertebrates and Fish - To Assess the Impact of Toxic Industrial Releases on a Small French River

Manon Lainé, Soizic Morin*, Juliette Tison-Rosebery

UR EABX, Irstea, Cestas cedex, France

Abstract

The River Luzou flows through a sandy substrate in the South West of France. According to the results of two assessment surveys, the Water Agency appraised that this river may not achieve the good ecological status by 2015 as required by the Water Framework Directive (2000/60/EC). This ecosystem is impacted by industrial effluents (organic matter, metals and aromatic compounds). In order to assess and characterize the impact, this study aimed to combine a set of taxonomic and non-taxonomic metrics for diatoms, macrophytes, macroinvertebrates and fish along the up- to downstream gradient of the river. Diversity metrics, biological indices, biological and ecological traits were determined for the four biological quality elements (BQE). Various quantitative metrics (biomass estimates) were also calculated for diatom communities. The results were compared to physicochemical analysis. Biological measurements were more informative than physicochemical analysis, in the context of the study. Biological responses indicated both the contamination of water and its intensity. Diversity metrics and biological indices strongly decreased with pollution for all BQE but diatoms. Convergent trait selection with pollution was observed among BQE: reproduction, colonization strategies, or trophic regime were clearly modified at impaired sites. Taxon size and relation to the substrate diverged among biological compartments. Multiple anthropogenic pollution calls for alternate assessment methods of rivers' health. Our study exemplifies the fact that, in the case of complex contaminations, biological indicators can be more informative for environmental risk, than a wide screening of contaminants by chemical analysis alone. The combination of diverse biological compartments provided a refined diagnostic about the nature (general mode of action) and intensity of the contamination.

Editor: Jose Luis Balcazar, Catalan Institute for Water Research (ICRA), Spain

Funding: This study was funded by the Adour-Garonne Water Agency. The funders had no role in study design, data collection and analysis, decision to publish, or preparation of the manuscript.

Competing Interests: The authors have declared that no competing interests exist.

* Email: soizic.morin@irstea.fr

Introduction

Rivers are a key component for the development of civilizations and, therefore, human-induced impacts on these ecosystems are diverse and always increasing. Since 2000 in Europe, the Water Framework Directive [1] requires water bodies to achieve good ecological status by 2015, and consider biology as the central element of the assessment. This Directive recommends using four biological quality elements (BQEs) – diatoms, macrophytes, benthic macroinvertebrates and fish – to assess the rivers' ecological status.

The River Luzou flows through a sandy substrate in the South West of France (Landes). From the results of two assessment surveys (2004 and 2006), the Water Agency appraised that this river may not achieve the good ecological status by 2015 [2]. The ecosystem is impacted by the effluent of an industrial plant producing rubber: the pollution is very diverse (mainly organic matter, metals and aromatic compounds including aniline), variable in composition over the year, and released in pulses (51 releases in 2010).

In this context of abundant, complex, toxic pollution, the question is how to characterize the ecological status of the river Luzou? In France, river assessment is based on several biological indices: IPR (Indice Poisson Rivière) for fish fauna [3–4], IBMR (Indice Biologique Macrophytes Rivières) for macrophytes [5], IBGN (Indice Biologique Global Normalisé) for benthic macro-invertebrates [6] and IBD (Indice Biologique Diatomées) for diatoms [7–8]. While these indices are mostly sensitive to trophic pollution, recent works in Europe focused on methods to assess multi-stress conditions: IPR+ [9], I2M2 [10], SPEARorg [11] or SPEARpest for phytosanitary disturbance [12]. Usually, the four BQEs are studied separately but it has already been demonstrated that to assess ecological status of rivers, they are complementary [13–16].

Nevertheless, even when combined, indices often lead to excessive reduction of environmental information [17], so the use of non-taxonomic measures such as biological and ecological traits provides new perspectives for biological assessment methods [18–19]. Traits represent qualitative and quantitative information related to the biology of organisms and their relationship with the environment. Traits describe taxon ecological preferenda, life cycle, morphology, physiology or behaviour [12,20–25].

In this context, we hypothesized that the assessment of the ecological status of the River Luzou according to the recommendations of the WFD, e.g. biological indices calculation and punctual physicochemical analysis of the water, would be insufficient to highlight the impact suffered by the studied ecosystem. In other words, we assumed that the combination of relevant taxonomic and non-taxonomic metrics for the four BQEs cited along the up- to downstream gradient of the River Luzou, is a consistent way to assess the impacts of the industrial releases on the ecosystem. We tested our hypothesis by sampling benthic macroinvertebrates, diatoms, macrophytes, fish and water at three sites on an up- to downstream gradient of the river, in different seasons, in 2009 and 2010. We compared the relevance of the different types of results, e.g. classical biological indices, taxonomic and non-taxonomic metrics, punctual physicochemical parameter values, to accurately monitor the changes in major ecosystem components.

Materials and Methods

Study area and sampling sites

The River Luzou (Landes, South West France) is a small river (28 km long) flowing through a sandy substrate, and characterized by low pH, low conductivity and low nutrient concentrations (Fig. 1). The industrial plant, situated about 18 km from the river source, introduces highly toxic pollution and in addition increases the river water temperature through use of water for cooling and tank washing processes. The wastewater is stored in a settling pool and overflow is released into the River Luzou directly downstream of the plant. Three sampling sites publicly accessible were selected along the study reach: a site called "Up" situated upstream of the factory (Lambert 93 coordinates: X385686, Y6312556) was considered as the reference station (i.e. unimpacted by the toxic effluents); sites "Down 1" (X387414, Y6312572) and "Down 2" (X389923, Y6310416) were situated about 500 m and 4500 m downstream of the factory, respectively. Benthic macroinvertebrates, diatoms, macrophytes, fish and water were sampled at the three sites once or several times, in different seasons, in 2009 and 2010 (see below for details). The water, macrophyte, macroinvertebrate, and phytobenthos sampling procedures were approved by the Adour-Garonne Water Agency.

Physicochemical parameters

Temperature, pH, conductivity and oxygen saturation were measured *in situ* with appropriate WTW probes. During diatom samplings at day 30, two litres of water were collected from the main flow area near the middle of the river and kept at 4°C for analysis within 24 h, according to AFNOR standardized protocols. The following parameters were analyzed by an accredited laboratory: total suspended particulate matter (SPM), biological oxygen demand (BOD), and concentrations of sulfates (SO_4^{2-}), ammonium (NH_4^+), nitrites (NO_2^-), nitrates (NO_3^-), Kjeldahl nitrogen (Nkjeldahl), phosphates (PO_4^{3-}), total phosphorus (PT), carbonates (HCO_3^-), calcium (Ca), chlorine (Cl), heavy metals (copper, lead, aluminium, iron, manganese, nickel, zinc, cadmium, mercury), cyanide, arsenic, pesticides (142 substances, including degradation products), benzene and nitrobenzene, toluene and 2-nitrotoluene.

The substances monitored were chosen because they are known to be present in the releases, as stated in the self-monitoring data from the plant itself. Parameters showing values below the detection limit for all upstream and downstream sites were removed from the analysis. For those occasionally below, half the value of the detection limit was used [26].

Additionally, maximal theoretical concentration of aniline (a target component of the toxic wastes released by the industrial site) was evaluated using industrial monitoring data from year 2010 and the QMNA5 value (monthly low flow value that may not occur more than once every 5 years).

Fish sampling

Ethics statement: All fish were properly collected and handled in an ethical manner, with all required permissions from the Adour-Garonne Water Agency. No other permissions were required for completion of this research, and this study does not include endangered or otherwise protected species.

Fish surveys were carried out by experienced fisheries staff of the departmental fishing federation AAPPMA64 (accredited by ECCEL Environnement) and all sampling procedures complied with the French and European Union legislation on animal welfare. Electric fishing was carried out at the minimum power settings needed to incapacitate the fish and thus no adverse impact was expected. The fish were handled with great care. This includes electrofishing and manipulations (counting, weighing and measurements) where fish were maintained in river water. After measurements, all fish were returned alive into the river.

All sites were electrofished (pulsed direct current waveform), during low flow period (September 2009) and according to the NF EN 14011 standard [27]. Electrofishing was authorized by the Adour Garonne Water Agency and performed by operators accredited by ECCEL Environnement. Fish were sorted and stored in a large basin, and then counted, measured and weighed, then released alive in the water. In the case of very numerous individuals, fish were counted and weighed in homogenous sets. The sampling area was systematically reported.

Macrophyte sampling

All sampling was carried out according the standard NF T90–395 [5] during vegetation periods (September 2009, July 2010). The areas covered by macrophyte beds and by each taxon were evaluated for each site. Taxa difficult to identify *in situ* were collected, packed and transported to the laboratory for determination.

Benthic macroinvertebrate sampling

Macroinvertebrates were collected at each of the 4 seasons (Autumn: September 2009, Winter: November 2009, Spring: April 2010 and Summer: July 2010). Macroinvertebrates were collected with a Surber sampler (mesh size 500 μm, sampling area 0.05 m²), the device required by the French standardized protocol XP T90–333 [28]. Micro-samples were taken at each site from among twelve mesohabitats defined as visually distinct units within the stream, considering apparent physical uniformity (*sensu* Armitage et al. 1995), and described by a combination of substrate types and current velocities. Mesohabitats were sampled in a hierarchical order according to the IBGN standard [6] to maximize the taxonomic richness of the faunal assemblage at the site scale, after gathering sample units. Macroinvertebrates were sorted and identified to the family level following standard XP T90–388 [29], except for some groups identified at a higher taxonomic level (i.e. *Oligochaeta*, *Bryozoa*, *Nematoda*, *Hydracarina*). The twelve micro-samples were pooled to finally constitute a single sample, used to calculate the IBGN index [6].

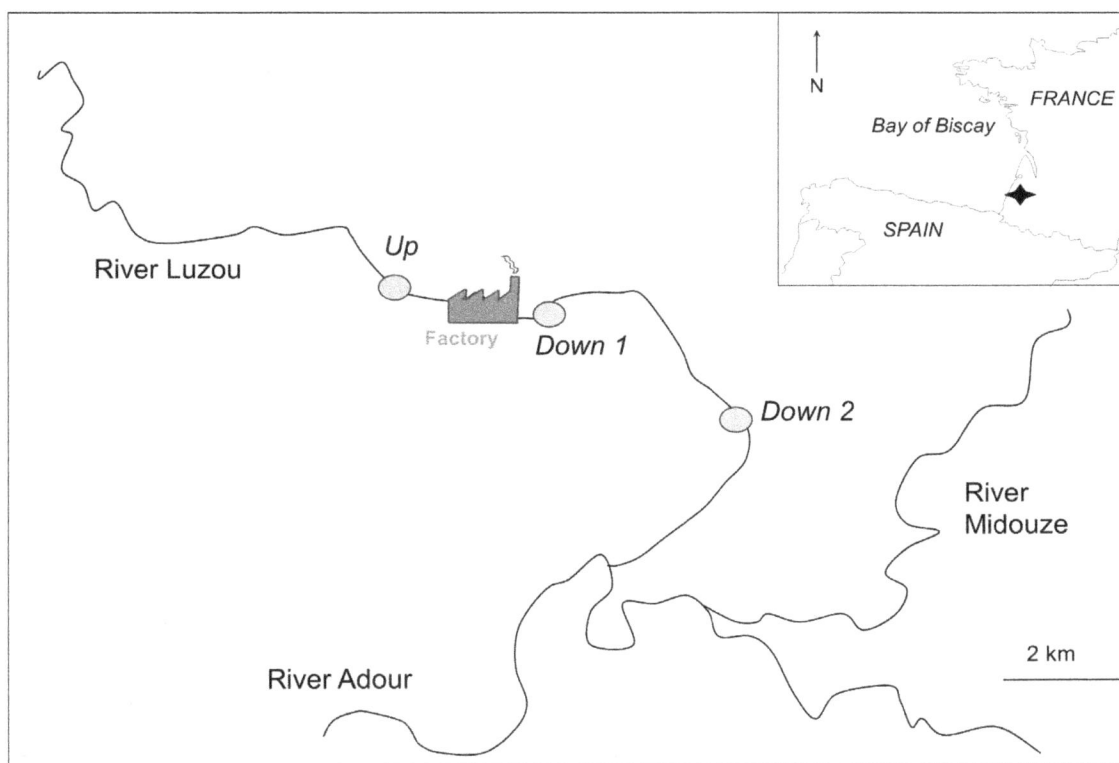

Figure 1. Study area and sampling sites.

Phytobenthos sampling

Phytobenthos was also sampled four times (Autumn: September 2009, Winter: November 2009, Spring: April 2010 and Summer: August 2010).

Quantitative measurements. Artificial substrates were immersed in the river to quantify dry weight, ash-free dry mass of biofilms, chlorophyll a concentrations, and the number of live and dead diatom cells. Six glass slides (total surface area reaching 300 cm^2) were placed in a rack at the three sites [30]. After 15 days immersion (t15) and 30 days immersion (t30), three slides were removed from the water and scraped into a standard volume of mineral water, to obtain three replicates per sampling date then separated into aliquots.

A 20-mL aliquot was used to determine the dry weight (DW) and ash-free dry mass (AFDM) of the biofilm, expressed as mg cm^{-2}, according to European standard NF EN 872 [31]. Ten millilitres of the suspension were filtered through a Whatman GF/C filter, then extracted with acetone for 24 h before spectrophotometric analyses. Chlorophyll a concentrations were calculated following Lorenzen [32]. A 5-mL aliquot was preserved with 0.5 mL of formalin solution for diatom cell density enumeration [live and dead, 33] and taxonomic identification.

Diatom community characterization. Diatom samples were collected from natural surfaces (pebbles or macrophytes, at t30) and artificial substrates (at t15 and t30), according to a standardized method NF T 90–354 [7]. For each slide, 400 valves were determined to the lowest taxonomical level possible. Diatom species were identified at 1000× magnification (Leitz DMRB light microscope), mainly according to Krammer and Lange-Bertalot [34] and Lange-Bertalot [35], by examining permanent slides of cleaned diatom frustules, digested in boiling H_2O_2 (30%) and HCl

(35%) and mounted in a high refractive index medium (Naphrax, Northern Biological Supplies Ltd., UK; RI = 1.74). A total of 234 diatom taxa were identified. The 107 taxa with abundances higher than five individuals (considering all samples) were used to describe community structure.

Taxonomic and indicial metrics

For the four biological compartments and for all sampling dates, specific richness (S), Shannon diversity (H) [36] and Pielou equitability (J) [37] indices were calculated, as well as the French biological indices used for biomonitoring: IBGN [6], IBD [7], IPR [3,38] and IBMR [5]. In addition, for benthic macroinvertebrates SPEARpesticides [12] and SPEARorganic [11] indices were determined thanks to an online application (http://www.systemecology.eu/spear/). Finally, for diatoms the polluosensitivity index IPS [39] and the occurrence of teratogenic forms were calculated based on the 234 taxa.

Biofilm-related quantitative metrics

Seasonal variations in biomass: dry weight, ash-free dry mass, chlorophyll a concentrations, and the number of live and dead diatoms were reduced by normalization using the mean values calculated on the reference site (Up) at each season.

Functional metrics

For each taxon within each biological compartment, different functional traits were listed from the literature. Each trait shows at least two modalities, and for each modality a score is assigned to the taxon. A Taxa × Traits table was obtained for each compartment. Fish traits are listed from Keith and Allardi [20]: trophic guilds, nesting substrates and position in the water column.

Affinity scores were based on a 0 ("no affinity") to 1 ("affinity") scale. Considering benthic macroinvertebrates, eleven biological traits and eleven ecological traits were determined from Tachet et al. [40]. Affinity scores were scaled from 0 ("no affinity") to 5 ("high affinity"). Diatom traits were derived from the Irstea database (https://hydrobio-dce.cemagref.fr/) for biovolumes and pioneer forms, Kelly et al. [41] for growth forms and Passy [22] for guilds. Affinity scores were scaled from 0 ("no affinity") to 1 ("affinity"). For macrophytes, floristic groups -phanerogams, algae, bryophytes and heterotrophs- (scores from 0 "no affinity" to 1 "affinity") were assigned to taxa and biological phanerogam traits (scores from 0 "no affinity" to 2 "strong affinity") were listed from Willby et al. [25].

Data analysis

Major differences between sampling sites considering physico-chemical parameters were investigated using Principal Component Analysis (PCA). Data were normalized and redundant variables identified through Spearman pairwise correlation tests. Among redundant variables, only one was kept. Concentrations of aniline were taken as supplementary variable, assuming that the upstream site (Up) was free of aniline.

Concerning biological data, diatom and macroinvertebrate community structures were described using PCA based on species relative abundances. Functional traits-related information was obtained from the tables Taxa × Samples and Taxa × Traits for each compartment. Taxa × Samples tables were expressed in abundances, except for macrophytes (percentage cover) and fish (biomass). In order to produce functional profiles for each trait (relative distribution of the information among the categories), the following process was applied: i) for a given site and for each category of traits, taxon scores were weighted by abundances (cover or biomass); ii) the sums of the weighted scores were then expressed as a relative abundance distribution (within a trait), giving the site trait profile. For diatoms, this analysis was based on the 107 dominant taxa.

Only diatoms and benthic macroinvertebrates enabled statistical analyses (four sampling dates). Taxonomic, indicial and functional metrics were compared between sites by Kruskal-Wallis tests. For macrophytes and fish (respectively two and one replicates), only simple visual comparisons were possible.

All analyses were performed using the R software, version 3.0.2 [42], with packages ade4 [43] for descriptive analyses of data and agricolae [44] for Kruskal-Wallis tests.

Results

Analysis of environmental data

Alkalinity and HCO_3^- giving redundant ecological information (Spearman test = 1), HCO_3^- was removed from the dataset. The following parameters, even though they were known to be released in the effluents, were systematically below the detection limit and were thus removed from the analysis: copper, lead, manganese, nickel, cadmium, mercury, cyanide, arsenic, benzene and nitrobenzene, toluene and 2-nitrotoluene. For the same reason, pesticides were also removed.

The environmental parameters finally kept for analysis and their contributions to axes 1 and 2 are listed in Table 1.

Axes 1 and 2 account for 70% of the total inertia (Fig. 2a). Axis 1 discriminates upstream from downstream conditions but does not clearly discriminate Down 1 from Down 2, while both Axes 1 and 2 separates winter samples from the others.

Downstream sites are characterized by higher concentrations of Zn, Nkjeldahl, SO_4^{2-}, NH_4^+, BOD5, higher conductivity, alkalinity and temperature (Fig. 2b).

Mean calculated concentration of aniline downstream was 0.44 µg L^{-1}. Aniline, represented a posteriori, shows a cos^2 equal to 0.42 on axis 1 and to 0.28 on axis 2.

Analysis of biological data

Upstream communities were typical of the Landes ecoregion. Minnow (Phoxinus phoxinus), gudgeon (Gobio gobio), stone loach (Barbatula barbatula), brook lamprey (Lampetra planeri) and eel (Anguilla anguilla) were present in the River Luzou. The benthic macroinvertebrate fauna (thirty taxa) is dominated by Chironomidae and Gammaridae. Acidophilous and neutrophilous diatom taxa from the genera Eunotia and Brachysira were abundant, associated with numerous Karayevia oblongella, Tabellaria flocculosa and Peronia fibula. The macrophytic community was mainly composed of phanerogams characteristic of oligotrophic and acidic waters: Myriophyllum alterniflorum and Potamogeton polygonifolium.

Taxonomical metrics and indices

Table 2 gathers information about differences between upstream and downstream conditions, according to Kruskal-Wallis tests. Taxonomic metrics based on fish, macrophytes and benthic macroinvertebrates showed similar responses to toxic pollution: a strong decrease in species richness, diversity and equitability was observed at station Down 1. Fish response was very marked, as only two minnow specimens were found at station Down 1 (considered as a null biomass), and a still low biomass was harvested at station Down 2. Macrophyte community structure was strongly shifted from a diversity of algae, bryophytes (e.g. Fontinalis antipyretica) and spermaphytes (in particular, Callitriche platycarpa), to a Sphaerotilus sp.-dominated community at Down 1, causing the dramatic decrease in taxonomic metrics observed. At Down 2, the diversity of taxonomic groups increased, and high percentages of algae (such as Cladophora sp.) were found. Taxonomic metrics for benthic macroinvertebrates were also significantly lower at stations Down 1 and Down 2, reflecting that macroinvertebrate species composition was mainly influenced by the up- to downstream gradient (Fig. 3a), more than by season (cold vs. warm waters). No significant differences between downstream and upstream sites were globally observed for diatoms, but the flora showed large seasonal variations (Fig. 3b).

Poor IPR, IBMR and IBGN scores classified station Down 1 in a "bad" ecological status, and station Down 2 in a "medium" to "poor" status (Table 2). The high abundance (up to 60%) at downstream sites of Achnanthidium minutissimum, a species considered as oligo- to mesotrophic by IBD and IPS, maintained good index scores whereas the SPEARorganic index differentiated upstream and downstream sites.

Diatom communities sampled on glass slides 30 days after immersion presented a strong decrease in live cell density and in chlorophyll a concentrations downstream. At station Down 1, dry weight and ash-free dry mass increased for communities sampled on glass slides 15 days after immersion, but diatom mortality was significantly higher (decrease in the ratio of live to dead cells).

Functional metrics

Diatoms. Even though Kruskall Wallis test results were not significant, some functional traits from communities sampled on natural substrates (pebbles) showed trends between upstream and downstream sites (Table 3). Pioneer species were more abundant at station Down 2 while "high profile" and "motile" species

Table 1. Average (±standard errors, n = 4) values and contributions of environmental parameters to axis 1 and axis 2 of the PCA performed on physicochemical parameters (see Figure 2).

Parameter	Code	Tox	Up	Down1	Down2	Axis 1	Axis 2
Aluminium (µg L⁻¹)	Al	100	306±163	322±213	266±141	-0.38	*0.51*
Aniline (µg L⁻¹)	Ani	2.2	b.d.l.	0.44	0.44	0.48	*0.19*
Calcium (mg L⁻¹)	Ca		5.5±0.3	7.5±0.5	8.4±0.9	0.48	0.19
Chloride (mg L⁻¹)	Cl		19.7±0.5	20.4±0.3	20.1±0.4	-0.01	*0.65*
Conductivity (µs cm⁻¹)	Cond.		123±4	175±26	176±22	0.88	0.06
Biological Oxygen Demand (mgO₂ L⁻¹)	DBO5		1.3±0.2	4.9±2.2	2.9±0.5	*0.60*	0.00
Total Iron (µg L⁻¹)	Fe.brut	300	287±62	340±64	326±58	0.00	*0.56*
Total manganese (µg L⁻¹)	Mn.brut		16±3	17±3	24±3	0.00	0.40
Total Suspended Matter (mg L⁻¹)	SPM		11.0±9.6	3.9±1.7	4.0±1.6	-0.22	0.20
Kjeldahl nitrogen (mg L⁻¹)	Nk		0.8±0.3	3.3±1.4	3.0±1.2	*0.82*	0.01
Ammonium (mg L⁻¹)	NH4		0.1±0.0	2.3±1.2	2.3±1.2	*0.85*	0.01
Nitrates (mg L⁻¹)	NO3		5.1±1.1	4.6±1.2	5.2±1.0	-0.47	0.46
Oxygen saturation (%)	O2sat		97.9±6.3	92.3±8.7	89.0±8.2	-0.46	0.08
pH	pH		6.8±0.3	6.5±0.3	6.1±0.4	0.09	*-0.8*
Total phosphorus (mg L⁻¹)	Ptot		0.03±0.00	0.03±0.01	0.02±0.01	-0.41	0.03
Sulfates (mg/ L⁻¹)	SO4		10.9±0.9	28.4±9.1	26.1±7.3	0.88	0.06
Temperature (°C)	Temp		12.4±0.9	12.7±1.4	12.7±1.3	*0.51*	-0.01
Alkalinity (°f)	TAC		0.9±0.1	1.4±0.3	1.5±0.4	*0.84*	-0.02
Total zinc (µg L⁻¹)	Zn.brut	30	5.8±0.6	21.0±4.7	17.3±0.6	*0.67*	0.14
Dissolved zinc (µg L⁻¹)	Zn.dis		4.8±1.2	13.5±2.4	13.8±1.4	0.18	0.39

Aniline data calculated *a posteriori* are provided for information, b.d.l. below detection limit. Toxicity benchmarks (Tox) are freshwater values from the Canadian Water Quality Guidelines for the Protection of Aquatic Life (http://www.pesticideinfo.org/). Italics indicate the variables significantly discriminant (cos²>0.50).

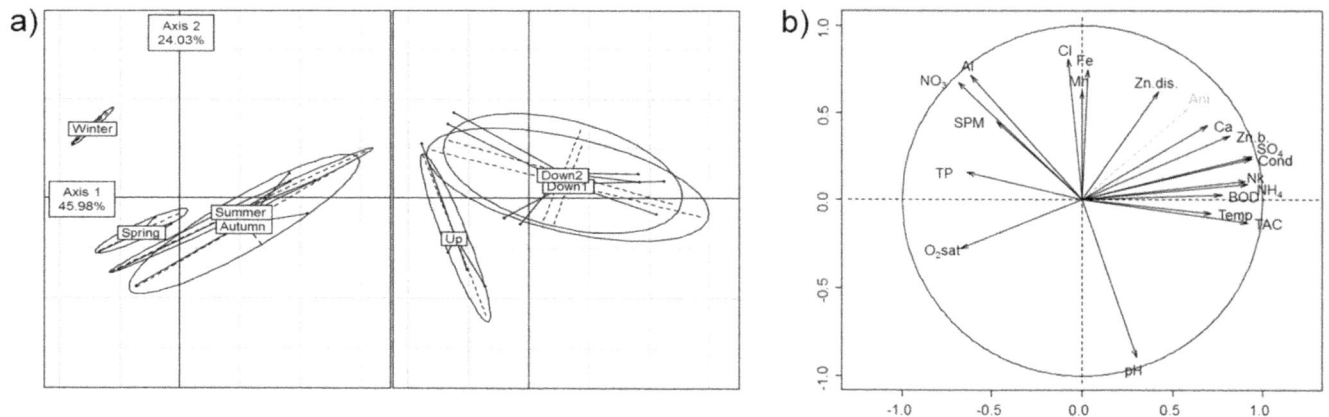

Figure 2. Principal Component Analysis based on physicochemistry. a) Principal Component Analysis performed on physicochemical data, with samples grouped by season (left panel) or by sampling site (right panel); b) Correlation circle. See Table 1 for abbreviations and contributions of environmental parameters on axis 1 and 2.

decreased. Downstream, high biovolume taxa regressed in favour of smaller ones (<99 μm³). No particular upstream-downstream pattern emerged from growth forms.

Benthic macroinvertebrates. A higher number of taxa showing more than one reproductive cycle per year, and/or an asexual way of reproduction was observed downstream, in addition to a greater number of taxa showing a high maximal size (Table 4). Taxa with tegumentary respiration, burrowers, interstitial or permanently attached were also favoured downstream. Taxa without any form of resistance decreased in favour of taxa able to produce cocoons. Active dispersal (aerial and aquatic) decreased while aquatic passive dispersion increased (station Down 1). Feeding behaviours were also modified downstream whereas absorbers and deposit feeders, eating detritus, fine sediments and microorganisms dominated. As a result shredders, filter-feeders and scrapers (only at station Down 1) eating plants (alive or dead) or living microinvertebrates decreased. Concerning ecological traits, polysaprobic and eutrophic taxa increased downstream while xenosaprobic, oligotrophic and mesotrophic taxa decreased.

Macrophytes. The different floristic group proportions were clearly modified downstream (Table 5). Station Down 1 presented more than 99% of heterotrophic forms, hence phanerogram-related traits were not reliable for its description. At station Down 2 the phanerograms did not recover the upstream reference status,

allowing the installation of filamentous algae. At this site, taxa were preferentially annual, with asexual reproduction mainly by fragmentation and sexual reproduction based on higher numbers of seeds, with entomophilous dispersion. These taxa showed larger emergent leaves, related to a higher morphological index (large size).

Fish. Functional metrics were not applicable in station Down 1, where only one species (2 individuals) was found. At station Down 2 the biomass and abundance of invertivorous species decreased in favour of omnivores (Table 6). Pelagic species became dominant over benthic ones.

Discussion

The importance of biology for the assessment of ecological status

PCA performed on environmental data discriminated upstream from downstream conditions, but sites Down 1 and Down 2 were not well distinguished, in contrast to results obtained with biological data (especially for macro-invertebrates). A high seasonal variation can also be noted, with different dilution conditions according to whether the flow was low or high.

Moreover, several toxicants known to be released in the plant effluents were not detected by physicochemical analysis. Thus,

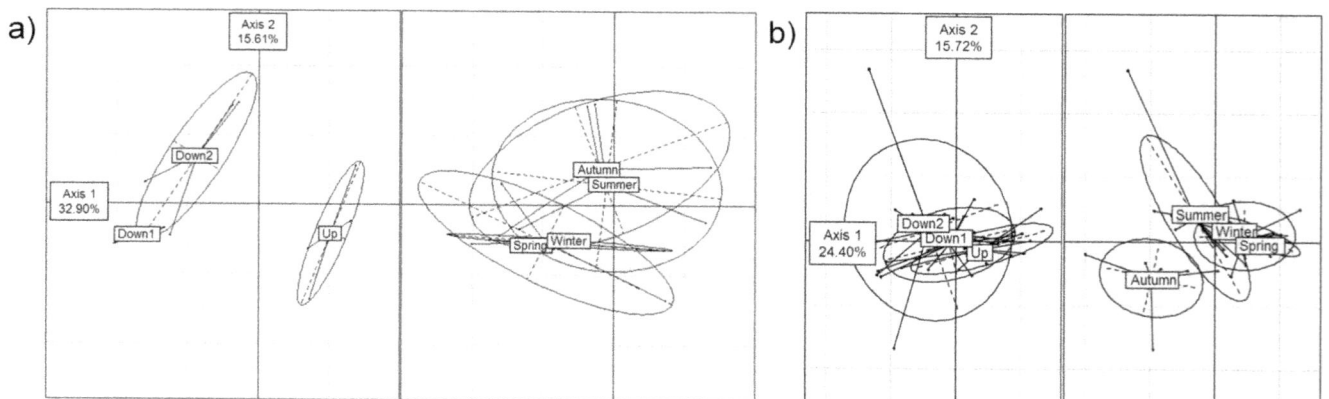

Figure 3. Contrasted responses of a) benthic macroinvertebrates and b) diatoms to seasonal and longitudinal changes.

Table 2. Taxonomic and quantitative metrics: calculation (mean and standard deviation) and differences between sites.

		Up	Down1	Down2	P value
Fish	S	6.00	*1.00*	8.00 *	-
	H	1.27	*0.00*	1.31 *	-
	J	0.70	*NA*	0.63	-
	Biomass	57.50	*0.00*	*12.00 **	-
	IPR	14.62	*52.28*	*22.49 **	-
Macrophytes	S	10.50±2.12	*4.00±1.41*	15.00±1.41	-
	H	1.22±0.07	*0.03±0.04*	2.02±0.03	-
	J	0.52±0.02	*0.02±0.02*	0.75±0.04	-
	IBMR	13.29±0.55	*3.22±3.49*	10.93±0.07 *	-
Macroinvertebrates	S	30.00±5.77	*10.00±3.37*	*18.50±3.42 **	0.008
	H	1.95±0.43	*0.56±0.17*	*1.03±0.45*	0.02
	J	0.57±0.10	*0.24±0.05*	*0.35±0.15*	0.03
	IBGN	13.00±2.16	*3.50±1.29*	*8.50±0.58*	0.007
	SPEARp	35.05±2.77	27.37±5.66	34.95±4.55	0.08
	SPEARo	-0.54±0.04	*-0.68±0.02*	*-0.62±0.08*	0.04
Diatoms - pebbles	S	40.25±4.92	44.00±8.87	33.75±13.99	0.5
	H	2.73±0.42	2.88±0.50	2.36±0.86	0.61
	J	0.74±0.09	0.76±0.09	0.67±0.17	0.73
	TER%	0.24±0.34	1.65±1.20	1.00±1.41	0.18
	IPS	17.4±1.47	16.85±1.36	17.45±1.07	0.87
	IBD	19.95±0.10	18.68±1.65	19.38±0.43	0.3
Diatoms - t30	S	36.00±7.39	*46.25±8.02*	*49.5±6.61*	0.03
	H	2.72±0.46	3.06±0.20	2.98±0.28	0.58
	J	0.76±0.09	0.80±0.03	0.77±0.07	0.87
	TER%	0.13±0.25	0.37±0.14	0.19±0.24	0.36
	IPS	17.85±2.02	17.88±0.85	15.73±3.60	0.49
	IBD	19.93±0.15	20.00±0.00	18.95±2.10	0.57
	Live	1.00±0.18	*0.55±0.37*	*0.33±0.25 **	0.0001
	Live/Dead	1.00±0.12	1.12±0.74	0.97±0.48	0.81
	DW	1.00±0.26	2.07±2.44	0.55±0.53	0.06
	AFDM	0.98±0.31	1.36±0.97	0.63±0.48	0.07
	Chlo.a	0.98±0.69	*0.38±0.58*	*0.31±0.28*	0.002
Diatoms - t15	S	42.00±7.53	48.00±7.87	45.25±11.00	0.72
	H	2.90±0.27	3.06±0.17	2.38±1.01	0.69
	J	0.78±0.07	0.79±0.05	0.62±0.23	0.38
	TER%	0.00±0.00	0.06±0.13	0.37±0.60	0.3
	IPS	18.50±0.90	17.80±1.05	15.23±2.55	0.07
	IBD	19.73±0.55	19.70±0.60	18.18±2.25	0.23
	Live	1.00±0.22	0.95±0.50	0.85±0.41	0.45
	Live/Dead	1.00±0.23	0.69±0.48	0.78±0.38	0.048
	DW	0.98±0.16	*2.88±1.87*	1.50±0.97	0.03
	AFDM	1.07±0.28	*3.43±3.23*	1.54±0.90	0.03
	Chlo.a	0.89±0.40	1.06±1.13	0.56±0.57	0.23

Italics indicate that the metric is significantly different from the reference station (Up), stars indicate significant differences between Down 1 and Down 2. Abbreviations: Specific richness (S), Shannon diversity (H), Pielou equitability (J), percentage of diatoms abnormal forms (TER%), dry weight (DW), ash-free dry mass (AFDM), chlorophyll *a* (Chlo.a), and biological indices based on macroinvertebrates (IBGN), fish (IPR), and diatoms (IBD and IPS). NA: Metric calculation not possible. Quantitative metrics for diatoms from artificial substrates (t15, t30) are normalized by Up values at each sampling date.

Table 3. Diatom-related traits: calculation (mean and standard deviation) and differences between sites.

	Trait	Modality	Up	Down1	Down2	P value
Diatoms - pebbles	Pioneer forms	Non-pioneer	95.78±4.31	75.66±13.27	56.67±29.22	0.09
		Pioneer	4.22±4.31	24.34±13.27	43.33±29.22	0.09
	Growth forms	Adnate	2.97±1.29	1.24±0.49	1.59±1.52	0.15
		Pedunculate	19.58±9.08	28.16±4.71	36.46±11.77	0.09
		Colonial	38.65±17.73	28.75±15.85	17.46±18.87	0.14
		Non-colonial	38.80±9.69	41.85±12.49	44.49±8.64	0.58
	Biovolumes ($\mu m3$)	<99	30±19.85	33.54±11.40	53.19±21.29	0.21
		100–299	25.44±5.06	31.53±9.61	23.12±3.79	0.29
		300–599	15.31±9.51	12.89±5.45	6.79±5.26	0.16
		600–1499	17.85±7.54	13.26±5.70	11.17±9.83	0.43
		>1500	11.40±5.27	8.79±6.00	5.73±3.97	0.5
	Passy guilds	Low profile	42.46±21.27	37.71±11.55	65.79±25.38	0.23
		High profile	21.17±14.07	24.72±14.72	16.33±16.40	0.43
		Mobile	24.56±8.73	28.94±9.52	*11.96±4.87* *	0.02
		Variable	3.98±5.34	3.63±1.75	1.82±1.41	0.43
Diatoms - t30	Pioneer forms	Non-pioneer	97.92±2.59	92.94±6.55	90.19±13.58	0.21
		Pioneer	2.08±2.59	7.06±6.55	9.81±13.58	0.21
	Growth forms	Adnate	3.36±2.00	3.77±1.97	2.81±1.54	0.59
		Pedunculate	15.95±10.15	20.15±9.08	18.75±8.16	0.79
		Colonial	46.09±21.86	44.63±16.44	35.63±21.39	0.66
		Non-colonial	34.60±14.60	31.46±8.68	42.81±14.93	0.43
	Biovolumes (μm^3)	<99	26.98±15.17	23.96±4.47	22.92±10.21	0.84
		100–299	24.99±9.13	25.97±3.33	31.79±17.19	0.98
		300–599	18.69±5.35	15.95±3.08	14.79±3.26	0.33
		600–1499	17.01±6.22	20.04±4.84	17.31±7.11	0.49
		>1500	12.33±4.20	14.08±1.79	13.20±5.09	0.84
	Passy guilds	Low profile	35.08±14.64	34.80±7.39	31.97±12.85	0.69
		High profile	30.57±18.84	38.90±10.87	28.95±17.18	0.39
		Mobile	21.91±12.33	15.28±4.55	30.28±13.84	0.33
		Variable	3.64±3.29	3.13±2.52	2.05±1.48	0.87
Diatoms - t15	Pioneer forms	Non-pioneer	91.74±11.74	92.09±7.82	80.8±24.64	0.98
		Pioneer	8.26±11.74	7.91±7.82	19.20±24.64	0.98
	Growth forms	Adnate	3.57±2.32	2.96±1.73	2.35±1.96	0.77
		Pedunculate	20.33±9.53	20.54±7.31	20.91±14.17	0.92
		Colonial	40.17±21.35	39.66±16.74	19.60±14.68	0.23
		Non-colonial	35.93±13.41	36.83±10.88	57.15±20.56	0.12
	Biovolumes (μm^3)	<99	23.99±9.55	21.69±7.10	27.39±21.94	0.98
		100–299	25.35±9.28	27.09±5.14	13.20±6.89	0.05
		300–599	15.53±4.34	13.89±3.68	36.45±33.16	0.55
		600–1499	23.32±3.35	24.52±5.88	15.25±8.95	0.12
		>1500	11.81±6.43	12.82±2.55	7.71±5.22	0.24
	Passy guilds	Low profile	34.38±11.26	29.46±9.31	32.32±22.83	0.87
		High profile	30.98±13.85	34.59±11.76	18.46±13.25	0.19
		Mobile	19.93±7.78	21.17±7.98	42.67±28.49	0.23
		Variable	3.90±1.30	3.60±0.84	2.23±2.11	0.43

Italics indicate that the metric is significantly different from the reference station (Up), stars indicate significant differences between Down 1 and Down 2.

Table 4. Macroinvertebrate-related traits: calculation (mean and standard deviation) and differences between sites.

Trait	Modality	Up	Down1	Down2	P value
Maximal size	≤0.25 cm	0.10±0.05	0.46±0.06	0.35±0.17	0.03
	>0.25–.5 cm	22.41±13.39	5.59±1.78	8.34±3.63	0.02
	>0.5–1 cm	25.86±3.14	12.7±3.93	21.34±13.96	0.07
	>1–2 cm	27.51±8.47	13.42±1.47	16.58±5.20	0.03
	>2–4 cm	13.74±4.98	9.29±0.26	9.95±1.26	0.24
	>4–8 cm	8.15±6.13	45.77±5.58	33.98±17.07	0.03
	>8 cm	2.24±1.69	12.76±1.55	9.47±4.76	0.03
Life span	≤1 year	61.84±9.98	18.95±9.6	37.52±29.15	0.07
	>1 year	38.16±9.98	81.05±9.6	62.48±29.15	0.07
Number of reproductive cycles per year	<1	1.31±0.80	0.40±0.21	1.44±0.99	0.17
	1	50.29±12.22	31.60±2.30	36.65±6.28	0.03
	>1	48.40±12.71	68±2.42	61.91±7.10	0.03
Aquatic stage	egg	27.05±7.23	28.24±2.91	26.00±4.43	0.87
	larva	39.06±1.53	36.77±1.39	40.66±5.25	0.16
	nymph	19.5±7.46	8.80±4.73	13.46±9.28	0.21
	adult	14.4±3.91	26.18±3.15	19.87±9.74	0.07
Reproduction	ovoviviparity	17.59±11.17	1.84±0.93	3.03±1.92	0.02
	isolated eggs, free	2.20±1.23	5.63±0.72	4.65±2.15	0.08
	isolated eggs, cemented	4.82±3.42	2.77±0.34	3.55±1.57	0.66
	clutches, cemented or fixed	51.35±19.01	56.97±3.75	52.65±7.01	0.66
	clutches, free	15.30±10.60	9.21±4.95	13.20±9.26	0.66
	clutches, in vegetation	0.45±0.36	0.00±0.01	0.21±0.21	0.06
	clutches, terrestrial	4.72±2.37	3.32±1.52	7.68±4.33	0.19
	asexual reproduction	3.56±2.73	20.26±2.47	15.03 ± 7.55	0.02
Dispersal	aquatic passive	46.23±5.32	80.11±6.02	67.06±18.51	0.04
	aquatic active	21.28±3.39	9.99±1.10	13.24±4.03	0.02
	aerial passive	12.96±4.70	5.93±3.12	10.03±7.12	0.19
	aerial active	19.53±6.53	3.96±1.86	9.67±7.61	0.03
Resistance forms	eggs, statoblasts	3.56±2.57	0.01±0.02	4.23±4.93	0.02
	cocoons	6.53±5.02	36.92±4.50	27.34±13.82	0.03
	housings against desiccation	0±0	0±0	0±0	0.11
	diapause or dormancy	6.14±2.01	2.30±1.14	4.79±3.02	0.09
	none	83.77±5.91	60.78±3.36	63.64±7.10	0.02
Respiration	tegument	49.45±12.09	90.58±2.47	79.90±12.31	0.02
	gill	42.28±9.52	9.01±2.45	18.56±12.40	0.01
	plastron	6.75±6.77	0.10±0.14	0.50±0.32 *	0.01
	spiracle	1.52±0.97	0.30±0.37	1.03±1.20	0.17
	hydrostatic vesicle	0±0	0±0	0±0	
Locomotion	flier	2.63±2.62	0.04±0.07	0.20±0.12 *	0.01
	surface swimmer	0.06±0.03	0.01±0.01	0.06±0.09	0.16
	full water swimmer	13.21±6.51	8.57±1.55	11.24±5.76	0.55
	crawler	41.60±8.07	7.52±3.65	17.59±13.13	0.02
	burrower	9.11±4.94	26.09±1.89	20.88±6.98	0.02
	interstitial	19.50±6.97	51.59±4.34	41.85±13.85	0.02
	temporarily attached	13.85±11.44	6.00±1.07	8.03±3.02	0.24
	permanently attached	0.04±0.03	0.17±0.02	0.16±0.03	0.02
Food	microorganisms	3.59±2.26	15.66±1.52	11.98±5.07	0.02
	detritus (<1 mm)	26.85±4.00	52.40±2.81	45.28±10.39	0.02
	dead plant (≥1 mm)	9.97±1.38	1.09±0.48	2.80±2.66	0.02

Table 4. Cont.

Trait	Modality	Up	Down1	Down2	P value
	living microphytes	32.22±8.85	23.28±0.40	24.64±2.43	0.05
	living macrophytes	5.26±1.89	1.42±0.68	2.64±2.12	0.04
	dead animal (≥1 mm)	2.90±1.53	0.37±0.19	1.06±1.24	0.06
	living microinvertebrates	10.13±5.33	2.08±1.11	3.52±2.48	0.04
	living macroinvertebrates	9.06±2.61	3.70±1.63	8.05±5.04	0.07
	vertebrates	0.02±0.04	0±0	0.03±0.03 *	0.04
Feeding habitats	absorber	2.58±1.96	14.78±1.80	10.95±5.53	0.03
	deposit feeder	19.1±9.57	64.99±5.44	52.16±16.83	0.02
	shredder	21.91±6.67	2.23±1.04	3.63±2.12	0.02
	scraper	27.79±11.74	8.61±1.66	14.82±9.42	0.04
	filter-feeder	13.66±15.90	2.22±1.22	5.39±3.98	0.11
	piercer	1.29±0.89	1.35±0.71	4.47±2.95	0.11
	predator	11.67±4.08	4.62±2.30	6.90±4.90	0.13
	parasite	1.99±1.40	1.20±0.65	1.68±1.19	0.77
Transverse distribution	river channel	25.78±13.91	11.33±1.10	13.81±1.89	0.03
	banks, connected side-arms	35.64±2.91	31.87±0.16	32.54±0.65 *	0.01
	ponds, pools, disconnected side-arms	8.46±4.10	11.17±0.44	10.85±0.49	0.74
	marshes, peat bogs	3.89±2.17	6.02±0.29	5.59±0.27	0.13
	temporary waters	7.87±3.67	6.15±1.23	7.53±2.87	0.79
	lakes	16.29±5.28	22.83±0.32	21.76±1.01	0.02
	groundwaters	2.07±1.48	10.64±1.30	7.92±4.00	0.03
Longitudinal distribution	crenon	10.64±1.26	12.32±0.23	12.10±0.37	0.02
	epirithron	16.62±2.67	13.79±0.12	14.44±0.50	0.02
	metarithron	17.34±3.19	14.75±0.22	15.19±0.26	0.23
	hyporithron	17.60±3.63	16.13±0.39	16.21±0.51	0.92
	epipotamon	13.95±1.34	16.91±0.40	15.86±1.27	0.03
	metapotamon	9.71±2.42	15.25±0.52	13.76±1.85	0.03
	estuary	3.84±2.21	1.15±0.49	1.61±0.96	0.04
	outside river system	10.31±4.15	9.71±1.35	10.81±2.42	0.87
Altitude	lowlands	54.64±5.20	67.68±3.27	61.11±9.16	0.04
	piedmont level	25.87±4.61	15.00±0.89	18.08±3.74	0.02
	alpine level	10.23±2.29	8.48±1.27	11.10±3.52	0.31
Substrate	flags/boulders/cobbles/pebbles	16.27±3.38	11.96±0.41	13.61±2.00	0.06
	gravel	12.88±2.44	17.38±1.29	15.14±3.33	0.14
	sand	11.03±2.04	12.81±0.24	11.96±1.20	0.39
	silt	6.00±1.83	8.65±0.29	7.58±1.49	0.07
	macrophytes	16.22±1.08	12.87±0.54	14.91±2.72	0.09
	microphytes	2.62±1.09	3.10±0.14	2.61±0.61	0.43
	twigs/roots	8.53±3.66	4.16±0.26	5.32±1.80	0.03
	organic detritus/litter	5.82±2.00	5.98±0.08	5.95±0.15	0.98
	mud	5.92±3.35	14.52±0.43	12.60±2.33	0.02
Current velocity	null	16.62±8.61	24.75±0.67	22.73±1.55	0.05
	slow (<25 cm/s)	31.16±3.62	32.91±0.55	32.47±1.72	0.58
	medium (25–50 cm</s)	33.93±7.92	26.26±0.38	27.57±1.20	0.01
	fast (>50 cm/s)	18.30±1.78	16.08±0.26	17.22±2.05	0.15
Trophic status	oligotrophic	37.28±6.57	31.68±0.75	32.73±0.76	0.02
	mesotrophic	42.33±1.69	38.78±0.25	38.66±0.27	0.02
	eutrophic	20.39±5.89	29.54±0.55	28.61±0.77	0.01
Salinity (preferences)	freshwater	85.41±7.13	89.05±1.75	87.01±4.7	0.79

Table 4. Cont.

Trait	Modality	Up	Down1	Down2	P value
	brackish water	14.59±7.13	10.95±1.75	12.99±4.7	0.79
Temperature	cold (<15°C)	21.34±4.37	25.92±0.48	23.81±2.46	0.29
	warm (>15°C)	10.53±2.91	*16.23±0.17*	14.71±2.10	0.03
	eurythermic	68.12±6.30	*57.85±0.51*	61.48±4.55	0.03
Saprobity	xenosaprobic	9.25±2.05	*3.25±1.08*	*5.42±3.36*	0.04
	oligosaprobic	27.68±1.76	23.67±0.83	25.40±2.44	0.07
	β-mesosaprobic	40.95±5.03	40.74±0.98	40.51±1.69	0.84
	α-mesosaprobic	18.47±4.05	*23.78±0.73*	21.45±2.62	0.03
	polysaprobic	3.64±2.14	*8.56±0.25*	*7.22±1.51*	0.01

Italics indicate that the metric is significantly different from the reference station (Up), stars indicate significant differences between Down 1 and Down 2.

such analysis does not seem to be particularly reliable in the case of the River Luzou: to characterize intermittent pollution, measurements should be performed exactly during the releases, or by high resolution analyses. The problem of toxicity related to cocktail effects and degradation compounds can also hardly be tackled in this way. In this context, biological related metrics are potentially more informative than chemical analyses. Seasonal variations in community structure (Fig. 3) were also observed, and the different metrics used allowed the nature, and intensity, of the pollution present in the River Luzou to be highlighted.

Temporal scales of biological responses

The seasonal variations observed in the PCA performed on environmental data were correlated to changes in diatom responses over time. Although quantitative measurements indicated decreasing diatom biomass from up- to downstream whatever the sampling season, the number of cells settled was up to 15 fold higher in warmer conditions compared to winter (data not shown). The seasonality in water contamination was highlighted by, e.g., higher percentages of teratologies (up to 3% on natural substrates) and of the species *Achnanthidium minutissimum* downstream in Autumn, indicating toxic pollution [45,46].

Diatoms have fast growth rates (from hours to days), and thus respond very quickly to variations in their environment. BQE with longer life span reflect more averaged water quality on different time scales. Seasonal patterns were thus less pronounced for macrophytes and macroinvertebrates that integrated global quality over the year. Ultimately, fish responses were expected to reveal environmental conditions on the longer term (years).

Taxonomic metrics reveal pollution intensity

Richness, diversity and equitability indices are classically used to evaluate ecological status of water and their decline is indicative of a disturbed environment [47]. Except for diatoms, these metrics clearly decreased for all the biological compartments studied, at station Down 1, which is in accordance with the literature [16,48–49]. They recovered to variable extents at Down 2, indicating weaker biological impact likely due to lower toxicant availability (by dilution and/or toxicant degradation). For diatoms, the phenomenon observed contrasted with the literature, rather reporting a decrease of these taxonomic metrics in toxic conditions [45,50]. For this BQE in our study, richness, diversity and equitability indices were not relevant to highlight the toxic impact of pollution (see below).

Biological indices rather indicate the nature of pollution

Biological indices except IBD and IPS drastically decreased downstream, due to the presence of numerous tolerant taxa for IBMR and IBGN (*Sphaerotilus* sp., *Oligochaeta* and *Chironomidae*), or to the lack of fish populations for IPR. SPEARorganic index being negatively correlated to toxicants (icides, surfactants, petrochemicals) [11], its decrease downstream was also consistent. In contrast IPS and IBD scores remained good due to the high proportion of *Achnanthidium minutissimum* (up to 60%). However, the occurrence of abnormal forms, mainly affecting *A. minutissimum*, clearly characterized downstream conditions. Above 1%, this rate of occurrence is considered to reflect the impact of toxic pollutants on diatom communities [45–46]. Moreover, recent works also suggested that, due to its pioneering character, *A. minutissimum* was indicative of toxic pollution [45]. Therefore, the use of diatom-based biological indices to highlight toxic pollution is not recommended, but a careful analysis of community composition (as used for index calculations) can also provide information regarding the nature of the contamination.

Additionally, quantitative metrics (decrease of chlorophyll *a* concentrations and diatom density downstream) were consistent with Morin et al. [45], and with Wang et al. [51] who demonstrated that a derivative of aniline could inhibit adhesion in certain diatom species. The percentage of live cells tended to decrease downstream, as already reported by Stevenson and Bahls [52] and Gillet et al. [53] who examined whether the percentage of live diatoms in periphyton communities could be used as a metric of human disturbance in streams and rivers.

Periphytic biomass (dry weight and ash-free dry mass) tended to increase downstream, whereas diatom cell numbers and chlorophyll a did not. This unexpected result can perhaps be related to the periphyton becoming more heterotrophic, which would be consistent with the massive development of *Sphaerotilus* sp. observed at Down 1 in the macrophytic community. One could also hypothesize that rivers from Landes ecoregion, characterized by naturally acidic waters and nutrient depletion, represent a particular ecosystem where artificial eutrophication (concomitant to industrial release) can enhance the periphyton growth and richness. Other recent works [54] reported reduced impacts on biomass of metal toxicity in acid-adapted biofilms. These phenomena may explain the typical quantitative response of phytobenthos towards toxic pollution.

Table 5. Macrophyte-related traits: calculation (mean and standard deviation) and differences between sites.

Trait	Modality	Up	Down1	Down2
Floristic group	Algae	3.63±2.84	0.02±0.03	50.84±33.98 *
	Heterotroph	0±0	99.58±0.53	0±0
	Bryophyte	35.00±5.67	0.36±0.51	35.55±27.43
	Phanerogam	61.38±2.83	0.03±0.05	13.61±6.54 *
Growth form	Anchored, floating leaves	33.23±0.06	NA	31.31±2.41
	Anchored, submerged leaves	33.47±0.01	NA	34.34±1.21
	Anchored, emergent leaves	4.25±3.27	NA	9.02±5.42
	Anchored, heterophylly	29.05±3.31	NA	25.32±4.21
Vertical shoot architecture	Single apical growth point	2.92±2.19	NA	0±0
	Single basal growth point	12.26±10.39	NA	14.98±1.26
	Multiple apical growth point	84.82±8.21	NA	85.02±1.26
Leaf type	Capillary	0.52±0.10	NA	0±0
	Entire	99.48±0.10	NA	100±0
Leaf area	Small (<1 cm^2)	78.01±9.83	NA	64.86±19.20
	Medium (1–20 cm^2)	9.02±2.27	NA	11.93±6.70
	Large (20–100 cm^2)	12.76±7.86	NA	20.01±7.96
	Extra-large (>100 cm^2)	0.21±0.30	NA	3.21±4.53
Morphology index (score)	3–5	44.91±1.89	NA	34.28±6.89
	6–7	46.70±3.12	NA	49.94±13.26
	8–9	7.93±5.19	NA	15.53±6.72
	10	0.45±0.18	NA	0.25±0.35
Mode of reproduction	Rhizome	7.40±3.34	NA	9.39±3.58
	Fragmentation	40.25±5.89	NA	36.49±4.78
	Stolons	5.67±4.53	NA	10.05±4.51
	Seeds	46.67±1.98	NA	44.06±3.31
Number of reproductive organs per year per individual	Medium (10–100)	60.85±4.79	NA	54.36±7.42
	High (100–1000)	39.15±4.79	NA	45.64±7.42
Perrenation	Annual	29.49±1.87	NA	34.74±15.65
	Biennial / Short lived perrenial	0.08±0.11	NA	0.89±1.26
	Perennial	70.43±1.98	NA	64.38±14.4
Gamete vector	Wind	43.93±2.65	NA	43.32±0.28
	Water	18.39±0.60	NA	16.69±2.58
	Insect	0.14±0.20	NA	4.03±4.82
	Self	37.54±1.85	NA	35.96±1.96
Body flexbility	Low (<45°)	0.08±0.11	NA	0±0
	Medium (45–300°)	32.81±0.12	NA	37.33±4.99
	High (>300°)	67.11±0.01	NA	62.67±4.99
Leaf texture	Soft	37.49±0.91	NA	37.09±0.81
	Rigid	3.47±0.92	NA	4.36±2.04
	Waxy	21.54±0.90	NA	21.46±0.42
	Non-waxy	37.49±0.91	NA	37.09±0.81
Period of production of reproductive organ	Early (March-May)	31.34±0.96	NA	28.45±3.97
	Mid (June-August)	36.07±2.94	NA	38.68±2.41
	Late (August-September)	31.57±1.28	NA	32.87±1.56
	Very late (post- September)	1.02±0.70	NA	0±0

Table 5. Cont.

Trait	Modality	Up	Down1	Down2
Fruit size	<1 mm	0±0	NA	0±0
	1–3 mm	87.74±10.39	NA	85.02±1.26
	>3 mm	12.26±10.39	NA	14.98±1.26

Italics indicate that the metric is significantly different from the reference station (Up), stars indicate significant differences between Down 1 and Down 2. NA: Metric calculation not possible.

Survival strategies under high pollution

Functional metrics illustrated how the ecosystem of the River Luzou changed under toxic pollution. According to the results, many modifications concerning the four BQE were convergent, like reproduction and colonization strategies, or the trophic regime. Southwood [55] wrote that those physiological adaptations were typically found in impaired sites, as they induce tolerance to harsh conditions.

First, when escaping pollution, as fish attempt to do, was not possible, resistant taxa became dominant (*Sphaerotilus* sp., *Oligochaeta* and *Chironomidae*), with production of cocoons as extreme resistance forms for macroinvertebrates.

Early colonizers with ruderal strategies [56] were also favoured, like the diatom *Achnanthidium minutissimum* which reached an abundance of up to 60% downstream. Morin et al. [57] already observed this high abundance of *A. minutissimum* under toxic conditions, implying a thinning of the biofilm by loss of high-profile taxa (according to Passy [22], high profile guild reaches a maximum in nutrient-rich sites and in conditions of low flow disturbance). *Achnanthidium minutissimum* was probably less disfavoured thanks to its small size reducing the exposure time to toxicants [58], or its adnate posture. Macrophyte communities evolved towards the predominance of annual taxa producing a great number of reproductive organs per year and per individual

(e.g. *Berula erecta*, *Scirpus fluitans* or *Sparganium erectum*), ensuring a rapid spread across the river. Macroinvertebrates' high colonization ability was provided by the combination of a greater number of reproductive cycles and dispersion by drift, those two strategies being already reported in the literature [59]. Drift is an important way of species dissemination and recolonisation of river systems by lotic macroinvertebrates [60], and is known to increase with chemical disturbance [61].

Finally, taxon sizes showed opposite trends. Diatom taxa with lower biovolumes were favoured, whereas bigger macroinvertebrate taxa, or macrophytes with large emergent leaves, increased downstream. This, however, resulted from the same strategy of organisms facing toxicants. Indeed, a decrease of diatom cell sizes are generally observed in environments exposed to toxic pollutions: reduction of size expresses increased vegetative multiplication, and reduced sexual reproduction (not measured here, but concordant with traits of other compartments: higher reproduction rates and asexual reproduction), and selection of smaller species (reducing uptake of toxicants into the cell).

Toxic pollution also drives indirect changes across the trophic web

Our results highlighted changes in feeding habits, which represent an important aspect of the community trophic structure

Table 6. Fish-related traits: calculation (mean and standard deviation) and differences between sites.

	Trait	Up	Down1	Down2
Biomass	Invertivorous	55.65	NA	25.00
	Omnivorous	44.35	NA	75.00
	Other	0.00	NA	0.00
	Phytophilic	0.00	NA	0.00
	Lithophilic	44.35	NA	100.00
	Mixed	55.65	NA	0.00
	Benthic	74.78	NA	16.70
	Pelagic	25.22	NA	83.33
Abundance	Invertivorous	21.29	NA	14.52
	Omnivorous	78.71	NA	85.48
	Other	0.00	NA	0.00
	Phytophilic	0.00	NA	0.00
	Lithophilic	78.71	NA	90.32
	Mixed	21.29	NA	9.68
	Benthic	43.78	NA	8.06
	Pelagic	56.22	NA	91.94

Italics indicate that the metric is significantly different from the reference station (Up). NA: Metric calculation not possible.

modifications. Summarizing, the reference ecosystem (Up) showed balanced communities composed of species typical of the Landes ecoregion [62]. The different compartments are driven by complex interactions. Primary producers provide food resources to primary consumers (like macroinvertebrates and fish feeding on phytobenthos) but also refuge and habitat and/or egg-laying substrates (especially macrophytes). The modifications observed in these communities (disappearance or sharp decline of phanerogams, biofilm thinning) represent a real impact on invertebrates and may explain the changes observed downstream. In particular, in such streams like the river Luzou where there is a poor diversification of abiotic substrates, macrophytes not only provide a food source, but also a shelter for invertebrates [63]. Macroinvertebrate structure and biomass are also modified by the presence or absence of predators. With high pollution (Down 1), biomass of primary producers decreased to be replaced by filamentous heterotrophs, affecting the subsequent components in a cascade. Aside from some potential direct toxicity (according to the toxicants mode of action), macroinvertebrates were probably driven by the resource, selecting absorbers and deposit feeders feeding on detritus and microorganisms in accordance with Archaimbault et al. [60] and Schultheis et al. [64]. Moreover, the decline of macroinvertebrate communities may have forced invertivorous macroinvertebrates and fish to leave the site. Further downstream (Down 2), primary producers tended to diversify, as well as primary consumers and predators. This partial recovery thus reflected both direct (reduction of toxic pressure, slight increase in nutrient availability) and indirect (return to a more balanced ecosystem) improvement of the ecosystem.

Conclusions

In conclusion, our study revealed that in a context of multiple contaminants, of pulse inputs, of complex cocktails, and/or of release of unknown substances, a combination of different biological measurements, from different aquatic communities, can be much more informative than punctual physicochemical analysis and single biotic indices.

References

1. European Parliament (2000) Directive 2000/60/EC of the European Parliament and of the Council establishing a framework for community action in the field of water policy. O.J.L327.
2. Rosebery J, Morin S, Chauvin C, Leibig H, Hupin C, et al. (2010) Bilan hydrobiologique du Luzou. Rapport pour l'Agence de l'Eau Adour-Garonne.
3. AFNOR (2011) Qualité de l'eau - Détermination de l'indice poissons rivières (IPR) - NF T90–344.
4. Oberdorff T, Pont D, Hugeny B, Porcher JP (2002) Development and validation of a fish-based index (FBI) for the assessment of "river health" in France.Freshwat. Biol. 47, 1720–1734.
5. AFNOR (2003) Qualité de l'eau - Détermination de l'indice biologique macrophytique en rivière (IBMR) - NF T90–395.
6. AFNOR (2004) Qualité de l'eau - Détermination de l'indice biologique globale normalisé (IBGN) - NF T90–350.
7. AFNOR (2007) Qualité de l'Eau - Détermination de l'Indice Biologique Diatomées (IBD) - NF T90–354.
8. Coste M, Boutry S, Tison-Rosebery J, Delmas F (2009) Improvements of the Biological Diatom Index (BDI): Description and efficiency of the new version (BDI-2006). Ecol. Indic. 9, 621–650.
9. Marzin A, Delaigue O, Logez M, Belliard J, Pont D (2014) Uncertainty associated with river health assessment in a varying environment: The case of a predictive fish-based index in France. Ecol Indic. 43: 195–204.
10. Mondy CP, Villeneuve B, Archaimbault V, Usseglio-Polatera P (2012) A new macroinvertebrate-based multimetric index (I2M2) to evaluate ecological quality of French wadeable streams fulfilling the WFD demands: A taxonomical and trait approach.Ecol. Indic. 18, 452–467.
11. Beketov MA, Liess M (2008) An indicator for effects of organic toxicants on lotic invertebrate communities: Independence of confounding environmental factors over an extensive river continuum.Environ. Pollut. 156, 980–987.

Under such strong and diverse anthropogenic pressure, a multicompartment approach allows the integrated observation of community trajectories towards adaptation, accounting for the complex biotic relationships in aquatic ecosystems. Colonization strategies, reproduction and trophic regimes seem to be key indicators of this adaptation.

A further step would be to continue this multi-compartment survey during restoration programmes, to identify the behaviour, and/or biological elements that tend to recover more rapidly. Data from the literature suggest that mobile organisms would have the greatest ability to recolonize sites after water quality improvement. The time necessary to reach complete recovery of the ecosystem when pollution ceases is also likely to be very variable among biological compartments and is an important component to be determined in the context of the implementation of the Water Framework Directive.

Acknowledgments

We would like to thank Sylvia Moreira, Gwilherm Jan (Irstea) for help in field sampling, Maryse Boudigues and Muriel Bonnet (Irstea) for physicochemical analyses, Hervé Liebig, Claire Hupin (ECCEL Environnement) and Vincent Renard (FLPPMA) for sampling and identification of macroinvetebrates and fish, Vincent Bertrin and Christian Chauvin (Irstea) for sampling and identification of macrophytes, as well as Evelyne Trichet (Irstea) for assistance. Peter Winterton is acknowledged for proofreading the manuscript. Finally, the authors would like to thanks the reviewers for helping them to improve the manuscript. This study has been carried out in the framework of the Cluster of Excellence COTE.

Author Contributions

Conceived and designed the experiments: SM JTR. Performed the experiments: SM JTR. Analyzed the data: ML SM JTR. Contributed reagents/materials/analysis tools: SM JTR. Wrote the paper: ML SM JTR.

12. Liess M, Von Der Ohe PC (2005) Analysing effects of pesticides on invertebrates communities in streams. Environ. Toxicol. Chem.24, 954–965.
13. Cellamare M, Morin S, Coste M, Haury J (2012) Ecological assessment of French Atlantic lakes based on phytoplankton, phytobenthos and macrophytes. Environ. Monit. Assess.184, 4685–4708.
14. Hering D, Johnson RK, Kramm S, Schmutz S, Szoszkiewicz K, et al. (2006) Assessment of European streams with diatoms, macrophytes, macroinvertebrates and fish: a comparative metric-based analysis of organism response to stress.Freshwat. Biol. 51, 1757–1785.
15. Johnson RK, Hering D, Furse MT, Verdonschot PFM (2006) Indicators of ecological change: comparison of the early response of four organism groups to stress gradients. Hydrobiologia 566, 139–152.
16. Marzin A, Archaimbault V, Belliard J, Chauvin C, Delmas F, et al. (2012) Ecological assessment of running waters: Do macrophytes, macroinvertebrates, diatoms and fish show similar responses to human pressures? Ecol. Indic. 23, 56–65.
17. Usseglio-Polatera P, Bournaud M, Richoux P, Tachet H (2000) Biological and ecological traits of benthic freshwater macroinvertebrates: relationships and definition of group with similar traits.Freshwat. Biol. 43, 175–205.
18. Bonada N, Prat N, Resh VH, Statzner B (2006) Developments in aquatic insect biomonitoring: A comparative analysis of recent approaches.Ann. Rev. Entomol. 51, 495–523.
19. Burkholder JM, Wetzel RG, Klomparens KL (1990) Direct comparison of phosphate uptake by adnate and loosely attached microalgae within an intact biofilm matrix.Appl. Microbiol. 56, 2882–2890.
20. Keith P, Allardi J (2001) Atlas des poissons d'eau douce de France. Patrimoines Naturels, Paris.
21. Morin S, Coste M, Delmas F (2008) A comparison of specific growth rates of periphytic diatoms of varying cell size under laboratory and field conditions. Hydrobiologia 614, 285–297.

22. Passy SI (2007) Diatom ecological guilds display distinct and predictable behavior along nutrient and disturbance gradients in running waters.Aquat. Bot. 86, 171–178.

23. Potapova M, Charles DF (2003) Distribution of benthic diatoms in U.S. rivers in relation to conductivity and ionic composition.Freshwat. Biol. 48, 1311–1328.

24. van Dam H, Mertens A, Sinkeldam J (1994) A coded checklist and ecological indicator values of freshwater diatoms from the Netherlands.Neth. J. Aquat. Ecol. 28, 117–133.

25. Willby NJ, Abernethy VJ, Demars BOL (2000) Attribute-based classification of European hydrophytes and its relationship to habitat utilization.Freshwat. Biol. 43, 43–74.

26. Helsel DR (1990) Less than obvious - statistical treatment of data below the detection limit.Environ. Sci. Technol. 24, 1766–1774.

27. AFNOR (2003) Qualité de l'eau - Échantillonnage des poissons à l'électricité - NF EN 14011.

28. AFNOR (2009) Qualité de l'eau - Prélèvement des macroinvertébrés aquatiques en rivières peu profondes - XP T90–333.

29. AFNOR (2010) Qualité de l'eau - Traitement au laboratoire d'échantillons contenant des macro-invertébrés de cours d'eau - XP T90–388.

30. Morin S, Vivas-Nogues M, Duong TT, Boudou A, Coste M, et al. (2007) Dynamics of benthic diatom colonization in a cadmium/zinc-polluted river (Riou-Mort, France). *Fundam. Appl.* Limnol.168, 179–187.

31. AFNOR (2005) Qualité de l'eau. Dosage des matières en suspension. Méthode par filtration sur filtre en fibres de verre - NF EN 872.

32. Lorenzen CJ (1967) Determination of chlorophyll and pheopigments: spectrophotometric equations.Limnol. Oceanogr. 12, 343–346.

33. Morin S, Proia L, Ricart M, Bonnineau C, Geiszinger A, et al. (2010) Effects of a bactericide on the structure and survival of benthic diatom communities. Vie Milieu 60, 109–116.

34. Krammer K, Lange-Bertalot H (1985–1991) Bacillariophyceae 1. Teil: Naviculaceae; 2. Teil: Bacillariaceae, Epithemiaceae, Surirellaceae; 3. Teil: Centrales, Fragilariaceae, Eunotiaceae; 4. Teil: Achnanthaceae. Kritische Ergänzungen zu *Navicula* (Lineolatae) und *Gomphonema. Süßwasserflora von Mitteleuropa. Band 2/1-4* (eds H Ettl, J Gerloff, H Heynig & D Mollenhauer). G. Fischer Verlag, Stuttgart.

35. Lange-Bertalot H (1995–2012) Iconographia Diatomologica. Annotated Diatom Micrographs, Vol. 1–23. Koeltz Scientific Books, Königstein.

36. Shannon CE (1948) A Mathematical Theory of Communication.Bell Syst. Tech. J. 27, 379–423.

37. Pielou EC (1975) Ecological Diversity. New York, Wiley.

38. Rogers C, Pont D (2005) Création de base de données thermiques devant servir au calcul de l'Indice Poisson normalisé. Université de Lyon I CSP.

39. Coste M (1982) Etude des méthodes biologiques d'appréciation quantitative de la qualité des eaux. Rapport Cemagref Q.E. Lyon-A.F. Bassin Rhône Méditerranée Corse.

40. Tachet H, Richoux P, Bournaud M, Usseglio-Polatera P (2010) *Invertébrés d'eau douce.* Systématique, biologie, écologie. CNRS Editions, Paris.

41. Kelly MG, Bennion H, Cox EJ, Goldsmith B, Jamieson J, et al. (2005) *Common freshwater diatoms of Britain and Ireland: an interactive key.* Environment Agency, Bristol.

42. R Development Core Team (2008) R: A language and environment for statistical computing. R Foundation for Statistical Computing, Vienna, Austria. ISBN 3-900051-07-0, Available: http://www.r-project.org

43. Thioulouse J, Chessel D, Dolédec S, Olivier JM (1997) ADE-4: a multivariate analysis and graphical display software.Stat. Comput. 7, 75–83.

44. De Mendiburu F (2007) *Agricolae: Statistical Procedures for Agricultural Research.* R package version 1.0-3. Available: http://tarwi.lamolina.edu.pe/~fmendiburu

45. Morin S, Cordonier A, Lavoie I, Arini A, Blanco S, et al. (2012) Consistency in diatom response to metal-contaminated environments.Handbook of Environ-

mental Chemistry (H. Guasch, A. Ginebreda & A. Geiszinger), Springer, Heidelberg. 19: 117–146.

46. Falasco E, Bona F, Badino G, Hoffmann L, Ector L (2009) Diatom teratological forms and environmental alterations: a review. Hydrobiologia 623, 1–35.

47. Hooper DU, Chapin III FS, Ewel JJ, Hector A, Inchausti P, et al. (2005) Effects of biodiversity on ecosystem functioning: A consensus of current knowledge. Ecol. Monogr. 75, 3–35.

48. Barbour MT, Gerritsen J, Snyder BD, Stribling JB (1999) *Rapid bioassessment protocols for use in streams and wadeable rivers: Periphyton, benthic macroinvertebrates and fish.* 2nd ed. Washington D.C., US Environmental Protection Agency, Office of Water.

49. Camargo JA, Gonzalo C, Alonso A (2011). Assessing trout farm pollution by biological metrics and indices based on aquatic macrophytes and benthic macroinvertebrates: A case study.Ecol. Indic. 11, 911–917.

50. Lavoie I, Lavoie M, Fortin C (2012) A mine of information: Benthic algal communities as biomonitors of metal contamination from abandoned tailings.Sci. Tot. Environ. 425, 231–241.

51. Wang YK, Stevenson RJ, Metzmeier L (2005) Development and evaluation of a diatom-based Index of Biotic Integrity for the Interior Plateau Ecoregion, USA.J. N. Am. Benthol. Soc. 24, 990–1008.

52. Stevenson RJ, Bahls L (1999) Periphyton protocols. Rapid bioassessment protocols for use in streams and wadeable rivers: Periphyton, benthic macroinvertebrates and fish, 2nd edn. (eds MT Barbour, J Gerritsen, BD Snyder & JB Stribling) Washington D.C., U.S. Environmental Protection Agency; Office of Water. 6, pp. 1–22.

53. Gillett ND, Pan Y, Manoylov KM, Stevenson RJ (2011) The role of live diatoms in bioassessment: a large-scale study of Western US streams. Hydrobiologia 665, 79–92.

54. Luís AT, Bonet B, Corcoll N, Almeida SFP, Ferreira da Silva E, et al. (2014) Experimental evaluation of the contribution of acidic pH and Fe concentration to the structure, function and tolerance to metals (Cu and Zn) exposure in fluvial biofilms. Ecotoxicology, in press.

55. Southwood TRE (1988) Tactics, strategies and templets. Oikos 52, 3–18.

56. Biggs B, Stevenson RJ, Lowe RL (1998) A habitat matrix conceptual model for stream periphyton.Arch. Hydrobiol. 143, 21–56.

57. Morin S, Duong TT, Dabrin A, Coynel A, Herlory O, et al. (2008) Long term survey of heavy metal pollution, biofilm contamination and diatom community structure in the Riou-Mort watershed, South West France.Environ. Pollut. 151, 532–542.

58. Khoshmanesh A, Lawson F, Prince IG (1997) Cell surface area as a major parameter in the uptake of cadmium by unicellular green microalgae.Chem. Eng. J. 65, 13–19.

59. Townsend CR, Hildrew AG (1976) Field Experiments on the Drifting, Colonization and Continuous Redistribution of Stream Benthos. J. Anim. Ecol 45, 759–772.

60. Archaimbault V, Rosebery J, Morin S (2010) Traits biologiques et écologiques, intérêt et perspectives pour la bio-indication des pollutions toxiques. Sciences Eaux & Territoires 1, 46–51.

61. Bournaud M, Maucet D, Chavanon G (1984) Méthode pratique de mesure de la dérive des macroinvertébrés dans un cours d'eau: application à la détection de perturbations du milieu.Bull. Ecol. 15, 199–209.

62. Tison J, Park YS, Coste M, Wasson JG, Ector L, et al. (2005) Typology of diatom communities and the influence of hydro-ecoregions: A study on the French hydrosystem scale.Wat. Res. 39, 3177–3188.

63. Cortelezzi A, Sierra MV, Gomez N, Marinelli C, Capitulo AR (2013) Macrophytes, epilithic biofilms, and invertebrates as biotic indicators of physical habitat degradation of lowland streams (Argentina). Environn. Monit. Asses. 185, 5801–5815.

64. Schultheis AS, Sanchez M, Hendricks AC (1997) Structural and functional responses of stream insects to copper pollution. Hydrobiologia 346, 85–93.

23

Higher Isolation of NDM-1 Producing *Acinetobacter baumannii* from the Sewage of the Hospitals in Beijing

Chuanfu Zhang[1⊙]**, Shaofu Qiu**[1]*⊙**, Yong Wang**[1⊙]**, Lihua Qi**[1⊙]**, Rongzhang Hao**[1⊙]**, Xuelin Liu**[1⊙]**, Yun Shi**[1⊙]**, Xiaofeng Hu**[1⊙]**, Daizhi An**[1⊙]**, Zhenjun Li**[2⊙]**, Peng Li**[1⊙]**, Ligui Wang**[1⊙]**, Jiajun Cui**[1]**, Pan Wang**[1]**, Liuyu Huang**[1]**, John D. Klena**[3]**, Hongbin Song**[1]*

1 Institute of Disease Control and Prevention, Academy of Military Medical Science, Beijing, People's Republic of China, **2** State Key Laboratory for Infectious Disease Prevention and Control, China Center of Disease Control and Prevention, Beijing, People's Republic of China, **3** United States Centers for Disease Control and Prevention, China –US Collaborative Program on Emerging and Re-emerging Infectious Diseases, Beijing, People's Republic of China

Abstract

Multidrug resistant microbes present in the environment are a potential public health risk. In this study, we investigate the presence of New Delhi metallo-β-lactamase 1 (NDM-1) producing bacteria in the 99 water samples in Beijing City, including river water, treated drinking water, raw water samples from the pools and sewage from 4 comprehensive hospitals. For the bla_{NDM}-1 positive isolate, antimicrobial susceptibility testing was further analyzed, and Pulsed Field Gel Electrophoresis (PFGE) was performed to determine the genetic relationship among the NDM-1 producing isolates from sewage and human, as well as the clinical strains without NDM-1. The results indicate that there was a higher isolation of NDM-1 producing *Acinetobacter baumannii* from the sewage of the hospitals, while no NDM-1 producing isolates were recovered from samples obtained from the river, drinking, or fishpond water. Surprisingly, these isolates were markedly different from the clinical isolates in drug resistance and pulsed field gel electrophoresis profiles, suggesting different evolutionary relationships. Our results showed that the hospital sewage may be one of the diffusion reservoirs of NDM-1 producing bacteria.

Editor: Raymond Schuch, Rockefeller University, United States of America

Funding: This work was supported in part by the National Science and technology Major Project of China (2012ZX10004-215, 2013ZX10004-218, 2013ZX10004-607, 2012ZX10004-801, 2013ZX10004-203, 2013ZX10004-217-002) and the National Natural Science Foundation of China (81000723, 81070969, 81171554). The funders had no role in study design, data collection and analysis, decision to publish, or preparation of the manuscript.

Competing Interests: The authors have declared that no competing interests exist.

* E-mail: hongbinsong@263.net (HBS); qiushf0613@hotmail.com (SFQ)

⊙ These authors contributed equally to this work.

Introduction

The increasing threat of antibiotic resistance in microbes affecting humans has been recognized as a challenge for treatment of clinical infection. The emergence and spread of pathogenic bacteria with broad spectrum antibiotic resistance pose real threats to the public health systems of any country. In 2009, a new metallo-β-lactamase gene (*bla*NDM-1), encoding the metallo-β-lactamase protein New Delhi metallo-β-lactamase 1 (NDM-1) with high carbapenemase activity and which can destroy carbapenem-type antibiotics, was first identified from a Swedish patient of Indian origin [1]. Carbapenems represent a last line of antibiotic defense for many infections; resistant organisms are capable of causing death in infected hosts. To date, infections associated with NDM-1 positive strains have been reported in several countries and district including U.K., U.S., Canada, Australia, France, Holland, India and China, Sweden, Sultanate of Oman, Kenya, Singapore, Bangladesh, Australia, Switzerland, France, Iraq, Norway, Singapore, Belgium, Montenegro, Germany, Pakistan, Italy, Japan, Spain [2]. In total, strains of multiple speciesof *bla*NDM-1 carrying bacteria, including *Shigella boydii* and *Vibrio cholerae*, have been identified worldwide [2–4]. The genes encoding NDM-1 are known to be carried on a plasmid, and it is suspected

horizontal gene transfer (HGT) promotes the exchange of resistance among Gram-negative organisms.

Sewage is a complex matrix composed of multiple components from many fecal sources. Discharged sewage from hospitals if improperly treated may contain pathogens and antibiotic residues that could lead to acute infections, or to the selection and spread of resistance through the HGT of genetically mobile resistance cassettes. Sewage is a hot spot of gene transfer between organisms [5]. In this way, the hospital may become a source of spread of the resistant bacteria. NDM-1 producing bacteria have been identified in water pools, sewage and tap water in New Delhi [4].

Acinetobacter baumannii is an important pathogen for hospital-acquired infections and widely distributed in a variety of environments in the hospital. Infections caused by *A. baumannii* expressing a broad drug-resistance spectrum have been frequently reported worldwide. The resistance patterns associated with the isolates has created great obstacles for clinical treatment. Additionally, strains are resistant to heat, ultraviolet and chemical sanitizers and thus not easily disinfected by routine sanitizers. Recently, 14% of *bla*NDM-1 carrying bacteria and many *A. junii* isolates and a couple of *A.baumanii* isolates were found from over 10,000 faecal samples in China [6]. Two *Acinetobacter johnsonii* strain carrying a blaNDM-1 plasmid was also isolated from Sewage of a hospital in China [7]. In addition, it was proved that NDM-1 is a recently made gene via the fusion of two resistance genes and that this event happened in Acinetobacter [8]. Beijing, the capital city of China, has an estimated population of 20 million, spread over a 16,410.54 km² area. Potable water is a limited resource to Beijing, therefore it is of prime importance to ensure the water systems of Beijing City remain pathogen-free. Surveillance of water systems that may be adversely affected by point sources of contamination, such as hospital effluent, is critical to the maintenance of safe water supplies. The aims of this study were to determine whether NDM-1 producing *A. baumannii* could be detected in hospital effluent, environmental sources, and clinical cases in Beijing and to determine the relationship between NDM-1 and non-NDM-1 producing *A. baumannii* isolates regardless of their source of origin.

Materials and Methods

Collection and Identification of Bacteria

The institutional review board of the Academy of Military Medical Sciences waived the need for written informed consent from the participants. This study is approved and authorized for each location by the Academy of Military Medical Sciences Review Board. There was no request for a specific permission according to Chinese law. A total of 119 water samples were gathered from river water ($n = 20$), treated drinking water samples ($n = 50$), sewage from 4 comprehensive hospitals ($n = 20$), raw water samples from the pools ($n = 9$) and community life sewage ($n = 20$) in Beijing from September to November 2010. For each sample, 500 ml of target water was collected and a 100 ml aliquot was centrifuged at $1000 \times g$ for 10 minutes at ambient temperature. After carefully decanting the supernatant, pellets were re-suspended with 1 ml Luria-Bertani (LB) liquid medium and 400 µl samples were seeded onto LB agar plates containing imipenem (10 µg/ml). All colonies on the culture plates were selected and identified by PCR as previously described [4] and was further sequenced for confirmation. All NDM-1 positive strains were identified using Vitek GNI+ cards (bioMérieux, France), and sequence analysis of the 16S rRNA gene. The primers used to amplify the 16s rRNA gene were 5'-TACCTTGTTACGACTT-3' and 5'- AGAGTTTGATCITGGA-3' [9]. The other *A. baumannii* isolates, as identified by Vitek GNI+ cards (bioMérieux, France) were isolated in our laboratory, between September 2009 and February 2010 (Table 1).

Antibiotic Susceptibility Testing

Bacterial susceptibility testing was carried out by the Kirby-Bauer method according to the Clinical Laboratory Standards Institute (CLSI) guidelines (2008) [10]. The antibiotic discs used were imipenem, meropenem, cefdazadime, cefotaxime, cefepime, gentamicin, tobramycin, tetracycline, ciprofloxacin, polymyxin B, chloramphenicol and tigecycline. The tested bacterium was picked up with sterile loop and suspended in peptone water and incubated at 37°C for 3 hours. The turbidity of the suspension was adjusted to 0.5 McFarland's standard, and the suspension was then spread on the surface of a LB agar plates using sterile cotton swab. The antimicrobial susceptibility test disc was placed on the agar. The plates were incubated at 37°C overnight. The zone of inhibition was measured and interpreted as per the CLSI guidelines.

Conjugation and Transformation Experiments and Plasmid Analysis

Conjugation transfer assay was performed in broth culture with *E. coli* J53 as the recipient. Donor was respectively mixed at a ratio of 1:3 with ten *A. baumannii* containing the *bla*NDM-1 gene from the sewage. Transconjugants were selected on MacConkey medium containing sodium azide (100 mg/mL) and ceftazidime (16 mg/mL). Plasmid DNA was respectively extracted from ten *A. baumannii* containing the *bla*NDM-1 gene from the sewage, and was transformed by electroporation into competent *E. coli* JM109. The transformants were selected on LB agar plates containing ceftazidime (16 mg/mL). Transformants and transconjugants were further confirmed by Vitek GNI+ cards (bioMérieux, France) and tested for antimicrobial susceptibility by the Kirby-Bauer method according to CLSI (2008). Plasmids were extracted from donor strain, recipient strain (*E. coli* J53 and *E. coli* JM109), transconjugant and transformant by the QIAGEN Large-Construct Kit and were further analyzed by specific PCR and sequencing for blaDNM-1.

PFGE

Pulsed-field gel electrophoresis (PFGE) of *bla*NDM-1 carrying *A. baumannii* isolates recovered from sewage, environmental sources, and clinical isolates with or without *bla*NDM-1 was carried out with a CHEF-Mapper XA PFGE system (Bio-Rad, USA) for 22 h at 6 V/cm and 14°C, with a pulse angle of 120° and pulse times from 5 to 20 s. The restriction endonuclease ApaI was used for in-situ digestion of intact *A. baumannii* genomic DNA. PFGE banding patterns were analyzed visually by using a Bio-Rad Gel Doc 2000

Table 1. *A. baumannii* isolates with or without *bla*NDM-1 from the environment, sewage and human.

Isolate	Source	Region	NDM-1	isolation by year
HHG8②, HHG8③	Environment	Beijing	+	2009
WJ3-2, WJ3-5, WJ0117, WJ0111, WJ0102①, WJ0135, WJ0102②, WJ0102, WJ0147, 3070341	Sewage	Beijing	+	2010
10051750green	Human	Xiamen	+	2009
44, 65, ICU-1, Evn-60, 25, Evn-59,	Human	Beijing	–	2009
10051442blue, 192, 270, 136, 372,104,	Human	Xiamen	–	2009
NJ35, NJ35-1, NJ87-1-1,	Human	Nanjing	–	2009
10092903, 10092908, 10092901, 10092902, 10092904, 10092907, 10092910, Evn-52, Evn-37, Evn-38, Evn-50, Evn-41, Evn-47, Evn-43, Evn-44	Environment	Beijing		2009

system. Genetic relationship was analyzed by the BioNumerics version 46.0 software.

Results

Higher NDM-1 Producing *A. baumannii* Isolation from the Sewage of the Hospitals

Ten isolates containing the *bla*NDM-1 gene were identified in sewage prior to disinfect from the general hospitals; this included an effluent sample from a hospital after disinfecting by chlorination before discharge and community life sewage (Table 2). No NDM-1 producing isolates were recovered from samples obtained from the river, drinking, or fishpond water and community life sewage. Biochemical identification and sequencing of 16S rRNA demonstrated that the ten isolates were *A. baumannii*.

Characterization of the Drug Resistance Profile of the *bla*NDM-1 Positive Isolates

Antimicrobial susceptibility testing was performed to determine the drug resistance profile of the *bla*NDM-1 positive isolates. Previous studies indicated that most clinical NDM-1 producing bacteria were resistant to all antibiotics except colistin and tigecycline [1,3]. Zhou et al [11] reported clinical isolates from a child patient in China were resistant to all β-lactams except aztreonam but sensitive to aminoglycosides and quinolones. In this study, they were sensitive to aminoglycosides, chloramphenicol, colistin and tigecycline that *A. baumannii* possessing *bla*NDM-1 were isolated from the hospital environment and the hospital sewage (Table 3), In addition, two isolates were sensitive to quinolones and six isolates were non-resistant (four susceptible and two intermediate) to tetracycline, respectively. The data indicates that sewage-associated isolates of *A. baumannii* have unique antibiotic resistance profiles from those reported for clinically-obtained *A. baumannii* harbouring *bla*NDM-1.

Plasmid Analysis

The conjugation and transformation experiments were performed to investigate whether the gene for NDM-1 in *A. baumannii* isolates recovered from sewage were located on plasmids and whether the transconjugant and transformant reduced susceptibility of the recipient strain towards antibiotics. Transconjugants and transformants were respectively randomly chosen, and the

*bla*NDM-1 gene was determined by PCR amplification and sequencing (see in Fig. S1). The result showed that the plasmids carrying blaNDM-1 from A. baumannii from the sewage were successfully transferred to E. coli J53 and *E. coli* JM109. Susceptibility tests revealed that both the transconjugant and transformant decreased susceptibility to imipenem, cefepime, ciprofloxacin, cefdazadime, cefotaxime as compared with the recipients *E. coli* J53 and JM109 (Figure 1). Interestingly, all the transconjugants and transformants stably maintained the *bla*NDM-1-containing plasmid after seven passages in the absence of selection pressure.

PFGE Analysis

To investigate the genetic relationship between *bla*NDM-1 carrying *A. baumannii* isolates recovered from sewage, environmental sources, and clinical isolates with or without *bla*NDM-1, PFGE was performed as previously described [12]. Among the 13 *bla*NDM-1 positive isolates recovered in this study, 11 PFGE patterns were observed (Figure 2). Two isolates, WJ3-2 and WJ3-5, had an indistinguishable *Apa* I-PFGE pattern among the ten sewage-recovered isolates of *A. baumannii*; two additional pairs (WJ0102, WJ0102-2; and WH0111 and WJ0102-1) were ≥94% similar. All six of these isolates were *bla*NDM-1-positive and had nearly identical phenotypic antimicrobial sensitivity profiles (both WJ0111 and WJ0102-1 were TE non-susceptible, WJ0111 was resistant and WJ0102-1 was recorded as intermediate). Collectively this data suggests these isolates came from a common source. Similarly, environmental isolates grouped into several distinct clusters. Furthermore, a large number of band variations were observed between the *bla*NDM-1 carrying isolates recovered from hospital sewage, and the *bla*NDM-1 carrying clinical and hospital environmental isolates. This variation was also evident in *A. baumannii* isolates without *bla*NDM-1. These results suggest that *A. baumannii* isolates in Beijing hospital system are genetically diverse overall, and that among the *bla*NDM-1 carrying *A. baumannii* isolates, several clones were detected.

Discussion

Sewage from hospitals has the potential to contain a large number of pathogens including parasite ova, pathogenic bacteria and viruses; therefore, sewage should be disinfected before discharge. Guardabassi et al [13] reported a large number of

Table 2. Recovery of carbapenem-resistant and *bla*NDM-1 positive *A. baumannii* isolates from sewage sources of four Beijing hospitals and community.

Sample ID	disinfection	Sample number	Carbapenem resistant isolates[a]	*bla*NDM-1 positive isolates[b]
A	Before disinfection	4	32	3
	After disinfection	1	5	1
B	Before disinfection	4	29	2
	After disinfection	1	3	0
C	Before disinfection	4	41	1
	After disinfection	1	1	0
D	Before disinfection	4	37	3
	After disinfection	1	2	0
E[c]	Before treatment	20	37	0

[a]Phenotypic resistance to carbapenem.
[b]A. baumannii only.
A,B,C,D: The hospital sewage; E: Community life sewage.

Table 3. Antibiotic resistance profiles of *A. baumannii* isolates carrying *bla*NDM-1 isolated from sewage.

Strains	Drug resistance§											
	IPM	MEC	CAZ	CTX	FEP	GM	TM	TE	CIP	PB	CHL	TGC
WJ0135	R	R	R	R	R	S	S	S	S	S	S	S
WJ0117	R	R	R	R	R	S	S	S	S	S	S	S
HHG8②	R	R	R	R	R	S	S	S	S	S	S	S
HHG8③	R	R	R	R	R	S	S	S	S	S	S	S
3070341	R	R	R	R	R	S	S	I	S	S	S	S
WJ0102①	R	R	R	R	R	S	S	I	S	S	S	S
WJ0102②	R	R	R	R	R	S	S	R	S	S	S	S
WJ0111	R	R	R	R	R	S	S	S	S	S	S	S
WJ0147	R	R	R	R	R	S	S	R	S	S	S	S
WJ0102	R	R	R	R	R	S	S	R	S	S	S	S
WJ3-2	R	R	R	R	R	S	S	S	R	S	S	S
WJ3-5	R	R	R	R	R	S	S	S	R	S	S	S

Abbreviations: IPM: imipenem (10 μg); MEM: meropenem (10 μg); CAZ: cefdazadime (30 μg); CTX: cefotaxime (30 μg); FEP: cefepime (30 μg); CN: gentamicin (10 μg); TOB: tobramycin (10 μg); TE: tetracycline (30 μg); CIP: ciprofloxacin (5 μg); PB: polymyxin B (300U); CHL: chloramphenicol (30 μg); TGC: tigecycline (15 μg); R: resistance; S: sensitivity; I: intermediate;
§According to CLSI guidelines.

bacteria remained in the sewage discharged from the hospitals, antibiotic production factories and livestock farms after the disinfection process. We identified ten *bla*NDM-1 carrying isolates from the water samples collected from four general hospitals in Beijing, China. Even though waste water disinfection was performed in all hospitals, a *bla*NDM-1 carrying isolate was still recovered from the treated sewage. Sewage that is not treated appropriately may pollute the water systems including surface water, ground water and drinking water, which may enhance the possibility of infection induced by *bla*NDM-1 positive strains in human body. Additionally, the *bla*NDM-1 gene is known to be transferred between bacterial genera [3]. These observations suggest an increase exposure of hospitalized patients to *bla*NDM-1 carrying microbes and imply the incidence of *bla*NDM-1-associated infections in hospital-acquired infections, caused directly by *A. baumannii* or a bacterial genera that acquired the resistance mechanism from *A. baumannii* after mixing in a

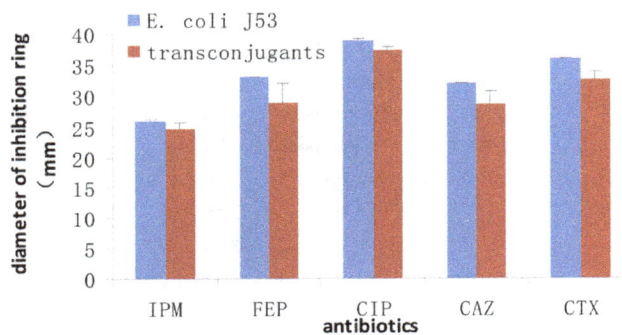

Figure 1. Antibiotic susceptibility testing of recipient strain *E. coli* J53 and transconjugants. IPM: imipenem; FEP: cefepime; CIP: ciprofloxacin; CAZ: cefdazadime; CTX: cefotaxime; The diameter of the conjugant to all the antibiotics reduced significantly, P<0.05.

permissive environment such as sewage discharge from hospitals, is likely to increase in China.

As previously reported, NDM-1-producing bacteria are resistant to β-lactamase antibiotics [3]. The NDM-1 positive isolates associated with sewage in this study were all sensitive to aminoglycosides, chloramphenicol, colistin and tigecycline. In addition, two isolates were sensitive to quinolones and six isolates were non-resistant to tetracycline, respectively. Our study indicated differences in drug resistance profiles between *bla*NDM-1-associated isolates obtained from sewage and those previously reported from clinical isolates. With the wide and excessive using of antibiotics in clinic, clinical strains quickly evolved into multi-drug resistant (MDR) or pan-drug resistant (PDR) bacteria. It contains a certain amount of antibiotic residues in the sewage, but the concentration of antibiotic is relatively lower than that in the clinic, which has to a certain extent slowed down the speed of evolution of the environmental resistant strains, and result in environmental resistant strains are still sensitive to some antibiotics. This may be one of the reasons for these differences. Even though no associated infection has been reported to date from these hospitals in Beijing, we can not neglect the potential threats of the drug resistance gene to the ecosystem and human lives.

The plasmids carrying the *bla*NDM gene successfully transferred to the recipient *E. coli* J53 and *E. coli* JM109, suggesting that this plasmid is mobile. The transconjugant and transformant decreased susceptibility to imipenem, cefepime, ciprofloxacin, cefdazadime, cefotaxime, and stably maintained the *bla*NDM-1-containing plasmid of *A. baumannii*, which displays the potential for the spread of *bla*NDM through plasmid transmission from A. baumannii to Enterobacteriaceae in the natural environment. To analyze the genetic diversity and evolutionary relationships among isolates, PFGE was performed. The results further support the genetic diversity of the *bla*NDM-1 carrying *A. baumannii*. Therefore, we hope to raise the attention of *bla*NDM-1-associated bacteria in the water systems of Beijing City. Currently, it is not clear to the reason of the difference of the drug resistance and genetic relatedness among the *bla*NDM-1 carrying obtained

PFGE-ApaI

PFGE-ApaI

Isolate		Source	Region	NDM-1
AB0100	HHG8②	Environment	Beijing	+
AB0101	HHG8③	Environment	Beijing	+
AB0003	10092903	Environment	Beijing	
AB0006	10092908	Environment	Beijing	
AB0001	10092901	Environment	Beijing	
AB0002	10092902	Environment	Beijing	
AB0004	10092904	Environment	Beijing	
AB0005	10092907	Environment	Beijing	
AB0007	10092910	Environment	Beijing	
AB0118	44	Human	Beijing	
AB0102	WJ3-2	Sewage	Beijing	+
AB0103	WJ3-5	Sewage	Beijing	+
AB0235	65	Human	Beijing	
AB0060	192	Human	Xiamen	
AB0107	WJ0117	Sewage	Beijing	+
AB0209	ICU-1	Human	Beijing	
AB0098	WJ0111	Sewage	Beijing	+
AB0105	WJ0102①	Sewage	Beijing	+
AB0068	270	Human	Xiamen	
AB0221	Env-60	Human	Beijing	
AB0008	10051442blue	Human	Xiamen	
AB0016	10051750green	Human	Xiamen	+
AB0067	136	Human	Xiamen	
AB0151	25	Human	Beijing	
AB0070	372	Human	Xiamen	
AB0220	Env-59	Human	Beijing	
AB0257	Env-52	Environment	Beijing	
AB0242	Env-37	Environment	Beijing	
AB0243	Env-38	Environment	Beijing	
AB0255	Env-50	Environment	Beijing	
AB0104	WJ0135	Sewage	Beijing	+
AB0083	NJ35	Human	Nanjing	
AB0084	NJ35-1	Human	Nanjing	
AB0086	NJ87-1-1	Human	Nanjing	
AB0099	WJ0102②	Sewage	Beijing	+
AB0106	WJ0102	Sewage	Beijing	+
AB0246	Env-41	Environment	Beijing	
AB0252	Env-47	Environment	Beijing	
AB0071	104	Human	Xiamen	
AB0248	Env-43	Environment	Beijing	
AB0249	Env-44	Environment	Beijing	
AB0109	3070341	Sewage	Beijing	+
AB0108	WJ0147	Sewage	Beijing	+

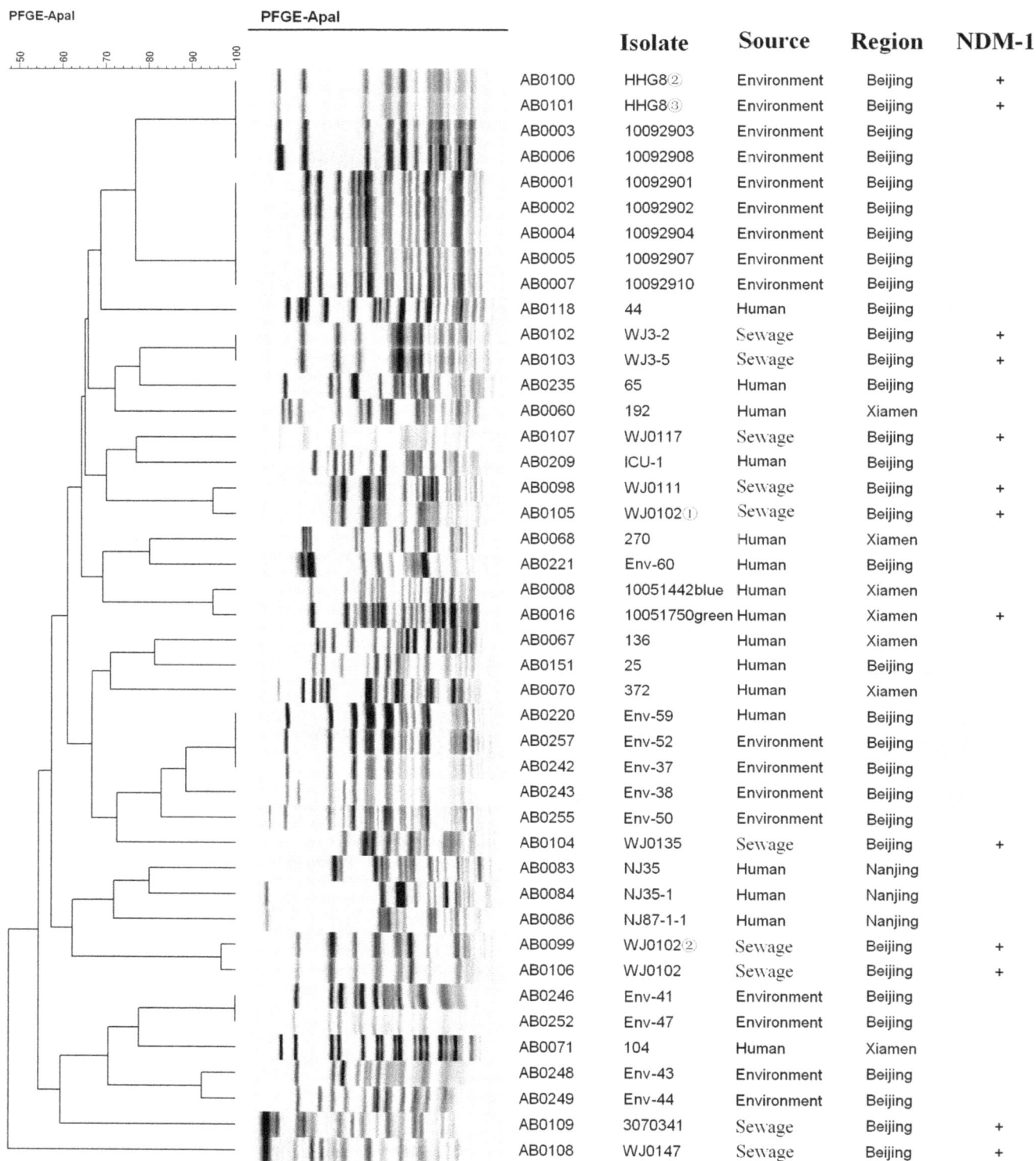

Figure 2. Dendrogram derived from PFGE patterns of ApaI-digested *A. baumanii* DNA. *A. baumannii* isolates were recovered from clinical cases (Beijing, Xiamen, and Nanjing) and the hospital environment (Beijing). Abbreviations: isolate: key number given to isolate in BioNumerics software (ABxxxx) and original isolate designation; source: location of recovery of isolate where environment could be (list sources here); region: city of isolation; NDM-1: "+" indicates an isolate containing *bla*NDM-1.

isolates from the sewage and the clinic, further in-depth genetic analysis must be performed.

Supporting Information

Figure S1 Identification of a mobile blaNDM-1 gene. Plasmids were extracted from donor strain, recipient strain *E. coli* J53 and *E. coli* JM109, transconjugants, transformants and PCR analysis of the blaNDM-1 gene. M: marker; 1-3: transconjugants; 4-5: transformants; 6: *E. coli* J53; 7: *E. coli* JM109; 8-9: WJ3-5,

WJ0135. The experimental results of other *A. baumannii* containing the blaNDM-1 gene from the sewage were similar to those.

Author Contributions

Conceived and designed the experiments: HBS CFZ SFQ YW RZH. Performed the experiments: CFZ XLL YS XFH DZA LHQ ZJL PW. Analyzed the data: CFZ JJC PL LGW LYH. Contributed reagents/materials/analysis tools: YS XFH DZA LHQ. Wrote the paper: CFZ SFQ HBS JDK.

References

1. Yong D, Toleman MA, Giske CG, Cho HS, Sundman K, et al. (2009) Characterization of a new metallo-beta-lactamase gene, *bla* (NDM-1), and a novel erythromycin esterase gene carried on a unique genetic structure in *Klebsiella pneumoniae* sequence type 14 from India. Antimicrob Agents Chemother 53: 5046–5054.
2. Bushnell G, Mitrani-Gold F, Mundy LM (2013) Emergence of New Delhi metallo-β-lactamase t ype 1-producing *Enterobacteriaceae* and non-*Enterobacteriaceae*: global case detection and bacterial surveillance. Int J Infect Dis 175: e325–333.
3. Kumarasamy KK, Toleman MA, Walsh TR, Bagaria J, Butt F, et al. (2010) Emergence of a new antibiotic resistance mechanism in India, Pakistan, and the UK: a molecular, biological, and epidemiological study. Lancet Infect Dis 10: 597–602.
4. Walsh TR, Weeks J, Livermore DM, Toleman MA (2011) Dissemination of NDM-1 positive bacteria in the New Delhi environment and its implications for human health: an environmental point prevalence study. Lancet Infect Dis 11: 355–362.
5. Kim S, Aga DS (2007) Potential ecological and human health impacts of antibiotics and antibiotic-resistant bacteria from wastewater treatment plants. J. Toxicol. Environ. Health B Crit Rev 10, 559–573.
6. Wang X, Liu W, Zou D, Li X, Wei X, et al. (2012) High rate of New Delhi metallo-β- lactamase 1-producing bacterial infection in China. Clin Infect Dis 56: 161–162.
7. Zong Z, Zhang X (2013) *bla*NDM-1-carrying *Acinetobacter johnsonii* detected in hospital sewage. J Antimicrob Chemother 68 (5): 1007–1010.
8. Toleman MA, Spencer J, Jones L, Walsh TR (2012) bla_{NDM-1} is a chimera likely constructed in *Acinetobacter baumannii*. Antimicrob Agents Chemother. 56: 2773–2776.
9. Vila J, Marcos MA, Jimenez de Anta MT (1996) A comparative study of different PCR-based DNA fingerprinting techniques for typing of the *Acinetobacter calcoaceticus- A. baumannii* complex. J Med Microbiol 44: 482–489.
10. Clinical and laboratory standards Institute (2008) Performance standards for antimicrobial susceptibility testing; eighteen information supplement M100-S18. CLSI, Wayne, PA.
11. Zhou Z, Guan R, Yang Y, Chen L, Fu J, et al. (2012) Identification of New Delhi metallo-β-lactamase gene (NDM-1) from a clinical isolate of *Acinetobacter junii* in China. Can J Microbiol 58: 112–115.
12. King SJ, Leigh JA, Heath PJ, Luque I, Tarradas C, et al. (2002) Development of a multilocus sequence typing scheme for the pig pathogen *Streptococcus suis*: identification of virulent clones and potential capsular serotype exchange. J Clin Microbiol 40: 3671–3680.
13. Guardabassi L, Lo Fo Wong DM, Dalsgaard A (2002) The effects of tertiary wastewater treatment on the prevalence of antimicrobial resistant bacteria. Water Res 36: 1955–1964.

PERMISSIONS

LIST OF CONTRIBUTORS

Rita Branco
IMAR, 3004-517 Coimbra, Portugal
Escola Universitária Vasco da Gama, Mosteiro de S. Jorge de Milréu, Estrada da Conraria, Castelo Viegas – Coimbra, Portugal

Armando Cristóvão
Center for Neuroscience and Cell Biology, University of Coimbra, Coimbra, Portugal
Department of Life Sciences, FCTUC, University of Coimbra, Coimbra, Portugal

Paula V. Morais
IMAR, 3004-517 Coimbra, Portugal
Department of Life Sciences, FCTUC, University of Coimbra, Coimbra, Portugal

Antonio Tovar-Sánchez and David Sánchez-Quiles
Department of Global Change Research, Mediterranean Institute for Advanced Studies (UIB-CSIC), Esporles, Balearic Island, Spain

Gotzon Basterretxea
Department of Ecology and Marine Resources, Mediterranean Institute for Advanced Studies (UIB-CSIC), Esporles, Balearic Island, Spain

Juan L. Benedé, Alberto Chisvert and Amparo Salvador
Department of Analytical Chemistry, Facultad de Química, Universitat de Valéncia, Burjassot, Valencia, Spain

Ignacio Moreno-Garrido and Julián Blasco
ICMAN-Instituto de Ciencias Marinas de Andalucía (CSIC), Puerto Real, Cádiz, Spain

Wen Liang, Xuefei Zhou and Yalei Zhang
State Key Laboratory of Pollution Control and Resources Reuse, Tongji University, Shanghai, China

Chaomeng Dai
State Key Laboratory of Pollution Control and Resources Reuse, Tongji University, Shanghai, China
College of Civil Engineering, Tongji University, Shanghai, China

Hongjun Wang, Youguang Liang, Sixin Li and Jianbo Chang
Key Laboratory of Ecological Impacts of Hydraulic-Projects and Restoration of Aquatic Ecosystem of Ministry of Water Resources, Institute of Hydroecology, Ministry of Water Resources and Chinese Academy of Sciences, Wuhan, P.R. China

Anja Henneberg, Heinz-R. Köhler, Diana Maier and Rita Triebskorn
Animal Physiological Ecology, University of Tübingen, Tübingen, Germany

Katrin Bender, Sabrina Giebner, Jörg Oehlmann, Ulrike Schulte-Oehlmann, Agnes Sieratowicz and Simone Ziebart
Department Aquatic Ecotoxicology, University of Frankfurt am Main, Frankfurt am Main, Germany

Ludek Blaha
Faculty of Science, RECETOX, Masaryk University, Brno, Czech Republic

Bertram Kuch
Institute for Sanitary Engineering, Water Quality and Solid Waste Management, University of Stuttgart, Stuttgart, Germany

Doreen Richter and Marco Scheurer
Water Technology Center Karlsruhe, Karlsruhe, Germany

Wenzhong Tang, Baoqing Shan, Chao Wang and Wenqiang Zhang
State Key Laboratory on Environmental Aquatic Chemistry, Research Center for Eco-Environmental Sciences, Chinese Academy of Sciences, Beijing, China

Jingguo Cui
Beijing Sound Environmental Engineering Co., Ltd., Beijing, China

Ilunga Kamika and Maggie N. B. Momba
Department of Environmental, Water and Earth Sciences, Faculty of Science, Tshwane University of Technology, Pretoria, Gauteng, South Africa

Markus Brinkmann, Henning Blenkle, Kerstin Bluhm, Sabrina Schiwy and Henner Hollert
Department of Ecosystem Analysis, Institute for Environmental Research, RWTH Aachen University, Aachen, Germany

Helena Salowsky and Andreas Tiehm
Department of Environmental Biotechnology, Water Technology Center, Karlsruhe, Germany

Joep F. Schyns and Arjen Y. Hoekstra
Twente Water Centre, University of Twente, Enschede, The Netherlands

Joel T. Kidgell, Rocky de Nys, Nicholas A. Paul and David A. Roberts
MACRO - The Centre for Macroalgal Resources and Biotechnology, and School of Marine and Tropical Biology, James Cook University, Townsville, Queensland, Australia

Yi Hu
Advanced Analytical Centre, James Cook University, Townsville, Queensland, Australia

Janne Laine
Tampere University Hospital, Department of Internal Medicine, Tampere, Finland
National Institute of Health and Welfare, Epidemiologic Surveillance and Response Unit, Helsinki, Finland

Jukka Lumio
University of Tampere, School of Medicine, Tampere, Finland, 4 Nokia Health Centre, Nokia, Finland

Salla Toikkanen, Mikko J. Virtanen and Markku Kuusi
National Institute of Health and Welfare, Epidemiologic Surveillance and Response Unit, Helsinki, Finland

Terhi Uotila
Tampere University Hospital, Department of Internal Medicine, Tampere, Finland

Markku Korpela
Tampere University Hospital, Department of Internal Medicine, Tampere, Finland
University of Tampere, School of Medicine, Tampere, Finland

Eila Kujansuu
Nokia Health Centre, Nokia, Finland

Department of Social Services and Health Care, Tampere, Finland

Binbin Wu, Guoqiang Wang, Jin Wu and Changming Liu
College of Water Sciences, Beijing Normal University, Key Laboratory of Water and Sediment Sciences, Ministry of Education, Beijing, China

Qing Fu
Chinese Research Academy of Environmental Sciences, Beijing, China

Takeki Hamasaki, Noboru Nakamichi, Kiichiro Teruya and Sanetaka Shirahata
Department of Bioscience and Biotechnology, Faculty of Agriculture, Kyushu University, Higashi-ku, Fukuoka, Japan

Magalie Canuel
Institut national de santé publique du Québec (INSPQ), Québec City, Canada

Belkacem Abdous
Centre de recherche du Centre hospitalier universitaire de Québec, Québec City, Canada
Département de médecine sociale et préventive de l9Université Laval, Québec City, Canada

Diane Bélanger
Centre de recherche du Centre hospitalier universitaire de Québec, Québec City, Canada,
Institut national de la recherche scientifique, Centre Eau Terre Environnement, Québec City, Canada

Pierre Gosselin
Institut national de santé publique du Québec (INSPQ), Québec City, Canada
Centre de recherche du Centre hospitalier universitaire de Québec, Québec City, Canada
Département de médecine sociale et préventive de l9Université Laval, Québec City, Canada
Institut national de la recherche scientifique, Centre Eau Terre Environnement, Québec City, Canada

Gill T. Braulik
Sea Mammal Research Unit, Scottish Oceans Institute, University of St. Andrews, St. Andrews, Fife, United Kingdom
World Wildlife Fund-Pakistan, Lahore, Pakistan
Wildlife Conservation Society, Zanzibar, United Republic of Tanzania

Masood Arshad and Uzma Noureen
World Wildlife Fund-Pakistan, Lahore, Pakistan

Simon P. Northridge
Sea Mammal Research Unit, Scottish Oceans Institute, University of St. Andrews, St. Andrews, Fife, United Kingdom

Grace Hwee Boon Ng and Zhiyuan Gong
Department of Biological Sciences, NUS Graduate School for Integrative Sciences and Engineering, National University of Singapore, Singapore and Computation and Systems Biology, Singapore-MIT Alliance, Singapore, Singapore

Yilong Wang, Bingdi Chen and Fangfang Guo
The Institute for Biomedical Engineering and Nano Science, Tongji University School of Medicine, Shanghai, P. R. China

Song Liang, Shuili Yu and Yulin Tang
State Key Laboratory of Pollution Control and Resource Reuse, College of Environmental Science and Engineering, Tongji University, Shanghai, P. R. China

Sher Bahadar Khan, Mohammed M. Rahman, Abdullah M. Asiri and Malik Abdul Rub
Center of Excellence for Advanced Materials Research (CEAMR), King Abdulaziz University, Jeddah, Saudi Arabia
Chemistry Department, Faculty of Science, King Abdulaziz University, Jeddah, Saudi Arabia

Kalsoom Akhtar
Division of Nano Sciences and Department of Chemistry, Ewha Womans University, Seoul, Korea

Alexander Golberg
Centre for Engineering in Medicine, Massachusetts General Hospital, Harvard Medical School, Shriners Burns Institute, Boston, Massachusetts, United States of America

Gregory Linshiz and Nathan J. Hillson
Fuels Synthesis Division, Joint BioEnergy Institute, Emeryville, California, United States of America
Physical BioSciences Division, Lawrence Berkeley National Labs Berkeley, California, United States of America
DOE Joint Genome Institute, Walnut Creek, California, United States of America

Ilia Kravets
Department of Computer Science, Technion Institute of Technology, Haifa, Israel

Nina Stawski
Fuels Synthesis Division, Joint BioEnergy Institute, Emeryville, California, United States of America
Physical BioSciences Division, Lawrence Berkeley National Labs Berkeley, California, United States of America

Martin L. Yarmush
Centre for Engineering in Medicine, Massachusetts General Hospital, Harvard Medical School, Shriners Burns Institute, Boston, Massachusetts, United States of America
Department of Biomedical Engineering, Rutgers University, New Jersey, United States of America

Robert S. Marks
Department of Biotechnology Engineering, The National Institute of Biotechnology in Negev, Ben Gurion University, Beer-Sheva, Israel

School of Materials Science and Engineering, Nanyang Technological University, Singapore
NRF CREATE program for Nanomaterials in Energy and Water Management, Singapore

Tania Konry
Department of Pharmaceutical Sciences, School of Pharmacy Bouvé College of Health Sciences, Northeastern University, Boston, Massachusetts, United States of America

Siang Yee Yeo and Beng Joo Reginald Thio
Engineering Product Development, Singapore University of Technology and Design, Singapore, Singapore

Siwon Choi, T. Alan Hatton and Patrick S. Doyle
Department of Chemical Engineering, Massachusetts Institute of Technology, Cambridge, Massachusetts, United States of America

Vivian Dien
Department of Materials Science and Engineering, Massachusetts Institute of Technology, Cambridge, Massachusetts, United States of America

Yoke Keow Sow-Peh
Science Department, Hwa Chong Institution (High School), Singapore, Singapore

Genggeng Qi
Kaust-Cornell Center for Energy and Sustainability, Cornell University, Ithaca, New York, United States of America

Wei Ouyang, Guanqing Cai and Fanghua Hao
School of Environment, State Key Laboratory of Water Environment Simulation, Beijing Normal University, Beijing, China

Siyang Chen
Marine Monitoring and Forecasting Center of Zhejiang, Hangzhou, China

Manon Lainé, Soizic Morin and Juliette Tison-Rosebery
UR EABX, Irstea, Cestas cedex, France

Chuanfu Zhang, Shaofu Qiu, Yong Wang, Lihua Qi, Rongzhang Hao, Xuelin Liu, Yun Shi, Xiaofeng Hu, Daizhi An, Peng Li, Ligui Wang, Jiajun Cui, Pan Wang, Liuyu Huang and Hongbin Song
Institute of Disease Control and Prevention, Academy of Military Medical Science, Beijing, People's Republic of China

Zhenjun Li
State Key Laboratory for Infectious Disease Prevention and Control, China Center of Disease Control and Prevention, Beijing, People's Republic of China

John D. Klena
United States Centers for Disease Control and Prevention, China –US Collaborative Program on Emerging and Re-emerging Infectious Diseases, Beijing, People's Republic of China

Index

www.ingramcontent.com/pod-product-compliance
Lightning Source LLC
Chambersburg PA
CBHW080521200326
41458CB00012B/4286